ENCYCLOPEDIA OF MATHEMATICS AND ITS APPLICATIONS

FOUNDED BY G.-C. ROTA

Editorial Board
R. S. Doran, M. Ismail, T.-Y. Lam, E. Lutwak, R. Spigler

Volume 90

Algebraic Combinatorics on Words

ENCYCLOPEDIA OF MATHEMATICS AND ITS APPLICATIONS

4   W. Miller, Jr. *Symmetry and separation of variables*
6   H. Minc *Permanents*
11 W. B. Jones and W. J. Thron *Continued fractions*
12 N. F. G. Martin and J. W. England *Mathematical theory of entropy*
18 H. O. Fattorini *The Cauchy problem*
19 G. G. Lorentz, K. Jetter and S. D. Riemenschneider *Birkhoff interpolation*
21 W. T. Tutte *Graph theory*
22 J. R. Bastida *Field extensions and Galois theory*
23 J. R. Cannon *The one-dimensional heat equation*
25 A. Salomaa *Computation and automata*
26 N. White (ed.) *Theory of matroids*
27 N. H. Bingham, C. M. Goldie and J. L. Teugels *Regular variation*
28 P. P. Petrushev and V. A. Popov *Rational approximation of real functions*
29 N. White (ed.) *Combinatorial geometries*
30 M. Pohst and H. Zassenhaus *Algorithmic algebraic number theory*
31 J. Aczel and J. Dhombres *Functional equations containing several variables*
32 M. Kuczma, B. Choczewski and R. Ger *Iterative functional equations*
33 R. V. Ambartzumian *Factorization calculus and geometric probability*
34 G. Gripenberg, S.-O. Londen and O. Staffans *Volterra integral and functional equations*
35 G. Gasper and M. Rahman *Basic hypergeometric series*
36 E. Torgersen *Comparison of statistical experiments*
37 A. Neumaier *Interval methods for systems of equations*
38 N. Korneichuk *Exact constants in approximation theory*
39 R. A. Brualdi and H. J. Ryser *Combinatorial matrix theory*
40 N. White (ed.) *Matroid applications*
41 S. Sakai *Operator algebras in dynamical systems*
42 W. Hodges *Model theory*
43 H. Stahl and V. Totik *General orthogonal polynomials*
44 R. Schneider *Convex bodies*
45 G. Da Prato and J. Zabczyk *Stochastic equations in infinite dimensions*
46 A. Bjorner, M. Las Vergnas, B. Sturmfels, N. White and G. Ziegler *Oriented matroids*
47 E. A. Edgar and L. Sucheston *Stopping times and directed processes*
48 C. Sims *Computation with finitely presented groups*
49 T. Palmer *Banach algebras and the general theory of *-algebras*
50 F. Borceux *Handbook of categorical algebra I*
51 F. Borceux *Handbook of categorical algebra II*
52 F. Borceux *Handbook of categorical algebra III*
54 A. Katok and B. Hasselblatt *Introduction to the modern theory of dynamical systems*
55 V. N. Sachkov *Combinatorial methods in discrete mathematics*
56 V. N. Sachkov *Probabilistic methods in discrete mathematics*
57 P. M. Cohn *Skew fields*
58 Richard J. Gardner *Geometric tomography*
59 George A. Baker, Jr. and Peter Graves-Morris *Padé approximants*
60 Jan Krajicek *Bounded arithmetic, propositional logic, and complex theory*
61 H. Gromer *Geometric applications of Fourier series and spherical harmonics*
62 H. O. Fattorini *Infinite dimensional optimization and control theory*
63 A. C. Thompson *Minkowski geometry*
64 R. B. Bapat and T. E. S. Raghavan *Nonnegative matrices and applications*
65 K. Engel *Sperner theory*
66 D. Cvetkovic, P. Rowlinson and S. Simic *Eigenspaces of graphs*
67 F. Bergeron, G. Labelle and P. Leroux *Combinatorial species and tree-like structures*
68 R. Goodman and N. Wallach *Representations of the classical groups*
69 T. Beth, D. Jungnickel and H. Lenz *Design theory volume I 2 ed.*
70 A. Pietsch and J. Wenzel *Orthonormal systems and Banach space geometry*
71 George E. Andrews, Richard Askey and Ranjan Roy *Special functions*
72 R. Ticciati *Quantum field theory for mathematicians*
76 A. A. Ivanov *Geometry of sporadic groups I*
78 T. Beth, D. Jungnickel and H. Lenz *Design theory volume II 2 ed.*
80 O. Stormark *Lie's structural approach to PDE systems*
81 C. F. Dunkl and Y. Xu *Orthogonal polynomials of several variables*
82 J. P. Mayberry *The foundations of mathematics in the theory of sets*
83 C. Foias, O. Manley, R. M. S. Rosa and R. Temam *Navier–Stokes equations and turbulence*
85 R. B. Paris and D. Kaminski *Asymptotics and Mellin-Barnes integrals*

ENCYCLOPEDIA OF MATHEMATICS AND ITS APPLICATIONS

Algebraic Combinatorics on Words

M. LOTHAIRE

CAMBRIDGE UNIVERSITY PRESS
Cambridge, New York, Melbourne, Madrid, Cape Town, Singapore,
São Paulo, Delhi, Dubai, Tokyo, Mexico City

Cambridge University Press
The Edinburgh Building, Cambridge CB2 8RU, UK

Published in the United States of America by Cambridge University Press, New York

www.cambridge.org
Information on this title: www.cambridge.org/9780521180719

© Cambridge University Press 2002

This publication is in copyright. Subject to statutory exception
and to the provisions of relevant collective licensing agreements,
no reproduction of any part may take place without the written
permission of Cambridge University Press.

First published 2002
First paperback edition 2010

A catalogue record for this publication is available from the British Library

Library of Congress Cataloguing in Publication data
    Lothaire, M.
      Algebraic combinatorics on words / M. Lothaire.
        p. cm – (Encyclopedia of mathematics ; 90)
      Includes bibliographical references and index.
      ISBN 0 521 81220 8
        1. Combinatorial analysis. 2. Word problems (Mathematics) I. Title.
      II. Encyclopedia of mathematics and its applications ; v. 90.
    QA164.L65 2002
    511'.6–dc21   2001037964

ISBN 978-0-521-81220-7 Hardback
ISBN 978-0-521-18071-9 Paperback

Cambridge University Press has no responsibility for the persistence or
accuracy of URLs for external or third-party internet websites referred to in
this publication, and does not guarantee that any content on such websites is,
or will remain, accurate or appropriate.

Contents

Preface ix

Chapter 1 Finite and Infinite Words 1

    1.0 Introduction 1
    1.1 Semigroups 1
    1.2 Words 3
    1.3 Automata 12
    1.4 Generating series 22
    1.5 Symbolic dynamical systems 26
    1.6 Unavoidable sets 33
        Problems 40
        Notes 44

Chapter 2 Sturmian Words 45

    2.0 Introduction 45
    2.1 Equivalent definitions 46
    2.2 Standard words 63
    2.3 Sturmian morphisms 83
        Problems 101
        Notes 108

Chapter 3 Unavoidable Patterns 111

    3.0 Introduction 111
    3.1 Definitions and basic properties 112
    3.2 Deciding avoidability: the Zimin algorithm 116
    3.3 Avoidability on a fixed alphabet 123
        Problems 131
        Notes 133

Chapter 4 Sesquipowers 135

- 4.0 Introduction 135
- 4.1 Bi-ideal sequences 136
- 4.2 Canonical factorizations 137
- 4.3 Sesquipowers and recurrence 141
- 4.4 Extensions of a theorem of Shirshov 144
- 4.5 Finiteness conditions for semigroups 149
- Problems 159
- Notes 162

Chapter 5 The Plactic Monoid 164

- 5.0 Introduction 164
- 5.1 Schensted's algorithm 165
- 5.2 Greene's invariants and the plactic monoid 167
- 5.3 The Robinson–Schensted–Knuth correspondence ... 170
- 5.4 Schur functions and the Littlewood–Richardson rule . 176
- 5.5 Coplactic operations 182
- 5.6 Cyclage and canonical embeddings 186
- Problems 192
- Notes 195

Chapter 6 Codes 197

- 6.0 Introduction 197
- 6.1 X-factorizations 198
- 6.2 Defect 204
- 6.3 More defect 209
- 6.4 A theorem of Schützenberger 221
- Problems 224
- Notes 227

Chapter 7 Numeration Systems 230

- 7.0 Introduction 230
- 7.1 Standard representation of numbers 231
- 7.2 Beta-expansions 236
- 7.3 U-representations 248
- 7.4 Representation of complex numbers 257
- Problems 263
- Notes 267

Contents vii

Chapter 8 Periodicity 269

    8.0 Introduction 269
    8.1 Periods in a finite word 270
    8.2 Local versus global periodicity 283
    8.3 Infinite words. 294
        Problems 306
        Notes 310

Chapter 9 Centralizers of Noncommutative Series and Polynomials 312

    9.0 Introduction 312
    9.1 Cohn's centralizer theorem 313
    9.2 Euclidean division and principal right ideals 315
    9.3 Integral closure of the centralizer 316
    9.4 Homomorphisms into $k[t]$ 319
    9.5 Bergman's centralizer theorem 320
    9.6 Free subalgebras and the defect theorem 322
    9.7 Appendix: some commutative algebra 324
        Notes 328

Chapter 10 Transformations on Words and q-Calculus 330

    10.0 Introduction 330
    10.1 The q-binomial coefficients 332
    10.2 The MacMahon Verfahren 334
    10.3 The insertion technique 337
    10.4 The (t,q)-factorial generating functions 339
    10.5 Words and biwords 341
    10.6 Commutations 343
    10.7 The two commutations 345
    10.8 The main algorithm 347
    10.9 The inverse of the algorithm 349
    10.10 Statistics on circuits 351
    10.11 Statistics on words and equidistribution properties . . . 353
        Problems 355
        Notes 363

Chapter 11 Statistics on Permutations and Words 365

    11.0 Introduction 365
    11.1 Preliminaries 366
    11.2 Words with a given shape 367
    11.3 Backsteps of permutations with a given shape 369

    11.4 Inversions of permutations with a given shape 372
    11.5 Lyndon factorization and cycles of permutations 375
    11.6 Major index of permutations with a given cyclic type . 378
          Problems 381
          Notes 385

Chapter 12 Makanin's Algorithm 387

    12.0 Introduction 387
    12.1 Words and word equations 389
    12.2 The exponent of periodicity 400
    12.3 Boundary equations 406
    12.4 Proof of Theorem 12.3.10 432
          Problems 435
          Notes 438

Chapter 13 Independent Systems of Equations 443

    13.0 Introduction 443
    13.1 Sets and equations 444
    13.2 The compactness property 447
    13.3 Independence of finite systems of equations 453
    13.4 Semigroups without the compactness property 458
    13.5 Semigroups with the compactness property 460
          Problems 468
          Notes 470

References 473

Index of Notation 497

General Index 499

Preface

Combinatorics on words is a field that has grown separately within several branches of mathematics, such as number theory, group theory or probability theory, and appears frequently in problems of theoretical computer science, as dealing with automata and formal languages.

A unified treatment of the theory appeared in Lothaire's *Combinatorics on Words*. Since then, the field has grown rapidly. This book presents new topics of combinatorics on words.

Several of them were not yet ripe for exposition, or even not yet explored, twenty years ago. The spirit of the book is the same, namely an introductory exposition of a field, with full proofs and numerous examples, and further developments deferred to problems, or mentioned in the Notes.

This book is independent of Lothaire's first book, in the sense that no knowledge of the first volume is assumed. In order to avoid repetitions, some results of the first book, when needed here, are explicitly quoted, and are only referred for the proof to the first volume.

This volume presents, compared with the previous one, two important new features. It is first of all a complement in the sense that it goes deeper in the same direction. For example, the theory of unavoidable patterns (Chapter 3) is a generalization of the theory of square-free words and morphisms. In the same way, the chapters on statistics on words and permutations (Chapters 10 and 11) are a continuation of the chapter on transformations on words of the previous volume. But this volume is also a complement in the sense that it presents aspects of combinatorics on words that had not been treated in the previous one. For example, the plactic monoid is presented here although it had not been mentioned at all in the previous volume. The same holds for several topics connected with symbolic dynamics, such as Sturmian words or beta-expansions.

Let us now describe in more detail the content of this volume. Most of the basic facts needed are given in Chapter 1, "Finite and Infinite Words", written by Jean Berstel and Dominique Perrin. This chapter also contains basic concepts on symbolic dynamical systems. Unavoidable sets of words are studied at the end of this chapter. They are considered again in Chapter 3.

Chapter 2, "Sturmian Words", written by Jean Berstel and Patrice Séébold, is a systematic exposition of a family of infinite words that have minimal complexity. These words share a number of extremal properties and can be defined in several quite different ways. After a treatment of these properties, morphisms preserving Sturmian words are characterized. A strong relationship to the continued fraction expansion of irrational numbers is established.

Chapter 3, "Unavoidable Patterns", is written by Julien Cassaigne. A pattern is unavoidable if there exist infinitely many words that do not encounter this pattern. This is the generalization of square-free words (covered in Lothaire's first book). The algorithm of Zimin for testing whether a pattern is unavoidable is given. When the alphabet is fixed, one gets a hierarchy of avoidability (squares are 3-avoidable but not 2-avoidable). Some results concerning this hierarchy are derived.

Chapter 4, "Sesquipowers", is written by Aldo De Luca and Stefano Varricchio. Sesquipowers can be defined by bi-ideal sequences. These sequences have interesting combinatorial properties, with links to recurrence and n-divisions. From these an improvement of an important theorem by Shirshov is obtained. Regularities of Coudrain and Schützenberger, and of Shirshov, can be presented in a unified way. Applications to finiteness conditions in semigroups are given.

Chapter 5, "The Plactic Monoid", is written by Alain Lascoux, Bernard Leclerc and Jean-Yves Thibon. The plactic monoid is an algebraic structure that takes into account most of the combinatorial properties of Young tableaux. The starting point of the theory is Schensted's algorithm. The defining relations of the plactic monoid were determined by D. Knuth. Applications include the Littlewood–Richardson rule, a combinatorial description of the Kostka–Foulkes polynomials, a noncommutative version of the Demazure character formula, and of the Schubert polynomials. Quite recently, combinatorics of Young tableaux were related to quantum groups.

Chapter 6, "Codes", written by Véronique Bruyère, is concerned with several kinds of codes, in relation to the so-called defect theorem. The defect effect still holds if the set is not an ω-code. A remarkable phenomenon appears when, for a finite code X, neither X nor its reversal $\widetilde{X}$ is an ω-code. In this case, the n elements of X can be expressed as a product of $n - 2$ words. The body of the chapter ends with a short and elementary proof of a result of Schützenberger stating that a finite maximal code X that is also an ω-code is a prefix code.

Chapter 7, "Numeration Systems", written by Christiane Frougny, deals with the various ways to write integers, reals, and complex numbers in positional number systems. Finite automata may exist to perform arithmetic operations, such as addition, and also to compute some standard representation. A special class of representations, called beta-expansions, has several interesting properties related to symbolic dynamical systems. Generalizing the notion of base leads to number systems with respect to a sequence of numbers, such as the Fibonacci numbers. Numeration systems for complex numbers, without sign, and without separating real and imaginary parts, are considered at the end of the chapter.

Chapter 8, "Periodicity", written by Filippo Mignosi and Antonio Restivo, considers periods of various kinds in finite and infinite words. Repetitions may be of rational (not only integer) order. The golden ratio appears to be an extremal value for periodicity in words. An important topic is the relation between local and global periodicity. Criteria for infinite words to be periodic are given next. Again, the golden ratio plays a central role.

The aim of Chapter 9, "Centralizers of Noncommutative Series and Polynomials", written by Christophe Reutenauer, is to give a self-contained proof of Cohn's and Bergman's centralizer theorems. These are analogues, in polynomials and series, of the well-known fact that two commuting words are powers of a third word. The proofs use noncommutative Euclidean division, the result that the centralizer of a noncommutative polynomial is integrally closed in its field of fractions, its embeddability in a one-variable polynomial ring, and a characterization of free subalgebras of a one-variable polynomial algebra. In addition, a defect theorem is shown to hold for two noncommutative polynomials.

Chapter 10, "Transformations on Words and q-Calculus", written by Dominique Foata and Guo-Niu Han, deals with statistics on words. There are several relevant statistics, such as the number of descents, of excedences, the major index, and the Denert statistics. MacMahon had already calculated the distributions of the early statistics. All calculations are presented here in a unified way. The second part is devoted to the derivation of an algorithm, which involves the introduction of commutation rules on biwords, and is useful for the construction of two bijections. The body of the chapter concludes with the proof of equidistribution properties.

Chapter 11, "Statistics on Permutations and Words", is written by Jacques Désarménien. It starts with the so-called shape of a word, computes statistics on shapes, considers inversion of permutations with a given shape. Lyndon words are related to cycles of permutations.

Chapter 12, "Makanin's Algorithm", written by Volker Diekert, is a self-contained exposition of the famous theorem of Makanin stating that it is decidable whether a set of equations in words has a solution. The first step towards Makanin's result is to bound the exponent of periodicity. Next, the problem is transformed to systems of boundary equations. This leads to a geometric reflection of the problem. An upper bound for the exponent of periodicity yields an upper bound on the length of convex chains. This in turn leads to an upper bound on the number of boundary equations. Then transformation rules are defined which either lead to a solution, or introduce additional boundary equations. Since their number is bounded, this procedure eventually stops.

Chapter 13, "Independent Systems of Equations", written by Tero Harju, Juhani Karhumäki and Wojciech Plandowski, is concerned with the existence of a notion of dimension for a set of words. A good example is the defect theorem already considered earlier. Another result is the compactness property (also known as Ehrenfeucht's conjecture) stating that every independent set of equations in words is finite. Existence of independent systems of equations, together with bounds on their size, is given. Although the problem generalizes in a natural way to all semigroups, the compactness property does not hold in all semigroups. Varieties of semigroups with that property are characterized in terms of ascending chains of congruences.

Each chapter of this book can be read independently of the others, in the sense that there is no logical dependence of the results of one chapter on those of another one. The introductory chapter (Chapter 1) is an exception, however. It contains definitions and results used in the rest of the volume and it has been designed as a reference for the other ones. Each of the chapters can be used for a separate graduate course or seminar. The necessary mathematical background does not exceed a general undergraduate level.

In the Problems sections at the ends of chapters, difficult ones are indicated with an asterisk or double asterisk.

A word about the process which gave rise to this volume. The authors of the previous volume have agreed to serve as a steering committee. The set of authors is to a large extent different from the previous one. On several occasions, the whole content of the book has been presented in seminars. This includes a special session of the Lotharingian seminar held in Bellagio in October 1996.

Finally, a third volume of combinatorics on words is in preparation. It is focused on applications. This includes natural language processing, text algorithms, fractals and tilings and bioinformatics.

The authors acknowledge helpful discussions and comments with a great number of colleagues. Among them are Laurence Bartz, Paul

Preface

Cohn, Clelia de Felice, Jeanne Devolder, Georges Hansel, Jacques Justin, Michel Koskas, Michel Latteux, Jean Mairesse, Yuri V. Matiyasevich, Anca Muscholl, Bruno Pettazoni, Gwénaël Richomme, Klaus U. Schulz, Stephanie van Willigenbourg.

Jean Berstel
Dominique Perrin

Marne-la-Vallée, May 16, 2001

CHAPTER 1

Finite and Infinite Words

1.0. Introduction

The aim of this chapter is to provide an introduction to several concepts used elsewhere in the book. It fixes the general notation on words used elsewhere. It also introduces more specialized notions of general interest. For instance, the notion of a uniformly recurrent word used in several other chapters is introduced here.

We start with the notation concerning finite and infinite words. We also describe the Cantor space topology on the space of infinite words.

We provide a basic introduction to the theory of automata. It covers the determinization algorithm, part of Kleene's theorem, syntactic monoids and basic facts about transducers. These concepts are illustrated on the classical combinatorial examples of the de Bruijn graph, and the Morse–Hedlund theorem.

We also consider the relationship with generating series, as a useful tool for the enumeration of words.

We introduce some basic concepts of symbolic dynamical systems, in relation with automata. We prove the equivalence between the notions of minimality and uniform recurrence. Entropy is considered, and we show how to compute it for a sofic system.

We also present a more specialized subject, namely unavoidable sets. This notion is easy to define but leads to interesting and significant results. In this sense, the last section of this chapter is a foretaste of the rest of the book.

1.1. Semigroups

As usual, $\mathbb{N}$, $\mathbb{Z}$, $\mathbb{Q}$, $\mathbb{R}$, $\mathbb{C}$ denote the sets of nonnegative integers, integers, rational, real and complex numbers respectively. We denote by Card X the cardinality of the set X.

A *semigroup* is a set equipped with a binary associative operation. The set of words over a given alphabet has an obvious semigroup structure for the concatenation of words. A subsemigroup is a subset closed under the operation. A semigroup *morphism* from a semigroup S into a semigroup T is a mapping $f: S \to T$ such that $f(uv) = f(u)f(v)$ for all $u, v \in S$.

A *monoid* M is a semigroup with a neutral element, i.e. an element ε such that $m\varepsilon = \varepsilon m = m$ for all $m \in M$. A *submonoid* of a monoid M is a subset of M closed under the operation and containing the neutral element of M. A monoid *morphism* $f: M \to N$ is a semigroup morphism such that $f(\varepsilon_M) = \varepsilon_N$.

Given two semigroups S and T, the set $S \times T$ is canonically equipped with a semigroup operation by setting $(s,t)(s',t') = (ss', tt')$. The semigroup $S \times T$ is the *direct product* of S and T. A subset of $S \times T$ is called a *relation* between (or over) S and T.

Let X and Y be two subsets of a semigroup S. The *product* of X and Y is the set
$$XY = \{xy \mid x \in X, y \in Y\}.$$

Given a set $X \subset S$, we denote by X^+ the subsemigroup generated by X, that is
$$X^+ = \{x_1 \cdots x_n \mid n \geq 1, \ x_i \in X\}.$$
The operation $X \mapsto X^+$ is called the *plus operation*. This unary operation should not be confused with the (binary) disjoint union. If S is a monoid, we also define
$$X^* = X^+ \cup \{\varepsilon\} \tag{1.1.1}$$
which is the submonoid generated by X. The operation $X \mapsto X^*$ is called the *star operation*.

A subset X of a semigroup S is *rational* if it can be obtained from the finite subsets of S by a finite number of the operations of union, product, and plus.

In a monoid M, the family of rational sets is closed under the star operation because of Formula (1.1.1). Actually, this family is also generated by the operations of union, product and star, because $X^+ = XX^*$.

A special case deserves a mention. A rational subset of a product semigroup is called a *rational relation*.

EXAMPLE 1.1.1. For any set Q, the set $2^{Q \times Q}$ of binary relations on Q is a monoid for the composition of relations. The identity relation is the neutral element. The set of partial functions from Q into Q is a submonoid of $2^{Q \times Q}$. The set of permutations of Q is a submonoid of the latter.

1.2. Words

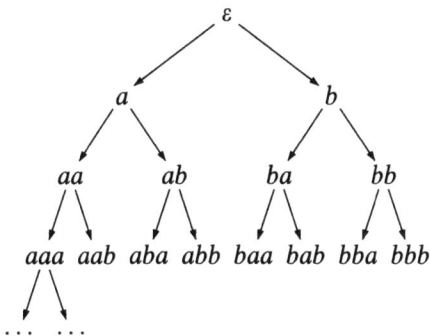

Figure 1.1. The tree of the free monoid.

EXAMPLE 1.1.2. For any finitely generated semigroup S, the set $D = \{(s,s) \mid s \in S\}$ is a rational relation called the *diagonal*.

1.2. Words

In this section, we first describe the (ordinary) finite words, before going to infinite one-sided and then two-sided words.

1.2.1. Finite words

We briefly introduce the basic terminology on words. A more detailed presentation can be found in Lothaire 1983. Let A be a set usually called the *alphabet*. We denote as usual by A^* the set of words over A and by ε the empty word. For a word w, we denote by $|w|$ the length of w. We use the notation $A^+ = A^* - \{\varepsilon\}$.

The set A^* is a monoid. Indeed, the concatenation of words is associative, and the empty word is a neutral element for concatenation. The set A^+ is usually called the *free semigroup* over A, while A^* is usually called the *free monoid*.

A word w is called a *factor* (resp. a *prefix*, resp. a *suffix*) of a word u if there exist words x, y such that $u = xwy$ (resp. $u = wy$, resp. $u = xw$). The factor (resp. the prefix, resp. the suffix) is *proper* if $xy \neq \varepsilon$ (resp. $y \neq \varepsilon$, resp. $x \neq \varepsilon$). The set of words over a finite alphabet A can be conveniently seen as a tree. Figure 1.1 represents $\{a, b\}^*$ as a binary tree. The vertices are the elements of A^*. The root is the empty word ε. The sons of a node x are the words xa for $a \in A$. Every word x can also be viewed as the path leading from the root to the node x. A word x is a prefix of a word y if it is an ancestor in the tree. We denote by Alph w the set of letters having at least one occurrence in the word w.

The set of factors of a word x is denoted by $F(x)$. We denote by $F(X)$ the set of factors of words in $X \subset A^*$. The *reversal* of a word $w = a_1 a_2 \cdots a_n$, where $a_1, \ldots, a_n$ are letters, is the word $\tilde{w} = a_n a_{n-1} \cdots a_1$. Similarly, for $X \subset A^*$, we denote $\tilde{X} = \{\tilde{x} \mid x \in X\}$. A *palindrome word* is a word w such that $w = \tilde{w}$. If $|w|$ is even, then w is a palindrome if and only if $w = x\tilde{x}$ for some word x. Otherwise w is a palindrome if and only if $w = xa\tilde{x}$ for some word x and some letter a.

An integer $p \geq 1$ is a *period* of a word $w = a_1 a_2 \cdots a_n$ where $a_i \in A$ if $a_i = a_{i+p}$ for $i = 1, \ldots, n-p$. The smallest period of w is called *the* period of w.

A word $w \in A^+$ is *primitive* if $w = u^n$ for $u \in A^+$ implies $n = 1$.

Two words x, y are *conjugate* if there exist words u, v such that $x = uv$ and $y = vu$. Thus conjugate words are just cyclic shifts of one another. Conjugacy is thus an equivalence relation. The conjugacy class of a word of length n and period p has p elements if p divides n and has n elements otherwise. In particular, a primitive word of length n has n distinct conjugates.

There are three order relations frequently used on words. We give the definition of each of them.

The *prefix order* is the partial order defined by $x \leq y$ if x is a prefix of y.

Two other orders, the *radix order* and the *lexicographic order*, are refinements of the prefix order which are defined for words over an ordered alphabet A. Both are total orders.

The *radix order* is defined by $x \leq y$ if $|x| < |y|$, or $|x| = |y|$ and $x = uax'$ and $y = uby'$ with a, b letters and $a \leq b$. If integers are represented in base k without leading zeros, then the radix order on their representations corresponds to the natural ordering of the integers. If we allow leading zeros, the same holds provided the words have the same length (which can always be achieved by padding).

For $k = 2$, the tree of words without leading zeros is given in Figure 1.2. The radix order corresponds to the order in which the vertices are met in a breadth-first traversal. The index of a word in the radix order is equal to the number represented by the word in base 2.

The *lexicographic order*, also called *alphabetic order*, is defined as follows. Given two words x, y, we have $x < y$ if x is a proper prefix of y or if there exist factorizations $x = uax'$ and $y = uby'$ with a, b letters and $a < b$. This is the usual order in a dictionary. Note that $x < y$ in the radix order if $|x| < |y|$, or $|x| = |y|$ and $x < y$ in the lexicographic order.

1.2. Words

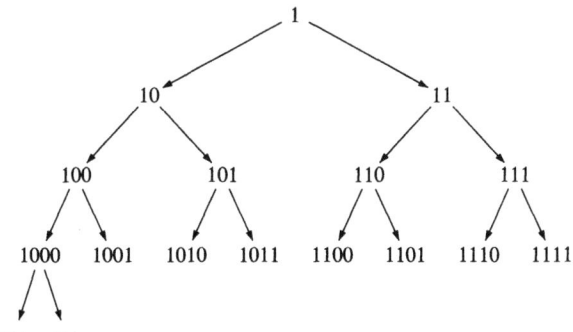

Figure 1.2. The tree of integers in binary notation.

A *Lyndon word* is a primitive word which is minimal for the lexicographic order in its conjugacy class. Thus, each nonempty word is a conjugate of a power of some Lyndon word.

The following result is known as Fine and Wilf's theorem (see Proposition 1.3.5 in Lothaire 1983 and Theorem 8.1.4). As usual, $\gcd(n,m)$ denotes the greatest common divisor of n and m.

PROPOSITION 1.2.1 (Fine and Wilf). *Let x, y be words, let $n = |x|$, $m = |y|$, $d = \gcd(n,m)$. If two powers x^p and y^q of x and y have a common prefix of length at least $n + m - d$, then x and y are powers of the same word.*

For $X, Y \subset A^*$, we say that the union of X and Y is *unambiguous* if $X \cap Y = \emptyset$. In this case, we write $X + Y$ as a notation equivalent to $X \cup Y$.

The *product* of X and Y is, as in any semigroup, the set $XY = \{xy \mid x \in X, y \in Y\}$. The product is said to be *unambiguous* if for each $z \in XY$ there is exactly one pair $(x, y) \in X \times Y$ such that $z = xy$. In particular, we define
$$X^0 = \{\varepsilon\}, \quad X^{n+1} = X^n X \ (n \geq 0).$$
Given a set $X \subset A^*$, the *star* of X is, as in any monoid, the set
$$X^* = \{x_1 \cdots x_n \mid n \geq 0, \ x_i \in X\} = \bigcup_{n \geq 0} X^n.$$

PROPOSITION 1.2.2. *Any submonoid M of A^* has a unique minimal generating set*
$$X = (M - \varepsilon) - (M - \varepsilon)^2.$$

Proof. First, we show that X generates M. Let $w \in M - \varepsilon$. If $w \notin (M - \varepsilon)^2$, then $w \in X$. Otherwise, $w = w'w''$, with $w', w'' \in M - \varepsilon$. By induction, $w', w'' \in X^*$, and thus $w \in X^*$. This shows that $M \subset X^*$. The converse inclusion is clear since $X \subset M$.

Let Y be a set of generators of M. Clearly, $X \subset Y^*$. Since no word of X is a product of two nonempty words of M, we have actually $X \subset Y$. This shows that X is minimal. ∎

We say that a subset X of A^+ is a *code* if there is no relation among the elements of X, i.e.
$$x_1 \cdots x_n = y_1 \cdots y_m$$
with $x_1, \ldots, x_n, y_1, \ldots, y_m \in X$ implies $n = m$ and $x_i = y_i$ for $i = 1, \ldots, n$. In this case, one has
$$X^* = \sum_{n \geq 0} X^n \tag{1.2.1}$$
since the sets X^n are pairwise disjoint. We say that the star operation is *unambiguous* on the set X if X is a code.

A *prefix code* is a set X of words such that none of its elements is a prefix of another one. A prefix code is clearly a code.

A set of words is *prefix-closed* if it contains the prefixes of all its elements. A set of words is *factor-closed* or *factorial* if it contains the factors of all its elements.

A function $f: A^* \to B^*$ is a *morphism* (also called a *substitution*) if $f(xy) = f(x)f(y)$ for all $x, y \in A^*$. A morphism is uniquely determined by its value on the alphabet. A morphism is *literal* if the image of a letter is a letter. It is *nonerasing* if the image of a letter is always a nonempty word.

A morphism $f: A^* \to B^*$ is injective if and only if f is injective on A and if $f(A)$ is a code (see Proposition 6.1.3 for a proof).

1.2.2. Infinite words

We denote by $A^{\mathbb{N}}$ the set of (right) infinite words. It is the set of sequences of symbols in A indexed by nonnegative integers. We also denote
$$A^{\infty} = A^* \cup A^{\mathbb{N}},$$
the set of finite or infinite words.

1.2. Words

For $x \in A^*$ and $y \in A^{\mathbb{N}}$, the product xy is well defined. This defines a left action of the semigroup A^* on the set $A^{\mathbb{N}}$ since $x(yz) = (xy)z$ for all $x, y \in A^*$ and $z \in A^{\mathbb{N}}$.

A finite word w is a *factor* of an infinite word x if $x = uwy$. The set of factors of x is denoted by $F(x)$ and the set of factors of length n is denoted by $F_n(x)$. For a subset X of $A^{\mathbb{N}}$, we denote by $F(X)$ the set of factors of words in X. An infinite word $s \in A^{\mathbb{N}}$ is a *suffix* of $x \in A^{\mathbb{N}}$ if there is a word p in A^* such that $x = ps$. The suffix is called *proper* if $p \neq \varepsilon$.

The *lexicographic order* has a simple expression for infinite words over an ordered alphabet, since $x < y$ if and only if $x = uax'$, $y = uby'$ for some word $u \in A^*$, some letters $a, b \in A$ with $a < b$ and $x', y' \in A^{\mathbb{N}}$.

For a set $X \subset A^*$, we denote by X^ω the set of all $x = x_0 x_1 x_2 \cdots$ with $x_i \in X - \varepsilon$. In particular, A^ω is the same as $A^{\mathbb{N}}$ and we use both notations without distinction. We also use the notation

$$X^\infty = X^* \cup X^\omega.$$

The *shift* function is the function $\sigma : A^{\mathbb{N}} \to A^{\mathbb{N}}$ defined by $\sigma(x_0 x_1 \cdots) = x_1 x_2 \cdots$.

Consider a finite or infinite word x over the alphabet A. The *complexity function* of x is the function that counts, for each integer $n \geq 0$, the number $P(x, n)$ of factors of length n in x:

$$P(x, n) = \mathrm{Card}(F_n(x)).$$

Clearly, $P(x, 0) = 1$ and $P(x, 1)$ is the number of letters appearing in x. If x is infinite, every factor can be extended to the right, whence $P(x, n) \leq P(x, n+1)$. Moreover,

$$P(x, n + m) \leq P(x, n) P(x, m)$$

since indeed $F_{n+m}(x) \subset F_n(x) F_m(x)$.

The set $A^{\mathbb{N}}$ is equipped with a distance defined as follows. For $x, y \in A^\omega$, we have $d(x, y) = 2^{-n}$ with

$$n = \min\{k \geq 0 \mid x_k \neq y_k\}$$

and the convention that $n = \infty$ and thus $d(x, y) = 0$ if $x = y$.

With respect to this distance, the set A^ω becomes a topological space, often called the *Cantor space*. A sequence $x^{(n)}$ of infinite words converges to y in this topology

$$y = \lim_{n \to \infty} x^{(n)},$$

if, for each index $i \in \mathbb{N}$, one has $x_i^{(n)} = y_i$ for large enough n.

For example, the sequence $x^{(n)} = a^n b^\omega$ converges to $y = a^\omega$.

A consequence of the definitions is that if a word w is a factor of $\lim_{n\to\infty} x^{(n)}$, then w is a factor of all but a finite number of the $x^{(n)}$.

A set of infinite words is (topologically) *closed* if it contains the limit of each convergent sequence of its elements.

The open sets are the complements of the closed sets. They happen to be also the sets of the form XA^ω for $X \subset A^*$ (Problem 1.2.3).

The following result is known as König's lemma.

PROPOSITION 1.2.3 (König's lemma). *If X is an infinite prefix-closed set of words over a finite alphabet A, there is an infinite word x having all its prefixes in X.*

Proof. There is a letter a_1 which is a prefix of an infinite number of elements of X. Similarly, there is a letter a_2 such that $a_1 a_2$ is a prefix of an infinite number of elements of X. Continuing this way, one obtains an infinite word $a_1 a_2 \cdots$ having all its prefixes in X. ∎

A set of words is *compact* if any sequence $x^{(n)}$ of infinite words of the set has a convergent sub-sequence.

PROPOSITION 1.2.4. *For a finite alphabet A, the space A^ω is compact.*

Proof. Consider a sequence $x^{(n)}$ of infinite words. Let X be the set of all prefixes of the words $x^{(n)}$. By König's lemma, there is an infinite word x which has all its prefixes in X.

For each $i > 0$, let u_i be the prefix of length i of x. Since $u_i \in X$, there is an integer n_i such that u_i is a prefix of $x^{(n_i)}$. Clearly, the sequence $x^{(n_i)}$ converges to x. ∎

A closed subset of $A^{\mathbb{N}}$ is also compact, as one may check, provided A is finite as above.

If $X_1 \supset X_2 \supset \cdots \supset X_n \supset \cdots$ is a decreasing sequence of nonempty closed subsets of $A^{\mathbb{N}}$ where A is finite, then their intersection X is nonempty. Consider indeed, for each n, a word x_n in X_n. By compactness, there is a sub-sequence converging to an infinite word x. Since the X_n are closed, x is in all of them, thus x is in X.

For any set $X \subset A^\infty$, we denote

$$X_* = X \cap A^*, \quad X_\omega = X \cap A^\omega.$$

A *binoid* over A is a subset M of A^∞ such that

$$M_* M \subset M, \quad (M_*)^\omega \subset M.$$

1.2. Words

Observe that in particular M_* is a submonoid of A^*. It is convenient to denote, for every set $X \subset A^\infty$,

$$X^\infty = X_*^\infty \cup (X_*)^* X_\omega = X_*^* \cup X_*^\omega \cup (X_*)^* X_\omega.$$

With this notation, a set $M \subset A^\infty$ is a binoid if and only if $M^\infty = M$. For any set $X \subset A^\infty$, the set X^∞ is a binoid called the binoid generated by X. It is also the intersection of all binoids containing X. A binoid M such that $M = X^\infty$ for some $X \subset A^*$ is called *finitary*.

EXAMPLE 1.2.5. Set $X = a \cup b^\omega$. Then $X^\infty = a^\infty \cup a^* b^\omega$. This binoid is not finitary.

PROPOSITION 1.2.6. *Any binoid M has a unique minimal generating set* $X = (M - \varepsilon) - (M_* - \varepsilon)(M - \varepsilon)$.

Proof. Observe first that X_* is the minimal generating set of M_* by Proposition 1.2.2.

To prove that X also generates M_ω, let $y \in M_\omega$. If $y \notin (M_*-\varepsilon)M$, then $y \in X$. Otherwise, $y = x_1 y_1$ for some $x_1 \in X_* - \varepsilon$ and $y_1 \in M_\omega$. Again, if $y_1 \notin (M_* - \varepsilon)M$, then $y_1 \in X$ and thus $y \in X^\infty$. Otherwise, $y_1 = x_2 y_2$ for some $x_2 \in X_* - \varepsilon$ and $y_2 \in M_\omega$. Thus $y = x_1 x_2 y_2$ is in X^∞. Continuing in this way, either we eventually obtain $y = x_1 x_2 \cdots x_n y_n \in X^\infty$, or $y = x_1 x_2 \cdots x_n \cdots \in (X_*)^\omega$. This proves that X generates M.

Let Y be a generating set of M. We know that $X_* \subset Y_*$. Let $x \in X_\omega$. Since $Y^\infty = Y_* Y \cup Y$, one has $x = yy'$ for some $y \in Y_* \cup \varepsilon$ and $y' \in Y$. By the definition of X, we have $y = \varepsilon$ and thus $x \in Y$. Thus $X \subset Y$. ∎

COROLLARY 1.2.7. *The minimal generating set of a finitary binoid M is the minimal generating set of the monoid M_*.* ∎

A word $x \in A^\omega$ is *periodic* if it is of the form $x = z^\omega$ for some $z \in A^+$. A word $x \in A^\omega$ is *eventually periodic* or *ultimately periodic* if it is of the form $x = yz^\omega$ for some $y, z \in A^+$. A word $x \in A^\omega$ is *aperiodic* if it is not eventually periodic.

A word x is periodic if and only if it is a proper suffix of itself or equivalently if $x = \sigma^p(x)$ for some $p > 0$.

A nonerasing morphism $f: A^* \to B^*$ defines a function, also called a morphism, from $A^\mathbb{N}$ to $B^\mathbb{N}$ by $f(a_0 a_1 \cdots a_n \cdots) = f(a_0) f(a_1) \cdots f(a_n) \cdots$.

A sequence $(u_n)_{n \geq 0}$ of finite words over an alphabet A *converges* to an infinite word x if every prefix of x is a prefix of all but a finite number of the words u_n. This word x is unique and is denoted by

$$x = \lim_{n \to \infty} u_n.$$

This definition can be related to the topology considered above (Problem 1.2.4).

As an example, the sequence $a^n b^n$ converges to a^ω. An important special case arises when every u_n is a prefix of u_{n+1}. Then the sequence converges, provided the lengths of the words u_n are unbounded. A special case of this is described in the following statement.

PROPOSITION 1.2.8. *Let h be a nonerasing morphism from A^* into itself, and let a be a letter such that $h(a) = as$ for some nonempty word s. Set, for $n \geq 0$,*

$$u_n = h^n(a), \quad v_n = h^n(s).$$

Then

(i) $u_{n+1} = u_n v_n$, *and in particular, u_n is a prefix of u_{n+1} for all $n \geq 0$,*
(ii) $u_{n+1} = a v_0 v_1 v_2 \cdots v_n$,
(iii) *the infinite word*

$$x = ash(s)h^2(s) \cdots h^n(s) \cdots \tag{1.2.2}$$

is the limit of the words u_n and x is a fixed point of h. Moreover, it is the unique fixed point of h starting with the letter a.

Proof. (i) $u_{n+1} = h^{n+1}(a) = h^n(h(a)) = h^n(as) = u_n v_n$.
Part (ii) holds for $n = 0$, and by induction

$$u_{n+1} = u_n v_n = a v_0 v_1 v_2 \cdots v_{n-1} v_n.$$

(iii) It is clear that x is the limit. Moreover,

$$h(x) = h(a)h(s)h^2(s)\cdots = x. \qquad \blacksquare$$

The word x of the proposition is also denoted by

$$x = h^\omega(a).$$

A word x obtained in this way is a *morphic* word.

We now develop two examples which will occur throughout the book.

EXAMPLE 1.2.9. The *Thue–Morse* infinite word t over the alphabet $A = \{0, 1\}$ is defined as the limit

$$t = \lim_{n \to \infty} u_n$$

where the sequences of words $(u_n)_{n \geq 0}$ and $(v_n)_{n \geq 0}$ are defined by

$$u_0 = 0, \qquad v_0 = 1,$$
$$u_{n+1} = u_n v_n, \qquad v_{n+1} = v_n u_n, \qquad n \geq 0.$$

1.2. Words

The first letters of t are

$t =$
01101001100101101001011001101001100101100110100101101001100$\cdots$.

The word t is actually a morphic word since $u_n = \mu^n(0)$ where μ is the morphism
$$\mu : \begin{array}{l} 0 \mapsto 01, \\ 1 \mapsto 10. \end{array}$$

The decomposition of x corresponding to Equation (1.2.2) is

$$t = 0\ 1\ 10\ 1001\ 10010110\ 1001011001101001 \cdots.$$

EXAMPLE 1.2.10. Let $A = \{a, b\}$. Let φ be the morphism defined by

$$\varphi : \begin{array}{l} a \mapsto ab, \\ b \mapsto a. \end{array}$$

The *Fibonacci* word is the infinite word $f = \varphi^\omega(a)$. The first letters of f are

$$f = abaababaabaababaababaabaababaabaab \cdots.$$

One has also $f = \lim_{n\to\infty} f_n$ where the sequence of words $f_n = \varphi^n(a)$ can also be defined by

$$f_0 = a, \ f_1 = ab, \ f_{n+2} = f_{n+1}f_n.$$

The sequence of lengths of the words f_n is the traditional sequence of Fibonacci numbers (see Example 1.4.2).

1.2.3. Two-sided infinite words

We denote by $A^{\mathbb{Z}}$ the set of two-sided infinite words on A, which is the set of sequences of symbols of A indexed by integers. For $x \in A^{\mathbb{Z}}$, the *shift* function is the function $\sigma : A^{\mathbb{Z}} \to A^{\mathbb{Z}}$ defined by $\sigma(x) = y$ with $y_n = x_{n-1}$ for $n \in \mathbb{Z}$. Observe that, contrary to the one-sided case, the shift is a one-to-one transformation on $A^{\mathbb{Z}}$. The *period* of $x \in A^{\mathbb{Z}}$ is the greatest common divisor of the integers $n > 1$ such that $\sigma^n(x) = x$. It is an integer or ∞. The terminology used for words or one-sided infinite words carries over. In particular, $F(x)$ denotes the set of (finite) factors of a word $x \in A^{\mathbb{Z}}$, and if $X \subset A^{\mathbb{Z}}$, we denote by $F(X)$ the set of factors of words in X.

The set $A^{\mathbb{Z}}$ is also equipped with a distance defined in a way quite analogous to the distance of A^{ω}. We define $d(x,y) = 2^{-n}$ where

$$n = \min\{k \geq 0 \mid x_k \neq y_k \text{ or } x_{-k} \neq y_{-k}\}$$

with the convention that $d(x,y) = 0$ when $x = y$.

This distance defines a topology on the set $A^{\mathbb{Z}}$ as in the one-sided case. There exists a two-sided version of König's lemma.

PROPOSITION 1.2.11. *For any infinite factorial set X of words over a finite alphabet, there exists a two-sided infinite word having all its factors in X.*

Proof. This is similar to the one-sided case. ∎

Again, the space $A^{\mathbb{Z}}$ is compact when A is finite.

For a set $X \subset A^+$, we denote by X^ς the closure under the shift of the set of all $w = (a_n)_{n \in \mathbb{Z}} \in A^{\mathbb{Z}}$ such that

$$\cdots a_{-1}a_0 = \cdots x_{-1}x_0, \quad a_1 a_2 \cdots = x_1 x_2 \cdots$$

with $x_n \in X$ for all $n \in \mathbb{Z}$. Observe that A^ς coincides with $A^{\mathbb{Z}}$. For a single word $x = a_1 a_2 \cdots a_n \in A^+$, the set x^ς is composed of the sequences of the form $\cdots a_n(a_1 a_2 \cdots a_n) a_1 a_2 \cdots$. Each word in x^ς has a period which divides n. The period is n if and only if the word x is primitive.

A literal morphism $f : A^* \to B^*$ defines a function, also called a literal morphism, from $A^{\mathbb{Z}}$ to $B^{\mathbb{Z}}$. It maps the word $x \in A^{\mathbb{Z}}$ to the word $y = f(x) \in B^{\mathbb{Z}}$ defined by $y_i = f(x_i)$.

1.3. Automata

1.3.1. Definitions

An *automaton* over the alphabet A is composed of a set Q of *states*, a set $E \subset Q \times A \times Q$ of *edges* or *transitions* and two sets $I, T \subset Q$ of *initial* and *terminal* states. For an edge $e = (p, a, q)$, the state p is the *origin*, a is the *label*, and q is the *end*.

The automaton is often denoted by $\mathcal{A} = (Q, E, I, T)$, or also (Q, I, T) when E is understood, or even $\mathcal{A} = (Q, E)$ if $Q = I = T$.

A *path* in the automaton $\mathcal{A}$ is a sequence

$$(p_0, a_1, p_1), (p_1, a_2, p_2), \ldots, (p_{n-1}, a_n, p_n)$$

1.3. Automata

of consecutive edges. Its label is the word $x = a_1 a_2 \cdots a_n$. The path *starts* at p_0 and *ends* at p_n. The path is often denoted by

$$p_0 \xrightarrow{x} p_n.$$

A path is *successful* if it starts in an initial state and ends in a terminal state. The set *recognized* by the automaton is the set of labels of its successful paths.

A state p is *accessible* if there is a path starting in an initial state and ending in p. It is *coaccessible* if there is a path starting in p and ending in a terminal state. An automaton is *trim* if every state is accessible and coaccessible.

An automaton is *unambiguous* if, for each pair of states p, q, and for each word w, there is at most one path from p to q labeled with w.

An automaton is *deterministic* if, for each state p and each letter a, there is at most one edge which starts at p and is labeled by a. This state is denoted by $p \cdot a$. Clearly, a deterministic automaton is unambiguous.

Given an automaton $\mathscr{A}$ and a state q of $\mathscr{A}$, the set of *first returns* to q is the set of labels of paths from q to q which do not pass another time through q. If $\mathscr{A}$ is unambiguous, then the set of first returns to a state q is a code. If $\mathscr{A}$ is deterministic, it is a prefix code.

EXAMPLE 1.3.1. Let $\mathscr{A}$ be the automaton shown in Figure 1.3. The set of first returns to state 1 is the prefix code $X = \{b, ab\}$ of Example 1.4.2.

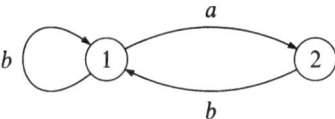

Figure 1.3. Golden mean automaton.

An automaton is *finite* if its set of states is finite. Since the alphabet is usually assumed to be finite, this means that the set of edges is finite.

A set of words X over A is *recognizable* if it can be recognized by a finite automaton.

PROPOSITION 1.3.2. *Every recognizable set can be recognized by a finite trim deterministic automaton having a unique initial state.*

Proof. Let $\mathscr{A} = (Q, E, I, T)$ be a finite automaton over A recognizing a set X. Let $\mathscr{B} = (\mathscr{R}, F, \{I\}, \mathscr{T})$ be the automaton defined as follows. Its states are the subsets

$$Q(u) = \{q \in Q \mid i \xrightarrow{u} q \text{ for some } i \in I\}$$

for all u in A^*. Since Q is finite, there is a finite number of subsets $Q(u)$. The edges of $\mathcal{B}$ are all triples

$$(Q(u), a, Q(ua)).$$

The set of terminal states is

$$\mathcal{T} = \{U \in \mathcal{R} \mid U \cap T \neq \emptyset\}.$$

It is easy to verify that $\mathcal{B}$ is trim, deterministic, and recognizes X. ∎

Let X be a subset of A^*. For $w \in A^*$, we define the *left* and *right* quotients

$$w^{-1}X = \{u \in A^* \mid wu \in X\}, \quad Xw^{-1} = \{u \in A^* \mid uw \in X\}.$$

The following relations hold for words v, w and a letter a:

$$(vw)^{-1}X = w^{-1}(v^{-1}X), \quad a^{-1}(XY) = (a^{-1}X)Y \cup (X \cap \varepsilon)a^{-1}Y.$$

The notation is extended as usual to sets by

$$X^{-1}Y = \bigcup_{x \in X} x^{-1}Y.$$

To every set $X \subset A^*$ is associated a deterministic automaton $\mathcal{A}(X)$ as follows. Its set of states $Q(X)$ is

$$Q(X) = \{w^{-1}X \mid w \in A^*\}.$$

Its initial state is X, its set of final states is $T(X) = \{S \in Q(X) \mid \varepsilon \in S\}$. Its transitions are defined for $S \in Q(X)$ and $a \in A$ by $S \cdot a = a^{-1}S$. The automaton $\mathcal{A}(X)$ is called the *minimal automaton* of X. It recognizes X because indeed

$$X \cdot w \in T(X) \iff \varepsilon \in w^{-1}X \iff x \in X.$$

An equivalence relation on A^* is *right regular* if $u \equiv v$ implies $ux \equiv vx$ for all $u, v, x \in A^*$.

PROPOSITION 1.3.3. *Let X be a subset of A^*. The following conditions are equivalent:*
(i) *X is recognizable,*

1.3. Automata

(ii) *the automaton $\mathcal{A}(X)$ is finite,*
(iii) *there exists a right regular equivalence relation of finite index on A^* for which X is a union of equivalence classes.*

Proof. (i) ⇔ (ii) First $\mathcal{A}(X)$ recognizes X. Conversely, let $\mathcal{A} = (Q, E, i, T)$ be a finite trim deterministic automaton recognizing X. For each state $q \in Q$, let $X_q = \{u \in A^* \mid q \cdot u \in T\}$. The set $Q(X)$ is equal to the set of the X_q for $q \in Q$. Indeed, for $w \in A^*$, we have $w^{-1}X = X_{i \cdot w}$.

(ii) ⇔ (iii) The equivalence relation defined by $u \equiv v$ if and only if $u^{-1}X = v^{-1}X$ satisfies the conditions. Conversely, if (iii) is satisfied, the set $Q(X)$ is finite. Indeed, $u \equiv v$ implies $u^{-1}X = v^{-1}X$, and thus the elements of $Q(X)$ are unions of equivalence classes. ∎

It can be shown (Problem 1.3.1) that $\mathcal{A}(X)$ is the (unique) smallest deterministic automaton recognizing X.

PROPOSITION 1.3.4. *A set of words X over the alphabet A is recognizable if and only if there exists a morphism $f : A^* \to S$ from A^* into a finite semigroup S such that $X = f^{-1}(f(X))$.*

Proof. Let $\mathcal{A} = (Q, E, I, T)$ be a finite automaton over A recognizing a set X. The set S of all binary relations over Q is a semigroup for the composition of relations. Define for each word w the relation $f(w)$ by

$$f(w) = \{(p, q) \in Q \times Q \mid p \xrightarrow{w} q\}.$$

It is easy to check that f is a semigroup morphism and that $X = f^{-1}(U)$, where $U = \{s \in S \mid s \cap I \times T \neq \emptyset\}$. Thus $X = f^{-1}(f(X))$.

Conversely, let $f : A^* \to S$ be a semigroup morphism satisfying the conditions of the statement. Define an automaton $\mathcal{A} = (S, E, f(\varepsilon), f(X))$ where $E = \{p, a, q) \in S \times A \times S \mid pf(a) = q\}$. It can be verified that this automaton recognizes the set X. ∎

A semigroup S is said to *recognize* a set X if there exists a morphism $f : A^* \to S$ such that $X = f^{-1}(f(X))$.

Let X be a set of words. The set of contexts of a word w is the set

$$C(w) = \{(x, y) \in A^* \times A^* \mid xwy \in X\}.$$

The *syntactic equivalence* of X is defined by $u \equiv v$ if and only if $C(u) = C(v)$. The syntactic equivalence is compatible with concatenation of words, and thus the quotient $A^*/\equiv$ is a semigroup. It is called the *syntactic semigroup* of X. It can be shown (Problem 1.3.2) that the syntactic semigroup of X is the smallest semigroup recognizing X. In particular, X is recognizable if and only if its syntactic semigroup is finite.

A subset X of A^* is *rational* if it can be obtained from the finite subsets of A by a finite number of the operations of union, product, and star.

A subset X of A^* is *unambiguously rational* if it can be obtained from the finite subsets of A by a finite number of the operations of unambiguous union, product, and star.

A well-known theorem of Kleene (see Notes) asserts that, over a finite alphabet, a set is rational if and only if it is recognizable. We prove here one direction of the equivalence in a slightly stronger form.

PROPOSITION 1.3.5. *Any recognizable set is unambiguously rational.*

Proof. Let $\mathcal{A} = (Q, i, T)$ be a finite deterministic automaton recognizing a set X with a unique initial state. For $p, q \in Q$, let $X_{p,q}$ be the set of nonempty words recognized by the automaton (Q, p, q). Then

$$X = \sum_{t \in T} X_{i,t} + \Delta_{i,T}$$

where $\Delta_{i,T} = \{\varepsilon\}$ if $i \in T$, and $\Delta_{i,T} = \emptyset$ otherwise. It is therefore enough to prove that each $X_{i,j}$ is unambiguously rational.

For $P \subset Q$ and $p, q \in Q$, we denote by $X_{p,P,q}$ the set of nonempty words that are labels of paths from p to q and which only pass through states in P (except perhaps at the beginning and the end). We prove that each $X_{p,P,q}$ is unambiguously rational by induction on the size of P. If $P = \emptyset$, then $X_{p,P,q}$ is a subset of the alphabet A. Set $Y_{p,q} = X_{p,P,q}$ and $Z_{p,q} = X_{p,P \cup \{r\},q}$ for some state $r \notin P$. We have the formula

$$Z_{p,q} = Y_{p,q} + Y_{p,r}(Y_{r,r}^*)Y_{r,q}.$$

The operations used in this formula are unambiguous. This proves the property by induction. ∎

1.3.2. Automata on infinite words

In this subsection, we introduce acceptance of infinite words by finite automata in Büchi's sense.

Let $\mathcal{A} = (Q, E, I, T)$ be a finite automaton over A. An infinite path is an infinite sequence

$$(p_0, a_0, p_1), (p_1, a_1, p_2), \ldots$$

1.3. Automata

of consecutive edges. Its label is the infinite word $x = a_0 a_1 \cdots$. The path is *successful* if $p_0 \in I$ and if $p_n \in T$ for infinitely many indices n.

The set of infinite words *recognized* by the automaton is the set of labels of successful infinite paths. An automaton used to recognize infinite words in this sense is frequently called a *Büchi automaton*.

EXAMPLE 1.3.6. Consider first the automaton given in the left part of Figure 1.4 with $I = \{1\}$ and $T = \{2\}$. It recognizes the set X of words having an infinite number of occurrences of a. The second automaton, given on the right, again with $I = \{1\}$ and $T = \{2\}$, recognizes the complement of X, namely the set of words with a finite number of occurrences of a.

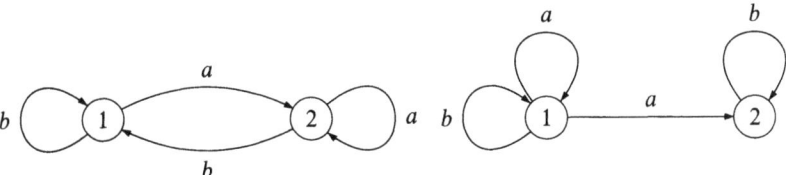

Figure 1.4. Büchi automata.

Observe that the complement of X is *not* obtained by simply complementing the set of terminal states in the first automaton.

1.3.3. Transducers

A *transducer* over the monoid $A^* \times B^*$ is composed of a set Q of *states*, a set $E \subset Q \times A^* \times B^* \times Q$ of *edges* and two sets $I, T \subset Q$ of *initial* and *terminal* states. For an edge $e = (p, x, y, q)$, the state p is the *origin*, x is the *input label*, y is the *output label*, and q is the *end*. Thus, a transducer is the same object as an automaton, except that the labels of the edges are pairs of words instead of letters.

A transducer is often denoted by $\mathcal{A} = (Q, E, I, T)$, or also (Q, I, T) when E is understood, or even $\mathcal{A} = (Q, E)$ if $Q = I = T$.

A *path* in the transducer $\mathcal{A}$ is a sequence

$$(p_0, x_1, y_1, p_1), (p_1, x_2, y_2, p_2), \ldots, (p_{n-1}, x_n, y_n, p_n)$$

of consecutive edges. Its input label is the word $x = x_1 x_2 \cdots x_n$, its output label is the word $y = y_1 y_2 \cdots y_n$. The path *starts* at p_0 and *ends* at p_n.

The path is often denoted by

$$p_0 \xrightarrow{x/y} p_n.$$

A path is *successful* if it starts in an initial state and ends in a terminal state. The set *recognized* by the transducer is the set of labels of its successful paths, which is actually a relation $R \subset A^* \times B^*$. The *function computed* by the transducer is the function f from A^* into the set of subsets of B^* associated to the relation R:

$$f(x) = \{y \in B^* \mid (x, y) \in R\}.$$

Thus a transducer can be seen as a machine computing nondeterministically output words from input words.

A transducer is *finite* if its set of states is finite. It can be shown that a subset of $A^* \times B^*$ is a rational relation if and only if it is the set recognized by a finite transducer.

EXAMPLE 1.3.7. The automaton of Figure 1.5 computes the identity function on $\{a, b\}^*$.

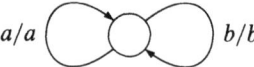

Figure 1.5. The transducer for the diagonal over $\{a, b\} \times \{a, b\}$.

EXAMPLE 1.3.8. The subset R of $a^* \times \{b, c\}^*$ defined as

$$R = (a^2, b^2)^* \cup (a^2, c^2)^*(a, c)$$

is a rational relation. Its elements have the form (a^n, d^n), with $d = b$ if n is even, and $d = c$ otherwise. The automaton of Figure 1.6 recognizes the relation R.

Let $\mathcal{A}$ be a transducer such that its edges are labeled by elements of $A \times B^*$. The *underlying input automaton* of $\mathcal{A}$ is obtained by omitting the output label of each edge.

The transducer $\mathcal{A}$ is *sequential* if the following conditions are satisfied:
(i) it has a unique initial state,
(ii) the underlying input automaton is deterministic,
(iii) every state is final.

1.3. Automata

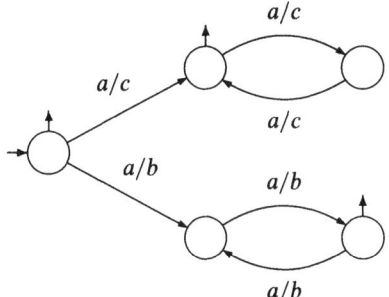

Figure 1.6. A transducer for the relation $(a^2, b^2)^* \cup (a^2, c^2)^*(a, c)$.

These conditions ensure that for each word $x \in A^*$, there is at most one word $y \in B^*$ such that (x, y) is recognized by $\mathcal{A}$. Thus, the function computed by $\mathcal{A}$ is a partial function from A^* into B^*.

EXAMPLE 1.3.9. The transducer of Example 1.3.7 is sequential. On the contrary, the transducer of Example 1.3.8 is not sequential. Actually, the function computed by this transducer is not computable by a sequential transducer. Indeed, one may verify that if f is a function computable by some sequential transducer, and if $f(xy)$ is defined, then $f(x)$ is a prefix of $f(xy)$.

A function f is *left sequential* (or *sequential* for short) if there is a sequential transducer which computes f. A function f is *right sequential* if the function $\tilde{f}$ defined by $y = \tilde{f}(x)$ if $\tilde{y} = f(\tilde{x})$ is left sequential. Thus, a right sequential function is a function computed by a sequential transducer operating from right to left.

A *subsequential transducer* $(\mathcal{A}, \omega)$ over $A^* \times B^*$ is a pair composed of a sequential transducer $\mathcal{A}$ over $A^* \times B^*$ with set of states Q, and of a function $\omega: Q \to B^*$. The function f computed by $(\mathcal{A}, \omega)$ is defined as follows. Let x be a word in A^*. The value $f(x)$ is defined if and only if there is a path $i \xrightarrow{x,y} q$ in $\mathcal{A}$ with input label x and starting in the initial state i. In this case, $f(x) = y\omega(q)$. Thus, the function ω is used to append a word to the output at the end of the computation.

A function computed by a subsequential transducer is a *left subsequential function*. *Right* subsequential functions are obtained by reversal. The following example shows that the successor of an integer in base 2 is a right subsequential function.

EXAMPLE 1.3.10. Let $A = \{0, 1\}$. For every word x in A^*, let $f(x)$ be the binary expansion of the successor of the integer represented by x in base 2, (most significant bit first). The pair $(\mathscr{A}, \omega)$, with $\mathscr{A}$ given in Figure 1.7, and $\omega(i) = 1$, $\omega(t) = \varepsilon$, computes the function $\tilde{f}$. Since $\mathscr{A}$ is sequential, f is right subsequential.

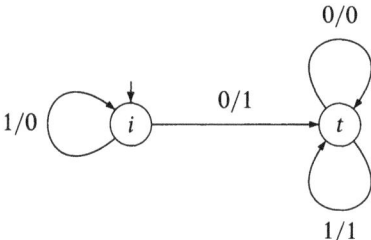

Figure 1.7. A transducer adding 1 in binary.

Any sequential transducer $\mathscr{A}$ over $A^* \times B^*$ defines in fact a function from A^∞ to B^∞. Indeed, for each x in A^∞, there is at most one y in B^∞ such that there is an infinite path starting in the initial state and labeled (x, y). A function $f : A^\infty \to B^\infty$ is *left sequential* if it can be computed by a sequential transducer.

1.3.4. Factor graphs

We now present a special family of automata which allows one to identify the occurrences of elements of a set X of words as factors of a given word.

For each $n \geq 1$, we define the de Bruijn graph of order n as the following labeled graph. The set of vertices is A^n and the set of edges is

$$E = \{(bs, a, sa) \mid a, b \in A, s \in A^{n-1}\}.$$

A word x is the label of a path from u to v if and only if v is the suffix of length n of ux. The de Bruijn graph of order 2 is given in Figure 1.8. For each two-sided infinite word x, there is a unique infinite path labeled by x. The set of vertices occurring in the path is the set $F_n(x)$ of factors of length n of x. The set of edges occurring in the path corresponds to the set $F_{n+1}(x)$.

More generally, consider an infinite word x. The *factor graph* $G_n(x)$ of order n is the labeled graph with vertex set $F_n(x)$ and edge set

$$E = \{(bs, a, sa) \mid a, b \in A, bsa \in F_{n+1}(x)\}.$$

1.3. Automata

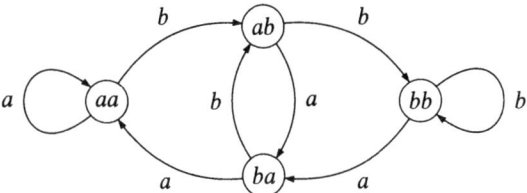

Figure 1.8. De Bruijn graph of order $n = 2$.

A word y is the label of a path from u to v in $G_n(x)$ if and only if $F_{n+1}(uy) \subset F_{n+1}(x)$ and v is the suffix of length n of uy. The de Bruijn graph is a particular case of a factor graph corresponding to a word x such that $F_{n+1}(x) = A^{n+1}$.

A factor p is called *conservative* if there is exactly one edge leaving p, it is *right special* otherwise.

EXAMPLE 1.3.11. Let t be the infinite Thue–Morse word (see Example 1.2.9). It is easily checked that 000 and 111 are not factors of t. The factor graph $G_3(t)$ is given in Figure 1.9.

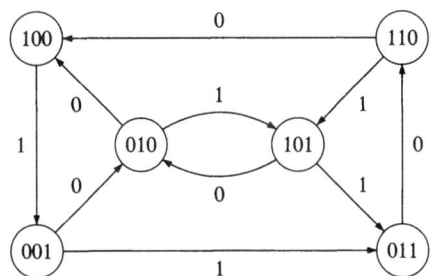

Figure 1.9. Factor graph of order 3 for the Thue–Morse word.

PROPOSITION 1.3.12. *A word y is the label of an infinite path starting at vertex p in $G_n(x)$ if and only if $F_{n+1}(py) \subset F(x)$.* ∎

Recall that the complexity function $P(x, n)$ of an infinite word x is defined by $P(x, n) = \text{Card}(F_n(x))$. The following result is a gap theorem. It shows that the complexity is either bounded or more than linear.

THEOREM 1.3.13. *Let x be an infinite word. The following are equivalent:*
  (i) *x is eventually periodic,*
  (ii) *$P(x, n) = P(x, n + 1)$ for some n,*
  (iii) *$P(x, n) < n + k - 1$ for some $n \geq 1$, where k is the number of letters appearing in x,*
  (iv) *$P(x, n)$ is bounded.*

Proof. (i)⇒(iv) Observe that if $x = uy^\omega$, then $P(x, n) \leq |uy|$.

The implication (iv)⇒(iii) is clear.

(iii)⇒(ii) Assume $P(x, m-1) < P(x, m)$ for $m = 0, \ldots, n$, then $P(x, n) \geq n - 1 + P(x, 1) = n - 1 + k$, a contradiction since $P(x, n)$ is a nondecreasing function of n.

(ii)⇒(i) Consider the factor graph $G_n(x)$. Since every factor of length n is a prefix of a factor of length $n+1$, there is at least one edge starting at each vertex. Since $P(x, n) = P(x, n + 1)$, there is exactly one edge leaving each vertex. This implies that the strongly connected components of the graph are simple circuits. Thus any infinite path will loop through a fixed circuit after a while, and consequently its label is eventually periodic. Since x is the label of a path, the claim is proved. ∎

There is a slightly stronger result that is sometimes useful to show that an infinite word is eventually periodic.

PROPOSITION 1.3.14. *Let x be an infinite word, let $n \geq 1$ and let c be the number of conservative factors of length n in x. If x has a factor of length $n + c$ whose factors of length n are all conservative, then x is eventually periodic.*

Proof. Let $w = a_1 a_2 \cdots a_{n+c}$ be a factor of length $n + c$ whose factors of length n are all conservative, and set $p_i = a_i \cdots a_{i+n-1}$ for $i = 0, \ldots, c$. In the factor graph $G_n(x)$, the path $\pi = (p_0, \ldots, p_c)$ is part of the path of x. Since there are only c conservative vertices, the path π contains a circuit, and since each vertex p_i has a unique outgoing edge, the path of x must stay in this circuit indefinitely. Thus x is eventually periodic. ∎

1.4. Generating series

For any set $X \subset A^*$, the *generating function* or *generating series* of X is the formal series
$$f_X(z) = \sum_{n \geq 0} u_n z^n$$

1.4. Generating series

where
$$u_n = \text{Card}(X \cap A^n).$$

PROPOSITION 1.4.1. *If X is a code, then $f_{X^*} = 1/(1 - f_X)$.*

Proof. If X is a code, every word in X^* has a unique decomposition as a product of words in X. This implies that
$$f_{X^n} = (f_X)^n$$
and thus, by Equation (1.2.1),
$$f_{X^*} = 1 + f_X + \cdots + f_{X^n} + \cdots = 1/(1 - f_X).\qquad\blacksquare$$

EXAMPLE 1.4.2. The set $X = \{b, ab\}$ is a prefix code. The series f_{X^*} is
$$f_{X^*}(z) = \frac{1}{1 - z - z^2}.$$

Let $(F_n)_{n \geq 0}$ be the sequence of Fibonacci numbers defined by $F_0 = 0$, $F_1 = 1$, and $F_{n+2} = F_{n+1} + F_n$. It follows from the recurrence relation that
$$\frac{z}{1 - z - z^2} = \sum_{n \geq 0} F_n z^n.$$

Consequently, $f_{X^*}(z) = \sum_{n \geq 0} F_{n+1} z^n$. It can also be proved by a combinatorial argument that the number of words of length n in X^* is F_{n+1}.

For any set $X \subset A^*$, we denote by ρ_X the radius of convergence of f_X. Since the coefficients u_n of f_X are bounded by $\text{Card}(A)^n$, one has $\rho_X \geq 1/\text{Card}(A)$.

PROPOSITION 1.4.3. *For any rational set X, the generating function $f_X(z)$ is a rational fraction.*

Proof. Let $X, Y \subset A^*$. If the union of X and Y is unambiguous, then $f_{X+Y} = f_X + f_Y$. Similarly, if the product of X and Y is unambiguous, then $f_{XY} = f_X f_Y$. Finally, if X is a code, then $f_{X^*} = 1/(1 - f_X)$ by Proposition 1.4.1.$\qquad\blacksquare$

The following lemma gives a method to compute the radius of convergence of an unambiguous star.

LEMMA 1.4.4. *Let $X \subset A^+$ be a code. If there is a real number α with $0 < \alpha < \rho_X$ such that $f_X(\alpha) = 1$, then $\rho_{X^*} = \alpha$ and f_{X^*} is unbounded on $(0, \alpha)$.*

Proof. The function $r \mapsto f_X(r)$ is continuous and increasing on $[0, \alpha]$. Thus $f_X(r)$ takes all values in $[0, 1]$ for $0 \le r \le \alpha$. Thus $f_{X^*}(r) = \sum(f_X(r))^n$ converges for $0 < r < \alpha$ and diverges for $r = \alpha$. In particular, $\rho_{X^*} = \alpha$ and f_{X^*} is unbounded on $(0, \alpha)$. ∎

The next lemma shows that the hypothesis of Lemma 1.4.4 is satisfied when the code is rational.

LEMMA 1.4.5. *Let $X \subset A^+$ be a nonempty rational set. For each real number $\alpha > 0$, there exist an real number β with $0 < \beta < \rho_X$ such that $f_X(\beta) = \alpha$.*

Proof. It suffices to prove that f_X is unbounded on the interval $(0, \rho_X)$. Indeed, the function $\beta \mapsto f_X(\beta)$ is a continuous increasing function on the interval $(0, \rho_X)$. By Proposition 1.3.5, any recognizable set is unambiguously rational. The proof is by induction on an unambiguous rational expression for X. The conclusion is clear if X is finite, i.e. if f_X is a polynomial. Assume that f_X is unbounded on $(0, \rho_X)$ and similarly f_Y is unbounded on $(0, \rho_Y)$. If the product XY is unambiguous, then $\rho_{XY} = \min\{\rho_X, \rho_Y\}$. Thus f_{XY} is unbounded on $(0, \rho_{XY})$. Similarly, if the sum $X + Y$ is unambiguous, then $\rho_{X+Y} = \min\{\rho_X, \rho_Y\}$. Thus f_{X+Y} is unbounded on $(0, \rho_{X+Y})$.

Finally, let X be a code such that f_X is unbounded on $(0, \rho_X)$. Let $0 < \beta < \rho_X$ be a real number such that $f_X(\beta) = 1$. The conclusion follows by Lemma 1.4.4. ∎

The following example shows that for a code X which is not rational, there may not exist any solution of the equation $f_X(r) = 1$.

EXAMPLE 1.4.6. The set of words on $A = \{a, b\}$ having an equal number of occurrences of a and b is a submonoid of A^* generated by a prefix code D. Since any word of D^* of length $2n$ is obtained by choosing n positions among $2n$, we have

$$f_{D^*}(z) = \sum_{n \ge 0} \binom{2n}{n} z^{2n}.$$

By a simple application of the binomial formula, we obtain

$$f_{D^*}(z) = (1 - 4z^2)^{-\frac{1}{2}}.$$

This follows indeed, using the simple identity

$$\binom{-\frac{1}{2}}{n} = \frac{1}{(-4)^n} \binom{2n}{n}.$$

1.4. Generating series

We have $f_D(z) = 1 - 1/f_{D^*}(z)$ and thus
$$f_D(z) = 1 - \sqrt{1 - 4z^2}.$$

Let D_a be the set of words of D which start with a and let D_b be the set of those which start with b. Then $D = D_a + D_b$ and $f_{D_a} = f_{D_b}$. Thus the prefix code $X = D_a$ satisfies
$$f_X = \frac{1 - \sqrt{1 - 4z^2}}{2}.$$

Since $z = 1/2$ is a singularity of $f_X(z)$, we have $\rho_X = 1/2$. However, $f_X(1/2) = 1/2$. Thus, $f_X([0, \rho_X)) = [0, \frac{1}{2})$.

The following result shows that the set of factors of a rational set X is not much "larger" than X itself.

PROPOSITION 1.4.7. *Let X be a rational set. Then*
$$\rho_X = \rho_{F(X)}.$$

Proof. Let
$$f_X(z) = \sum_{n \geq 0} a_n z^n, \qquad f_{F(X)}(z) = \sum_{n \geq 0} b_n z^n.$$

Since $X \subset F(X)$, we have $a_n \leq b_n$.

Let $\mathcal{A}$ be a finite automaton recognizing X with set of states Q. For each state q, there are words u_q and v_q and an initial state i_q and a terminal state t_q and a path $i_q \xrightarrow{u_q} q \xrightarrow{v_q} t_q$. Let w be a word of length n in $F(X)$. There exist two words u_q and v_p such that $u_q w v_p$ belongs to X. Thus
$$F(X) \subset \bigcup_{p,q \in Q} u_q^{-1} X v_p^{-1}.$$

It follows that
$$a_n \leq b_n \leq a_n + a_{n+1} + \cdots + a_{n+k+\ell}$$
where k is the maximal length of the words u_q and ℓ is the maximal length of the words v_p. This shows that the series $f_X(z)$ and $f_{F(X)}(z)$ have the same radius of convergence. ∎

The following example shows that Proposition 1.4.7 may be false when X is not rational.

EXAMPLE 1.4.8. Let $A = \{a, b\}$ and let $X = \{a^n b w \mid |w| = n\}$. This is a prefix code and $f_X(z) = \sum_{n \geq 1} 2^n z^{2n+1} = \frac{z}{1 - 2z^2}$. Thus $\rho_X = \sqrt{2}/2$. However, $F(X) = A^*$ and thus $\rho_{F(X)} = 1/2 < \rho_X$.

1.5. Symbolic dynamical systems

In this section, we present some basic concepts of symbolic dynamics. The alphabets considered in this section are finite.

1.5.1. Definitions

A *two-sided infinite word* $z \in A^{\mathbb{Z}}$ *avoids* a set of words $X \subset A^*$ if no factor of z is in X. We denote by S_X the set of all $y \in A^{\mathbb{Z}}$ which avoid X.

A *symbolic dynamical system* is a subset S of $A^{\mathbb{Z}}$ of the form S_X for some $X \subset A^+$.

A symbolic dynamical system S is defined by the set X of words that it avoids. Since $F(S)$ is, by definition, the complement of X in A^*, the set S is also determined by the set $F(S)$ of its factors. In particular, if S and T are two dynamical systems such that $F(S) = F(T)$, then $S = T$.

PROPOSITION 1.5.1. *A subset of $A^{\mathbb{Z}}$ is a symbolic dynamical system if and only if it is closed for the topology and invariant under the shift.*

Proof. It is clear that a symbolic dynamical system is both closed and invariant. Conversely, let $S \subset A^{\mathbb{Z}}$ be a closed and invariant set. Let $X = A^+ - F(S)$ be the set of words that do not appear as a factor in any of the words of S. We prove that $S = S_X$. It follows from the definition of X that $y \in S_X$ if and only if $F(y) \subset F(S)$. This shows that $S \subset S_X$. Conversely, let $y \in S_X$. For each integer n, let $w_n = y_{-n} \cdots y_{n-1} y_n$. Since $w_n \in F(y)$, there is a word $y^{(n)} \in S$ such that $w_n \in F(y^{(n)})$. Since S is shift-invariant, we can suppose that $w_n = y^{(n)}_{-n} \cdots y^{(n)}_{n}$. This implies that the sequence $y^{(n)}$ converges to y. Since S is closed, we obtain $y \in S$ and this concludes the proof. ∎

The system is denoted by S or (S, σ) to emphasize the role of the shift σ and it is also called a *subshift*.

For example, $(A^{\mathbb{Z}}, \sigma)$ itself is a symbolic dynamical system, often called the *full shift*.

As a less trivial example, let us consider the following subshift.

EXAMPLE 1.5.2. Let S be the set of two-sided infinite words on $A = \{a, b\}$ such that a symbol a is always followed by a symbol b. Since $S = S_{\{aa\}}$, it is a subshift often called the *golden mean* subshift.

Let $h: A^{\mathbb{Z}} \to B^{\mathbb{Z}}$ be a literal morphism, with A finite. For any subshift S of $A^{\mathbb{Z}}$, the set $T = h(S)$ is a subshift of $B^{\mathbb{Z}}$. Indeed, T is clearly

1.5. Symbolic dynamical systems

shift-invariant. It is also closed: Consider a sequence (y_n) of elements of T converging to some $y \in B^{\mathbb{Z}}$. Let (x_n) be a sequence of elements of S such that $y_n = h(x_n)$. Since S is compact, there is a sub-sequence (x_{n_i}) of the (x_n) which converges to some x in S. Then $y = h(x)$ and thus y is in T.

Conversely, if T is a subshift of $B^{\mathbb{Z}}$, then it is easy to see that $h^{-1}(T)$ is a subshift of $A^{\mathbb{Z}}$, even if A is infinite.

A subshift $S \subset A^{\mathbb{Z}}$ is of *finite type* if $S = S_X$ for some *finite* set $X \subset A^+$. As an example, the golden mean subshift is of finite type.

Let S be a subshift, and let $I(S) = A^+ - F(S)$ be the set of words avoided by S. Let $X(S)$ be the set of elements of $I(S)$ which are minimal for the factor ordering (i.e. which have no proper factor in $I(S)$). Then $S = S_{X(S)}$ and S is of finite type if and only if $X(S)$ is finite.

EXAMPLE 1.5.3. Let $G = (Q, E)$ be a finite graph. The set of two-sided infinite paths in G is a subshift of finite type. Indeed, the set $X(S)$ consists of the set of pairs of nonconsecutive edges. This subshift is called the *edge-shift* of G.

A subshift S is *sofic* if $S = S_X$ for some *rational* set $X \subset A^+$. As above, a subshift S is sofic if and only if $X(S)$ is a rational set.

It is clear that a subshift of finite type is sofic. The converse is not true, as shown by the following examples.

EXAMPLE 1.5.4. Let $S \subset A^{\mathbb{Z}}$ be the set of two-sided infinite words on $A = \{a, b\}$ that contain at most one b. Then $X(S) = ba^*b$. Thus S is sofic. Since $X(S)$ is infinite, S is not a subshift of finite type.

EXAMPLE 1.5.5. Let $S \subset A^{\mathbb{Z}}$ be the set of two-sided infinite words on $A = \{a, b\}$ such that the number of occurrences of a between two consecutive b is even. S is called the *even subshift*. This system is sofic, since $S = S_X$ for $X = ba(aa)^*b$. Since every proper factor of an element of X is in $F(S)$, we have $X = X(S)$. Since X is infinite, S is not a subshift of finite type.

Let $\mathcal{A} = (Q, E)$ be a finite automaton. Let $S \subset A^{\mathbb{Z}}$ be the set of labels of all two-sided infinite paths in $\mathcal{A}$. We say that it is the subshift *recognized* by $\mathcal{A}$. Any sofic subshift is obtained in this way:

PROPOSITION 1.5.6. *Let S be a subshift. The following conditions are equivalent:*
  (i) *S is sofic;*

(ii) *S is recognizable by a finite automaton;*
(iii) *F(S) is recognizable.*

Proof. (i) ⇒ (ii) Let X be a rational set such that $S = S_X$. Let $\mathscr{A}$ be a finite trim automaton recognizing the rational set $A^* - A^*XA^*$. Let S' be the subshift recognized by $\mathscr{A}$. We claim that $F(S') = F(S)$. Indeed, a word $w \in F(S')$ is the label of some path in $\mathscr{A}$. Since $\mathscr{A}$ is trim, there exist words u, v such that uwv is recognized by $\mathscr{A}$. Thus, $w \in F(S)$. Conversely, by compactness, any label of a path in $\mathscr{A}$ is a factor of the label of a two-sided infinite path. Thus $S = S'$. The implications (ii) ⇒ (iii) and (iii) ⇒ (i) are clear. ∎

EXAMPLE 1.5.7. The golden mean subshift of Example 1.5.2 is recognized by the golden mean automaton given in Figure 1.3. The subshift of Example 1.5.4 is recognized by the automaton given in Figure 1.10.

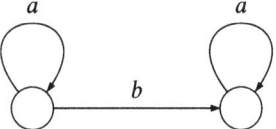

Figure 1.10. One b automaton.

The subshift of Example 1.5.5 is recognized by the even a automaton given in Figure 1.11.

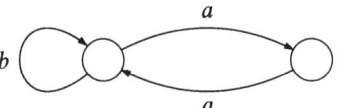

Figure 1.11. Even a automaton.

A consequence of Proposition 1.5.6 is that a subshift S is sofic if and only if the set $F(S)$ of its factors is rational. Indeed, if S is sofic, $F(S)$ is recognized by any automaton recognizing S.

The class of sofic subshifts is closed under image and inverse image by a literal morphism h. Indeed, if S is a sofic system and if $T = h(S)$, then $F(T) = h(F(S))$, and thus $F(T)$ is a rational set. Conversely, if T

1.5. Symbolic dynamical systems

is a sofic system, then $S = h^{-1}(T)$, then $F(S) = h^{-1}(F(T))$. Thus again, $F(S)$ is a rational set.

As an example, let S be a sofic subshift recognized by an automaton $\mathcal{A}$. The set S is the image of the edge-shift of $\mathcal{A}$ under the literal morphism that maps each edge to its label.

A subshift S is *irreducible* if for all $u, v \in F(S)$, there is a word $w \in F(S)$ such that uwv is in $F(S)$.

The subshift of Example 1.5.4 is not irreducible since b appears at most once in a word. On the contrary, the subshift of Example 1.5.5 and the golden mean system are irreducible, as a consequence of the following result.

PROPOSITION 1.5.8. *A sofic subshift is irreducible if and only if it can be recognized by a strongly connected automaton.*

Proof. To prove that the condition is sufficient, let $\mathcal{A}$ be a strongly connected automaton recognizing S. Let u, v be words in $F(S)$, and consider two paths $p \xrightarrow{u} q$ and $r \xrightarrow{v} s$ in $\mathcal{A}$. Since $\mathcal{A}$ is strongly connected, there is a path from q to r labeled by some word w. Thus, uwv is the label of some path in $\mathcal{A}$ and $uwv \in F(S)$.

Conversely, we consider a trim deterministic automaton recognizing the set $F(S)$. Let $\mathcal{A}'$ be an automaton that is a strongly connected component of $\mathcal{A}$ without edges leaving this component. We prove that $\mathcal{A}'$ recognizes S. It suffices to prove that any word in $F(S)$ is the label of some path in $\mathcal{A}'$. Let $w \in F(S)$. Let u be the label of a path from the initial state i of $\mathcal{A}$ to some state p of $\mathcal{A}'$. Since S is irreducible, there is a word v such that $uvw \in F(S)$. Since $\mathcal{A}$ is deterministic, there is a path starting from p labeled with vw. This path is in $\mathcal{A}'$. Thus, w is the label of a path in $\mathcal{A}'$. ∎

The notions introduced above can also be formulated in the context of one-sided infinite words. A one-sided symbolic system, or one-sided subshift, is a set $S \subset A^{\mathbb{N}}$ which is both closed and invariant. Equivalently, it is the set of right-infinite sequences that appear in a subshift. We shall usually work with two-sided subshifts because two-sided shifts take into account both the past and the future. An exception will be made in Subsection 1.5.2 concerning the notion of recurrence.

1.5.2. Recurrence and minimality

In this subsection, we concentrate on a special kind of symbolic dynamical systems: the smallest system containing a given infinite word. It is more appropriate to present it in the one-sided case. We define $S(x) = \{y \in$

$A^\mathbb{N} \mid F(y) \subset F(x)\}$ where $F(x)$ denotes the set of factors of x. The set $S(x)$ is the smallest subshift containing x. Indeed, $S(x) = S_{A^*-F(x)}$. This shows that $S(x)$ is a subshift. Moreover, if $x \in S_X$ for some $X \subset A^+$, then $X \subset A^* - F(x)$ and thus $S_X \supset S_{A^*-F(x)} = S(x)$.

A one-sided infinite word $x \in A^\mathbb{N}$ is said to be *recurrent* if any factor occurring in x has an infinite number of occurrences. It obviously suffices, for x to be recurrent, that any prefix of x has a second occurrence in x.

It is easy to verify that x is recurrent if and only if the subshift $S(x)$ is irreducible. Indeed, if $S(x)$ is irreducible, then for any prefix u of x there is a v such that $uvu \in F(x)$ and thus u has a second occurrence. Conversely, if x is recurrent then for any $u, v \in F(x)$, v has an occurrence following any occurrence of u and thus there is a word w such that $uwv \in F(x)$.

A word $x \in A^\mathbb{N}$ is said to be *uniformly recurrent* if every block of x appears infinitely often at bounded distance, in other terms if, for every word $w \in F(x)$, there exists an integer r such that w is a factor of every word in $F_r(x)$.

A periodic word is obviously uniformly recurrent. We shall see another example below (Example 1.5.10).

These notions are strongly related to that of a *minimal subshift*, i.e. a subshift $S \subset A^\mathbb{N}$ such that $T \subset S$ implies $T = \emptyset$ or $T = S$.

The following result is one of the earliest in symbolic dynamics.

THEOREM 1.5.9. *Let $x \in A^\mathbb{N}$ be a one-sided infinite word. The following conditions are equivalent:*
  (i) *x is uniformly recurrent;*
  (ii) *$S(x)$ is minimal.*

Proof. (i) $\Rightarrow$ (ii) Let $S \subset S(x)$ be a subshift and let $y \in S$. Then $S(y) \subset S$. Since $y \in S(x)$, we have $F(y) \subset F(x)$ by the definition of $S(x)$. Let $w \in F(x)$. Since x is uniformly recurrent, w must appear in every long enough factor of x. If v is a factor of y of this length, then since it is also a factor of x, it admits w as a factor. Hence $w \in F(y)$. This shows that $F(x) = F(y)$ and this implies that $S(y) = S = S(x)$.

(ii) $\Rightarrow$ (i) For every $y \in S(x)$, one has $F(x) = F(y)$, since $S(x)$ is minimal. For a given block w of x, we define $i_w(y)$ to be the function assigning to $y \in S(x)$ the least integer i such that $y = uwz$ with $|u| = i$. Since i_w is continuous and $S(x)$ is compact, i_w is bounded. Let w be a block of $x = uwy$. Since $y \in S(x)$, $w \in F(y)$ and thus w has a second occurrence in x at a distance equal to $|w| + i_w(y)$, hence bounded. ∎

1.5. Symbolic dynamical systems

EXAMPLE 1.5.10. The Thue–Morse word $t = \mu^\omega(0)$ is uniformly recurrent. Indeed, 000 and 111 are not in $F(t)$. Thus successive occurrences of 0 or 1 are separated by at most two symbols. It follows that any block of t appears at bounded distance since it has to appear in some $\mu^k(0)$ or $\mu^k(1)$, and successive occurrences of blocks are again separated by at most two blocks. The system $S(t)$ is known as the *Morse minimal set*.

We used in the proof of Theorem 1.5.9 a possible variant of the definition of a uniformly recurrent word: for all $n > 0$ there is an $m > n$ such that any factor of length n appears in any factor of length m. This condition can be used as a definition for a uniformly recurrent two-sided infinite word. It also leads to the definition of a function $r_x(n)$ called the *recurrence index* of x. We let $r_x(n) = m$ if m is the smallest possible integer such that any factor of length n appears in any factor of length m. It is well defined for all integers n if and only if x is uniformly recurrent.

THEOREM 1.5.11. *Every nonempty subshift contains a uniformly recurrent word.*

Proof. Let S be a nonempty subshift. We define a decreasing sequence (H_n) of subshifts of S as follows. Let $H_0 = S$. Suppose that H_{n-1} is already defined. Let H_n be the set of elements of H_{n-1} which have a minimal number of factors of length n. Each H_n is a subshift. Let H be the intersection of the H_n. Since $A^\mathbb{N}$ is compact, any decreasing sequence of closed subsets has a nonempty intersection. Thus H is nonempty. Let x be an infinite word in H and let $S(x)$ be the smallest subshift containing x. Then $S(x)$ is clearly minimal and thus, x is uniformly recurrent. ∎

As an application of this result, we mention

PROPOSITION 1.5.12. *For any infinite set L of words over a finite alphabet, there is a uniformly recurrent infinite word x such that $F(x) \subset F(L)$.*

Proof. Let S be the subshift avoiding $A^* - F(L)$. Then $F(S) \subset F(L)$ and $F(S)$ is not empty by König's lemma. Any uniformly recurrent word x in S satisfies $F(x) \subset F(L)$. ∎

1.5.3. Entropy

Let S be a nonempty subshift, and let $F = F(S)$ be the set of factors of S. The *entropy* of S is the number

$$h(S) = -\log(\rho_F).$$

Clearly, $0 \le h(S) \le \log k$, where $k = \operatorname{Card} A$. It is also clear that, for any subshifts S, T, if $S \subset T$, then $h(S) \le h(T)$.

A subshift S is said to be *coded* if there is a prefix code X such that $F(S) = F(X^*)$. In this case, we say that X is a code for S. Any sofic system is coded. Indeed, if $\mathscr{A}$ is a deterministic strongly connected automaton recognizing S, the code of first returns to some state of $\mathscr{A}$ is a code for S. The following example shows that the notion of a coded subshift is more general than the notion of a sofic system.

EXAMPLE 1.5.13. Let $X = \{a^n b^n \mid n \ge 1\}$. The system S coded by X is not sofic. Let us indeed suppose the contrary and let $\mathscr{A}$ be a finite automaton recognizing S. For each $n \ge 1$, since $(a^n b^n)^\zeta$ is included in S, there is in $\mathscr{A}$ a cycle labeled by some power of $a^n b^n$. These cycles cannot be all disjoint since the automaton is finite. This clearly gives infinite words which are not in S.

The following theorem gives a method to compute the entropy of a sofic system.

THEOREM 1.5.14. *Let S be a sofic subshift and let X be a rational code for S. Then*
$$h(S) = -\log r$$
where r is the unique positive solution of the equation $f_X(z) = 1$.

Proof. By Proposition 1.4.7, we have $\rho_{X^*} = \rho_{F(X^*)}$. By Lemma 1.4.4, $\rho_{X^*} = r$. ∎

An alternative method of computing r is to use the fact that $1/r$ is the maximal eigenvalue of the matrix associated to any unambiguous automaton recognizing S. Let, in fact, M be a matrix with real coefficients. The spectral radius ρ of M is the maximal modulus of its eigenvalues. One has (see Gantmacher 1960 for example)
$$\rho = \limsup \sqrt[n]{\|M^n\|}$$
where $\|M\|$ is any norm of the matrix M.

Let now S be a sofic system, and let $\mathscr{A} = (Q, E)$ be an unambiguous automaton recognizing S. Let M be the $Q \times Q$ matrix with integer coefficients defined by
$$M_{p,q} = \operatorname{Card}\{a \in A \mid (p, a, q) \in E\}.$$

Let us choose the particular norm equal to the sum of moduli of all coefficients. Then the number s_n of factors of length n appearing in S

satisfies $s_n \leq \|M^n\| \leq c \cdot s_n$ for some constant c. Thus the entropy of S is $\log \rho$.

EXAMPLE 1.5.15. Let S be the even subshift recognized by the automaton of Figure 1.11. We have $X = \{aa, b\}$ and $f_X(z) = z + z^2$. Thus $r = 1/\tau$ where τ is the golden mean. Accordingly, the maximal eigenvalue of the matrix
$$M = \begin{pmatrix} 1 & 1 \\ 1 & 0 \end{pmatrix}$$
is τ and the entropy of S is $h(S) = \log \tau$.

EXAMPLE 1.5.16. Let $X = \{a^n b^n \mid n \geq 1\}$ and let S be the subshift coded by X, as in Example 1.5.13. We have $F(X^*) = QX^*P$ where P (Q) is the set of proper left (right) factors of words of X. Actually $P = a^+X \cup \varepsilon$ since the nonempty words of P have the form $a^n b^m$ for $0 \leq m < n$. Similarly, $Q = Xb^+ \cup \varepsilon$. Thus $\rho_{F(X^*)} = \min\{\rho_Q, \rho_{X^*}, \rho_P\} = \rho_{X^*}$. Since $f_X(z) = \sum_{n \geq 1} z^{2n}$ we have $f_X(z) = \frac{z^2}{1-z^2}$. The equation $f_X(r) = 1$ has the solution $r = \sqrt{2}/2$. Hence
$$h(S) = \log \sqrt{2}.$$

1.6. Unavoidable sets

Unavoidable sets are sets of words X such that any infinite word has a factor in X. The purpose of this section is to present several properties of unavoidable sets. The main result is that, for each integer k, there is an explicit description of the unavoidable sets of cardinality k.

We start with several equivalent definitions of unavoidable sets and some elementary properties.

In this section, all alphabets are supposed to be finite and to contain at least two letters.

1.6.1. Definitions and elementary properties

Recall from Section 1.5 that a two-sided infinite word $z \in A^{\mathbb{Z}}$ avoids a set of words $X \subset A^*$ if no factor of z is in X. The set of all $y \in A^{\mathbb{Z}}$ which avoid X is denoted by S_X.

A set X of words over an alphabet A is called *unavoidable* (over A) if the set S_X is empty.

EXAMPLE 1.6.1. The set A^n is unavoidable for all $n \geq 0$.

EXAMPLE 1.6.2. The set $X = \{a, bb\}$ is unavoidable over $\{a, b\}$. Indeed, any two-sided infinite word over $\{a, b\}$ either contains an a or is reduced to b^{ζ}.

PROPOSITION 1.6.3. *A set X of words over A is unavoidable if and only if the set $A^* - A^*XA^*$ is finite.*

Proof. Assume first that X is unavoidable. Arguing by contradiction, suppose that $Y = A^* - A^*XA^*$ is infinite. By König's lemma, there is a two-sided infinite word y with all its factors in Y. Consequently, y is in S_X, a contradiction.

Conversely, if Y is finite, any two-sided infinite word has a factor in X. ∎

PROPOSITION 1.6.4. *Any unavoidable set X contains a finite unavoidable set.*

Proof. Let d be the maximal length of the words in the finite set $Y = A^* - A^*XA^*$. Let Z be the set of words in X of length at most $d+1$. Every word of length $d+1$ has a factor in X which actually is in Z. Thus Z is unavoidable. ∎

A set containing an unavoidable set is again unavoidable. It is therefore natural to consider minimal unavoidable sets.

Minimal unavoidable sets contained in a given unavoidable set are not necessarily unique. Indeed the set $\{aa, ab, ba, bb\}$ contains both $\{aa, ab, bb\}$ and $\{aa, ba, bb\}$, which both are easily seen to be unavoidable and minimal.

The following example shows the existence of minimal unavoidable sets of arbitrary size $n \geq k$ on an alphabet with $k \geq 2$ letters.

EXAMPLE 1.6.5. Let first $A = \{a, b\}$. For each $n \geq 2$, the set

$$X = \{aa, aba, abba, \ldots, ab^{n-2}a, b^{n-1}\}$$

is a minimal unavoidable set with n elements. Indeed, any infinite word avoiding b^{n-1} has a block of the form $ab^i a$ with $i < n - 1$. This shows that X is unavoidable. For each $0 \leq i < n$, the infinite word $(ab^i)^{\zeta}$ has only one factor in X, namely $ab^i a$ for $i < n - 1$ and b^{n-1} for $i = n - 1$. This shows that X is minimal.

Let now A be an alphabet with $k \geq 3$ letters. We use two symbols $a, b \in A$ to build as above a minimal unavoidable set X having size $n - k + 2$. The set $X \cup A - \{a, b\}$ is a minimal unavoidable set of size n.

1.6. Unavoidable sets

It is worth observing that if X is a finite unavoidable set over A, then A has to be finite. Indeed, for each letter $a \in A$, some a^n is in X and thus $\mathrm{Card}(X) \geq \mathrm{Card}(A)$.

The following result gives an equivalent formulation of the definition of finite unavoidable sets which will be used in what follows.

PROPOSITION 1.6.6. *A finite set X of words is unavoidable if and only if every periodic two-sided infinite word has a factor in X.*

Proof. The condition is clearly necessary. For the converse, we argue by contradiction. Let x be a two-sided infinite word avoiding X. Let n be the maximal length of the words in X. Consider two disjoint occurrences of a factor u of length n in x. Then there is a factor $y = uzu$ of x for some word z. Each word in $(uz)^{\zeta}$ is a periodic two-sided infinite word avoiding X. ∎

It is worth observing that the proposition becomes false when X is infinite. Indeed, the set of squares over a three-letter alphabet is a counterexample because it is avoidable, but every periodic word contains a square.

To check in practice that a given finite set X is unavoidable, there are two possible algorithms.

The first one consists in computing a graph $G = (P, E)$, where P is the set of prefixes of X and E is the set of pairs (p, s) for which there is a letter $a \in A$ such that s is the longest suffix of pa which is in P.

PROPOSITION 1.6.7. *A finite set X is unavoidable if and only if every cycle in G contains a vertex in X.*

Proof. For each integer $n \geq 0$, and vertices $u, v \in P$, there is a path of length n from u to v if and only if there exists a word y of length n such that v is the longest suffix of uy in P. This can be proved by induction on n. It follows that there is a path of length n from ε to a vertex $x \in X$ if and only if $AX \cap A^n \neq \emptyset$. ∎

EXAMPLE 1.6.8. For $X = \{a, bb\}$, the word graph is given in Figure 1.12. By inspection, the set X is unavoidable.

The second algorithm is sometimes easier to write down by hand. Say that a set Y of words is obtained from a finite set of words X by an *elementary derivation* if

(i) either there exist words $u, v \in X$ such that u is a proper factor of v, and
$$Y = X - v,$$

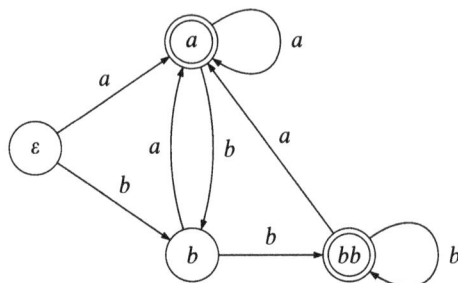

Figure 1.12. The graph for $X = \{a, bb\}$.

(ii) or there exists a word $x = ya \in X$ with $a \in A$ such that, for each letter $b \in A$, there is a suffix z of y such that $zb \in X$, and

$$Y = X - x + y.$$

A *derivation* is a sequence of elementary derivations. We say that Y is derived from X if Y is obtained from X by a derivation.

EXAMPLE 1.6.9. Let $X = \{aaa, b\}$. Then we have the derivations

$$X \to \{aa, b\} \to \{a, b\} \to \{\varepsilon, b\} \to \{\varepsilon\}$$

where the first three arrows follow case (ii) and the last one case (i).

The following result shows in particular that if Y is derived from X, then X is unavoidable if and only if Y is unavoidable.

PROPOSITION 1.6.10. *If Y is derived from X, then $S_X = S_Y$.*

Proof. It is enough to consider the case of an elementary derivation. In the first case where $Y = X - v$, where v has a factor in X, then clearly $S_X = S_Y$. In the second case, we clearly have $S_Y \subset S_X$ since Y is obtained by replacing an element of X by one of its factors. Conversely, assume by contradiction the existence of some $s \in S_X - S_Y$. The only possible factor of s in Y is y. Let b be the letter following y in s. Then s has a factor in X, namely zb where z is the suffix of y such that $zb \in X$ whose existence is granted by the definition of a derivation. This is a contradiction. ∎

The notion of a derivation gives a practical method to check whether a set is unavoidable. We have indeed the following result.

1.6. Unavoidable sets

PROPOSITION 1.6.11. *A finite set X is unavoidable if and only if there is a derivation from X to the set $\{\varepsilon\}$.*

Proof. Let $X \neq \{\varepsilon\}$ be unavoidable. We prove the existence of a derivation to $\{\varepsilon\}$ by induction on the sum $l(X)$ of the lengths of words in X. If $\varepsilon \in X$, we may derive $\{\varepsilon\}$ from X. Thus assume $\varepsilon \notin X$, and let w be a word of maximal length avoiding X. For each $b \in A$ there is a word $x_b = zb \in X$ which is a suffix of wb. Let $x_a = ya$ be the longest of the words x_b. Then the hypotheses of case (ii) are satisfied and thus there is a derivation from X to a set Y with $l(Y) < l(X)$. The converse is clear by Proposition 1.6.10. ∎

In practice, there is a shortcut which is useful in performing derivations. It is described in the following transformation from X to Y:

(iii) there is a word y such that $ya \in X$ for each $a \in A$ and

$$Y = X - \sum_{a \in A} ya + y.$$

It is clear that such a set Y can be derived from X and thus, we do not change the definition of derivations by adding case (iii) to the definition of elementary derivations. We use this new definition in the following example.

EXAMPLE 1.6.12. Let $X = \{aaa, aba, abb, bbb\}$. We have the following sequence of derivations (with the symbol a in the word $x = ya$ underlined at each step):

$$\{aaa, ab\underline{a}, ab\underline{b}, bbb\} \to \{aa\underline{a}, ab, bbb\}$$
$$\to \{a\underline{a}, a\underline{b}, bbb\} \to \{a, bb\underline{b}\} \to \{a, b\underline{b}\}$$
$$\to \{\underline{a}, \underline{b}\} \to \{\varepsilon\}.$$

Derivations could of course be performed on the left rather than on the right (see Problem 1.5.3).

1.6.2. The structure theorem

We will now see how one can describe the unavoidable sets with a fixed number of elements. Our aim is to give a description of this family in a parametric form $\{x_1^{k_1}, \ldots, x_m^{k_m}\}$ for some words $x_1, \ldots, x_m$ and integers $k_1, \ldots, k_m$.

To avoid confusion, we call a *family* any set of subsets of A^*. We will thus speak of the family of unavoidable sets. An *n-subset* is a set with n elements.

An *n-section* of a family $\mathscr{F} = \{X_1, \ldots, X_m\}$ is a set X of n words of A^* containing at least one element of each $F(X_i)$, for $i = 1, \ldots, m$. We denote by $\sec_n(X_1, \ldots, X_m)$ the family of n-sections of $\mathscr{F}$.

EXAMPLE 1.6.13. The family of 2-sections of the family $\mathscr{F} = \{\{a\}, b^*\}$ is composed of the sets $\{\varepsilon, w\}$ for $w \in A^+$, and of the sets $\{a, b^n\}$ for $n \geq 1$.

A subset X of A^* is *simple* if it is of the form $X = \{u\}$ or $X = u^*$ for some word u. A family $\mathscr{F}$ is *simple* if it is composed of simple subsets of A^*.

A family $\mathscr{X}$ of n-subsets of A^* has finite *dimension* if it is the union of the n-sections of a finite number of simple families, i.e.

$$\mathscr{X} = \bigcup_{i=1}^{k} \sec_n(X_{i,1}, \ldots, X_{i,m(i)}) \qquad (1.6.1)$$

where the $X_{i,j}$ are simple sets. The dimension of $\mathscr{X}$ is the minimum value of the maximum of the $m(i)$ for all representations of $\mathscr{X}$ in the form (1.6.1).

EXAMPLE 1.6.14. The family $\mathscr{F} = \{\{a\}, b^*\}$ of Example 1.6.13 is simple. Thus, the family of its 2-sections has dimension 2.

EXAMPLE 1.6.15. The family of one-element sets has infinite dimension. Indeed otherwise there would exist a finite number of words such that every word is a factor of a power of one of these words.

THEOREM 1.6.16. *For each integer $n \geq 0$, the family of unavoidable sets having n elements has dimension n.*

Proof. Let Y be a subset of A^*. Let $\mathscr{R}_n(Y)$ be the family of n-subsets X of A^* such that $X \cup Y$ is unavoidable. We prove by induction on n that, for every Y, the family $\mathscr{R}_n(Y)$ has dimension at most n.

The claim holds for $n = 0$ since the family $\mathscr{R}_0(Y)$ is formed of the empty set, and thus has dimension 0.

Let us suppose that the result holds for $\mathscr{R}_{n-1}(Y)$ for any set Y. Let now Y be any set, and consider $\mathscr{R}_n(Y)$. We distinguish two cases.

Case 1. There exist $n + 1$ words $u_1, \ldots, u_{n+1}$ such that the sets u_i^ζ are disjoint and all their elements avoid Y. Let T be the set of words which

1.6. Unavoidable sets

are factors of at least two u_i^ζ. The set T is finite. Indeed, if u and v are any two primitive words which are not conjugate, then by Fine and Wilf's theorem, any word in $F(u^\zeta) \cap F(v^\zeta)$ has length at most $|u| + |v| - 2$.

Any set $X \in \mathcal{R}_n(Y)$ contains an element of T. Indeed, each element of u_i^ζ has a factor in X and since there are $n+1$ words u_i, two of them have a common factor in X. If $t \in X$, we may write $X \cup Y$ as $Z \cup (Y \cup \{t\})$ with $Z = X - \{t\}$. This shows that

$$\mathcal{R}_n(Y) = \bigcup_{t \in T} \{Z + t \mid Z \in \mathcal{R}_{n-1}(Y + t)\}.$$

By the induction hypothesis, each family $\mathcal{R}_{n-1}(Y + t)$ has dimension at most $n - 1$. Since T is finite, this implies that $\mathcal{R}_n(Y)$ has dimension at most n.

Case 2. The set of two-sided infinite periodic words avoiding Y is contained in a union $u_1^\zeta \cup \cdots \cup u_k^\zeta$ with $k \le n$. Then $\mathcal{R}_n(Y)$ is the family of n-sections of the family $\{u_1^*, \ldots, u_k^*\}$.

This proves that the family of unavoidable sets of cardinality n has dimension at most n. This results indeed from the case $Y = \emptyset$ in the above argument.

The proof that the dimension is exactly n relies on the existence of minimal unavoidable sets of arbitrary size (see Example 1.6.5).

We argue by contradiction and suppose that the family of unavoidable sets with n elements has dimension less than n. Let X be a minimal unavoidable set with n elements. By assumption, X is an n-section of some simple family $\mathcal{F}$ with less than n elements, and all n-sections of $\mathcal{F}$ are unavoidable. There is some $x \in X$ such that $Y = X - x$ in an $(n-1)$-section of $\mathcal{F}$. Then, for any $x' \in A^* - X$, the set $Y + x'$ is an n-section of $\mathcal{F}$, and thus it is unavoidable. It is clear that this implies that Y itself is unavoidable, a contradiction. ∎

EXAMPLE 1.6.17. Let $\mathcal{U}_n$ denote the family of unavoidable sets with n elements on $A = \{a, b\}$. We give below a list of finite sets $\mathcal{F}_n$ of simple families for $n \le 4$ such that $\mathcal{U}_n$ is the set of n-sections of the elements of $\mathcal{F}_i$ for $i \le n$.

Let first $\mathcal{F}_1$ be reduced to $\{\varepsilon\}$ (we identify in this example $\{w\}$ with w). Next, $\mathcal{F}_2$ is composed of the two families

$$\{a, b^*\}, \quad \{a^*, b\}.$$

The family $\mathcal{F}_3$ is composed of the simple families

$$\{aa, bb, (ab)^*\}, \quad \{aa, b^*, bab\}, \quad \{a^*, b^*, ab\}$$

and the two additional ones obtained by exchanging a and b.

The family $\mathscr{F}_4$ is composed of the simple families

$\{aa, bbb, babbab, (ab)^*\}$, $\{aa, bbb, babab, (abb)^*\}$, $\{aa, b^*, babab, bbabb\}$,
$\{aaa, b^*, bab, baab\}$, $\{a^*, b^*, bab, baa\}$, $\{a^*, b^*, bab, aab\}$

and the six additional ones obtained by exchanging a and b.

Problems

Section 1.1

1.1.1 Consider a binary operation on the set $\mathbb{N} \times \mathbb{N}$ defined by

$$(i,j)(k,\ell) = \begin{cases} (i+k-j, \ell) & \text{if } j \leq k, \\ (i, j-k+\ell) & \text{otherwise}. \end{cases}$$

Show that this operation is associative, with neutral element $(0,0)$. The set $\mathbb{N} \times \mathbb{N}$ equipped with this operation is called the *bicyclic monoid*.

Section 1.2

1.2.1 Show that w is primitive if and only if its period is not a proper divisor of its length.

1.2.2 Let A be an alphabet with k elements. Let X be a subset of A^* such that $F(X) \neq A^*$. Show that $\rho_X > 1/k$. (*Hint*: Take $w \notin F(X)$. If w is a letter, then $X \subset (A - w)^*$ and thus $\rho_X > 1/(k-1)$. In the general case, show that the result holds for each set $X_i = \{x \in X \mid |x| \equiv i \pmod{|w|}\}$.)

1.2.3 Show that a set of infinite words is open for the topology if it is of the form XA^ω for some $X \subset A^*$.

1.2.4 Let A be an alphabet, and let $\$$ be a letter not in A. Any word w over A can be viewed as the infinite word $w\$^\omega \in (A \cup \$)^\mathbb{N}$. Show that a sequence (u_n) of words over A converges to an infinite word x if and only if it is not ultimately constant and if the sequence $u_n\$^\omega$ converges to x in the topological space $(A \cup \$)^\mathbb{N}$.

1.2.5 Consider a closed curve in the plane which is *normal*, i.e. has only finitely many self-intersections and these are transverse double points. Label the intersections with distinct symbols from an alphabet A. The *Gauss code* of the curve is the word obtained as the successive intersection points met by proceeding along the

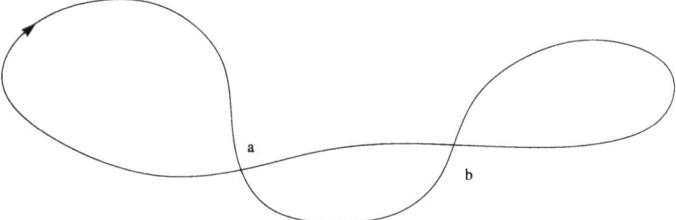

Figure 1.13. A closed curve.

curve and noting each crossing point label as it is traversed. The word obtained is really a conjugacy class.

For example, the Gauss code of the curve of Figure 1.13 is *abba*. Prove that every symbol appears exactly twice in a Gauss code. For each symbol $a \in A$, let $\Delta(a)$ be the set of symbols occurring exactly once between two occurrences of a. Prove that in a Gauss code, the cardinality of $\Delta(a)$ is even for each $a \in A$. Show that this condition is not sufficient for a word with two occurrences of each symbol to be a Gauss code. (*Hint*: Consider the word *abcadcedbe*.)

Section 1.3

1.3.1 Let $\mathscr{A} = (Q, E, i, T)$ be a finite trim deterministic automaton with a unique initial state recognizing a set X. Show that there is a function f from Q onto $Q(X)$ such that $f(i) = X$ and $f(q \cdot a) = f(q) \cdot a$ for all $q \in Q$ and $a \in A$. Derive from this that $\mathscr{A}(X)$ is the unique deterministic automaton recognizing X having a minimal number of states.

1.3.2 Show that the syntactic semigroup of X is the smallest semigroup recognizing X in the sense that, for every semigroup S recognizing X, there exists a morphism from S onto the syntactic semigroup of X. Show that X is recognizable if and only if its syntactic semigroup is finite.

Let $\mathscr{A}(X)$ be the minimal automaton of X. Define a semigroup morphism f from A^+ into $Q(X)^{Q(X)}$ by $f(w): q \mapsto q \cdot w$. Show that the semigroup $f(A^+)$ is isomorphic to the syntactic semigroup of X.

1.3.3 Set $A = \{1, 2, 3\}$, and consider the following function $f: A^* \to A^*$.

A word $x \in A^*$ can be written as
$$x = a_1^{i_1} a_2^{i_2} \cdots a_n^{i_n}$$
with $i_j \geq 1$ and $a_i \neq a_{i+1}$ for $1 \leq i < n$. Let X be the set of words x for which $i_j \leq 3$ for $j = 1, \ldots, n$. Define a function f on X by
$$f(x) = i_1 a_1 i_2 a_2 \cdots i_n a_n.$$
The iterates of f on the letter 1 are
$$\begin{array}{c} 1 \\ 1\ 1 \\ 2\ 1 \\ 1\ 2\ 1\ 1 \\ 1\ 1\ 1\ 2\ 2\ 1 \\ 3\ 1\ 2\ 2\ 1\ 1 \end{array}$$

1. Show that $f(x) \in X$.
2. Show that f is both left subsequential and right subsequential.

1.3.4 Use the de Bruijn graph of order n over a q-letter alphabet to prove the existence of an infinite word of period q^{n+1} having all words of length $n+1$ as its factors. (*Hint:* Use the fact that in the de Bruijn graph, there exists a circuit using each edge exactly once (such a circuit is called Eulerian).)

Section 1.4

1.4.1 Let $f : A^* \to A^*$ be a morphism such that for any two letters $a, b \in A$ there is at least one occurrence of b in $f(a)$. Then any infinite word x such that $f(x) = x$ is uniformly recurrent (Martin 1971).

1.4.2 Show that a recurrent one-sided infinite word x over a finite alphabet is uniformly recurrent if and only if any infinite set of factors of x contains two elements which are factors one of the other.

1.4.3 Let (u_n) be a sequence of positive real numbers such that $u_{n+m} \leq u_n + u_m$. Show that $\lim_{n \to \infty}(u_n/n)$ exists and is equal to $\inf_{n \to \infty}(u_n/n)$.

Section 1.5

1.5.1 Let X be a finite unavoidable set over A. Let n be the maximal length of the elements of X and let $k = \mathrm{Card}(A)$. Show that any word of length $k^n + n - 1$ has a factor in X.

Problems

1.5.2 For a given set $X \subset A^+$, we define an order on A^* by requiring that for all $u, v \in A^*$ and $x \in X$, we have

$$uv < uxv.$$

Prove the following generalization of Higman's theorem: the above defined order is a well-quasi-order if and only if the set X is unavoidable.

1.5.3 A *two-sided derivation* is obtained by adding in the definition of a derivation the following case:

(ii') there exists a word $x = ay \in X$ with $a \in A$ such that, for each letter $b \in A$, there is a prefix z of y such that $bz \in X$, and
$$Y = X - x + y.$$

Show that for any finite set X there is a two-sided derivation from X to the set Y of words avoided by S_X which are minimal for the factor ordering.

1.5.4 A set $X \subset A^+$ is called *separable* if for every $x \in X$ there is a periodic two-sided infinite word s such that $F(s) \cap X = \{x\}$. Show that a finite set X is a subset of a minimal unavoidable set if and only if it is separable. (*Hint*: To prove that the condition is sufficient, choose for each $x \in X$ a periodic infinite word s_x such that $F(s_x) \cap X = \{x\}$. Let $S(x)$ be the closure of s_x under the shift. Since

$$S = \bigcup_{x \in X} S(x)$$

is a subshift of finite type, there is a finite set I such that $S = S_I$. Then $X + I$ is unavoidable and there is a minimal unavoidable set Y such that $X \subset Y \subset X + I$.)

1.5.5 An unavoidable set $X \subset A^+$ is called *irreducible* if it is minimal and if, for x in X and for every proper factor y of x, the set $X - x + y$ is not minimal.
   1. Show that there is, up to symmetries (exchange of a and b and reversal), only one irreducible unavoidable set with four elements, namely $\{aa, aba, abb, bbb\}$.
   2. Show that for each integer n, there are only finitely many irreducible sets with n elements on a given alphabet.

(*Hint*: Use Theorem 1.6.16 and Dickson's lemma (see e.g. Problem 6.1.2 in Lothaire 1983).)

1.5.6 Let $X = \{aaa, bbbb, abbbab, abbab, abab, bbaabb, baabaab\}$. Verify that X is unavoidable.

Show that for any word x in X and any letter a in A, the sets

$$Y = X - x + xa \quad \text{and} \quad Y' = X - x + ax$$

are avoidable.

Notes

There are numerous references concerning automata and formal languages including detailed expositions of Kleene's theorem, see e.g. Hopcroft and Ullman 1979, Eilenberg 1974. The original reference to de Bruijn graphs is de Bruijn 1946. See also Problem 1.3.4 and van Lint and Wilson 1992 for a reference to Eulerian circuits. Theorem 1.3.13 is due to Coven and Hedlund 1973.

A good reference on symbolic dynamics is the book of Lind and Marcus 1995. Our presentation here follows closely the survey by Béal and Perrin 1997. Theorems 1.5.9, 1.5.11 are classical results due to Morse and Hedlund (Morse and Hedlund 1938).

Unavoidable sets seem to appear for the first time in Schützenberger 1964. In this paper, he proves an asymptotic estimate of the minimal cardinality $c(k, n)$ of an unavoidable set of words of fixed length n on a k-letter alphabet.

Further papers on unavoidable sets appeared later (see Choffrut and Culik 1984, Crochemore, Lerest, and Wender 1983).

Theorem 1.6.16 is due to Rosaz 1998. Laurent Rosaz has also obtained many other interesting results. Problem 1.5.5 appears in Rosaz 1998. The example of Problem 1.5.6 is due to Rosaz (Rosaz 1998). It answers in the negative a conjecture which had been formulated by Ehrenfeucht and asking whether for any unavoidable set X there are a word x in X and a letter a in A such that $X + xa - x$ is still unavoidable. Problem 1.5.4 is a result again due to Rosaz 1995.

The Gauss codes of Problem 1.2.5 were introduced by Gauss in Gauss 1900. Several characterizations of Gauss codes has been given (see Treybig 1968, Marx 1969, Lovasz and Marx 1976, Rosenstiehl 1976). For a history of the subject, see Grünbaum 1972 or Rosenstiehl 1999.

Problem 1.3.3 describes a transformation studied by J. H. Conway under the name of "audioactive decay" (Conway 1987). A number of amazing results were obtained by Conway (see also Vardi 1991, Ekhad and Zeilberger 1997)

Problem 1.5.2 is due to Ehrenfeucht, Hausler, and Rozenberg 1983a. It is a generalization of the famous Higman theorem (see Lothaire 1983).

CHAPTER 2

Sturmian Words

2.0. Introduction

Sturmian words are infinite words over a binary alphabet that have exactly $n + 1$ factors of length n for each $n \geq 0$. It appears that these words admit several equivalent definitions, and can even be described explicitly in arithmetic form. This arithmetic description is a bridge between combinatorics and number theory. Moreover, the definition by factors makes Sturmian words define symbolic dynamical systems. The first detailed investigations of these words were done from this point of view. Their numerous properties and equivalent definitions, and also the fact that the Fibonacci word is Sturmian, have led to a great development, under various terminologies, of the research.

The aim of this chapter is to present basic properties of Sturmian words and of their transformation by morphisms. The style of exposition relies basically on combinatorial arguments.

The first section is devoted to the proof of the Morse–Hedlund theorem stating the equivalence of Sturmian words with the set of balanced aperiodic words and the set of mechanical words of irrational slope. We also mention several other formulations of mechanical words, such as rotations and cutting sequences. We next give properties of the set of factors of one Sturmian word, such as closure under reversal, the minimality of the associated dynamical system, the fact that the set depends only on the slope, and we give the description of special words.

In the second section, we give a systematic exposition of standard pairs and standard words. We prove the characterization by the double palindrome property, describe the connection with Fine and Wilf's theorem. Then, standard sequences are introduced to connect standard words to characteristic Sturmian words. The relation to Beatty sequences is in the Problems. This section also contains the enumeration formula for finite Sturmian words. It ends with a short description of frequencies.

The third section starts by proving that the monoid of Sturmian morphisms is generated by three well-known morphisms. Then, standard morphisms are investigated. A description of all Sturmian morphisms in terms of standard morphisms is given next. The section ends with the characterization of those algebraic numbers that yield fixed points by standard morphisms.

Some Problems are just exercises, but most contain additional properties of Sturmian words, with appropriate references. It is difficult to trace back many of the properties of Sturmian words, because of the scattered origins, terminology and notation. When we quote a reference in the Notes section, we are only relatively certain that it is the source of the result.

In this chapter, words will be over a binary alphabet $A = \{0, 1\}$.

2.1. Equivalent definitions

This section is devoted to the proof of a theorem (Theorem 2.1.13) stating the equivalence of three properties, all defining what we call Sturmian words. We start by defining Sturmian words to have minimal complexity among aperiodic infinite words. We first prove that Sturmian words are exactly the aperiodic balanced words. We then introduce so-called mechanical words and prove that these yield another characterization of Sturmian words. Other formulations of the mechanical definition, by rotation and cutting sequences, are given in the second subsection. The third subsection contains several properties concerning the set of factors of a Sturmian word.

2.1.1. Complexity and balance

The *complexity function* of an infinite word x over some alphabet A was defined in Chapter 1. It is the function that counts, for each integer $n \geq 0$, the number $P(x, n)$ of factors of length n in x:

$$P(x, n) = \text{Card}(F_n(x)).$$

A *Sturmian* word is an infinite word s such that $P(s, n) = n + 1$ for any integer $n \geq 0$. According to Theorem 1.3.13, Sturmian words are aperiodic infinite words of minimal complexity. Since $P(s, 1) = 2$, any Sturmian word is over two letters. A *right special* factor of a word x is a word u such that $u0$ and $u1$ are factors of x. Thus, s is a Sturmian word if and only if it has exactly one right special factor of each length.

A suffix of a Sturmian word is a Sturmian word.

2.1. Equivalent definitions

EXAMPLE 2.1.1. We show that the *Fibonacci* word

$$f = 01001010010010100101001001010011 \cdots$$

defined in Chapter 1 is Sturmian. It will be convenient, in this chapter, to start the numeration of finite Fibonacci words differently, and to set $f_{-1} = 1$, $f_0 = 0$.

Since $f = \varphi(f)$, it is a product of words 01 and 0. Thus, the word 11 is not a factor of f and consequently $P(f, 2) = 3$. The word 000 is not a factor of $\varphi(f)$, since otherwise it is a prefix of some $\varphi(x)$ for a factor x of f, and x has to start with 11.

To show that f is Sturmian, we prove that f has exactly one right special factor of each length.

We start by showing that for all words x, either $0x0$ or $1x1$ is not a factor of f. This is clear if x is the empty word and if x is a single letter. Arguing by induction on the length, assume that $0x0$ and $1x1$ are in $F(f)$. Then x starts and ends with 0, and $x = 0y0$ for some y. Since $00y00$ and $10y01$ have to be factors of $\varphi(f)$, there exists a factor z of f such that $\varphi(z) = 0y$. Moreover, $00y0 = \varphi(1z1)$ and $010y01 = \varphi(0z0)$, showing that $1z1$ and $0z0$ are factors of f. This is a contradiction because $|z| \leq |\varphi(z)| < |x|$.

We show now that f has at most one right special factor of each length. Assume indeed that u and v are right special factors of the same length, and let x be the longest common suffix of u and v. Then the four words $0x0$, $0x1$, $1x0$, $1x1$ are factors of f, which contradicts our previous observation.

To show that f has at least one right special factor of each length, we use the relation

$$f_{n+2} = g_n \tilde{f}_n \tilde{f}_n t_n \qquad (n \geq 2) \tag{2.1.1}$$

where $g_2 = \varepsilon$ and for $n \geq 3$

$$g_n = f_{n-3} \cdots f_1 f_0, \quad t_n = \begin{cases} 01 & \text{if } n \text{ is odd,} \\ 10 & \text{otherwise.} \end{cases}$$

Observe that the first letter of $\tilde{f}_n$ is the opposite of the first letter of t_n. This proves that $\tilde{f}_n$ is a right special factor for each $n \geq 2$. Since a suffix of a right special factor is itself a right special factor, this proves that right special factors of any length exist.

Equation (2.1.1) is proved by induction. Indeed, $f_4 = \varepsilon(010)(010)10$ and $f_5 = 0(10010)(10010)01$. Next, is it easily checked by induction that

$$\varphi(\tilde{u})0 = 0(\varphi(u))^\sim \tag{2.1.2}$$

for any word u. It follows that $\varphi(\tilde{f}_n t_n) = 0\tilde{f}_{n+1} t_{n+1}$ and since $\varphi(g_n)0 = g_{n+1}$, one gets (2.1.1).

We now start to give another description of Sturmian words, namely as balanced words. The *height* of a word x is the number $h(x)$ of letters equal to 1 in x. Given two words x and y of the same length, their *balance* $\delta(x, y)$ is the number

$$\delta(x, y) = |h(x) - h(y)|.$$

A set of words X is *balanced* if

$$x, y \in X, \; |x| = |y| \; \Rightarrow \; \delta(x, y) \leq 1.$$

A finite or infinite word is itself balanced if the set of its factors is balanced.

PROPOSITION 2.1.2. *Let X be a factorial set of words. If X is balanced, then for all $n \geq 0$,*

$$\mathrm{Card}(X \cap A^n) \leq n + 1.$$

Proof. The conclusion is clear for $n = 0, 1$, and it holds for $n = 2$ because X cannot contain both 00 and 11. Arguing by contradiction, let $n \geq 3$ be the smallest integer for which the statement is false. Set $Y = X \cap A^{n-1}$ and $Z = X \cap A^n$. Then $\mathrm{Card}(Y) \leq n$ and $\mathrm{Card}(Z) \geq n + 2$. For each $z \in Z$, its suffix of length $n-1$ is in Y. By the pigeon-hole principle, there exist two distinct words $y, y' \in Y$ such that all four words $0y, 1y, 0y', 1y'$ are in Z. Since $y \neq y'$ there exists a word x such that $x0$ and $x1$ are prefixes of y and y'. But then, both $0x0$ and $1x1$ are words in X, showing that X is unbalanced. ∎

The argument used in the proof can be refined as follows.

PROPOSITION 2.1.3. *Let X be a factorial set of words. The set X is unbalanced if and only if there exists a palindrome word w such that $0w0$ and $1w1$ are in X.*

Proof. The condition is clearly sufficient. Conversely, assume that X is unbalanced. Consider two words $u, v \in X$ of the same length n such that $\delta(u, v) \geq 2$, and take them of minimal length. The first letters of u and v are distinct, and so are the last letters. Assuming that u starts with 0 and v with 1, there are factorizations $u = 0wau'$ and $v = 1wbv'$ for some words w, u', v' and letters $a \neq b$. In fact $a = 0$ and $b = 1$ since otherwise $\delta(u', v') = \delta(u, v)$, contradicting the minimality of n. Thus, again by minimality, $u = 0w0$ and $v = 1w1$.

2.1. Equivalent definitions

Assume next that w is not a palindrome. Then there are a prefix z of w and a letter a such that za is a prefix of w, $\tilde{z}$ is a suffix of w but $a\tilde{z}$ is not a suffix of w. Then of course $b\tilde{z}$ is a suffix of w, where b is the other letter. This gives a proper prefix $0za$ of u and a proper suffix $b\tilde{z}1$ of v. If $a = 0$ and $b = 1$, then $\delta(0z0, 1\tilde{z}1) = 2$, contradicting the minimality of n. But then $u = 0z1u''$ and $v = v''1\tilde{z}0$ for two words with $\delta(u'', v'') = \delta(u, v)$, again contradicting the minimality. Thus w is a palindrome. ∎

REMARK 2.1.4. In the proof that the Fibonacci word f is Sturmian given in Example 2.1.1, we actually started by showing that f is balanced.

THEOREM 2.1.5. *Let x be an infinite word. The following conditions are equivalent:*
  (i) *x is Sturmian,*
  (ii) *x is balanced and aperiodic.*

Proof. If x is aperiodic, then $P(x, n) \geq n + 1$ for all n by Theorem 1.3.13. If x is balanced, then by Proposition 2.1.2, $P(x, n) \leq n + 1$ for all n. Thus x is Sturmian.

To prove the converse, we assume x is Sturmian and unbalanced, and show that x is eventually periodic. Since x is unbalanced, there is a palindrome word w such that $0w0, 1w1$ are factors of x. This shows that w is right special. Set $n = |w| + 1$. Since x is Sturmian, there is a unique right special factor of length n, which is either $0w$ or $1w$. We suppose that $0w$ is right special, so $1w$ is not, and $0w1$ is a factor of x and $1w0$ is not.

Any occurrence of $1w$ in x is followed by the letter 1. Let v be a word of length $n - 1$ such that $u = 1w1v$ is in $F(x)$. The word u has length $2n$. We prove that all factors of length n of u are conservative. In view of Proposition 1.3.14, x is eventually periodic.

To show the claim, it suffices to prove that the only right special factor of length n, that is $0w$, is not a factor of u. Assume the contrary. Then there exist factorizations $w = s0t, v = yz, w = t1y$.

				u			
1		w		1		v	
			0	w			
1	s	0	t	1	y	z	

Since w is a palindrome, the first factorization implies $w = \tilde{t}0\tilde{s}$, and the letter following the prefix t in w is both a 0 and a 1. ∎

The *slope* of a nonempty word x is the number $\pi(x) = \frac{h(x)}{|x|}$.

EXAMPLE 2.1.6. The height of $x = 0100101$ is 3, and its slope is $3/7$. The word x can be drawn on a grid by representing a 0 (resp. a 1) as a horizontal (resp. a diagonal) unit segment. This gives a polygonal line from the origin to the point $(|x|, h(x))$, and the line from the origin to this point has slope $\pi(x)$. See Figure 2.1.

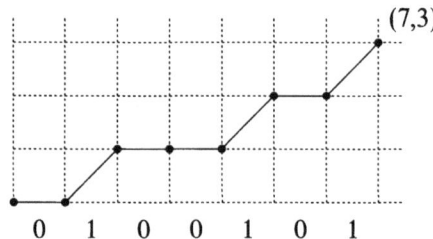

Figure 2.1. Height and slope of the word 0100101.

It is easily checked that

$$\pi(xy) = \frac{|x|}{|xy|}\pi(x) + \frac{|y|}{|xy|}\pi(y).$$

PROPOSITION 2.1.7. *A factorial set of words X is balanced if and only if, for all $x, y \in X$, $x, y \neq \varepsilon$,*

$$|\pi(x) - \pi(y)| < \frac{1}{|x|} + \frac{1}{|y|}. \qquad (2.1.3)$$

Proof. Assume first that (2.1.3) holds. For $x, y \in X$ of the same length, the equation gives

$$|h(x) - h(y)| < 2$$

showing that X is balanced.

Conversely, assume that X is balanced, and let x, y be in X. If $|x| = |y|$, then (2.1.3) holds. Assume $|x| > |y|$, and set $x = zt$, with $|z| = |y|$. Arguing by induction on $|x| + |y|$, we have

$$|\pi(t) - \pi(y)| < \frac{1}{|t|} + \frac{1}{|y|}$$

2.1. Equivalent definitions

and since X is balanced, $|h(z) - h(y)| \leq 1$, whence $|\pi(z) - \pi(y)| \leq \frac{1}{|y|}$.
Next,

$$\pi(x) - \pi(y) = \frac{|z|}{|x|}\pi(z) + \frac{|t|}{|x|}\pi(t) - \pi(y)$$

$$= \frac{|z|}{|x|}\bigl(\pi(z) - \pi(y)\bigr) + \frac{|t|}{|x|}\bigl(\pi(t) - \pi(y)\bigr);$$

thus

$$|\pi(x) - \pi(y)| < \frac{1}{|x|} + \frac{|t|}{|x|}\left(\frac{1}{|y|} + \frac{1}{|t|}\right) = \frac{1}{|x|} + \frac{1}{|y|}. \quad \blacksquare$$

COROLLARY 2.1.8. *Let x be an infinite balanced word, and for each $n \geq 1$, let x_n be the prefix of length n of x. The sequence $(\pi(x_n))_{n\geq 1}$ converges for $n \to \infty$.*

Proof. Indeed, (2.1.3) shows that $(\pi(x_n))_{n\geq 1}$ is a Cauchy sequence. $\blacksquare$

The limit
$$\alpha = \lim_{n\to\infty} \pi(x_n)$$
is the *slope* of the infinite word x.

EXAMPLE 2.1.9. To compute the slope of an infinite balanced word, it suffices to compute the limit of the slopes of an increasing sequence of prefixes (or even factors, as shown by the next proposition). For the Fibonacci infinite word, the slopes of the finite Fibonacci words f_n are easily computed. Indeed, $|f_n| = F_n$ and $h(f_n) = F_{n-2}$, whence

$$\pi(f) = \lim_{n\to\infty} \frac{F_{n-2}}{F_n} = \frac{1}{\tau^2},$$

where $\tau = (1 + \sqrt{5})/2$.

PROPOSITION 2.1.10. *Let x be an infinite balanced word with slope α. For every nonempty factor u of x, one has*

$$|\pi(u) - \alpha| \leq \frac{1}{|u|}. \tag{2.1.4}$$

More precisely, one of the following holds: either

$$\alpha|u| - 1 < h(u) \leq \alpha|u| + 1 \quad \text{for all } u \in F(x) \tag{2.1.5}$$

or

$$\alpha|u| - 1 \leq h(u) < \alpha|u| + 1 \quad \text{for all } u \in F(x) \tag{2.1.6}$$

Of course, the inequalities in (2.1.5) and (2.1.6) are strict if α is irrational.

Proof of proposition. Let x_n be the prefix of length n of x. Given some ε, consider n_0 such that for all $n \geq n_0$,
$$|\pi(x_n) - \alpha| \leq \varepsilon.$$

Then, using (2.1.3),
$$|\pi(u) - \alpha| \leq |\pi(u) - \pi(x_n)| + |\pi(x_n) - \alpha| < \frac{1}{|u|} + \frac{1}{n} + \varepsilon.$$

For $n \to \infty$ and then $\varepsilon \to 0$, the inequality follows. Equation (2.1.4) means that
$$\alpha|u| - 1 \leq h(u) \leq \alpha|u| + 1.$$

If the second claim were wrong, there would exist u, v in $F(x)$ such that $\alpha|u| - 1 = h(u)$ and $\alpha|v| + 1 = h(v)$. But then $|\pi(u) - \pi(v)| = 1/|u| + 1/|v|$, in contradiction with (2.1.3). ∎

PROPOSITION 2.1.11. *Let x be an infinite balanced word. The slope α of x is a rational number if and only if x is eventually periodic.*

Proof. If $x = uy^\omega$, then
$$\pi(uy^n) = \frac{h(u) + nh(y)}{|u| + n|y|} \to \pi(y)$$

for $n \to \infty$, showing that the slope is rational.

For the converse, we suppose that (2.1.5) holds. The other case is symmetric. The slope of x is a rational number $\alpha = q/p$ with q and p relatively prime. By (2.1.5), any factor u of x of length p has height q or $q + 1$. There are only finitely many occurrences of factors of length p and height $q + 1$, since otherwise there is a factor $w = uzv$ of x with $|u| = |v| = p$ and $h(u) = h(v) = q + 1$. In view of (2.1.5)
$$2 + 2q + h(z) = h(uzv) \leq 1 + \alpha p + \alpha|z| + \alpha p = 1 + 2q + \alpha|z|$$

whence $h(z) \leq \alpha|z| - 1$, in contradiction with (2.1.5).

By the preceding observation, there is a factorization $x = ty$ such that every word in $F_p(y)$ has the same height. Consider now an occurrence azb of a factor in y of length $p + 1$, with a and b letters. Since $h(az) = h(zb)$, one has $a = b$. This means that y is periodic with period p. Consequently, x is eventually periodic. ∎

2.1. Equivalent definitions

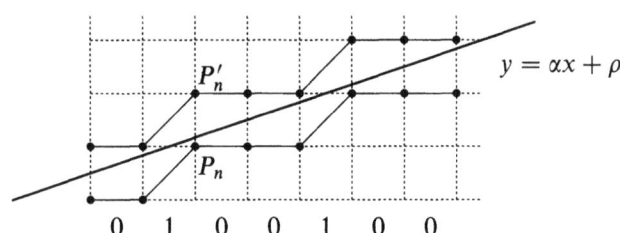

Figure 2.2. Mechanical words associated with the line $y = \alpha x + \rho$.

2.1.2. Mechanical words, rotations

Given two real numbers α and ρ with $0 \leq \alpha \leq 1$, we define two infinite words

$$s_{\alpha,\rho} : \mathbb{N} \to A, \quad s'_{\alpha,\rho} : \mathbb{N} \to A$$

by

$$s_{\alpha,\rho}(n) = \lfloor \alpha(n+1) + \rho \rfloor - \lfloor \alpha n + \rho \rfloor,$$
$$s'_{\alpha,\rho}(n) = \lceil \alpha(n+1) + \rho \rceil - \lceil \alpha n + \rho \rceil \quad (n \geq 0).$$

It is easy to check that $s_{\alpha,\rho}(n)$ and $s'_{\alpha,\rho}(n)$ are indeed in $\{0, 1\}$. The word $s_{\alpha,\rho}$ is the *lower mechanical word* and $s'_{\alpha,\rho}$ is the *upper mechanical word* with *slope* α and *intercept* ρ. (This slope will be shown in a moment to be the same as the slope of a balanced word.) It is clear that if $\rho - \rho'$ is an integer, then $s_{\alpha,\rho} = s_{\alpha,\rho'}$ and $s'_{\alpha,\rho} = s'_{\alpha,\rho'}$. Thus we may assume $0 \leq \rho < 1$ or $0 < \rho \leq 1$ (both will be useful).

The terminology stems from the following graphical interpretation (see Figure 2.2). Consider the straight line with equation $y = \alpha x + \rho$. The points with integer coordinates just below this line are $P_n = (n, \lfloor \alpha n + \rho \rfloor)$. Two consecutive points P_n and P_{n+1} are joined by a straight line segment that is horizontal if $s_{\alpha,\rho}(n) = 0$ and diagonal if $s_{\alpha,\rho}(n) = 1$.

The same observation holds for the points $P'_n = (n, \lceil \alpha n + \rho \rceil)$ located just above the line.

Clearly,
$$s_{0,\rho} = s'_{0,\rho} = 0^\omega, \quad s_{1,\rho} = s'_{1,\rho} = 1^\omega.$$

Let $0 < \alpha < 1$. Since $1 + \lfloor \alpha n + \rho \rfloor = \lceil \alpha n + \rho \rceil$ whenever $\alpha n + \rho$ is not an integer, one has $s_{\alpha,\rho} = s'_{\alpha,\rho}$ except when $\alpha n + \rho$ is an integer for some $n \geq 0$. In this case (see Figure 2.3),

$$s_{\alpha,\rho}(n) = 0, \quad s'_{\alpha,\rho}(n) = 1$$

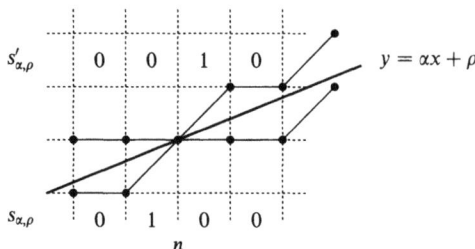

Figure 2.3. Mechanical words with an integral point.

and, if $n > 0$,
$$s_{\alpha,\rho}(n-1) = 1, \quad s'_{\alpha,\rho}(n-1) = 0.$$

Thus, if α is irrational, $s_{\alpha,\rho}$ and $s'_{\alpha,\rho}$ differ by at most one factor of length 2. A mechanical word is *rational* or *irrational* according as its slope is rational or irrational.

A special case deserves consideration, namely when $0 < \alpha < 1$ and $\rho = 0$. In this case, $s_{\alpha,0}(0) = \lfloor \alpha \rfloor = 0$, $s'_{\alpha,0}(0) = \lceil \alpha \rceil = 1$, and if α is irrational
$$s_{\alpha,0} = 0 c_\alpha, \quad s'_{\alpha,0} = 1 c_\alpha$$

where the infinite word c_α is called the *characteristic* word of α.

REMARK 2.1.12. The condition $0 \le \alpha \le 1$ in the definition of mechanical words is not a restriction, but a simplification. One could indeed use the same definition of $s_{\alpha,\rho}$ without any condition on α. Since $\lfloor \alpha \rfloor \le s_{\alpha,\rho}(n) \le 1 + \lfloor \alpha \rfloor$, the numbers $s_{\alpha,\rho}(n)$ then can have the two values k and $k+1$ where $k = \lfloor \alpha \rfloor$. Thus the words $s_{\alpha,\rho}$ and $s'_{\alpha,\rho}$ are over the two-letter alphabet $\{k, k+1\}$. This alphabet can be transformed back into $\{0, 1\}$ by using the formula
$$s_{\alpha,\rho}(n) = \lfloor \alpha(n+1) + \rho \rfloor - \lfloor \alpha n + \rho \rfloor - \lfloor \alpha \rfloor.$$

Mechanical words can be interpreted in several other ways. Consider again a straight line $y = \beta x + \rho$, for some $\beta > 0$ not restricted to be less than 1, and ρ not restricted to be positive. Consider the intersections of this line with the lines of the grid with nonnegative integer coordinates. We get a sequence $Q_0, Q_1, \ldots$ of intersection points. We call $Q_n = (x_n, y_n)$ *horizontal* if y_n is an integer, and *vertical* if x_n is an integer. If both are integers, we insert before Q_n a sibling Q_{n-1} of Q_n with the same coordinates, and we agree that the first is horizontal and the second is

2.1. Equivalent definitions

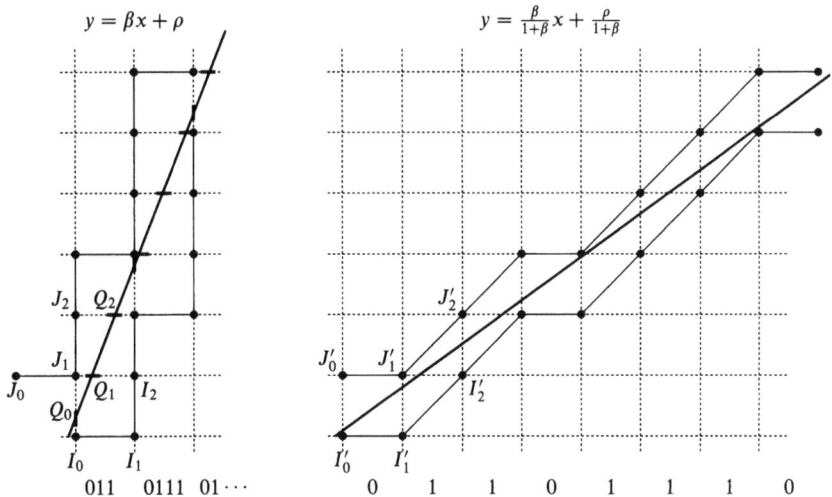

Figure 2.4. Cutting sequence and corresponding mechanical sequence.

vertical (or vice versa, but we always make the same choice). In Figure 2.4, Q_0 is vertical, because ρ is positive.

Writing a 0 for each vertical point and a 1 for each horizontal point, we obtain an infinite word $K_{\beta,\rho}$ that is called the (lower) *cutting sequence* (with the other choice for labeling siblings, one gets an upper cutting sequence $K'_{\beta,\rho}$).

To each $Q_n = (x_n, y_n)$, we associate a point $I_n = (u_n, v_n)$ with integer coordinates. The point I_n is the point below (below and to the right of) Q_n if Q_n is vertical (horizontal). Formally,

$$(u_n, v_n) = \begin{cases} (\lceil x_n \rceil, y_n - 1) & \text{if } Q_n \text{ is horizontal,} \\ (x_n, \lfloor y_n \rfloor) & \text{if } Q_n \text{ is vertical.} \end{cases}$$

Similar points J_n are defined above the line (see Figure 2.4). It is easy to check that $u_n + v_n = n$ for $n \geq 0$, and that

$$K_{\beta,\rho}(n) = v_{n+1} - v_n = 1 + u_n - u_{n+1}.$$

In the special case $\rho = 0$ and β irrational, we again get the same infinite word up to the first letter. There is a word C_β such that

$$K_{\beta,0} = 0C_\beta, \qquad K'_{\beta,0} = 1C_\beta.$$

Observe that Q_n is horizontal if and only if
$$1 + v_n \leq u_n\beta + \rho < 1 + \rho + v_n \tag{2.1.7}$$
and Q_n is vertical if and only if
$$v_n \leq u_n\beta + \rho < 1 + v_n. \tag{2.1.8}$$
We now check that
$$K_{\beta,\rho} = s_{\beta/(1+\beta),\rho/(1+\beta)}.$$
Indeed, the transformation $(x, y) \mapsto (x + y, x)$ of the plane maps the line $y = \beta x + \rho$ to $y = \beta/(1+\beta)x + \rho/(1+\beta)$, and a point $I_n = (u_n, v_n)$ to $I'_n = (n, v_n)$. It remains to show that
$$v_n = \left\lfloor \frac{\beta}{1+\beta}n + \frac{\rho}{1+\beta} \right\rfloor. \tag{2.1.9}$$
Using $u_n + v_n = n$, we get from (2.1.7) that
$$v_n + 1/(1+\beta) \leq \beta/(1+\beta)n + \rho/(1+\beta) < 1 + v_n$$
and from (2.1.8) that
$$v_n \leq \beta/(1+\beta)n + \rho/(1+\beta) < v_n + 1/(1+\beta).$$
Thus, (2.1.9) holds for horizontal and for vertical steps. Thus, cutting sequences are just another formulation of mechanical words.

Mechanical words can also be generated by rotations. Let $0 < \alpha < 1$. The *rotation* of angle α is the mapping $R = R_\alpha$ from $[0, 1)$ into itself defined by
$$R(z) = \{z + \alpha\}$$
where $\{z\} = z - \lfloor z \rfloor$ is the fractional part of z. Iterating R, one gets
$$R^n(\rho) = \{n\alpha + \rho\}.$$
Moreover, a straightforward computation shows that
$$\lfloor (n+1)\alpha + \rho \rfloor = 1 + \lfloor n\alpha + \rho \rfloor \iff \{n\alpha + \rho\} \geq 1 - \alpha.$$
Thus, defining a partition of $[0, 1)$ by
$$I_0 = [0, 1 - \alpha), \quad I_1 = [1 - \alpha, 1),$$
one gets
$$s_{\alpha,\rho}(n) = \begin{cases} 0 & \text{if } R^n(\rho) \in I_0, \\ 1 & \text{if } R^n(\rho) \in I_1. \end{cases} \tag{2.1.10}$$

2.1. Equivalent definitions

It will be convenient to identify $[0, 1)$ with the torus (or the unit circle). For $0 \le b < a < 1$, the set $[a, 1] \cup [0, b)$ is considered as an interval denoted by $[a, b)$. Then, for any subinterval I of $[0, 1)$, the sets $R(I)$ and $R^{-1}(I)$ are always intervals (even when overlapping the point 0).

As an example of the use of rotations, consider a word $w = b_0 b_1 \cdots b_{m-1}$, with $b_0, b_1, \ldots$ letters. We want to know whether w is a factor of some $s_{\alpha,\rho} = a_0 a_1 \cdots$, with $a_0, a_1, \ldots$ letters. By (2.1.10), $a_{n+k} = b_i$ if and only if $R^{n+i}(\rho) \in I_{b_i}$, or equivalently, if and only if $R^n(\rho) \in R^{-i}(I_{b_i})$. Thus, for $n \ge 0$,

$$w = a_n a_{n+1} \cdots a_{n+m-1} \iff R^n(\rho) \in I_w \qquad (2.1.11)$$

where I_w is the interval

$$I_w = I_{b_0} \cap R^{-1}(I_{b_1}) \cap \cdots \cap R^{-m+1}(I_{b_{m-1}}).$$

The interval I_w is nonempty if and only if w is a factor of $s_{\alpha,\rho}$. Observe that this property is independent of ρ, and thus the words $s_{\alpha,\rho}$ and $s_{\alpha,\rho'}$ have the same set of factors. A combinatorial proof will be given later (Proposition 2.1.18).

Mechanical words are quite naturally defined as two-sided infinite words. However, it appears that several properties, such as Theorem 2.1.13 below, only hold with some restrictions (see Problem 2.1.1).

THEOREM 2.1.13. *Let s be an infinite word. The following are equivalent:*
  (i) *s is Sturmian;*
  (ii) *s is balanced and aperiodic;*
  (iii) *s is irrational mechanical.*

The proof will be a simple consequence of two lemmas. In the proofs, we will use several times the formula

$$x' - x - 1 < \lfloor x' \rfloor - \lfloor x \rfloor < x' - x + 1.$$

LEMMA 2.1.14. *Let s be a mechanical word with slope α. Then s is balanced of slope α. If α is rational, then s is purely periodic. If α is irrational, then s is aperiodic.*

Proof. Let $s = s_{\alpha,\rho}$ be a lower mechanical word. The proof is similar for upper mechanical words. The height of a factor $u = s(n) \cdots s(n+p-1)$ is the number $h(u) = \lfloor \alpha(n+p) + \rho \rfloor - \lfloor \alpha n + \rho \rfloor$, thus

$$\alpha|u| - 1 < h(u) < \alpha|u| + 1. \qquad (2.1.12)$$

This implies $\lfloor \alpha|u| \rfloor \le h(u) \le 1 + \lfloor \alpha|u| \rfloor$, and shows that $h(u)$ takes only two consecutive values, when u ranges over the factors of a fixed length of s. Thus, s is balanced. Moreover, by (2.1.12)

$$|\pi(u) - \alpha| < \frac{1}{|u|}.$$

Thus $\pi(u) \to \alpha$ for $|u| \to \infty$ and α is the slope of s as it was defined for balanced words. This proves the first statement.

If α is irrational, the word s is aperiodic by Proposition 2.1.11. If $\alpha = q/p$ is rational, then $\lfloor \alpha(n+p) + \rho \rfloor = q + \lfloor \alpha n + \rho \rfloor$, for all $n \ge 0$. Thus $s(n+p) = s(n)$ for all n, showing that s is purely periodic. ∎

LEMMA 2.1.15. *Let s be a balanced infinite word. If s is aperiodic, then s is irrational mechanical. If s is purely periodic, then s is rational mechanical.*

Proof. In view of Corollary 2.1.8, s has a slope, say α. Denote by h_n the height of the prefix of length n of s.

For every real number τ, one at least of the following holds: $h_n \le \lfloor \alpha n + \tau \rfloor$ for all n; or $h_n \ge \lfloor \alpha n + \tau \rfloor$ for all n. Indeed, suppose on the contrary there exist a real number τ and two integers $n, n+k$ such that $h_n < \lfloor \alpha n + \tau \rfloor$ and $h_{n+k} > \lfloor \alpha(n+k) + \tau \rfloor$ (or the symmetric relation). This implies that $h_{n+k} - h_n \ge 2 + \lfloor \alpha(n+k) + \tau \rfloor - \lfloor \alpha n + \tau \rfloor > 1 + \alpha k$, in contradiction with (2.1.4).

Set

$$\rho = \inf\{\tau \mid h_n \le \lfloor \alpha n + \tau \rfloor \text{ for all } n\}.$$

By Proposition 2.1.10, one has $\rho \le 1$, and $\rho < 1$ if α is irrational. Observe that for all $n \ge 0$

$$h_n \le \alpha n + \rho \le h_n + 1 \qquad (2.1.13)$$

since otherwise there is an integer n such that $h_n + 1 < \alpha n + \rho$, and setting $\sigma = h_n + 1 - \alpha n$, one has $\sigma < \rho$ and $\alpha n + \sigma = h_n + 1 > h_n$, in contradiction with the definition of ρ.

If s is aperiodic, then α is irrational by Proposition 2.1.11, and $\alpha n + \rho$ is an integer for at most one n. By (2.1.13), either $h_n = \lfloor \alpha n + \rho \rfloor$ for all n, and then $s = s_{\alpha,\rho}$, or $h_n = \lfloor \alpha n + \rho \rfloor$ for all but one n_0, and $h_{n_0} + 1 = \alpha n_0 + \rho$. In this case, one has $h_n = \lceil \alpha n + \rho - 1 \rceil$ for all n and $s = s'_{\alpha,\rho-1}$.

If $s = u^\omega$ is purely periodic with period $|u| = p$, then $\alpha = q/p$ with $q = h(u) = h_p$. Again $h_n = \lfloor \alpha n + \rho \rfloor$ if $\alpha n + \rho$ is never an integer (this depends on ρ).

If $h_n = \alpha n + \rho$ for some n, we claim that $h_n = \lfloor \alpha n + \rho \rfloor$ for all n. Assume the contrary. Then by (2.1.13), $1 + h_m = \alpha m + \rho$, for some m and

2.1. Equivalent definitions

we may assume $n < m < n + p$. Consider the words $y = s(n+1) \cdots s(m)$ and $z = s(m+1) \cdots s(n+p)$. Then $\pi(y) = (h_m - h_n)/(m-n) = \alpha - 1/|y|$ and $\pi(z) = (h_{n+p} - h_m)/(n+p-m) = \alpha + 1/|z|$, whence $|\pi(y) - \pi(z)| = 1/|y| + 1/|z|$, in contradiction with Proposition 2.1.7. Similarly, if $1 + h_n = \alpha n + \rho$ for some n, then $h_n = \lceil \alpha n + \rho \rceil$ for all n. ∎

Proof of Theorem 2.1.13. We know already by Theorem 2.1.5 that (i) and (ii) are equivalent. Assume that s is irrational mechanical. Then s is balanced and aperiodic by Lemma 2.1.14. Conversely, if s is balanced and aperiodic, then by Lemma 2.1.15 s is irrational mechanical. ∎

EXAMPLE 2.1.16. To show that a balanced infinite word is not always mechanical when the slope is rational (so the converse is false in Lemma 2.1.14), consider the infinite balanced word 01^ω. It is not a mechanical word. Indeed, it has slope 1, and all mechanical words $s_{1,\rho}$ are equal to 1^ω.

Let us consider mechanical words with rational slope in some more detail. For a rational number $\alpha = p/q$ with $0 \leq \alpha \leq 1$ and p, q relatively prime, the infinite words $s_{\alpha,0}$ and $s'_{\alpha,0}$ are purely periodic. Define finite words
$$t_{p,q} = a_0 \cdots a_{q-1}, \quad t'_{p,q} = a'_0 \cdots a'_{q-1}$$
by
$$a_i = \left\lfloor (i+1)\frac{p}{q} \right\rfloor - \left\lfloor i\frac{p}{q} \right\rfloor, \quad a'_i = \left\lceil (i+1)\frac{p}{q} \right\rceil - \left\lceil i\frac{p}{q} \right\rceil.$$

Clearly, $t_{p,q}$ and $t'_{p,q}$ have height p. They are primitive words because $(p,q) = 1$. In particular, $t_{0,1} = 0$ and $t_{1,1} = 1$. These words are called *Christoffel words*. In any case, $s_{p/q,0} = t_{p,q}^\omega$ and $s'_{p/q,0} = t'_{p,q}{}^\omega$. Moreover, if $0 < p/q < 1$, the word $t_{p,q}$ starts with 0 and ends with 1 (and $t'_{p,q}$ starts with 1 and ends with 0). There is a word $z_{p,q}$ such that

$$t_{p,q} = 0 z_{p,q} 1, \quad t'_{p,q} = 1 z_{p,q} 0. \tag{2.1.14}$$

The word $z_{p,q}$ is easily seen to be a palindrome. Later, we will see that these words, called *central words*, have remarkable combinatorial properties.

The following result deals with finite words.

PROPOSITION 2.1.17. *A finite word w is a factor of some Sturmian word if and only if it is balanced.*

Proof. Clearly a factor of a Sturmian word is balanced. For the converse, consider a balanced word w, and define

$$\alpha' = \max\{\pi(u) - 1/|u|\}, \qquad \alpha'' = \min\{\pi(u) + 1/|u|\}$$

where the maximum and the minimum are taken over all nonempty factors u of w. Since w is balanced, one gets from Proposition 2.1.10 that

$$\pi(u) - 1/|u| < \pi(v) + 1/|v|$$

for all nonempty factors u and v of w. Thus $\alpha' < \alpha''$.

Take any irrational number α with $\alpha' < \alpha < \alpha''$. Then by construction, for every nonempty factor u of w,

$$|\pi(u) - \alpha| < 1. \tag{2.1.15}$$

Let w_n be the prefix of length n of w. By (2.1.15), there exists a real ρ_n such that

$$h(w_n) = n\alpha + \rho_n, \quad |\rho_n| < 1.$$

Moreover, for $n > m$, setting $w_n = w_m u$, one gets $h(w_n) - h(w_m) = h(u) = (n-m)\alpha + (\rho_n - \rho_m)$, showing that $|\rho_n - \rho_m| < 1$. Set

$$\rho = \max\{\rho_n \mid 1 \le n \le |w|\}.$$

Then

$$n\alpha + \rho \ge h(w_n) = n\alpha + \rho + (\rho_n - \rho) > n\alpha + \rho - 1$$

whence $h(w_n) = \lfloor n\alpha + \rho \rfloor$. This proves that w is a prefix of the Sturmian word $s_{\alpha,\rho}$. ∎

2.1.3. The factors of one Sturmian word

The aim of this subsection is to give properties of the set of factors of a single Sturmian word.

PROPOSITION 2.1.18. *Let s and t be Sturmian words.*
  (i) *If s and t have same slope, then $F(s) = F(t)$.*
  (ii) *If s and t have distinct slopes, then $F(s) \cap F(t)$ is finite.*

Proof. Let α be the common slope of s and t. By Proposition 2.1.10, every factor u of s satisfies

$$|\pi(u) - \alpha| < \frac{1}{|u|}$$

2.1. Equivalent definitions

(indeed, equality is impossible because α is irrational). Next, for every factor v of t,
$$|\pi(v) - \alpha| < \frac{1}{|v|}.$$
Let $X = F(s) \cup F(t)$. The set X is factorial. It is also balanced since
$$|\pi(u) - \pi(v)| \le |\pi(u) - \alpha| + |\pi(v) - \alpha| < \frac{1}{|u|} + \frac{1}{|v|}.$$
In view of Proposition 2.1.2
$$\mathrm{Card}(X \cap A^n) \le n + 1$$
for every n. Thus $F(s) = X = F(t)$.

Let now α be the slope of s and β be the slope of t. We may suppose that $\beta > \alpha$. For any factor u of s such that $(\beta - \alpha) \ge 2/|u|$, one has $\pi(u) - \alpha > -1/|u|$ by Proposition 2.1.10 whence $\pi(u) - \beta = (\pi(u) - \alpha) + (\beta - \alpha) \ge 1/|u|$ showing that u is not a factor of t. ∎

PROPOSITION 2.1.19. *The set $F(s)$ of factors of a Sturmian word s is closed under reversal.*

Proof. Set $\tilde{F}(s) = \{\tilde{x} \mid x \in F(s)\}$. The set $X = F(s) \cup \tilde{F}(s)$ is balanced. In view of Proposition 2.1.2, $\mathrm{Card}(X \cap A^n) \le n + 1$, for each n, and since $\mathrm{Card}(F(s) \cap A^n) = n + 1$, one has $X = F(s)$. Thus $\tilde{F}(s) = F(s)$. ∎

We now compare Sturmian words, with respect to their slope and intercept. The lexicographic order defined in Chapter 1 extends to infinite words as follows, with the assumption that $0 < 1$. Given two infinite words $x = a_0 \cdots a_n \cdots$ and $y = b_0 \cdots b_n \cdots$, we say that x is *lexicographically less* than y, and we write $x < y$, if there is an integer n such that $a_i = b_i$ for $i = 0, \ldots, n-1$ and $a_n = 0$, $b_n = 1$.

PROPOSITION 2.1.20. *Let $0 < \alpha < 1$ be an irrational number and let ρ, ρ' be real numbers with $0 \le \rho, \rho' < 1$. Then*
$$s_{\alpha, \rho} < s_{\alpha, \rho'} \iff \rho < \rho'.$$

Proof. Since α is irrational, the set of fractional parts $\{\alpha n\}$ for $n \ge 0$ is dense in the interval $[0, 1)$. Thus $\rho < \rho'$ if and only if there exists an integer $n \ge 1$ such that $1 - \rho' \le \{\alpha n\} < 1 - \rho$, and this is equivalent to $\lfloor \alpha n + \rho' \rfloor = 1 + \lfloor \alpha n + \rho \rfloor$. If n is the smallest integer for which this equality holds, then $s_{\alpha, \rho}(n-1) = 0$ and $s_{\alpha, \rho'}(n-1) = 1$ and $s_{\alpha, \rho'}(k) = s_{\alpha, \rho}(k)$ for $k < n - 1$. ∎

Observe that this proposition does not hold for rational slopes, since indeed $s_{0,\rho} = 0^\omega$ for all ρ.

LEMMA 2.1.21. *Let $0 < \alpha, \alpha' < 1$ be irrational numbers and let ρ, ρ' be real numbers. Any of the equalities $s_{\alpha,\rho} = s_{\alpha',\rho'}$, $s_{\alpha,\rho} = s'_{\alpha',\rho'}$, $s'_{\alpha,\rho} = s'_{\alpha',\rho'}$ implies $\alpha = \alpha'$ and $\rho \equiv \rho' \pmod{1}$.*

Proof. Any of the equalities implies that $\alpha = \alpha'$ because equal words have the same slope. Next, $s_{\alpha,\rho} = s_{\alpha,\rho'}$ implies $\rho \equiv \rho' \pmod{1}$ by the preceding proposition. Finally, consider the equality $s_{\alpha,\rho} = s'_{\alpha,\rho'}$. If $\alpha n + \rho'$ is not an integer for all $n \geq 1$, then $s'_{\alpha,\rho'} = s_{\alpha,\rho'}$ and the conclusion holds. Otherwise, let n be the unique integer such that $\alpha n + \rho'$ is an integer. Then $s_{\alpha,\rho+(1+n)\alpha} = s'_{\alpha,\rho'+(1+n)\alpha}$, showing again that $\rho \equiv \rho' \pmod{1}$. ∎

Sturmian words with intercept 0 have many interesting properties. We observed already that, for an irrational number $0 < \alpha < 1$, the words $s_{\alpha,0}$ and $s'_{\alpha,0}$ differ only in their first letter, and that

$$s_{\alpha,0} = 0c_\alpha, \quad s'_{\alpha,0} = 1c_\alpha$$

where c_α is the *characteristic word* of slope α. Equivalently,

$$c_\alpha = s_{\alpha,\alpha} = s'_{\alpha,\alpha}.$$

The following proposition states a combinatorial characterization of characteristic words among Sturmian words.

PROPOSITION 2.1.22. *For every Sturmian word s, either $0s$ or $1s$ is Sturmian. A Sturmian word s is characteristic if and only if $0s$ and $1s$ are both Sturmian.*

Proof. The first claim follows from the fact that $s_{\alpha,\rho-\alpha} = as_{\alpha,\rho}$, for some $a \in \{0,1\}$.

If $s = s_{\alpha,\alpha} = s'_{\alpha,\alpha}$ is the characteristic word of slope α, then $0s = s_{\alpha,0}$ and $1s = s'_{\alpha,0}$ are Sturmian.

Conversely, the Sturmian words $0s$ and $1s$ have same slope, say α. Denote by ρ and ρ' their intercepts. Then their common shift s has intercept $\rho + \alpha = \rho' + \alpha$, and by Lemma 2.1.21, $\rho \equiv \rho' \pmod{1}$ and we may take $0 \leq \rho = \rho' < 1$. Thus $0s = s_{\alpha,\rho}$ and $1s = s'_{\alpha,\rho}$. Assume $\rho > 0$. The first letter of $0s$ is $0 = \lfloor \alpha + \rho \rfloor - \lfloor \rho \rfloor = \lfloor \alpha + \rho \rfloor$ and the first letter of $1s$ is $1 = \lceil \alpha + \rho \rceil - \lceil \rho \rceil$. Then $2 = \lceil \alpha + \rho \rceil$, a contradiction. Thus $\rho = 0$. ∎

We are now able to describe right special factors.

2.2. Standard words

PROPOSITION 2.1.23. *The set of right special factors of a Sturmian word is the set of reversals of the prefixes of the characteristic word of same slope.*

Call a factor w of a Sturmian word s *left special* if both $0w$ and $1w$ are factors of s. Clearly, w is left special if and only if $\tilde{w}$ is right special. Thus the proposition states that the set of left special factors of a Sturmian word is the set of prefixes of the characteristic word of the same slope.

Proof of proposition. Let s be a Sturmian word of slope α. By Proposition 2.1.22, the infinite words $0c_\alpha$ and $1c_\alpha$ are Sturmian and clearly have slope α. Thus
$$F(s) = F(c_\alpha) = F(0c_\alpha) = F(1c_\alpha)$$
by Proposition 2.1.18. Consequently, for each prefix p of c_α, $0p$ and $1p$ are factors of s. Since $F(s)$ is closed under reversal, this shows that $\tilde{p}$ is right special. Thus $\tilde{p}$ is the unique right special factor of length $|p|$. ∎

EXAMPLE 2.1.24. Consider again the Fibonacci word f. We have seen in Example 2.1.1 that its right special factors are the reversals of its prefixes. Thus each prefix of f is left special. This shows that $F(f) = F(0f) = F(1f)$. Consequently, f is characteristic of slope $1/\tau^2$.

PROPOSITION 2.1.25. *The dynamical system generated by a Sturmian word is minimal.*

Proof. Let s be a Sturmian word, and let x be an infinite word such that $F(x) \subset F(s)$. Clearly, x is balanced. Also, x has the same irrational slope as s. Thus x is aperiodic and therefore is Sturmian. By Proposition 2.1.18(1), $F(x) = F(s)$. This shows that s and x generate the same dynamical system. ∎

Observe that Proposition 2.1.18(ii) is a consequence of Proposition 2.1.25. Indeed, the intersection of two distinct minimal dynamical systems is the trivial system.

2.2. Standard words

This section is concerned with a family of finite words that are basic bricks for constructing characteristic Sturmian words, in the sense that every characteristic Sturmian word is the limit of a sequence of standard words. This will be shown in Subsection 2.2.2.

2.2.1. Standard words and palindrome words

After basic definitions, we give two characterizations of standard words. The first is by a special decomposition into palindrome words (Theorem 2.2.4), the second (Theorem 2.2.11) by an extremal property on the periods of the word that is closely related to Fine and Wilf's theorem. We give then a "mechanical" characterization of central and standard words (Proposition 2.2.15). We end with an enumeration formula for standard words.

Consider two functions Γ and Δ from $\{0,1\}^* \times \{0,1\}^*$ into itself defined by

$$\Gamma(u,v) = (u, uv), \quad \Delta(u,v) = (vu, v).$$

The set of *standard pairs* is the smallest set of pairs of words containing the pair $(0,1)$ and closed under Γ and Δ. A *standard word* is any component of a standard pair.

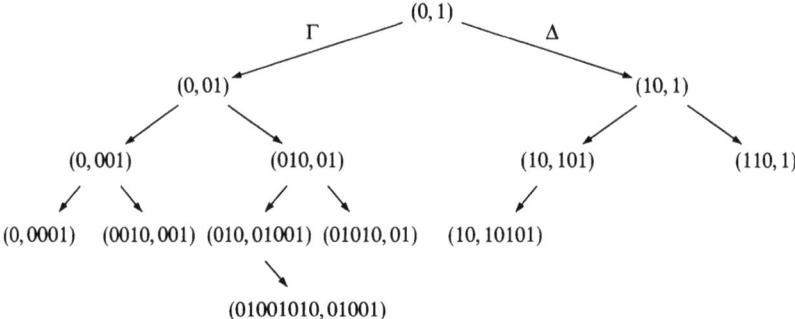

Figure 2.5. The tree of standard pairs.

EXAMPLE 2.2.1. Figure 2.5 shows the beginning of the tree of standard pairs. Considering the leftmost and rightmost paths, one gets the pairs

$$(0, 0^n 1), \quad (1^n 0, 1) \quad (n \geq 1).$$

Next to them are the pairs

$$(0(10)^n, 01), \quad (10, (10)^n 1) \quad (n \geq 1).$$

These are the pairs with one component of length 1 or 2.

Finite Fibonacci words are standard, since $(f_0, f_{-1}) = (0, 1)$, and for $n \geq 1$, $(f_{2n+2}, f_{2n+1}) = \Delta\Gamma(f_{2n}, f_{2n-1})$.

2.2. Standard words

Every standard word which is not a letter is a product of two standard words which are the components of some standard pair. The next proposition states some elementary facts.

PROPOSITION 2.2.2. *Let $r = (x, y)$ be a standard pair.*
  (i) *If $r \ne (0, 1)$ then one of x, y is a proper prefix of the other.*
  (ii) *If x (resp. y) is not a letter, then x ends with 10 (resp. y ends with 01).*
  (iii) *Only the last two letters of xy and yx are different.*

Proof. We prove the last claim by induction on $|xy|$. Assume indeed that $xy = p01$ and $yx = p10$. Then $\Gamma(r) = (x, xy)$ and $xxy = xp01$, $(xy)x = x(yx) = xp10$, so the claim is true for $\Gamma(r)$. The same holds for $\Delta(r)$. ∎

Every standard pair is obtained in a unique way from $(0, 1)$ by iterated use of Γ and Δ. Indeed, if (x, y) is a standard pair, then it is an image through Γ (resp. Δ) if and only if $|x| < |y|$ (resp. $|x| > |y|$). Thus, there is a unique product $W = \Lambda_1 \circ \cdots \circ \Lambda_n$, with $\Lambda_i \in \{\Gamma, \Delta\}$, such that

$$(x, y) = W(0, 1).$$

Consider two matrices

$$L = \begin{pmatrix} 1 & 0 \\ 1 & 1 \end{pmatrix}, \quad R = \begin{pmatrix} 1 & 1 \\ 0 & 1 \end{pmatrix}$$

and define a morphism μ from the monoid generated by Γ and Δ into the set of 2×2 matrices by

$$\mu(\Gamma) = L, \quad \mu(\Delta) = R,$$

and $\mu(\Lambda_1 \circ \cdots \circ \Lambda_n) = \mu(\Lambda_1) \cdots \mu(\Lambda_n)$. If $(x, y) = W(0, 1)$, then a straightforward induction shows that

$$\mu(W) = \begin{pmatrix} |x|_0 & |x|_1 \\ |y|_0 & |y|_1 \end{pmatrix}. \tag{2.2.1}$$

Observe that every matrix $\mu(W)$ has determinant 1. Thus if (x, y) is a standard pair,

$$|x|_0 |y|_1 - |x|_1 |y|_0 = 1, \tag{2.2.2}$$

showing that the entries in the same row (column) of $\mu(W)$ are relatively prime. From (2.2.2), one gets

$$h(y)|x| - h(x)|y| = 1 \tag{2.2.3}$$

(recall that $h(w) = |w|_1$ is the *height* of w). This shows also that $|x|$ and $|y|$ are relatively prime. A simple consequence is the following property.

PROPOSITION 2.2.3. *A standard word is primitive.*

Proof. Let w be a standard word which is not a letter. Then $w = x$ or $w = y$ for some standard pair (x, y). From (2.2.3), one gets that $h(w)$ and $|w|$ are relatively prime. This implies that w is primitive. ∎

The operations Γ and Δ can be explained through three morphisms E, G, D on $\{0, 1\}^*$ which we introduce now. These will be used also in what follows. Let

$$E : \begin{matrix} 0 \mapsto 1 \\ 1 \mapsto 0 \end{matrix}, \quad G : \begin{matrix} 0 \mapsto 0 \\ 1 \mapsto 01 \end{matrix}, \quad D : \begin{matrix} 0 \mapsto 10 \\ 1 \mapsto 1 \end{matrix}.$$

It is easily checked that $E \circ D = G \circ E = \varphi$. We observe that, for every morphism f,

$$\Gamma(f(0), f(1)) = (fG(0), fG(1)), \quad \Delta(f(0), f(1)) = (fD(0), fD(1)).$$

For $W = \Lambda_1 \circ \cdots \circ \Lambda_n$, with $\Lambda_i \in \{\Gamma, \Delta\}$, define $\hat{W} = \hat{\Lambda}_n \circ \cdots \circ \hat{\Lambda}_1$, with $\hat{\Gamma} = G$, $\hat{\Delta} = D$. Then

$$W(0, 1) = (\hat{W}(0), \hat{W}(1)). \tag{2.2.4}$$

Standard words have the following description.

THEOREM 2.2.4. *A word w is standard if and only if it is a letter or there exist palindrome words p, q and r such that*

$$w = pab = qr \tag{2.2.5}$$

where $\{a, b\} = \{0, 1\}$. Moreover, the factorization $w = qr$ is unique if $q \neq \varepsilon$.

EXAMPLE 2.2.5. The word 01001010 is standard (see Figure 2.5) and

$$01001010 = (010010)10 = (010)(01010).$$

We start the proof with a lemma of independent interest.

LEMMA 2.2.6. *If a primitive word is a product of two nonempty palindrome words, then this factorization is unique.*

Proof. Let w be a primitive word and assume $w = pq = p'q'$ for palindrome words p, q, p', q'. We suppose $|p| > |p'|$, so that $p = p's(= \tilde{s}p')$, $sq = q'(= q\tilde{s})$ for some nonempty word s. Thus $\tilde{s}p'q = pq = p'q' = p'q\tilde{s}$, showing that $p'q$ and $\tilde{s}$ are powers of some word z. But then $w = pq = \tilde{s}p'q = z^n$ for some $n \geq 2$, contradicting primitivity. ∎

Observe that (2.2.5) implies the following relations.

2.2. Standard words

LEMMA 2.2.7. *If $w = pab = qr$ for palindrome words p, q, r, and letters $a \neq b$, then one of the following holds:*
  (i) *$r = \varepsilon$, $p = (ba)^n b$, $q = (ba)^{n+1} b = w$ for some $n \geq 0$;*
  (ii) *$r = b$, $p = a^n$, $q = a^{n+1}$, $w = a^{n+1} b$ for some $n \geq 0$;*
  (iii) *$r = bab$, $p = b^{n+1}$, $q = b^n$, $w = b^{n+1} ab$ for some $n \geq 0$;*
  (iv) *$r = basab$, $p = qbas$, $w = qbasab$ for some palindrome word s.* ∎

We need another lemma.

LEMMA 2.2.8. *Let x, y be words with $|x|, |y| \geq 2$. The pair (x, y) is a standard pair if and only if there exist palindrome words p, q, r such that*

$$x = p10 = qr \quad \text{and} \quad y = q01 \qquad (2.2.6)$$

or

$$x = q10 \quad \text{and} \quad y = p01 = qr. \qquad (2.2.7)$$

Proof. Assume that (2.2.6) holds (the other case is symmetric). If r is the empty word, then by the previous lemma

$$(x, y) = ((01)^{n+1} 0, (01)^{n+1} 001) = \Gamma((01)^{n+1} 0, 01)$$

showing that the pair (x, y) is standard.

If $r = 0$, then $(x, y) = (1^n 0, 1^n 01) = \Gamma(1^n 0, 1)$, and if $r = 010$, then $(x, y) = (0^n 10, 0^n 1) = \Delta(0, 0^n 1)$.

Thus, we may assume that $r = 01s10$ for some palindrome word s. By (2.2.6), if follows that y is a prefix of x, so $x = yz$ for some word z. We show that (z, y) is standard. From $p = q01s = s10q$ it follows that $q \neq s$. Assume $|q| < |s|$ (the other case is symmetric). Then $s = qt$ for some word t, and the equation $p = qt10q$ shows that the word $r' = t10$ is a palindrome. Thus

$$y = q01, \quad z = qr' = s10$$

and (z, y) satisfies (2.2.6).

Conversely, let (x, y) be a standard pair, and assume $(x, y) = \Gamma(x, z)$, that is $y = xz$. If z is a letter, then $(x, z) = (1^n 0, 1)$ for some $n \geq 1$ and

$$x = q10, \ y = p01 = qr$$

for $q = 1^{n-1}$, $p = 1^n$, $r = 101$.

Thus we may assume that for some palindrome words p, q, r, either

$$x = p10 = qr, \quad z = q01$$

or
$$x = q10, \quad z = p01 = qr.$$

In the first case,
$$x = p10, \quad y = xz = (qrq)01 = p(10q01).$$

In the second case,
$$x = q10, \quad y = xz = q(10p01) = (qrq)01$$

because $10p = rq$. Thus (2.2.7) holds. ∎

Proof of Theorem 2.2.4. Let w be a standard word, $|w| \geq 2$. Then there exists a standard pair (x, y) such that $w = xy$ (or symmetrically $w = yx$). If $x = 0$, then $y = 0^n 1$ for some $n \geq 0$, and $xy = 0^{n+1} 1$ has the desired factorization. A similar argument holds for $y = 1$. Otherwise, either (2.2.6) or (2.2.7) of Lemma 2.2.8 holds. In the first case, $xy = p(10q01) = qrq01$ and in the second case, $xy = q(10p01) = qrq01$ because $10p = rq$. The factorization is unique by Lemma 2.2.6 because a standard word is primitive.

Conversely, if $w = p10 = qr$ (or $w = p01 = qr$) for palindrome words p, q, r, then by Lemma 2.2.8, the word w is a component of some standard pair, and thus is a standard word. ∎

A word w is *central* if $w01$ (or equivalently $w10$) is a standard word. As we shall see, central words indeed play a central role.

COROLLARY 2.2.9. *A word is central if and only if it is in the set*

$$0^* \cup 1^* \cup (P \cap P10P)$$

where P is the set of palindrome words. The factorization of a central word w as $w = p10q$ with p, q palindrome words is unique.

Observe that $P \cap P10P = P \cap P01P$.

Proof of corollary. Let $w \in 0^* \cup 1^* \cup (P \cap P10P)$. By the previous characterization, $w01$ is a standard word, so w is central. Conversely, if $w01$ is standard, then w is a palindrome and $w01 = qr$ for some palindrome words q and r. Either $w \in 0^* \cup 1^*$, or by Lemma 2.2.7, $r = \varepsilon$ and $w = (10)^n 1$ for some $n \geq 1$, or $w = q10s$ for some palindrome s, as required. ∎

As a simple consequence, we obtain the following.

2.2. Standard words

COROLLARY 2.2.10. *A palindrome prefix (suffix) of a central word is central.*

Proof. We consider the case of a prefix. Let p be a central word. If $p \in 0^* \cup 1^*$, the result is clear. Let x be a standard word such that $x = pab$, with $\{a,b\} = \{0,1\}$. Then $x = yz$ for a standard pair (y,z) or (z,y). Set $y = qba$ and $z = rab$, where q, r are central words. Then $p = q\bar{b}ar = rabq$ and by symmetry we may assume that $|r| < |q|$.

Let w be a palindrome prefix of p. If $|w| \le |q|$, the result holds by induction. If $w = qb$ then w is a power of b. Thus set $w = qbat$ where t is a prefix of r. Since r is a prefix of q, the word t is a prefix of q, and since $w = \tilde{t}abq$, one has $t = \tilde{t}$. Thus, by Corollary 2.2.9, $w = qbat$ is central. ∎

The next characterization relates central words to periods in words. Recall from Chapter 1 that given a word $w = a_1 \cdots a_n$, where $a_1, \ldots, a_n$ are letters, an integer k is a *period* of w if $k \ge 1$ and $a_i = a_{i+k}$ for all $1 \le i \le n-k$. Any integer $k \ge n$ is a period with this definition.

An integer k with $1 \le k \le |w|$ is a period of w if and only if there exist words x, y, and z such that

$$w = xy = zx, \quad |y| = |z| = k.$$

Fine and Wilf's theorem states that if a word w has two periods k and ℓ, and $|w| \ge k + \ell - \gcd(k, \ell)$, then $\gcd(k, \ell)$ is also a period of w. In particular, if k and ℓ are relatively prime, and $|w| \ge k + \ell - 1$, then w is the power of a single letter. The bound is sharp, and the question arises how to describe the words w of length $|w| = k + \ell - 2$ having periods k and ℓ. This is the object of the next theorem.

THEOREM 2.2.11. *A word w is central if and only if it has two periods k and ℓ such that $\gcd(k, \ell) = 1$ and $|w| = k + \ell - 2$. Moreover, if $w \notin 0^* \cup 1^*$, and $w = p10q$ with p, q palindrome words, then $\{k, \ell\} = \{|p| + 2, |q| + 2\}$ and the pair $\{k, \ell\}$ is unique.*

The proof will show that any word w having two periods k and ℓ such that $\gcd(k, \ell) = 1$ and $|w| = k + \ell - 2$ is over an alphabet with at most two letters.

Proof of theorem. Let w be a central word. Then $w01$ is a standard word, and there is a standard pair (x, y) such that $w01 = xy$. If $x = 0$ or $y = 1$, then w is a power of 0 or 1, respectively, and w has periods $k = 1$ and $\ell = |w| + 1$. Otherwise, $x = p10$ and $y = q01$ for some palindrome words p, q, and $w = p10q = q01p$ has two periods $k = |x|$ and $\ell = |y|$ which are relatively prime by Equation (2.2.3). Assume that w has also periods $\{k', \ell'\}$, with $k' + \ell' - 2 = |w|$. We may suppose $k < k' < \ell' < \ell$. Since $k + \ell' - 1 \le |w|$, Fine and Wilf's theorem applies. So w has also the period $d = \gcd(k, \ell')$. Similarly, w has also the period $d' = \gcd(k, k')$. So it has the period $\gcd(d, d') = 1$. This proves that the pair $\{k, \ell\}$ is unique.

Conversely, if w is a power of a letter, the result is trivial. Thus we assume that w contains two distinct letters. Since $k, \ell \ne 1$, we assume $2 \le k < \ell$.

Since w has period k, there is a word x of length $|x| = \ell - 2$ that is both a prefix and a suffix of w. Similarly, there is a word y of length $|y| = k - 2$ that is both a prefix and a suffix of w. Consequently, there exist words u and v, both of length 2, such that

$$w = yux = xvy.$$

We prove by induction on $|w|$ that x, y, w are palindrome words, that u and v are composed of distinct letters, and that no other letters than those of u appear in w (that is w is over an alphabet of two letters).

If $k = 2$, then y is the empty word. Thus $ux = xv$, and ℓ is odd. Therefore $u = ab$, $v = ba$, $x = (ab)^n a$, $w = (ab)^{n+1} a$ for letters $a \ne b$ and some $n \ge 0$. The result holds in this case.

If $k = \ell - 1$, then $x = ya = by$ for letters a and b. But then $a = b$ and w is a power of a letter, a case that we have excluded.

Thus we assume $k \le \ell - 2$. Then yu is a prefix of x. Define z by $yuz = x$. Then

$$x = yuz = zvy$$

showing that x has periods $|yu| = k$ and $|uz| = \ell - k$. Since $\gcd(k, \ell - k) = 1$ and $|x| = k + (\ell - k) - 2$, we get by induction that x is a palindrome, and that its prefix of length $k - 2$, that is y, and its suffix of length $\ell - k - 2$, that is z, are also palindromes. Moreover, $u = ab$ for letters $a \ne b$, and $\tilde{u} = v$ because $yuz = z\tilde{u}y = zvy$. Also, the word x (and y, and therefore also w) is composed only of a's and b's. Thus w is central. ∎

Theorem 2.2.11 associates, to every central word of length m, a pair $\{k, \ell\}$ of relatively prime integers such that $k + \ell - 2 = m$. We now show that, for each pair $\{k, \ell\}$ of relatively prime integers, there indeed exists a central word of length $k + \ell - 2$ and periods k and ℓ.

Let h, m be relatively prime integers with $1 \le h < m$. Define a word

$$z_{h,m} = a_1 a_2 \cdots a_{m-2} \qquad (a_n \in \{0, 1\})$$

by

$$a_n = \left\lfloor (n+1)\frac{h}{m} \right\rfloor - \left\lfloor n\frac{h}{m} \right\rfloor.$$

2.2. Standard words

These words have already been mentioned in our discussion of rational mechanical words (Equation (2.1.14)). Each word $z_{h,m}$ has length $m-2$ and height $h-1$.

PROPOSITION 2.2.12. *For every pair $1 \le h < m$ of relatively prime integers, the word $z_{h,m}$ is central. It has the periods k and ℓ where $k + \ell = m$ and $kh \equiv 1 \pmod{m}$.*

Proof. Define k by $1 \le k \le m-1$, and set $kh = 1 + \lambda m$. Observe that k exists because h and m are relatively prime. Let $\ell = m - k$. Then $\ell h \equiv -1 \pmod{m}$, and ℓ is the unique integer in the interval $[0, \ldots, m-1]$ with this property. Next

$$\left\lfloor (n+k)\frac{h}{m} \right\rfloor = \lambda + \left\lfloor \frac{nh+1}{m} \right\rfloor.$$

Since $nh \not\equiv -1 \pmod{m}$ for $1 \le n \le \ell - 1$, it follows that

$$\left\lfloor \frac{nh+1}{m} \right\rfloor = \left\lfloor \frac{nh}{m} \right\rfloor \quad (1 \le n \le \ell - 1).$$

Consequently, $a_{n+k} = a_n$ for $1 \le n \le \ell - 2$. A similar argument holds when k is replaced by ℓ and -1 is changed into 1.

Assume that some integer d divides k and ℓ. Then d divides also m. But k and ℓ are relatively prime to m, so $d = 1$ and $\gcd(k, \ell) = 1$. This proves, by Theorem 2.2.11, that $z_{h,m}$ is central. ∎

EXAMPLE 2.2.13. The words $z_{1,m} = 0^{m-2}$ and $z_{m-1,m} = 1^{m-2}$ are central. In particular, $z_{1,2} = \varepsilon$.

EXAMPLE 2.2.14. For $h = 5$, $m = 18$, one gets $z_{5,18} = 0010001001000100$, a word of length 16. By inspection, one finds the periods 7 and 11. The previous proposition allows one to compute them, since $11 \cdot 5 \equiv 1 \pmod{18}$.

PROPOSITION 2.2.15. *Let h, m be relatively prime integers with $1 \le h < m$. There exist exactly two standard words of height h and length m, namely $z_{h,m}10$ and $z_{h,m}01$. These words are balanced.*

Proof. By Proposition 2.2.12, the words $z_{h,m}10$ and $z_{h,m}01$ are standard words of height h and length m. They are factors of the Sturmian words $s_{h/m,0}$ and $s'_{h/m,0}$ and therefore are balanced. We prove that there exists

only one standard word of height h and length m ending in 10. Assume there are two, say w and w'. Then
$$w = xy, \quad w' = x'y'$$
for some standard pairs (x, y), (x', y'). By Formula (2.2.3),
$$h(x)|y| - h(y)|x| = 1, \quad h(x')|y'| - h(y')|x'| = 1.$$
Since $m = |x| + |y|$ and $h = h(x) + h(y)$, this gives
$$h(x)m - |x|h = 1, \quad h(x')m - |x'|h = 1$$
whence
$$(h(x) - h(x'))m = (|x'| - |x|)h.$$
Since $\gcd(m, h) = 1$, m divides $|x'| - |x|$. Thus $|x| = |x'|$, that is $x = x'$ and $y = y'$. ∎

Recall that Euler's *totient function* ϕ is defined for $m \geq 1$ as the number $\phi(m)$ of positive integers less than m and relatively prime to m

COROLLARY 2.2.16. *The number of standard words of length m is $2\phi(m)$, the number of central words of length m is $\phi(m+2)$, where ϕ is Euler's totient function.* ∎

2.2.2. Standard sequences and characteristic words

In this subsection, we use particular morphisms that will also be considered in the next subsection. Three of them, namely E, G, and D, were already introduced earlier. Here, these morphisms are used to relate standard words to characteristic words, and both to the continued fraction expansion of the slope of a characteristic word. Consider the morphisms

$$E : \begin{matrix} 0 \mapsto 1 \\ 1 \mapsto 0 \end{matrix}, \quad \varphi : \begin{matrix} 0 \mapsto 01 \\ 1 \mapsto 0 \end{matrix}, \quad \tilde{\varphi} : \begin{matrix} 0 \mapsto 10 \\ 1 \mapsto 0 \end{matrix}.$$

From these, we get other morphisms, denoted by G, $\tilde{G}$, D, $\tilde{D}$ and defined by

$$G = \varphi \circ E : \begin{matrix} 0 \mapsto 0 \\ 1 \mapsto 01 \end{matrix}, \quad \tilde{G} = \tilde{\varphi} \circ E : \begin{matrix} 0 \mapsto 0 \\ 1 \mapsto 10 \end{matrix},$$

$$D = E \circ \varphi : \begin{matrix} 0 \mapsto 10 \\ 1 \mapsto 1 \end{matrix}, \quad \tilde{D} = E \circ \tilde{\varphi} : \begin{matrix} 0 \mapsto 01 \\ 1 \mapsto 1 \end{matrix}.$$

Of course, $\varphi = G \circ E = E \circ D$ and $\tilde{\varphi} = \tilde{G} \circ E = E \circ \tilde{D}$.

2.2. Standard words

LEMMA 2.2.17. *For any real number ρ, the following relations hold:*
$E(s_{\alpha,\rho}) = s'_{1-\alpha,1-\rho}$ *and* $E(s'_{\alpha,\rho}) = s_{1-\alpha,1-\rho}$.

Proof. For $n \geq 0$,

$$s'_{1-\alpha,1-\rho}(n) = \lceil (1-\alpha)(n+1) + 1 - \rho \rceil - \lceil (1-\alpha)n + 1 - \rho \rceil$$
$$= 1 - (\lceil -\alpha n - \rho \rceil - \lceil -\alpha(n+1) - \rho \rceil) = 1 - s_{\alpha,\rho}(n)$$

because $-\lceil -r \rceil = \lfloor r \rfloor$ for every real number r. This proves the first equality, and the second is symmetric. ∎

LEMMA 2.2.18. *Let $0 < \alpha < 1$. For $0 \leq \rho < 1$,*

$$G(s_{\alpha,\rho}) = s_{\frac{\alpha}{1+\alpha},\frac{\rho}{1+\alpha}}, \quad \tilde{G}(s_{\alpha,\rho}) = s_{\frac{\alpha}{1+\alpha},\frac{\rho+\alpha}{1+\alpha}}, \quad \varphi(s_{\alpha,\rho}) = s'_{\frac{1-\alpha}{2-\alpha},\frac{1-\rho}{2-\alpha}}$$

and for $0 < \rho \leq 1$,

$$G(s'_{\alpha,\rho}) = s'_{\frac{\alpha}{1+\alpha},\frac{\rho}{1+\alpha}}, \quad \tilde{G}(s'_{\alpha,\rho}) = s'_{\frac{\alpha}{1+\alpha},\frac{\rho+\alpha}{1+\alpha}}, \quad \varphi(s'_{\alpha,\rho}) = s_{\frac{1-\alpha}{2-\alpha},\frac{1-\rho}{2-\alpha}}.$$

Proof. Let $s = a_0 a_1 \cdots a_n \cdots$ be an infinite word, the a_i being letters. An integer n is the index of the kth occurrence of the letter 1 in s if $a_0 \cdots a_n$ contains k letters 1 and $a_0 \cdots a_{n-1}$ contains $k-1$ letters 1. If $s = s_{\alpha,\rho}$ and $0 \leq \rho < 1$, this means that

$$\lfloor \alpha(n+1) + \rho \rfloor = k, \quad \lfloor \alpha n + \rho \rfloor = k - 1$$

which implies $\alpha n + \rho < k \leq \alpha(n+1) + \rho$, that is

$$n = \left\lceil \frac{k-\rho}{\alpha} - 1 \right\rceil.$$

Similarly, if $s = s'_{\alpha,\rho}$ and $0 < \rho \leq 1$, then

$$\lceil \alpha(n+1) + \rho \rceil = k+1, \quad \lceil \alpha n + \rho \rceil = k$$

and $n = \left\lfloor \frac{k-\rho}{\alpha} \right\rfloor$.

Set $G(s_{\alpha,\rho}) = b_0 b_1 \cdots b_i \cdots$, with $b_i \in \{0,1\}$. Since every letter 1 in $s_{\alpha,\rho}$ is mapped to 01 in $G(s_{\alpha,\rho})$, the prefix $a_0 \cdots a_n$ of $s_{\alpha,\rho}$ (where n is the index of the kth letter 1) is mapped onto the prefix $b_0 b_1 \cdots b_{n+k}$ of $G(s_{\alpha,\rho})$. Thus the index of the kth letter 1 in $G(s_{\alpha,\rho})$ is

$$n + k = \left\lceil \frac{k - \frac{\rho}{1+\alpha}}{\frac{\alpha}{1+\alpha}} - 1 \right\rceil.$$

This proves the first formula.

Next, we observe that, for any infinite word x, one has
$$G(x) = 0\tilde{G}(x).$$
Indeed, the formula $G(w)0 = 0\tilde{G}(w)$ is easily shown to hold for finite words w by induction. Furthermore, if a Sturmian word $s_{\alpha,\rho}$ starts with 0 and setting $s_{\alpha,\rho} = 0t$, one gets $t = s_{\alpha,\alpha+\rho}$. Altogether $\tilde{G}(s_{\alpha,\rho}) = s_{\alpha/(1+\alpha),(\rho+\alpha)/(1+\alpha)}$ for $0 \le \rho < 1$. The proof of the other formula is similar. Finally, since $\varphi = G \circ E$, $\varphi(s_{\alpha,\rho}) = G(s'_{1-\alpha,1-\rho}) = s'_{(1-\alpha)/(2-\alpha),(1-\rho)/(2-\alpha)}$. ∎

COROLLARY 2.2.19. *For any Sturmian word s, the infinite words $E(s)$, $G(s)$ $\tilde{G}(s)$, $\varphi(s)$, $\tilde{\varphi}(s)$, $D(s)$ $\tilde{D}(s)$ are Sturmian.* ∎

Formulas similar to those of Lemma 2.2.18 hold for $\tilde{\varphi}, D, \tilde{D}$ (Problem 2.2.6). Recall that the characteristic word of irrational slope α is defined by
$$c_\alpha = s_{\alpha,\alpha} = s'_{\alpha,\alpha}.$$
The previous lemmas imply

COROLLARY 2.2.20. *For any irrational α with $0 < \alpha < 1$, one has*
$$E(c_\alpha) = c_{1-\alpha}, \quad G(c_\alpha) = c_{\alpha/(1+\alpha)}.$$
∎

For $m \ge 1$, define a morphism θ_m by
$$\theta_m : \begin{array}{l} 0 \mapsto 0^{m-1}1 \\ 1 \mapsto 0^{m-1}10 \end{array}.$$
It is easily checked that
$$\theta_m = G^{m-1} \circ E \circ G.$$

COROLLARY 2.2.21. *For $m \ge 1$, one has $\theta_m(c_\alpha) = c_{1/(m+\alpha)}$.*

Proof. Since $E \circ G(c_\alpha) = c_{1/(1+\alpha)}$, the formula holds for $m = 1$. Next, $G(c_{1/(k+\alpha)}) = c_{1/(1+k+\alpha)}$, so the claim is true by induction. ∎

We use this corollary for connecting continued fractions to characteristic words. Recall that every irrational number γ admits a unique expansion as a continued fraction
$$\gamma = m_0 + \cfrac{1}{m_1 + \cfrac{1}{m_2 + \cfrac{1}{\cdots}}} \tag{2.2.8}$$

2.2. Standard words

where $m_0, m_1, \ldots$ are integers, $m_0 \geq 0$, $m_i > 0$ for $i \geq 1$. If (2.2.8) holds, we write
$$\gamma = [m_0, m_1, m_2, \ldots].$$

The integers m_i are called the *partial quotients* of γ. If the sequence (m_i) is eventually periodic, and $m_i = m_{k+i}$ for $i \geq h$, this is reported by overlining the purely periodic part, as in
$$\gamma = [m_0, m_1, m_2, \ldots, m_{h-1}, \overline{m_h, \ldots, m_{h+k-1}}].$$

Let $\alpha = [0, m_1, m_2, \ldots]$ be the continued fraction expansion of an irrational α with $0 < \alpha < 1$. If, for some β with $0 < \beta < 1$,
$$\beta = [0, m_{i+1}, m_{i+2}, \ldots]$$
we agree to write
$$\alpha = [0, m_1, m_2, \ldots, m_i + \beta].$$

COROLLARY 2.2.22. *If* $\alpha = [0, m_1, m_2, \ldots, m_i + \beta]$ *for some irrational* α *and* $0 < \alpha, \beta < 1$, *then*
$$c_\alpha = \theta_{m_1} \circ \theta_{m_2} \circ \cdots \circ \theta_{m_i}(c_\beta).\qquad\blacksquare$$

Let $(d_1, d_2, \ldots, d_n, \ldots)$ be a sequence of integers, with $d_1 \geq 0$ and $d_n > 0$ for $n > 1$. To such a sequence, we associate a sequence $(s_n)_{n \geq -1}$ of words by
$$s_{-1} = 1, \quad s_0 = 0, \quad s_n = s_{n-1}^{d_n} s_{n-2} \quad (n \geq 1). \tag{2.2.9}$$

The sequence $(s_n)_{n \geq -1}$ is a *standard sequence*, and the sequence $(d_1, d_2, \ldots)$ is its *directive sequence*. Observe that if $d_1 > 0$, then any s_n ($n \geq 0$) starts with 0; on the contrary, if $d_1 = 0$, then $s_1 = s_{-1} = 1$, and s_n starts with 1 for $n \neq 0$. Every s_{2n} ends with 0, every s_{2n+1} ends with 1.

EXAMPLE 2.2.23. The directive sequence $(1, 1, \ldots)$ gives the standard sequence defined by $s_n = s_{n-1} s_{n-2}$, that is the sequence of finite Fibonacci words. Observe that the directive sequence $(0, 1, 1, \ldots)$ results in the sequence of words obtained from Fibonacci words by exchanging 0 and 1.

Every standard word occurs in some standard sequence, and every word occurring in a standard sequence is a standard word. This results by induction from the fact that, for $s_n = s_{n-1}^{d_n} s_{n-2}$, one has
$$(s_n, s_{n-1}) = \Delta^{d_n}(s_{n-2}, s_{n-1}), \quad (s_{n-1}, s_n) = \Gamma^{d_n}(s_{n-1}, s_{n-2}).$$

Thus
$$(s_{2n}, s_{2n-1}) = \Delta^{d_{2n}} \circ \Gamma^{d_{2n-1}} \circ \cdots \circ \Gamma^{d_1}(0, 1),$$
$$(s_{2n}, s_{2n+1}) = \Gamma^{d_{2n+1}} \circ \Delta^{d_{2n}} \circ \Gamma^{d_{2n-1}} \circ \cdots \circ \Gamma^{d_1}(0, 1).$$

By Equation (2.2.4), this gives the expressions
$$s_{2n} = G^{d_1} \circ D^{d_2} \circ \cdots \circ D^{d_{2n}}(0) = G^{d_1} \circ \cdots \circ D^{d_{2n}} \circ G^{d_{2n+1}}(0),$$
$$s_{2n+1} = G^{d_1} \circ D^{d_2} \circ \cdots \circ D^{d_{2n+2}}(1) = G^{d_1} \circ \cdots \circ D^{d_{2n}} \circ G^{d_{2n+1}}(1).$$

PROPOSITION 2.2.24. *Let $\alpha = [0, 1 + d_1, d_2, \ldots]$ be the continued fraction expansion of some irrational α with $0 < \alpha < 1$, and let (s_n) be the standard sequence associated to $(d_1, d_2, \ldots)$. Then every s_n is a prefix of c_α and*
$$c_\alpha = \lim_{n \to \infty} s_n.$$

Proof. By definition, $s_n = s_{n-1}^{d_n} s_{n-2}$ for $n \geq 1$. Define morphisms h_n by
$$h_n = \theta_{1+d_1} \circ \theta_{d_2} \circ \cdots \circ \theta_{d_n}.$$

We claim that
$$s_n = h_n(0), \quad s_n s_{n-1} = h_n(1), \quad n \geq 1.$$

This holds for $n = 1$ since $h_1(0) = 0^{d_1} 1 = s_1$ and $h_1(1) = 0^{d_1} 10 = s_1 s_0$. Next, for $n \geq 2$,
$$h_n(0) = h_{n-1}(\theta_{d_n}(0)) = h_{n-1}(0^{d_n-1} 1) = s_{n-1}^{d_n-1} s_{n-1} s_{n-2} = s_n$$

and
$$h_n(1) = h_{n-1}(0^{d_n-1} 10) = s_n s_{n-1}.$$

For any infinite word x, the infinite word $h_n(x)$ starts with s_n because both $h_n(0)$ and $h_n(1)$ start with s_n. Thus, setting $\beta_n = [0, d_{n+1}, d_{n+2}, \ldots]$, one has $c_\alpha = h_n(c_{\beta_n})$ by Corollary 2.2.22 and thus c_α starts with s_n. This proves the first claim. The second is an immediate consequence. ∎

It is easily checked that
$$\theta_{1+d_1} \circ \theta_{d_2} \circ \cdots \circ \theta_{d_r} = G^{d_1} \circ E \circ G^{d_2} \circ E \circ \cdots \circ G^{d_r} \circ E \circ G$$
$$= \begin{cases} G^{d_1} \circ D^{d_2} \circ \cdots \circ D^{d_r} \circ G & \text{if } r \text{ is even,} \\ G^{d_1} \circ D^{d_2} \circ \cdots \circ D^{d_r} \circ D \circ E & \text{otherwise.} \end{cases}$$

EXAMPLE 2.2.25. The directive sequence for the Fibonacci word is $(1, 1, \ldots)$. The corresponding irrational is $1/\tau^2 = [0, 2, 1, 1, \ldots]$, and indeed the infinite Fibonacci word is the characteristic word of slope $1/\tau^2$.

2.2. Standard words

EXAMPLE 2.2.26. Since $1/\tau = [0,1,1,1,\ldots]$, the corresponding standard sequence is $s_1 = 1$, $s_2 = 10$, $s_3 = 101,\ldots$. The sequence is obtained from the Fibonacci sequence by exchanging 0's and 1's, in concordance with Lemma 2.2.17, since indeed $1/\tau + 1/\tau^2 = 1$.

EXAMPLE 2.2.27. Consider $\alpha = (\sqrt{3} - 1)/2 = [0,2,1,2,1,\ldots]$. The directive sequence is $(1,1,2,1,2,1,\ldots)$, and the standard sequence starts with $s_1 = 01$, $s_2 = 010$, $s_3 = 01001001$, $\ldots$, whence

$$c_{(\sqrt{3}-1)/2} = 010010010100100100101001001001 \cdots.$$

Due to the periodicity of the development, we get for $n \geq 2$ that $s_{n+2} = s_{n+1}^2 s_n$ if n is odd, and $s_{n+2} = s_{n+1} s_n$ if n is even.

COROLLARY 2.2.28. *Every standard word is a prefix of some characteristic word.* ∎

Thus, every standard word is left special.

COROLLARY 2.2.29. *A word is central if and only if it is a palindrome prefix of some characteristic word.*

Proof. A central word is a prefix of some standard word, so also of some characteristic word. Conversely, a palindrome prefix of a characteristic word is a prefix of any sufficiently long word in its standard sequence, so also of some sufficiently long central word. Thus the result follows from Corollary 2.2.10. ∎

Proposition 2.2.24 has several interesting consequences. The relation to fixpoints is left to subsection 2.3.6. We focus on two properties, first the powers that may appear in a Sturmian word, and then the computation of the number of factors of Sturmian words.

Let x be an infinite word. For $w \in F(x)$, the *index* of w in x is the greatest integer d such that $w^d \in F(x)$, if such an integer exists. Otherwise, w is said to have infinite index.

PROPOSITION 2.2.30. *Every nonempty factor of a Sturmian word s has finite index in s.*

Proof. Assume the contrary. There exist a Sturmian word s and a nonempty factor u of s such that u^n is a factor of s for every $n \geq 1$. Consequently, the periodic word u^ω is in the dynamical system generated by s. Since this system is minimal, $F(s) = F(u^\omega)$, a contradiction. ∎

An infinite word x has *bounded index* if there exists an integer d such that every nonempty factor of x has an index less than or equal to d.

THEOREM 2.2.31. *A Sturmian word has bounded index if and only if the continued fraction expansion of its slope has bounded partial quotients.*

We start with a lemma.

LEMMA 2.2.32. *Let $(s_n)_{n\geq -1}$ be the standard sequence of the characteristic word c_α, with $\alpha = [0, 1 + d_1, d_2, \ldots]$. For $n \geq 3$, the word $s_n^{1+d_{n+1}}$ is a prefix of c_α, and $s_n^{2+d_{n+1}}$ is not a prefix. If $d_1 \geq 1$, this holds also for $n = 2$.*

EXAMPLE 2.2.33. For the Fibonacci word $f = 0100101001001\cdots$, we have $s_n = f_n$ and $d_n = 1$ for all n. The lemma claims that for $n \geq 2$, the word f_n^2 is a prefix of the infinite word f, and that f_n^3 is not. As an example, $f_2^2 = 010010$ is a prefix and $f_2^3 = 010010010$ is not. Observe also that $f_1^2 = 0101$ is not a prefix of f.

Proof of lemma. We show that for $n \geq 3$ (and for $n \geq 2$ if $d_1 \geq 1$), one has
$$s_{n-1}s_n = s_n t_{n-1}, \quad \text{with } t_n = s_{n-1}^{d_n-1} s_{n-2} s_{n-1}.$$

Indeed
$$s_{n-1}s_n = s_{n-1}s_{n-1}^{d_n} s_{n-2} = s_{n-1}^{d_n} s_{n-2}^{d_{n-1}} s_{n-3} s_{n-2}$$
$$= s_{n-1}^{d_n} s_{n-2} s_{n-2}^{d_{n-1}-1} s_{n-3} s_{n-2} = s_n t_{n-1}$$

provided $d_{n-1} \geq 1$. Observe that t_{n-1} is not a prefix of s_n, since otherwise $s_n = t_{n-1}u$ for some word u, and $s_{n-1}s_n u = s_n^2$ and s_n is not primitive.

Clearly, $s_{n+1}s_n$ is a prefix of the characteristic word c_α. Since
$$s_{n+1}s_n = s_n^{d_{n+1}} s_{n-1} s_n = s_n^{1+d_{n+1}} t_{n-1}$$

the word $s_n^{1+d_{n+1}}$ is a prefix of c_α, and since t_{n-1} is not a prefix of s_n, the word $s_n^{2+d_{n+1}}$ is not a prefix of c_α. ∎

Proof of Theorem 2.2.31. Since a Sturmian word has the same factors as the characteristic word of the same slope, it suffices to prove the result for characteristic words. Let c be the characteristic word of slope $\alpha = [0, 1 + d_1, d_2, \ldots]$. Let $(s_n)_{n \geq -1}$ be the associated standard sequence.

To prove that the condition is necessary, observe that $s_n^{d_{n+1}}$ is a prefix of c for each $n \geq 1$. Consequently, if the sequence (d_n) of partial quotients is unbounded, the infinite word c has factors of arbitrarily great exponent.

Conversely, assume that the partial quotients (d_n) are bounded by some D and arguing by contradiction, suppose that c has unbounded index. Let r be some integer such that $F(c)$ contains a primitive word of length r with index greater than $D + 4$. Among those words, let w be a

2.2. Standard words

word of length r of maximal index. Let $d+1$ be the index of w. Then $d \geq D+3$. The proof is in three steps.

(1) The characteristic word c has prefixes of the form w^d, with $d \geq D+3$. Indeed, if w^{d+1} is a prefix of c, we are done. Otherwise, consider an occurrence of w^{d+1}. Set $w = za$ with a a letter, and let b be the letter preceding the occurrence of w^{d+1}. If $b = a$, replace w by az and proceed. The process will stop after at most $|w| - 1$ steps either because a prefix of c is obtained, or because otherwise w would occur in c to the power $d+2$. Thus, we may assume $b \neq a$. Thus $b(za)^{d+1}$ is a factor of c. This implies that $a(za)^d$ and $b(za)^d$ are factors, so w^d is a right special factor, and therefore it is a prefix of c.

(2) If w^d is a prefix of the characteristic word c, then w is one of the standard words s_n. Indeed, set $e = d-2$, so that $e \geq D+1$. Let n be the greatest integer such that s_n is a prefix of w^{e+1}. Then w^{e+1} is a prefix of $s_{n+1} = s_n^{d_{n+1}} s_{n-1}$, thus also of $s_n^{1+d_{n+1}}$. This shows that

$$(1+D)|w| \leq (1+e)|w| \leq (1+d_{n+1})|s_n| \leq (1+D)|s_n|$$

whence $|w| \leq |s_n|$. Now, since both w^{e+2} and $s_n^{1+d_{n+1}}$ are prefixes of c, one is a prefix of the other. If w^{e+2} is the shorter one, then $|w^{e+2}| = |w^{e+1}| + |w| \geq |s_n| + |w|$. Thus, w^{e+2} and $s_n^{1+d_{n+1}}$ share a common prefix of length $\geq |s_n| + |w|$. Consequently, w and s_n are powers of the same word, and since they are primitive, they are equal.

If $s_n^{1+d_{n+1}}$ is the shorter one then, since $(1+e)|w| \leq (1+d_{n+1})|s_n|$,

$$|s_n^{1+d_{n+1}}| = |s_n| + d_{n+1}|s_n| \geq |s_n| + \frac{d_{n+1}}{1+d_{n+1}}(1+e)|w| \geq |s_n| + |w|$$

and the same conclusion holds.

(3) It follows that s_n^{1+e} is a prefix of c and, since $e \geq D+1 \geq d_{n+1}+1$, also $s_n^{2+d_{n+1}}$ is a prefix of c, contradicting Lemma 2.2.32. ∎

We conclude this subsection with the computation of the number of factors of Sturmian words. Another characterization of central words will help. Recall that a finite word is balanced if and only if it is a factor of some Sturmian word. Moreover, every balanced word w, as a factor of some uniformly recurrent infinite word, can be extended to the right and to the left, that is wa and bw are balanced for some letters a, b.

PROPOSITION 2.2.34. *For any word w, the following are equivalent:*
  (i) *the word w is central;*
  (ii) *the words $0w0$, $0w1$, $1w0$, $1w1$ are balanced;*
  (iii) *the words $0w1$ and $1w0$ are balanced.*

Proof. (i) ⇒ (ii) The words $w01$ and $w10$ are standard, and therefore are prefixes of some characteristic words c and c'. By Proposition 2.1.22 the four infinite words $0c$, $1c$, $0c'$ and $1c'$ are Sturmian, and consequently their prefixes $0w0$, $0w1$, $1w0$, $1w1$ are balanced. The implication (ii) ⇒ (iii) is trivial.

(iii) ⇒ (i) We prove first that w is a palindrome word. Assume the contrary. Then there are words u,v, v' and letters $a \ne b$ such that $w = uav = v'b\tilde{u}$. But then $awb = auavb = av'b\tilde{u}b$ has factors aua and $b\tilde{u}b$ with heights satisfying $|h(aua) - h(b\tilde{u}b)| = 2$, a contradiction.

Let c be a characteristic word such that $0w1 \in F(c)$. Since $F(c)$ is closed under reversal (Proposition 2.1.19), and w is a palindrome, $1w0 \in F(c)$, showing that w is a right special factor of c. Thus its reversal (that is w itself) is a prefix of c. In view of Corollary 2.2.29, the word w is central. ∎

Words satisfying condition (ii) are sometimes called *strictly bispecial*.

We now want to count the number of balanced words of length n. We need a lemma.

LEMMA 2.2.35. *Let w be a word. If $w0$ and $w1$ are balanced, then there is a letter a such that $aw0$ and $aw1$ are balanced.*

Before giving the proof, let us observe that there seems to be a difference, for a word w, between being right special and having both extensions $w0$ and $w1$ balanced. Indeed, a word w can only be right special with respect to some Sturmian word s that contains both factors $w0$ and $w1$. On the contrary, if $w0$ and $w1$ are balanced, then there exist Sturmian words x and y such that $w0 \in F(x)$ and $w1 \in F(y)$, but x and y need not be the same. In fact, one can show (Problem 2.2.7) that the two notions coincide.

Proof of Lemma 2.2.35. Since $w0$ and $w1$ are factors of Sturmian words, there exist letters a and b such that $aw0$ and $bw1$ are balanced. If $a = b$, we get the claim. If $a = 1$ and $b = 0$, then w is central by Proposition 2.2.34, and therefore is balanced. Thus suppose $a = 0, b = 1$. Then $0w0$ and $1w1$ are balanced, but neither $1w0$ nor $0w1$ is. According to Proposition 2.1.3, there exists a palindrome word u such that $1u1$ and $0u0$ are factors of $1w0$. However, since $1w$ and $w0$ are balanced, $1u1$ is a prefix of $1w0$ and $0u0$ is a suffix of $1w0$. Thus there exist words p, s such that $1w0 = 1u1s0 = 1p0u0$, whence $w = u1s = p0u$. Similarly, there exist words u', p', s' such that $w = u'0s' = p'1u'$. We may assume $|u| < |u'|$ and set $u' = u1x = y0u$ for some words x, y. Then $w = y0u0s' = p'1u1x$, showing that w is unbalanced, a contradiction. ∎

2.2. Standard words

THEOREM 2.2.36. *The number of balanced words of length n is*

$$1 + \sum_{i=1}^{n}(n+1-i)\phi(i)$$

where ϕ is Euler's totient function.

Proof. Let $R(n)$ be the set of words w of length n such that $0w$ and $1w$ are balanced, and set $r(n) = \text{Card } R(n)$. Then $r(0) = 1 = \phi(1)$ and

$$r(n+1) = r(n) + \phi(n+2).$$

Indeed, for each $w \in R(n)$, one has $0w \in R(n+1)$ or $1w \in R(n+1)$ by Lemma 2.2.35, and both $0w, 1w \in R(n+1)$ if and only if $w \in R(n)$ and $0w1$ and $1w0$ are balanced, that is if and only if w is central, by Proposition 2.2.34. Thus $r(n+1) - r(n)$ is the number of central words of length n, which in turn is $\phi(n+2)$ by Corollary 2.2.16. It follows that

$$r(n) = \sum_{i=1}^{n+1}\phi(n).$$

Let $g(n)$ be the number of balanced words of length n. Then

$$g(n+1) = g(n) + r(n)$$

since for each balanced word w, the word $w0$ or $w1$ is balanced, and both are balanced if and only if $w \in R(n)$. Since $g(0) = 1$, it follows that

$$g(n) = 1 + \sum_{k=0}^{n-1} r(k) = 1 + \sum_{k=0}^{n-1}\sum_{i=1}^{k+1}\phi(i)$$

$$= 1 + \sum_{k=1}^{n}\sum_{i=1}^{k}\phi(i) = 1 + \sum_{i=1}^{n}(n+1-i)\phi(i)$$

as required. ∎

2.2.3. Frequencies

Let x be an infinite word. Recall from Chapter 1 that the *factor graph* $G_n(x)$ of order n is the graph with vertex set $F_n(x)$ and domain $F_{n+1}(x)$. A triple (p, a, s) is an edge if and only if $pa = bs \in F_{n+1}(x)$ for some letter b.

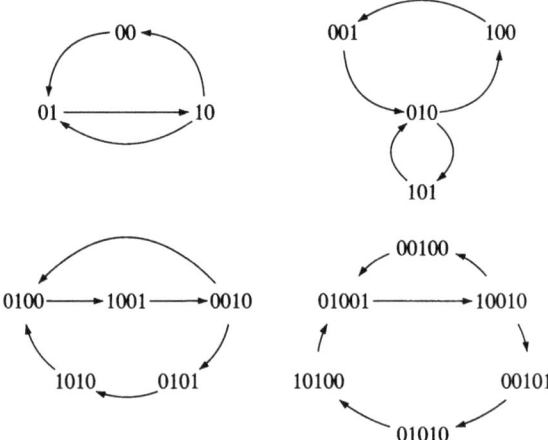

Figure 2.6. Factor graphs for the Fibonacci word.

If x is a Sturmian word, then there is exactly one vertex in $G_n(x)$ with out-degree 2. This is the right special factor d_n of length n. The edges leaving d_n are $(d_n, 0, d_{n-1}0)$ and $(d_n, 1, d_{n-1}1)$, because d_{n-1} is a suffix of d_n. Similarly, there is exactly one vertex with in-degree 2. This is the left special factor g_n of length n. Let a be the letter such that $g_n = g_{n-1}a$. Then the edges entering g_n are $(0g_{n-1}, a, g_n)$ and $(1g_{n-1}, a, g_n)$. Observe that $d_n = g_n$ if and only if d_n is a palindrome word. See Figure 2.6 for the word graphs of the Fibonacci word.

The factor graph of order n of a Sturmian word x is composed of three paths. The first is from g_n to d_n, both vertices included. This path is never empty. There are two other paths, from d_n to g_n, one through vertex $d_{n-1}0$, the other through $d_{n-1}1$. We consider that the endpoints d_n and g_n are not part of these paths. Then such a path may be empty. This happens if and only if $d_{n-1}0 = g_n$ or $d_{n-1}1 = g_n$ which in turn is the case if and only if $d_{n-1} = g_{n-1}$ because g_{n-1} is a prefix of g_n.

Let $s = s_{\alpha,\rho}$ be a Sturmian word of slope α. We have seen how to associate to s a rotation R on the unit circle. Also (Formula (2.1.11)), a word w is a factor of s if and only if the interval I_w of the unit circle is nonempty. Moreover, an integer $n \geq 0$ is the starting index of an occurrence of w in s if and only if $R^n(\rho) \in I_w$.

Let $\mu_N(w)$ be the number of occurrences of w in the prefix of length $N + |w| - 1$ of s. This is exactly the number of integers n, with $0 \leq n < N$, such that $R^n(\rho) \in I_w$. It is known from number theory that the numbers $R^n(\rho)(n \geq 1)$ are uniformly distributed in the interval $[0,1)$. As a consequence, the limit

$$\mu(w) = \lim_{N \to \infty} \mu_N(w)$$

always exists and is equal to the length of the interval I_w. The number $\mu(w)$ is the *frequency* of w in s. Of course, $\mu(w) = 0$ if and only if $w \notin F(s)$. It is easily seen that, for any word w, one has $\mu(0w) + \mu(1w) = \mu(w)$ and symmetrically $\mu(w) = \mu(w0) + \mu(w1)$.

THEOREM 2.2.37. *Let s be a Sturmian word. For each n, the frequencies of the factors of length n take at most three values. If they take three values, then one is the sum of the two others.*

LEMMA 2.2.38. *Let s be a Sturmian word. Let (p, a, q) be an edge in $G_n(s)$. If p is not right special and q is not left special, then $\mu(p) = \mu(q)$.*

Proof. There exists a letter b such that $pa = bq \in F_{n+1}(s)$. Since $pb, aq \notin F_{n+1}$, one has $\mu(p) = \mu(pa) = \mu(bq) = \mu(q)$. ∎

Proof of Theorem 2.2.37. By the lemma, the frequencies are constant on each of the three paths in the factor graph $G_n(s)$. Thus there are at most three frequencies. Assume that none of the three paths in the factor graph is empty. According to our discussion, this happens if and only if $d_{n-1} \neq g_{n-1}$. Moreover, the frequencies are those of any set of vertices taken in the paths, e.g. $\mu(d_n)$, $\mu(d_{n-1}0)$, and $\mu(d_{n-1}1)$. Set $d_n = 0d_{n-1}$. Since d_{n-1} is not left special, $1d_{n-1}$ is not a factor of s. Thus

$$\mu(d_n) = \mu(0d_{n-1}) = \mu(d_{n-1}) = \mu(d_{n-1}0) + \mu(d_{n-1}1)$$

showing the second part of the theorem. ∎

2.3. Sturmian morphisms

All morphisms will be endomorphisms of $\{0,1\}^*$. The identity morphism id and the morphism E that exchanges the letters 0 and 1 will be called *trivial* morphisms.

A morphism f is *Sturmian* if $f(s)$ is a Sturmian word for every Sturmian word s. Since an erasing morphism can never be Sturmian, all morphisms considered here are assumed to be nonerasing. The trivial morphisms id and E are Sturmian. The set of Sturmian morphisms is closed under composition, and consequently is a submonoid of the monoid of endomorphisms of $\{0,1\}^*$.

2.3.1. A set of generators

The main result of this section is the characterization of Sturmian morphisms (Theorem 2.3.7). Consider the morphisms

$$\varphi : \begin{matrix} 0 \mapsto 01 \\ 1 \mapsto 0 \end{matrix}, \qquad \tilde{\varphi} : \begin{matrix} 0 \mapsto 10 \\ 1 \mapsto 0 \end{matrix}.$$

Recall from Chapter 1 that the morphism φ generates the infinite Fibonacci word $f = \varphi(f) = 010010100100101001010\cdots$.

PROPOSITION 2.3.1. *The morphisms E, φ and $\tilde{\varphi}$ are Sturmian.*

Proof. This follows from Corollary 2.2.19. ∎

We shall see below that every Sturmian morphism is a composition of these three morphisms. The following property gives a converse of Proposition 2.3.1.

PROPOSITION 2.3.2. *Let x be an infinite word.*
 (i) *If $\varphi(x)$ is Sturmian then x is Sturmian.*
 (ii) *If $\tilde{\varphi}(x)$ is Sturmian and x starts with the letter 0, then x is Sturmian.*

Proof. Let x be an infinite word. If $\varphi(x)$ or $\tilde{\varphi}(x)$ is Sturmian, then x is clearly aperiodic. Arguing by contradiction, let us suppose that x is not balanced and suppose that $0v0$ and $1v1$ are both factors of x.

Clearly, $\varphi(0v0) = 01\varphi(v)01$, $\varphi(1v1) = 0\varphi(v)0$ and every occurrence of $\varphi(1v1)$ in $\varphi(x)$ is followed by the letter 0. Consequently $1\varphi(v)01$ and $0\varphi(v)00$ are both factors of $\varphi(x)$ which is not balanced.

Next, if x does not start with 1, then either $01v1$ or $11v1$ is a factor of x. But $\tilde{\varphi}(0v0)$ contains the factor $10\tilde{\varphi}(v)1$, and $\tilde{\varphi}(01v1)$ and $\tilde{\varphi}(11v1)$ both contain the factor $00\tilde{\varphi}(v)0$. Consequently, $\tilde{\varphi}(x)$ is not balanced. ∎

COROLLARY 2.3.3. *Let x be an infinite word and let f be a morphism that is a composition of E and φ. If $f(x)$ is Sturmian then x is Sturmian.* ∎

EXAMPLE 2.3.4. We give an example of a non-Sturmian word x starting with 1 and such that $\tilde{\varphi}(x)$ is Sturmian. Let f be the Fibonacci word. The infinite word $11f$ is not Sturmian because it contains both 00 and 11 as factors. However, since f is a characteristic word, the infinite word $0f$ is Sturmian. Consequently $\tilde{\varphi}(\varphi(0f)) = \tilde{\varphi}(01f) = 100\tilde{\varphi}(f)$ is Sturmian. Thus $00\tilde{\varphi}(f)$ is also Sturmian and, since $00 = \tilde{\varphi}(11)$, $\tilde{\varphi}(11f)$ is Sturmian.

2.3. Sturmian morphisms

Let us denote by St the submonoid of the monoid of endomorphisms obtained by composition of E, φ and $\tilde{\varphi}$ in any number and order. St is called the *monoid of Sturm* and by Proposition 2.3.1 all its elements are Sturmian. A first step to the converse is the following.

LEMMA 2.3.5. *Let f and g be two morphisms and let x a Sturmian word. If $f \in St$ and $f \circ g(x)$ is a Sturmian word, then $g(x)$ is a Sturmian word.*

Proof. Let x be a Sturmian word and g a morphism. It suffices to prove the conclusion for $f = E$, $f = \varphi$ and $f = \tilde{\varphi}$.

Set $y = g(x)$. If $E(y)$ is a Sturmian word then y is also a Sturmian word and, by Proposition 2.3.2, this also holds if $\varphi(y)$ is a Sturmian word. It remains to prove that if $\tilde{\varphi}(y)$ is a Sturmian word then so is y.

Suppose that y is not a Sturmian word. Observe that y is aperiodic, since otherwise $\tilde{\varphi}(y)$ is eventually periodic, thus it is not Sturmian. Thus $y = g(x)$ is not balanced and contains two factors $0v0$ and $1v1$ which are factors of images of some factors of x. The Sturmian word x is recurrent, thus $1v1$ occurs infinitely often in y, which implies that $01v1$ or $11v1$ is a factor of y. Since $\tilde{\varphi}(0v0) = 10\tilde{\varphi}(v)10$ and $\tilde{\varphi}(1v1) = 0\tilde{\varphi}(v)0$, both $10\tilde{\varphi}(v)1$ and $00\tilde{\varphi}(v)0$ are factors of $\tilde{\varphi}(y)$ and thus $\tilde{\varphi}(y)$ is not balanced. A contradiction. ∎

COROLLARY 2.3.6. *Let $f \in St$ and g be a morphism. The morphism $f \circ g$ is Sturmian if and only if g is Sturmian.*

Proof. Assume first that g is Sturmian. Since f is a composition of E, φ and $\tilde{\varphi}$, the morphism $f \circ g$ is Sturmian by Proposition 2.3.1.

Conversely, if $f \circ g$ is Sturmian, then for every Sturmian word x, the infinite word $f \circ g(x)$ is Sturmian and, by Lemma 2.3.5, the infinite word $g(x)$ is Sturmian. This means that g is Sturmian. ∎

A morphism f is *locally Sturmian* if there exists at least one Sturmian word x such that $f(x)$ is a Sturmian word.

THEOREM 2.3.7. *Let f be a morphism. The following three conditions are equivalent:*
  (i) $f \in St$;
  (ii) f *is Sturmian*;
  (iii) f *is locally Sturmian*.

The equivalence of (i) and (ii) means that the monoid of Sturm is exactly the monoid of Sturmian morphisms.

The *length* of a morphism f is the number $\|f\| = |f(0)| + |f(1)|$. The proof of Theorem 2.3.7 is based on the following fundamental lemma.

LEMMA 2.3.8. *Let f be a nontrivial morphism. If f is locally Sturmian then $f(0)$ and $f(1)$ both start or both end with the same letter.*

Proof. Let f be a nontrivial morphism and suppose that $f(0)$ and $f(1)$ do not both start or both end with the same letter.

Suppose $f(0)$ starts with the letter 0. Then $f(1)$ starts with the letter 1. If $f(0)$ ends with 1 then $f(1)$ ends with 0. But in this case $f(01)$ contains a factor 11 and $f(10)$ contains a factor 00. Thus the image of any Sturmian word contains the two factors 00 and 11 which means that f is not locally Sturmian.

Otherwise $f(0) \in 0A^*0 \cup \{0\}$ and $f(1) \in 1A^*1 \cup \{1\}$, and we prove the result by induction on $\|f\|$.

If $\|f\| = 3$, then $f(a) = cc$ and $f(b) = d$ for letters a, b, c, d, $a \neq b$, and since any Sturmian word x contains the two factors a^{n+1} and ba^nb for some integer n, $f(x)$ contains $(cc)^{n+1}$ and $d(cc)^n d$ and thus is not Sturmian.

Arguing by contradiction, suppose that $\|f\| \geq 4$ and f is locally Sturmian. Let x be a Sturmian word such that $f(x)$ is Sturmian (such a word exists because f is locally Sturmian) and suppose that x contains the factor 00 (the case where x contains 11 is clearly the same). Since $f(0)$ starts and ends with 0, $f(x)$ also contains 00. Consequently, since the infinite word $f(x)$ is balanced, neither $f(0)$ nor $f(1)$ contains the factor 11.

Since x is Sturmian, x does not contain 11 and there is an integer $m \geq 1$ such that every block of 0's between two consecutive occurrences of 1 is either 0^m or 0^{m+1}.

The word $f(0)$ does not contain the factor 00. Indeed, otherwise $f(0) = u00v$ and $f(1) = r1 = 1s$ for some words u, v, r, s. Since 0^{m+1} and $10^m 1$ are factors of w, the words $f(0^{m+1})$ and $f(10^m 1)$ are factors of $f(x)$. But

$$f(0^{m+1}) = u00vf(0^{m-1})u00v = uw_1v, \quad f(10^m 1) = r1f(0^{m-1})u00v1s = rw_2s$$

for suitable w_1, w_2, and one has $|w_1| = |w_2|$ and $\delta(w_1, w_2) = 2$, a contradiction.

Consequently $f(0) = (01)^n 0$ for some integer $n \geq 0$.

Since $10^m 1$ and $10^{m+1} 1$ are factors of x, the infinite word $f(x)$ contains the two factors $10^m 1$ and $10^{m+1} 1$ if $n = 0$, and the two factors 101 and 1001 if $n \neq 0$. Set $p = m$ if $n = 0$, and $p = 1$ if $n \neq 0$. Then in both cases, $f(x)$ contains the factors $10^p 1$ and $10^{p+1} 1$, and in both cases $1 \leq p \leq m$.

Since $f(1)$ does not contain the factor 11, there exist an integer $k \geq 0$, and integers $m_1, \ldots, m_k \in \{0, 1\}$, such that

$$f(1) = 10^{p+m_1} 10^{p+m_2} 1 \cdots 10^{p+m_k} 1.$$

2.3. Sturmian morphisms

Consider a new alphabet $B = \{a, b\}$ and two morphisms $\rho, \eta : B^* \to A^*$:

$$\rho : \begin{matrix} a \mapsto 0 \\ b \mapsto 0^p 1 \end{matrix}, \qquad \eta : \begin{matrix} a \mapsto (01)^n 0 \\ b \mapsto 0^p 1 \end{matrix}.$$

We show that there exists a word u over B such that $f(\rho(b)) = \eta(bub)$.

(1) If $n = 0$, set $u = a^{m_1} b a^{m_2} b \cdots b a^{m_k}$. Since $f(1) \neq 1$, one has $f(1) = 1\eta(u)0^p 1$. Thus $f(\rho(b)) = f(0^p 1) = \eta(bub)$.

(2) If $n \neq 0$ and $m_1 = \cdots = m_k = 0$, set $u = b^{k+n-1}$. Since $f(1) = (10)^k 1$, one gets $\eta(u) = (01)^{k+n-1}$ and $f(\rho(b)) = f(01) = \eta(bub)$.

(3) Otherwise $n \neq 0$ and $m_i = 1$ for at least one integer i, $1 \leq i \leq k$. Thus there exist integers $t \geq 2, n_1, \ldots, n_t$ such that

$$f(1) = 1(01)^{n_1} 0 (01)^{n_2} 0 \cdots (01)^{n_{t-1}} 0 (01)^{n_t}.$$

Since $f(01)$ starts with $(01)^{n+1}$, one has $n_1 \geq 0$, $n_i \geq n$ for $2 \leq i \leq t-1$ and $n_t \geq 1$. Set $u = b^{n_1} a b^{n_2 - n} a \cdots b^{n_{t-1} - n} a b^{n_t - 1}$. Then, again, $f(\rho(b)) = f(01) = \eta(bub)$.

Define a morphism $g : B^* \to B^*$ by

$$g : \begin{matrix} a \mapsto a \\ b \mapsto bub \end{matrix}.$$

Then $f \circ \rho = \eta \circ g$. Since $m \geq p$, by deleting if necessary some letters at the beginning of x, one may suppose that x starts with $0^p 1$. It follows that there exists a (unique) infinite word x' over B such that $\rho(x') = x$.

Thus there exists a (unique) infinite word y' over B such that

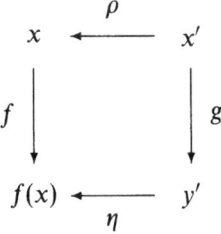

Identifying a with 0 and b with 1, one has $\rho = (\varphi \circ E)^p$. If $n = 0$ then $\eta = \rho$. If $n \neq 0$ then $p = 1$, so $\eta = \varphi \circ E \circ (E \circ \varphi)^n$. Thus since x and $f(x)$ are Sturmian, the words x' and y' are Sturmian by Corollary 2.3.3. Consequently the morphism g is locally Sturmian.

However, the words $g(0)$ and $g(1)$ do not start or end with the same letter and $3 \leq \|g\| < \|f\|$. By induction, g is not locally Sturmian, a contradiction. The lemma is proved. ∎

Proof of Theorem 2.3.7. It is easily seen that (i) ⇒ (ii) and (ii) ⇒ (iii).

So let us suppose that f is a locally Sturmian morphism. The property is straightforward if $f = \text{id}$ or $f = E$. Thus we assume $\|f\| \geq 3$.

Let x be a Sturmian word such that $f(x)$ is also a Sturmian word. Since $f(x)$ is balanced, it contains only one of the two words 00 and 11.

Suppose that $f(x)$ contains 00. From Lemma 2.3.8, the words $f(0)$ and $f(1)$ both start or both end with 0. Consider first the case where $f(0)$ and $f(1)$ both start with 0. Then $f(0), f(1) \in \{0, 01\}^+$ and there exist two words u and v such that $f(0) = \varphi(u)$ and $f(1) = \varphi(v)$. Define a morphism g by $g(0) = u$ and $g(1) = v$. Then $f = \varphi \circ g$ and, by Lemma 2.3.5, $g(x)$ is a Sturmian word. Next, $\|f\| = \|g\| + |uv|_0$ and $|uv|_0 > 0$. Otherwise, $f(0) = \varphi(u)$ and $f(1) = \varphi(v)$ would contain only 0 and $f(x) = 0^\omega$ would not be Sturmian. Thus $\|g\| < \|f\|$ and the result follows by induction.

If $f(0)$ and $f(1)$ both end with 0, the same argument holds with $\tilde{\varphi}$ instead of φ, and if $f(x)$ contains 11 then $E \circ f$ is of the same height and contains 00. ∎

We give here only one property of the monoid St which shows how decide whether a morphism is Sturmian by trying to decompose it over $\{E, \varphi, \tilde{\varphi}\}$. Other properties will be seen in Subsection 2.3.3 and in the problem section.

COROLLARY 2.3.9. *The monoid of Sturm is left and right unitary, i.e. for all morphisms f and g*
  (i) *if $f \circ g \in St$ and $f \in St$ then $g \in St$,*
  (ii) *if $f \circ g \in St$ and $g \in St$ then $f \in St$.*

Proof. Let f and g be two morphisms such that $f \circ g \in St$. Let x be a Sturmian word. Then $f \circ g(x)$ is a Sturmian word.

  (i) If $f \in St$ then by Lemma 2.3.5, $g(x)$ is a Sturmian word. Consequently g is locally Sturmian and, by Theorem 2.3.7, $g \in St$.

  (ii) If $g \in St$ then $g(x)$ is a Sturmian word. Thus f is locally Sturmian and by Theorem 2.3.7, $f \in St$. ∎

From this property we deduce an algorithm to decide whether a morphism is Sturmian. Indeed, if f is a nontrivial Sturmian morphism then f decomposes as $f = g \circ \sigma$, where g is Sturmian by Corollary 2.3.9 and where σ is one of the eight morphisms in $\{\varphi, \varphi \circ E, E \circ \varphi, E \circ \varphi \circ E, \tilde{\varphi}, \tilde{\varphi} \circ E, E \circ \tilde{\varphi}, E \circ \tilde{\varphi} \circ E\}$. According to σ, one gets the following factorizations of $f(0)$ and $f(1)$:

  $g(0) = f(1)$ and $f(0) = f(1)u$ with $u = g(1)$ if $\sigma = \varphi$;
  $g(0) = f(1)$ and $f(0) = uf(1)$ with $u = g(1)$ if $\sigma = \tilde{\varphi}$;
  $g(1) = f(1)$ and $f(0) = f(1)u$ with $u = g(0)$ if $\sigma = E \circ \varphi$;

2.3. Sturmian morphisms

$g(1) = f(1)$ and $f(0) = uf(1)$ with $u = g(0)$ if $\sigma = E \circ \tilde{\varphi}$;
$g(0) = f(0)$ and $f(1) = f(0)u$ with $u = g(1)$ if $\sigma = \varphi \circ E$;
$g(0) = f(0)$ and $f(1) = uf(0)$ with $u = g(1)$ if $\sigma = \tilde{\varphi} \circ E$;
$g(1) = f(0)$ and $f(1) = f(0)u$ with $u = g(0)$ if $\sigma = E \circ \varphi \circ E$;
$g(1) = f(0)$ and $f(1) = uf(0)$ with $u = g(0)$ if $\sigma = E \circ \tilde{\varphi} \circ E$.

PROPOSITION 2.3.10. *A morphism f is Sturmian if and only if, with f as input, the algorithm below ends with $g = $ id or E. In this case, the output h is a decomposition of f over $\{E, \varphi, \tilde{\varphi}\}$.* ∎

```
Algorithm:
    input:  f morphism;
    output: h morphism;
    local:  g morphism;
begin
    g ← f;
    h ← id;
    while one of the two words g(0) and g(1) is a proper prefix
              or a proper suffix of the other
        do if g(1) = g(0)u then
                g(1) ← u;  h ← φ ∘ E ∘ h
           else if g(1) = ug(0) then
                g(1) ← u;  h ← φ̃ ∘ E ∘ h
           else if g(0) = g(1)u then
                g(0) ← u;  h ← E ∘ φ ∘ h
           else {g(0) = ug(1)}
                g(0) ← u;  h ← E ∘ φ̃ ∘ h;
    if g = E then h ← E ∘ h
end.
```

Observe that $f(0)$ may be both a proper prefix and a proper suffix of $f(1)$ (or vice versa). In this case, there are two decompositions of f over $\{E, \varphi, \tilde{\varphi}\}$. These are obtained in the algorithm by inverting the order in the tests. We shall see in Subsection 2.3.3 that these are all decompositions (not containing E^2) of a given Sturmian morphism over $\{E, \varphi, \tilde{\varphi}\}$.

2.3.2. Standard morphisms

In this subsection it will be convenient to consider unordered standard pairs. An *unordered standard pair* is a set $\{x, y\}$ such that either (x, y) or (y, x) is a standard pair.

In particular, if $\{x, y\}$ is an unordered standard pair then $\{E(x), E(y)\}$ is an unordered standard pair. On the contrary, $\{\tilde{\varphi}(x), \tilde{\varphi}(y)\}$ is never an unordered standard pair because $\tilde{\varphi}(x)$ and $\tilde{\varphi}(y)$ end with the same letter (Proposition 2.2.2).

Consequently, Sturmian morphisms that are compositions of E and φ are an interesting special case. Because of the following proposition, a morphism is called *standard* if it is a composition of E and φ.

PROPOSITION 2.3.11. *A morphism f is standard if and only if $\{f(0), f(1)\}$ is an unordered standard pair.*

Proof. Assume first that f is standard and, arguing by induction on $\|f\|$, suppose that $\{f(0), f(1)\}$ is an unordered standard pair. If $g = f \circ E$, then $\{g(0), g(1)\} = \{f(0), f(1)\}$ is an unordered standard pair. If $g = f \circ \varphi$, then $\{g(0), g(1)\} = \{f(0)f(1), f(0)\}$ is also an unordered standard pair.

Conversely, assume that $\{f(0), f(1)\}$ is an unordered standard pair, and that $|f(0)| > |f(1)|$. Then $f(0) = f(1)v$ for some word v, and $\{v, f(1)\}$ is an unordered standard pair. By induction, there is a standard morphism g such that $\{g(0), g(1)\} = \{v, f(1)\}$. If $g(0) = f(1)$ and $g(1) = v$ then $f = g \circ \varphi$, in the other case $f = g \circ E \circ \varphi$. Thus f is standard. ∎

The set of standard morphisms is interesting because these morphisms are closely related to characteristic words (recall that an infinite word x is characteristic if and only if $0x$ and $1x$ are Sturmian words), as will appear in a moment.

A morphism f is *characteristic* if $f(x)$ is a characteristic word for every characteristic word x, and it is *locally characteristic* if there exists a characteristic word x such that $f(x)$ is a characteristic word.

The following theorem is an analogue of Theorem 2.3.7 for standard morphisms.

THEOREM 2.3.12. *Let f be a morphism. The following conditions are equivalent:*
  (i) *f is standard;*
  (ii) *f is characteristic;*
  (iii) *f is locally characteristic.*

To prove this result we need the following lemma.

LEMMA 2.3.13. *Let x be an infinite word;*
  (i) *x is characteristic if and only if $E(x)$ is characteristic;*
  (ii) *x is characteristic if and only if $\varphi(x)$ is characteristic.*

2.3. Sturmian morphisms

Proof. This is a consequence of Corollary 2.2.20 and Proposition 2.3.2. ∎

Proof of Theorem 2.3.12. The implication (ii) ⇒ (iii) is obvious and the implication (i) ⇒ (ii) is an immediate consequence of Lemma 2.3.13.

Let f be a locally characteristic morphism. Then f is locally Sturmian and by Theorem 2.3.7, it is a composition of E, φ and $\tilde{\varphi}$. We show that no occurrence of $\tilde{\varphi}$ appears in the decomposition of f, by induction on $\|f\|$.

If $\|f\| = 2$ then $f = \text{id}$ or $f = E$ and the result holds.

Assume $\|f\| \geq 3$ and let x be a characteristic word such that $f(x)$ is characteristic.

If x contains 11 as a factor then we can replace x by $E(x)$ which is also a characteristic word (Lemma 2.3.13) and consider $f \circ E$ instead of f, and if $f(x)$ contains 11 as a factor then we can consider $E \circ f$ instead of f. Since $\|f\| = \|f \circ E\| = \|E \circ f\|$, we may suppose that x and $f(x)$ both contain the factor 00 (and thus none contains the factor 11).

Since x and $f(x)$ are characteristic, both $1x$ and $1f(x)$ are Sturmian, and thus both x and $f(x)$ start with the letter 0, and thus $f(0)$ also starts with 0.

If $f(1)$ starts with 1 then, by Lemma 2.3.8, $f(0)$ and $f(1)$ end with the same letter. If this letter is a 1 then 11 is a factor of $f(01)$ and thus of $f(x)$ which is impossible. So $f(0)$ and $f(1)$ both end with the letter 0. Let $r \geq 1$ be such that x starts with $0^r 1$. Since $0x$ is Sturmian, x contains $0^{r+1} 1$ and then 10^{r+1} as a factor. Consequently $1f(0^r)1$ is a prefix of $1f(x)$ and $0f(0^r)0$ is a factor of $f(x)$. A contradiction.

Thus, $f(1)$ starts with 0 and since $f(0)$ and $f(1)$ do not contain 11 as a factor, $f(0) \in \{01,0\}^+$ and $f(1) \in \{01,0\}^+$. Consequently there exists a morphism g such that $f = \varphi \circ g$ with $\|g\| < \|f\|$. But $\varphi \circ g(x)$ is characteristic, thus $g(x)$ is characteristic (Lemma 2.3.13) and, by induction, $g \in \{E, \varphi\}^*$. So f is standard. ∎

2.3.3. A presentation of the monoid of Sturm

In this subsection, it will be convenient to write the composition of morphisms as a concatenation (so we will write fg instead of $f \circ g$).

Let $G = \varphi E$ and $\tilde{G} = \tilde{\varphi} E$. Clearly, the monoid of Sturm St is also generated by E, G and $\tilde{G}$.

THEOREM 2.3.14. *The monoid of Sturm has the presentation*

$$E^2 = \text{id}, \tag{2.3.1}$$

$$GEG^k E\tilde{G} = \tilde{G}E\tilde{G}^k EG, \quad k \geq 0. \tag{2.3.2}$$

Formula (2.3.2) can be rewritten, in terms of the generators φ and $\tilde{\varphi}$, as

$$\varphi(\varphi E)^k E\tilde{\varphi} = \tilde{\varphi}(\tilde{\varphi} E)^k E\varphi, \quad k \geq 0.$$

Proof of theorem. We consider words over the alphabet $\{E, G, \tilde{G}\}$. For each word W over $\{E, G, \tilde{G}\}$, denote by f_W the Sturmian morphism defined by composing the letters of W. Two words W and W' are *equivalent* if $f_W = f_{W'}$. The words W and W' are *congruent* ($W \sim W'$) if one can obtain one from the other by a repeated application of (2.3.1) and (2.3.2) viewed as rewriting rules (i.e. if W and W' are in the same equivalence class of the congruence generated by (2.3.1) and (2.3.2)).

We prove that equivalent words are congruent (the converse is clear). Let W, W' be equivalent words. The proof is by induction on $|WW'|$. We may assume that W and W' do not contain E^2. Since $E, G, \tilde{G}$ are injective, we may also assume that W and W' do not start with the same letter. Observe that if W starts with φ or $\tilde{\varphi}$, then $|f_W(01)|_1 < |f_W(01)|_0$ and if W starts with $E \circ \varphi$ or $E \circ \tilde{\varphi}$, then $|f_W(01)|_1 > |f_W(01)|_0$. Consequently W starts with E if and only if W' starts with E, so we suppose that neither does. Finally, since $G\tilde{G} \sim \tilde{G}G$, we may assume that one of W and W' starts with $G^n E$ and the other with $\tilde{G}^p E$ with $n \neq 0$ and $p \neq 0$. Thus

$$W = \tilde{G}^{r_1} E \tilde{G}^{r_2} G^{s_2} E \cdots E \tilde{G}^{r_q} G^{s_q},$$
$$W' = G^{s'_1} E \tilde{G}^{r'_2} G^{s'_2} E \cdots E \tilde{G}^{r'_{q'}} G^{s'_{q'}}$$

with $r_1, s'_1 \geq 1$, $r_i, s_i, r'_i, s'_i \geq 0$, and $r_i + s_i \geq 1$ for $2 \leq i < q$, $r'_j + s'_j \geq 1$ for $2 \leq j < q'$.

Observe first that $f_{W'}(0)$ and $f_{W'}(1)$ both start with the letter 0 (because G does).

Next, $s_2 = 0$. Indeed, otherwise W is congruent to a word starting with $\tilde{G}^{r_1} EG$, and since $\tilde{G}^{r_1} EG(0)$ and $\tilde{G}^{r_1} EG(1)$ both start with the letter 1, W' is not equivalent to W.

If $s_i = 0$ for $i = 3, \ldots, q$, then $W = \tilde{G}^{r_1} E \tilde{G}^{r_2} E \cdots E \tilde{G}^{r_q}$, and $f_W(0)$ or $f_W(1)$ starts with the letter 1, according to whether q is even or odd. Thus, there is a smallest $i \geq 3$ such that $s_i \geq 1$. Then W is congruent to a word starting with

$$U = \tilde{G}^{r_1} E \tilde{G}^{r_2} E \cdots E \tilde{G}^{r_{i-2}} E \tilde{G}^{r_{i-1}} EG.$$

If i is even, then $f_U(0)$ and $f_U(1)$ start with the letter 1. Thus i is odd, and using (2.3.2), U is congruent to

$$U' = \tilde{G}^{r_1} E \tilde{G}^{r_2} E \cdots E \tilde{G}^{r_{i-2}-1} GEG^{r_{i-1}} E\tilde{G}$$

2.3. Sturmian morphisms

and eventually U is congruent to

$$G\tilde{G}^{r_1-1}EG^{r_2}E\tilde{G}^{r_3}E\cdots E\tilde{G}^{r_{i-2}}EG^{r_{i-1}}E\tilde{G}.$$

Thus W' and some word congruent to W start with the same letter. By induction, they are congruent. ∎

As a corollary, we obtain a presentation of the monoid of standard morphisms.

COROLLARY 2.3.15. *The only nontrivial identity in the monoid of standard morphisms generated by E and φ is $E^2 = \mathrm{id}$.* ∎

2.3.4. Conjugate morphisms

In this subsection, we characterize Sturmian morphisms by standard morphisms. The main notion is a special kind of *conjugacy* relation for morphisms.

Let f and g be morphisms. The morphism g is a *right conjugate* of f, in symbols $f \triangleleft g$, if there is a word w such that

$$f(x)w = wg(x), \quad \text{for all words } x \in A^*. \tag{2.3.3}$$

This implies that the words $f(x)$ and $g(x)$ are conjugate, and moreover all pairs $(f(x), g(x))$ share the same "sandwich" word w. It suffices, for (2.3.3) to hold, that

$$f(a)w = wg(a), \quad \text{for all letters } a \in A, \tag{2.3.4}$$

since by induction $f(xa)w = f(x)f(a)w = f(x)wg(a) = wg(xa)$. Observe that if (2.3.4) holds for a nonempty word w, then all words $f(a)$ for $a \in A$ start with the same letter. Right conjugacy is a preorder over the set of all morphisms over A. Indeed, if $f(x)w = wg(x)$ and $g(x)v = vh(x)$, then $f(x)wv = wg(x)v = wvh(x)$.

EXAMPLE 2.3.16. The morphism $\tilde{\varphi}$ is a right conjugate of φ since $\varphi(0)0 = 010 = 0\tilde{\varphi}(0)$ and $\varphi(1) = \tilde{\varphi}(1) = 0$. Observe that φ is not a right conjugate of $\tilde{\varphi}$ since $\tilde{\varphi}(0)$ and $\tilde{\varphi}(1)$ do not start with the same letter.

This example shows that right conjugacy is not a symmetric relation. However, one has the following formulas.

LEMMA 2.3.17. *Let f, g, f', g' be morphisms.*
  (i) *If $f \triangleleft g$ and $f' \triangleleft g$, then $f \triangleleft f'$ or $f' \triangleleft f$.*

(ii) If $f \triangleleft g$ and $f \triangleleft g'$, then $g \triangleleft g'$ or $g' \triangleleft g$.
(iii) If $f \triangleleft g$ and $f' \triangleleft g'$, then $f \circ f' \triangleleft g \circ g'$.

Proof. We start with the first implication. If $f(x)w = wg(x)$ and $f'(x)v = vg(x)$, then for suitable x, the word $g(x)$ is longer than v and w. Thus w is a suffix of v or vice versa. Assume $v = zw$. Then $zf(x) = f'(x)z$. The second is symmetric.

For the third, assume $f(x)w = wg(x)$ for all words x. For any morphism h, $h(f(x)w) = h(f(x))h(w) = h(w)h(g(x))$, and consequently $h \circ f \triangleleft h \circ g$. Also $f(h(x))w = wg(h(x))$, showing that $f \circ h \triangleleft g \circ h$. Thus, if $f \triangleleft g$ and $f' \triangleleft g'$, then $f \circ f' \triangleleft g \circ f' \triangleleft g \circ g'$. ∎

The next result states that the monoid of Sturm is the closure under right conjugacy of the monoid of standard morphisms.

PROPOSITION 2.3.18. *A morphism is Sturmian if and only if it is a right conjugate of some standard morphism.*

Proof. We show first that a Sturmian morphism is a right conjugate of some standard morphism. Let g be a Sturmian morphism, and consider a decomposition
$$g = h_1 \circ h_2 \circ \cdots \circ h_n$$
with $h_1, \ldots, h_n \in \{E, \varphi, \tilde{\varphi}\}$. If none of the h_i is equal to $\tilde{\varphi}$, then g is standard. Otherwise, consider the smallest i such that $h_i = \tilde{\varphi}$. Then $g = g' \circ \tilde{\varphi} \circ g''$, for $g' = h_1 \circ \cdots \circ h_{i-1}$ and $g'' = h_{i+1} \circ \cdots \circ h_n$. By induction, g'' is a right conjugate of some standard morphism f'', and since $\varphi \triangleleft \tilde{\varphi}$ and by Lemma 2.3.17, $g' \circ \varphi \circ f'' \triangleleft g$, with $g' \circ \varphi \circ f''$ a standard morphism.

Conversely, let f be a standard morphism, and let g be a right conjugate of f. Then there is a word w such that $f(x)w = wg(x)$ for every word x. It follows that, for any infinite word s, one has $f(s) = wg(s)$. If s is a Sturmian word, then $g(s)$ is a Sturmian word, and g is a Sturmian morphism. ∎

We start an explicit description of the right conjugates of a standard morphism with the following observation.

PROPOSITION 2.3.19. *Right conjugate standard morphisms are equal.*

Proof. Let f and f' be two standard morphisms, and assume $f \triangleleft f'$. There is a word w such that
$$f(0)w = wf'(0), \quad f(1)w = wf'(1). \tag{2.3.5}$$

Set $x = f(0)$, $y = f(1)$, and $x' = f'(0)$, $y' = f'(1)$. Then $|x| = |x'|$ and $|y| = |y'|$. Next, by Proposition 2.3.11, $\{x, y\}$ and $\{x', y'\}$ are unordered

2.3. Sturmian morphisms

standard pairs. If $\{x, y\} = \{0, 1\}$, then $\{x, y\} = \{x', y'\}$ and $f = f'$. Otherwise, the words xy, yx, $x'y'$ and $y'x'$ are standard words with the same height and length by (2.3.5), and moreover $xy \neq yx$, $x'y' \neq y'x'$ by Proposition 2.2.2. In view of Proposition 2.2.15, there exist exactly two standard words of this height and length. Thus $xy = x'y'$, or both $xy = y'x'$ and $yx = x'y'$. In the first case, $f = f'$. In the second case, assume $|x| \leq |y|$. Then x is a prefix of y, and the equation $yx = x'y'$ shows that $x = x'$. Thus $f = f'$ in this case also. ∎

We now show a way to construct all Sturmian morphisms from standard morphisms.

As in Lothaire 1983, Section 1.3, we use the permutation γ over A^+ defined by $\gamma(ax) = xa$, $a \in A$, $x \in A^*$. Two words x, y are conjugate if and only if $y = \gamma^i(x)$ for some $0 \leq i < |x|$.

Let f be a standard morphism. For $0 \leq i \leq \|f\|-1$, define a morphism f_i by $f_i(01) = \gamma^i(f(01))$ and $|f_i(0)| = |f(0)|$.

EXAMPLE 2.3.20. Let f be the morphism defined by $f(0) = 01010$, $f(1) = 01$. The corresponding seven morphisms are

$$\begin{aligned}
f_0 &: 0 \mapsto 01010, & 1 &\mapsto 01, \\
f_1 &: 0 \mapsto 10100, & 1 &\mapsto 10, \\
f_2 &: 0 \mapsto 01001, & 1 &\mapsto 01, \\
f_3 &: 0 \mapsto 10010, & 1 &\mapsto 10, \\
f_4 &: 0 \mapsto 00101, & 1 &\mapsto 01, \\
f_5 &: 0 \mapsto 01010, & 1 &\mapsto 10, \\
f_6 &: 0 \mapsto 10101, & 1 &\mapsto 00.
\end{aligned}$$

It is easily checked that all morphisms except f_6 are Sturmian and are right conjugates of f.

PROPOSITION 2.3.21. *Let f be a nontrivial standard morphism. The right conjugates of f are the morphisms f_i, for $0 \leq i \leq \|f\| - 2$.*

This means that the morphism $f_{\|f\|-1}$ is never Sturmian (in the example above, this was f_6).

Proof of proposition. Let g be a right conjugate of f. Then $f(01)w = wg(01)$ for some word w, so $g = f_i$ for some i.

For the converse, we show first that $f_i(0)$ and $f_i(1)$ start with the same letter if and only if $0 \leq i \leq \|f\| - 3$. Indeed, set $x = f(0)$, $y = f(1)$, $x' = f_i(0)$ and $y' = f_i(1)$, and set $n = |x| = |x'|$. The word $x'y'$ is a factor of $xyxy$, thus there exists a nonempty word t of length i such that $xyxy$ starts with $tx'y'$. The first letter of x' is the $(i+1)$th letter of xy. The first letter of y' is the $(n+i+1)$th letter of xyx, i.e. the $(i+1)$th letter of yx. Since $\{x, y\}$ is an unordered standard pair, only the last two letters of the words xy and yx are different, by Proposition 2.2.2. Consequently the first letter of x' is equal to the first letter of y' if and only if $i + 1 \le \|f\| - 2$.

For any i with $0 \le i \le \|f\| - 3$, set $f_i(0) = au$, $f_i(1) = av$ for a letter a and words u, v. Then $f_{i+1}(0) = ua$, $f_{i+1}(1) = va$. Thus $f_i(0)a = af_{i+1}(0)$, $f_i(1)a = af_{i+1}(1)$, showing that $f_i \triangleleft f_{i+1}$, whence $f \triangleleft f_{i+1}$. ∎

PROPOSITION 2.3.22. *Let g be a Sturmian morphism. There exists a unique standard morphism f such that $f \triangleleft g$. This standard morphism is obtained from any decomposition of g in elements of $\{E, \varphi, \tilde{\varphi}\}$ by replacing all the occurrences of $\tilde{\varphi}$ by φ.*

Proof. Let g be a Sturmian morphism, and let f be obtained from a decomposition of g in elements of $\{E, \varphi, \tilde{\varphi}\}$ by replacing all the occurrences of $\tilde{\varphi}$ by φ. Since f is a composition of E and φ, f is standard. Moreover, since $\varphi \triangleleft \tilde{\varphi}$, one has $f \triangleleft g$ by repeated application of Lemma 2.3.17(iii).

Moreover if there exists a standard morphism f' such that $f' \triangleleft g$ then by Lemma 2.3.17, one has $f' \triangleleft f$ or $f \triangleleft f'$. By Proposition 2.3.19, $f = f'$ which proves that f is unique. ∎

2.3.5. Automorphisms of the free group

Consider two letters $\bar{0}, \bar{1}$ not in $A = \{0, 1\}$. The free monoid $A^* = \{0, 1, \bar{0}, \bar{1}\}^*$ is equipped with an involution by defining $\bar{\bar{a}} = a$ for $a \in A$, and $\overline{uv} = \bar{v}\bar{u}$. The *free group* $F(A)$ over $A = \{0, 1\}$ is the quotient of the free monoid A^* under the congruence relation generated by $0\bar{0} \equiv \bar{0}0 \equiv 1\bar{1} \equiv \bar{1}1 \equiv \varepsilon$. A word in A^* without factors of the form $0\bar{0}, \bar{0}0, 1\bar{1}, \bar{1}1$ is *reduced*. Every word in A^* is equivalent to a unique reduced word. If w is reduced, so is $\bar{w}$. The free group can be viewed as the set of reduced words. The product of two elements in $F(A)$ is the reduced word equivalent to the concatenation of the reduced words corresponding to the group elements, and the inverse of an element in $F(A)$ represented by w is $\bar{w}$. An element in $F(A)$ has a *length*. It is the length of its corresponding reduced word.

In this subsection, we give a characterization of Sturmian morphisms in terms of automorphisms of the free group $F(A)$.

Any morphism f on A is extended in a natural way to an endomorphism on $F(A)$, by defining $f(\bar{0}) = \overline{f(0)}$, $f(\bar{1}) = \overline{f(1)}$. It follows that $f(\bar{w}) = \overline{f(w)}$ for any $w \in F(A)$. Conversely, consider an endomorphism f of $F(A)$. It is called *positive* if the (reduced) words $f(0)$ and $f(1)$ are

2.3. Sturmian morphisms

words over A, that is do not contain any barred letter. An endomorphism f that is a bijection is an *automorphism*. Its inverse is denoted by f^{-1}.

The morphisms E, φ and $\tilde{\varphi}$ are extended to $F(A)$ by

$$E : \begin{array}{c} 0 \mapsto 1 \\ 1 \mapsto 0 \\ \bar{0} \mapsto \bar{1} \\ \bar{1} \mapsto \bar{0} \end{array}, \quad \varphi : \begin{array}{c} 0 \mapsto 01 \\ 1 \mapsto 0 \\ \bar{0} \mapsto \bar{1}\bar{0} \\ \bar{1} \mapsto \bar{0} \end{array}, \quad \tilde{\varphi} : \begin{array}{c} 0 \mapsto 10 \\ 1 \mapsto 0 \\ \bar{0} \mapsto \bar{0}\bar{1} \\ \bar{1} \mapsto \bar{0} \end{array}.$$

They are automorphisms, and their inverses are given by

$$E^{-1} = E, \quad \varphi^{-1} : \begin{array}{c} 0 \mapsto 1 \\ 1 \mapsto \bar{1}0 \end{array}, \quad \tilde{\varphi}^{-1} : \begin{array}{c} 0 \mapsto 1 \\ 1 \mapsto 0\bar{1} \end{array}.$$

It follows that every Sturmian morphism is a (positive) automorphism of $F(A)$. The converse also holds.

THEOREM 2.3.23. *The positive automorphisms of $F(A)$ are exactly the Sturmian morphisms.*

The theorem states that the three morphisms $E, \varphi, \tilde{\varphi}$ are a set of generators of the monoid of positive automorphisms. The full automorphism group of a free group is a well-known object (see Notes). In particular, sets of generators can be expressed in terms of so-called Nielsen transformations. In the present case, the morphisms

$$\begin{array}{cccc} 0 \mapsto 0 & 0 \mapsto \bar{0} & 0 \mapsto 01 & 0 \mapsto 0 \\ 1 \mapsto \bar{1} \end{array}, \begin{array}{c} \\ 1 \mapsto 1 \end{array}, \begin{array}{c} \\ 1 \mapsto 1 \end{array}, \begin{array}{c} \\ 1 \mapsto 10 \end{array}$$

generate the automorphism group of $F(A)$. The two last morphisms are $E \circ \tilde{\varphi}$ and $\tilde{\varphi} \circ E$.

We first prove a special case of the theorem.

PROPOSITION 2.3.24. *Let f be a positive automorphism of $F(A)$. If the words $f(0)$ and $f(1)$ do not end with the same letter, then f is a standard Sturmian morphism.*

Proof. Let f be a positive automorphism of $F(A)$. We may assume $|f(0)| \le |f(1)|$. We suppose first that $f(0)$ is not a prefix of $f(1)$. There exist words u, v_0, v_1 over A such that v_0 and v_1 start with different letters and $f(0) = uv_0$ and $f(1) = uv_1$. Since $f(0)$ and $f(1)$ do not end with the same letter, the words v_0 and v_1 also end with different letters. The images of reduced words of length 2 under f are uv_auv_b, $uv_a\bar{v}_b\bar{u}$, $\bar{v}_av_b$, $\bar{v}_a\bar{u}\bar{v}_b\bar{u}$. Each of these words is reduced because v_0 and v_1 start and end with different letters. It follows that for any reduced word w of length at least 2, the reduced word $f(w)$ has length at least 2. Consider now any letter $a \in A$. Since $|f(f^{-1}(a))| = 1$, it follows that $|f^{-1}(a)| = 1$, that is f is either the identity or E. Thus f is Sturmian.

Next, if $f(0)$ is a prefix of $f(1)$, there exists a word u such that $f(1) = f(0)u$. Define a morphism g by $g(0) = f(0)$ and $g(1) = u$. Then $f = g \circ \varphi \circ E$. Since f is a bijection, g is also a bijection. By induction on $\|g\|$, the morphism g is a standard Sturmian morphism, and so is f. ∎

Proof of Theorem 2.3.23. Let g be a positive automorphism. The words $g(01)$ and $g(10)$ are different because g is a bijection. They have the same length. Let u be their longest common suffix. There exist words v_0, v_1 over A of the same length such that $g(01) = v_0 u$, $g(10) = v_1 u$ and v_0, v_1 do not end with the same letter. Since for letters $a \ne b$, $g(aba) = v_a u g(a) = g(a) v_b u$, the words $u g(a)$ end with u. Define a morphism f by $f(a) = u g(a) \bar{u}$ for $a \in \{0, 1\}$. Then $f(w) = u g(w) \bar{u}$ for all w in $F(A)$. Since $u g(a)$ ends with u for $a \in \{0, 1\}$, the morphism f is positive.

Since g is a bijection, f is also a bijection. Moreover $f(01) = u v_0$ and $f(10) = u v_1$ end with different letters and since f is positive, $f(0)$ and $f(1)$ also end with different letters. By Proposition 2.3.24, f is a standard Sturmian morphism. Now $f(0)u = ug(0)$ and $f(1)u = ug(1)$ which means that g is a right conjugate of f. Consequently, by Proposition 2.3.18, g is a Sturmian morphism. ∎

2.3.6. Fixed points

In this subsection, we make use of Theorem 2.3.12 to describe those characteristic words that are fixed points of standard morphisms. As an example, we know from Chapter 1 that the morphism φ fixes the infinite Fibonacci word f.

We say that a morphism h *fixes* an infinite word x if $h(x) = x$. In this case, x is a *fixed point* of h. Every infinite word is fixed by the identity, and no infinite word is fixed by E.

For the description of characteristic words which are fixed points of morphisms, we introduce a special set of irrational numbers. A *Sturm number* is a number α that has a continued fraction expansion of one of the following kinds:

(i) $\alpha = [0, 1, a_0, \overline{a_1, \ldots, a_k}]$, with $a_k \ge a_0$,
(ii) $\alpha = [0, 1 + a_0, \overline{a_1, \ldots, a_k}]$, with $a_k \ge a_0 \ge 1$.

Observe that (i) implies $\alpha > 1/2$, and (ii) implies $\alpha < 1/2$. More precisely, α has an expansion of type (i) if and only if $1 - \alpha$ has an expansion of

2.3. Sturmian morphisms

type (ii). Consequently, α is a Sturm number if and only $1-\alpha$ is a Sturm number.

As an example, $1/\tau = [0,\overline{1}]$ is covered by the first case (for $k=1$ and $a_k = a_0 = 1$), and $1/\tau^2 = [0,2,\overline{1}]$ is covered by the second case.

We shall give later (Theorem 2.3.26) a simple algebraic description of Sturm numbers. There is also a simple combinatoric characterization of these numbers (Problem 2.3.4).

THEOREM 2.3.25. *Let $0 < \alpha < 1$ be an irrational number. The characteristic word c_α is a fixed point of some nontrivial morphism if and only if α is a Sturm number.*

Proof. Let
$$\alpha = [0, m_1, m_2, \ldots]$$
be the continued fraction expansion of α, and suppose that $f(c_\alpha) = c_\alpha$ for some morphism f. In view of Theorem 2.3.12, the morphism f is standard. Thus, f is a product of E and G, and is not a power of E. Also, f is not a proper power of G, because a morphism G^n with $n \geq 1$ fixes only the infinite word 0^ω. Thus (we write composition as concatenation), f has the form
$$f = G^{n_1} E G^{n_2} \cdots E G^{n_k} E G^{n_{k+1}}$$
for some $k \geq 1$, $n_1, n_{k+1} \geq 0$, and $n_2, \ldots, n_k \geq 1$. We use the morphisms $\theta_m = G^{m-1} E G$ for $m \geq 1$ and the fact (Corollary 2.2.21) that
$$\theta_m(c_\alpha) = c_{1/(m+\alpha)}.$$

There are three cases.

(a) Suppose first that $n_{k+1} > 0$. Then
$$f = \theta_{n_1+1} \theta_{n_2} \cdots \theta_{n_k} G^{n_{k+1}-1}.$$
Since f fixes c_α, this implies
$$[0, m_1, m_2, \ldots] = [0, 1+n_1, n_2, \ldots, n_k, n_{k+1}-1+m_1, m_2, \ldots]$$
which in turn gives $m_1 = 1 + n_1$, $m_2 = n_2$, ..., $m_k = n_k$, $m_{k+1} = n_{k+1} - 1 + m_1 = n_{k+1} + n_1$, and $m_j = m_{j+k}$ for $j \geq 2$. Thus
$$\alpha = [0, 1+n_1, \overline{n_2, \ldots, n_{k+1}+n_1}], \quad \text{with } n_1 \geq 0, n_2, \ldots, n_{k+1} \geq 1. \quad (2.3.6)$$

(b) Suppose now that $n_{k+1} = 0$, and consider the morphism $f' = EfE$. From $c_\alpha = f(c_\alpha)$, it follows that $f'(Ec_\alpha) = Ec_\alpha$, that is $f'(c_\beta) = c_\beta$ for $\beta = 1 - \alpha$. Now
$$f' = EG^{n_1} E G^{n_2} \cdots E G^{n_k}$$
where $n_1 \geq 0$ and $n_2, \ldots, n_k \geq 1$. There are two subcases.

(b.1) If $n_1 = 0$, then $k \geq 3$ and
$$f' = G^{n_2} \cdots EG^{n_k} = \theta_{n_2+1} \cdots \theta_{n_{k-1}} G^{n_k-1}$$
whence, as above, $\beta = [0, 1+n_2, \overline{n_3, \ldots, n_{k-1}, n_2 + n_k}]$ and since $n_2 \geq 1$,
$$\alpha = 1 - \beta = [0, 1, n_2, \overline{n_3, \ldots, n_{k-1}, n_2 + n_k}] \quad \text{with } n_2, \ldots, n_k \geq 1. \quad (2.3.7)$$

(b.2) If $n_1 \geq 1$, then
$$f' = EG^{n_1} \cdots EG^{n_k} = \theta_1 \theta_{n_1} \cdots \theta_{n_{k-1}} G^{n_k-1}$$
whence as above $\beta = [0, 1, \overline{n_1, \ldots, n_{k-1}, n_k}]$ and
$$\alpha = 1 - \beta = [0, 1 + n_1, \overline{n_2, n_3, \ldots, n_k, n_1}] \quad \text{with } n_1, \ldots, n_k \geq 1. \quad (2.3.8)$$

To show that Equations (2.3.6)–(2.3.8) exactly describe Sturm numbers, observe that Equation (2.3.6) with $n_1 = 0$ corresponds, in the definition of Sturm numbers, to case (i) with $a_k = a_0$, that Equation (2.3.6) with $n_1 > 0$ corresponds to case (ii) with $a_k > a_0$, that Equation (2.3.7) is equivalent to case (i) with $a_k > a_0$ and that Equation (2.3.8) is case (ii) with $a_k = a_0$.

The proof that a Sturm number indeed yields a fixed point is exactly the reverse of the preceding one. ∎

Sturm numbers have a simple algebraic description. Clearly, a Sturm number α is quadratic irrational, that is a solution of some equation
$$x^2 + px + q = 0$$
with rational coefficients p, q. The other solution of this equation is the *conjugate* of α, denoted by $\bar{\alpha}$, and satisfies $\alpha\bar{\alpha} = q$. It is easy to prove that the conjugate of $1 - \alpha$ is $1 - \bar{\alpha}$, and that the conjugate of $1/\alpha$ is $1/\bar{\alpha}$.

THEOREM 2.3.26. *A quadratic irrational α with $0 < \alpha < 1$ is a Sturm number if and only if $1/\bar{\alpha} < 1$.*

We need some facts from number theory. A quadratic irrational number γ is said to be *reduced* if $\gamma > 1$ and $-1 < \bar{\gamma} < 0$. This is equivalent to $1 > 1/\gamma > 0$ and $1/\bar{\gamma} < -1$. It is known that
(i) the continued fraction of a quadratic irrational γ is purely periodic if and only if γ is reduced,
(ii) if γ is reduced and $\gamma = [\overline{a_1, \ldots, a_n}]$, then $-1/\bar{\gamma} = [\overline{a_n, \ldots, a_1}]$.

Proof of Theorem 2.3.26. The condition $1/\bar{\alpha} < 1$ is equivalent to $\bar{\alpha} \notin [0,1]$. This in turn is equivalent to $1 - \bar{\alpha} \notin [0,1]$. Thus $\bar{\alpha}$ satisfies the condition if and only if $1 - \bar{\alpha}$ does. Consequently, it suffices to prove the equivalence for $0 < \alpha < 1/2$. We have to prove that $1/\bar{\alpha} < 1$ if and only if
$$\alpha = [0, 1 + a_0, \overline{a_1, \ldots, a_k}], \quad \text{with } a_k \geq a_0 \geq 1.$$
Let first α be a Sturm number with $0 < \alpha < 1/2$. Then
$$\alpha = \frac{1}{1 + a_0 + \frac{1}{\gamma}}, \quad \text{with } \gamma = [\overline{a_1, \ldots, a_k}], \quad a_k \geq a_0 \geq 1. \quad (2.3.9)$$
Thus γ is reduced, and since $-1/\bar{\gamma} = [\overline{a_k, \ldots, a_1}] > a_k$, it follows from (2.3.9) that
$$1/\bar{\alpha} = 1 + a_0 + 1/\bar{\gamma} < 1 + a_0 - a_k \leq 1.$$
Conversely, let $0 < \alpha < 1/2$ be a quadratic irrational with $1/\bar{\alpha} < 1$. Since $2 < 1/\alpha$, write
$$1/\alpha = 1 + a_0 + 1/\gamma \quad (2.3.10)$$
where $a_0 = \lfloor 1/\alpha - 1 \rfloor \geq 1$ and $1 < 1/\gamma < 1$. From $1/\bar{\alpha} < 1$ and the conjugate of (2.3.10), one gets
$$1/\bar{\gamma} < -a_0 \leq -1.$$
Thus γ is reduced, and writing $\gamma = [\overline{a_1, \ldots, a_k}]$, one gets
$$a_0 < -1/\bar{\gamma} = [\overline{a_k, \ldots, a_1}] < a_k + 1,$$
whence $a_k \geq a_0 \geq 1$ and
$$\alpha = \frac{1}{1 + a_0 + \frac{1}{\gamma}} = [0, 1 + a_0, \overline{a_1, \ldots, a_k}]. \quad \blacksquare$$

Problems

Section 2.1

2.1.1 We consider two-sided infinite words over $\{0,1\}$ of complexity $n+1$.
1. Show that the word x defined by $x(k) = 1$ for $k \geq 0$, and $x(k) = 0$ for $k < 0$ has $n+1$ factors of length n for each $n \geq 0$.
2. Let $z \notin 0^* \cup 1^*$ be a central word with period k and ℓ, and set $w = p10q$ where p and q are palindrome words with $k = |p|$,

$\ell = |q|$. Define two (one-sided) infinite words $x = (10q)^\omega$ and $y = (01p)^\omega$. Then the two-sided infinite word $\tilde{y}zx$ has $n+1$ factors of length n for each $n \geq 1$. (These are the only two-sided infinite words with complexity $n+1$; see Coven and Hedlund 1973.)

2.1.2 Let x be an infinite word which contains infinitely many occurrences of 0 and of 1. The *cell*-condition for x is the following: for any words w, w' such that $|w|_0 = |w'|_0$ and $0w0, 0w'0 \in F(x)$, one has $||w| - |w'|| \leq 1$, and the same condition with 0 and 1 exchanged. Show that x is balanced if and only if x satisfies the cell-condition. (Morse and Hedlund 1940. A proof consists in considering the word y such that $x = G(y)$.)

2.1.3 Let x be an infinite word. For $n \geq 1$, let X_n be the set of factors of x starting with 0, ending with 0, and containing exactly n occurrences of the letter 0. Define similarly Y_n, replacing 0 by 1. Show that x is Sturmian if and only if $\text{Card}(X_n) = \text{Card}(Y_n) = n$ for every n (Richomme 1999a).

2.1.4 Show that a word w is unbalanced if and only if it admits a factorization $w = xauay b\tilde{u}bz$ for words u, x, y, z and letters $a \neq b$. Use this characterization to prove that the set of unbalanced words is a context-free language. (Dulucq and Gouyou-Beauchamps 1990, see also Mignosi 1991, 1990.)

Section 2.2

2.2.1 Show that for any standard word $w \neq 0, 1$, there is only one standard pair (x, y) such that $w = xy$ or $w = yx$.

2.2.2 Define sequences of words $(A_n)_{n \geq 0}$ and $(B_n)_{n \geq 0}$ by

$$A_0 = a, \quad B_0 = b$$

and

$$R_1 : \begin{matrix} A_{n+1} = A_n \\ B_{n+1} = A_n B_n \end{matrix}, \quad \text{and} \quad R_2 : \begin{matrix} A_{n+1} = B_n A_n \\ B_{n+1} = B_n \end{matrix}.$$

The R_i's are called *Rauzy's rules* (see Rauzy 1985).
1. Show that, provided each of the rules R_i is applied infinitely often, the sequences A_n and B_n converge to the same infinite word which is characteristic.
2. Show that conversely every characteristic word is obtained in this way.

2.2.3 Let $0 \leq h \leq m$ be integers with $(h, m) = 1$. The lower and upper Christoffel words $t_{h,m}$ and $t'_{h,m}$ are defined by $t_{0,1} = t'_{0,1} = 0$, $t_{1,1} = t'_{1,1} = 1$, and $t_{h,m} = 0 z_{h,m} 1$, $t'_{h,m} = 1 z_{h,m} 0$ if $m \geq 2$. These are exactly the words defined in Subsection 2.1.2.

1. Show that if $h'm - m'h = 1$, then

$$t_{h,m} t'_{h',m'} = t_{h+h', m+m'}, \quad t'_{h',m'} t'_{h,m} = t'_{h+h', m+m'}.$$

2. For $1 \leq h < m$ and $(h, m) = 1$, show that there exist integers m', h' with $0 \leq h' \leq m' < m$, $h' < h$ such that $m'h - h'm = 1$, and

$$t_{h,m} = t_{h',m'} t_{h-h', m-m'}.$$

3. Define $\sigma_{h,m} = z_{h,m} 10$, $\sigma'_{h,m} = z_{h,m} 01$. Show that

$$\sigma'_{h,m} \sigma_{h',m'} = \sigma_{h+h', m+m'}, \quad \sigma_{h,m} \sigma'_{h',m'} = \sigma'_{h+h', m+m'}.$$

Show that the pairs of standard words are $(0, 1)$ and all the pairs $(\sigma_{h,m}, \sigma'_{h,m})$, for $h'm - hm' = 1$.

2.2.4 Consider a function Δ' from $\{0, 1\}^*$ into itself defined by $\Delta'(u, v) = (uv, v)$. The family of *Christoffel pairs* is the smallest set of pairs of words containing $(0, 1)$ and closed under Γ and Δ'. A standard pair and a Christoffel pair are *corresponding* if the sequence of Γ's and Δ's for the standard pair is the same as the sequence of Γ's and of (Δ')'s for the Christoffel pair.

1. Let (u, v) be a standard pair and let (u', v') be the corresponding Christoffel pair. Show that if $u = p10$, then $u' = 0p1$ and if $v = q01$, the $v' = 0q1$.

2. Show that the components of Christoffel pairs are exactly the lower Christoffel words. (See Borel and Laubie 1993.)

2.2.5 Christoffel words and Lyndon words.

1. Show that every lower Christoffel word is a Lyndon word.

2. Show that a balanced word is a Lyndon word if and only if it is a Christoffel word (Berstel and De Luca 1997).

3. Any lower Christoffel word w which is not a letter admits a unique factorization $w = xy$, where (x, y) is a Christoffel pair. Show that this factorization is the standard Lyndon factorization (Borel and Laubie 1993).

2.2.6 Show that, for $0 \leq \rho < 1$,

$$\tilde{\varphi}(s_{\alpha,\rho}) = s'_{\frac{1-\alpha}{2-\alpha}, \frac{2-\alpha-\rho}{2-\alpha}}, \quad D(s_{\alpha,\rho}) = s_{\frac{1}{2-\alpha}, \frac{1-\alpha+\rho}{2-\alpha}}, \quad \tilde{D}(s_{\alpha,\rho}) = s_{\frac{1}{2-\alpha}, \frac{\rho}{2-\alpha}}.$$

Show that for $0 < \rho \le 1$,

$$\tilde{\varphi}(s'_{\alpha,\rho}) = s'_{\frac{1-\alpha}{2-\alpha}, \frac{2-\alpha-\rho}{2-\alpha}}, \quad D(s'_{\alpha,\rho}) = s'_{\frac{1}{2-\alpha}, \frac{1-\alpha+\rho}{2-\alpha}}, \quad \tilde{D}(s'_{\alpha,\rho}) = s'_{\frac{1}{2-\alpha}, \frac{\rho}{2-\alpha}}.$$

(See Parvaix 1997.)

2.2.7 The aim of this problem is to prove that if w is a word such that $w0$ and $w1$ are balanced, then w is a right special factor of some Sturmian word.
Let w be a word such that $w0$ and $w1$ are balanced.
1. Show that if w is a palindrome, then w is central.
2. Show that if $w = uap$, with a a letter and p a palindrome, then pa is a prefix of some characteristic word.
3. Show that w is always a suffix of a central word.
4. Show that w is a right special factor of some Sturmian word.

(See De Luca 1997c.)

2.2.8 Let $\alpha = [0, 1 + d_1, d_2, \ldots]$ be the continued fraction expansion of the irrational α, let (s_n) be the associated standard sequence, and define $(t_n)_{n \ge -1}$ by

$$t_{-1} = 1, \quad t_0 = 0, \quad t_n = t_{n-1}^{d_n - 1} t_{n-2} t_{n-1} \quad (n \ge 1).$$

1. Show that $t_0 t_1 \cdots t_n = s_n \cdots s_1 s_0$.
2. Show the follow product formula: $c_\alpha = t_0 t_1 \cdots t_n \cdots$. (Brown 1993)

2.2.9 Let $\alpha = [0, 1 + d_1, d_2, \ldots]$ be the continued fraction expansion of the irrational α, let (s_n) be the associated standard sequence. Let w be a standard word that is a prefix of the characteristic word c_α. Show that there is an integer n such that $w = s_n^k s_{n-1}$ for some $1 \le k \le d_{n+1}$.

2.2.10 Let $\alpha = [0, 1 + d_1, d_2, \ldots]$ be the continued fraction expansion of the irrational α, let (s_n) be the associated standard sequence. Define three sequences of words by $(u_n)_{n \ge -1}$, $(v_n)_{n \ge -1}$ and $(w_n)_{n \ge -1}$,

$$u_{-1} = v_{-1} = w_{-1} = 1, \quad u_0 = v_0 = w_0 = 0$$

and

$$\begin{aligned}
u_{2n} &= u_{2n-2}(u_{2n-1})^{d_{2n}} \quad (n \ge 1), \\
u_{2n+1} &= (u_{2n})^{d_{2n+1}} u_{2n-1} \quad (n \ge 0), \\
v_{2n} &= (v_{2n-1})^{d_{2n}} v_{2n-2} \quad (n \ge 1), \\
v_{2n+1} &= v_{2n-1}(v_{2n})^{d_{2n+1}} \quad (n \ge 0), \\
w_n &= w_{n-2}(w_{n-1})^{d_n} \quad (n \ge 1).
\end{aligned}$$

Problems

1. Show that
$$0c_\alpha = \lim_{n\to\infty} u_n, \qquad 1c_\alpha = \lim_{n\to\infty} v_n,$$
$$01c_\alpha = \lim_{n\to\infty} w_{2n}, \qquad 10c_\alpha = \lim_{n\to\infty} w_{2n+1}.$$

2. Define a sequence $(p_n)_{n\geq -1}$ by $p_{-1} = 0^{-1}$, $p_0 = 1^{-1}$ and
$$p_{2n} = p_{2n-2}(10\pi_{2n-1})^{d_{2n}} \quad (n \geq 1),$$
$$p_{2n+1} = (p_{2n}10)^{d_{2n+1}}p_{2n-1} \quad (n \geq 0).$$

Show that the words p_n, for $n \geq 1$ are palindromes, and
$$s_{2n} = p_{2n}10, \qquad u_n = 0p_n 1, \qquad w_{2n} = 01p_{2n},$$
$$s_{2n+1} = p_{2n+1}01, \qquad v_n = 1p_n 0, \qquad w_{2n+1} = 10p_{2n+1}.$$

2.2.11 **A number system associated with a directive sequence.**
Let $\alpha = [0, 1+d_1, d_2, \ldots]$ be the continued fraction of the irrational α, and (s_n) be the associated standard sequence. Define integers by
$$q_{-1} = 1, \quad q_0 = 1, \quad q_n = d_n q_{n-1} + q_{n-2} \quad (n \geq 1).$$

Then of course $|s_n| = q_n$.

1. Show that any integer $m \geq 0$ can be written in the form
$$m = z_h q_h + \cdots + z_0 q_0 \quad (0 \leq z_i \leq d_{i+1}). \quad (2.4.1)$$

2. Show that every integer $0 \leq m \leq q_{h+1} - 1$ admits a unique such representation provided
$$z_i = d_{i+1} \implies z_{i-1} = 0 \quad (1 \leq i \leq h).$$

3. Show that if $m = z_h q_h + \cdots + z_0 q_0$ is as in Equation (2.4.1), then the prefix of c_α of length m has the form $s_h^{z_h} \cdots s_0^{z_0}$ (see Fraenkel 1985, 1982, Brown 1993 and the references cited there).

2.2.12 A *Beatty sequence* is a set $B = \{\lfloor rn \rfloor \mid n \geq 1\}$ for some irrational number $r > 1$ (it is a spectrum).
1. Let $\alpha = 1/r$, and let $c_\alpha = a_1 a_2 \cdots$ be the characteristic word of slope α. Show that $B = \{k \mid a_k = 1\}$.
2. Two Beatty sequences B and B' are *complementary* if B and B' form a partition of $\{1, 2, \ldots\}$. Show that the sets $\{\lfloor rn \rfloor \mid n \geq 1\}$ and $\{\lfloor r'n \rfloor \mid n \geq 1\}$ are complementary if and only if $1/r + 1/r' = 1$. (Use 1, see Beatty 1926)

2.2.13 Write $x < y$ if x is lexicographically less that y. Show that for any irrational characteristic word c, the word $0c$ is lexicographically smaller than all its proper suffixes, and $1c$ is lexicographically greater than all its proper suffixes. (Borel and Laubie 1993)

2.2.14 Define a mapping $C: \{0,1\}^* \to \{0,1\}^*$ by $C(\varepsilon) = \varepsilon$ and $C(ax) = xa$ for $a \in \{0,1\}$. This is just a cyclic permutation. Let $\alpha = [0, 1+d_1, d_2, \ldots]$ be the continued fraction of the irrational α, and (s_n) be the associated standard sequence.
1. Show that for $n \geq 0$, the words $C^{-1}(s_{2n})$ and $C^{|s_{2n}|-1}(s_{2n+1})$ are Lyndon words. (Borel and Laubie 1993, Melançon 1996)
2. Set $\ell_n = C^{|s_{2n}|-1}(s_{2n+1})$. Show that $c_\alpha = \ell_0^{d_2} \ell_1^{d_4} \cdots \ell_n^{d_{2n+2}} \cdots$ and that the sequence ℓ_n is a lexicographically strictly decreasing sequence.

2.2.15 Let $\alpha = [0, 1+d_1, d_2, \ldots]$ be the continued fraction of the irrational α, and (s_n) be the associated standard sequence.
1. Show that s_n^2 is a factor of c_α for every $n \geq 1$.
Since s_n is primitive, every factor of c_α of length $|s_n|$ except one is a conjugate of s_n. This is the *singular word*, denoted by w_n. For the Fibonacci word, the singular words are 00, 101, 00100, 10100101, ...
2. Let p_n be the palindrome prefix of s_n of length $|s_n| - 2$. Show that $w_n = a_n p_n a_n$, where $a_n = 0$ if n is odd, and $a_n = 1$ if n is even.
3. Show that the Fibonacci word is the product of 01 and its singular words: $f = 01(00)(101)(00100) \cdots$. (See Wen and Wen 1994b.)

2.2.16 To compute all conjugates of s_n, define sequences $(w_h)_{0 \leq h \leq n}$ of words parameterized by sequences of integers $z_0, \ldots, z_{n-1}$ with $0 \leq z_h \leq d_{h+1}$ by $w_{-1} = 1$, $w_0 = 0$ and $w_{h+1} = w_h^{d_{h+1}-z_h} w_{h-1} w_h^{z_h}$, $0 \leq h < n$.
1. Show that $w_n = C^k(s_n)$, where $k = \sum_{h=0}^{n-1} q_h z_h$.
2. Show that one gets all conjugates exactly once. (See Chuan 1997.)

2.2.17 Sturmian words and palindromes.
1. Let s be a Sturmian word. Show that $F(s)$ contains exactly one palindrome word of even length, and two palindrome words of odd length for each nonnegative integer.
2. Show that conversely, if $F(s)$ contains exactly one palindrome word of even length, and two palindrome words of odd length for each nonnegative integer, then s is Sturmian (Droubay and Pirillo 1999).

2.2.18 Sturmian words and decimation.
Let $1 \leq k \leq m$ be integers with $m \geq 2$. Let x be an infinite word with infinitely many 0's and 1's. The transformation $M_{k,m}$ deletes in x every 0 except those occurring at positions congruent to k modulo m. The transformation $D_{k,m}$ operates in the same way on 1's. For example, $M_{3,4}$, applied to

$$0100\,101001\mathit{0}010100\,101001\mathit{0}010100\,1001\cdots,$$

keeps only the italicized letter 0, and gives the word

$$101110110111011011\cdots.$$

1. Give a geometric argument (by cutting sequences) showing that $M_{k,m}(s)$ and $D_{k,m}(s)$ are Sturmian for Sturmian words.
2. Give explicit formulas for $M_{k,m}(s_{\alpha,\rho})$ and $D_{k,m}(s_{\alpha,\rho})$ similar to those of Problem 2.2.6.
3. Show that $M_{m,m} \circ D_{m,m}(c) = c$ for every characteristic word c. 4. Show that conversely, if $M_{m,m} \circ D_{m,m}(s) = s$ for every m, then the infinite word s is balanced. (Justin and Pirillo 1997; the explicit formulas are in Parvaix 1998.)

Section 2.3

2.3.1 For integers $m \geq 1, r \geq 1$, set

$$w_{m,r} = 0^{m-1}1(0^{m+1}1)^{r+1}0^m1(0^{m+1}1)^r0^m1,$$
$$w'_{m,r} = 0^m1(0^m1)^{r+1}0^{m+1}1(0^m1)^r0^{m+1}1.$$

In particular, $w_{1,1} = 10^210^21010^2101$ is a word of length 14. Any Sturmian word contains one and only one word from the set

$$\Omega = \{w_{m,r}, w'_{m,r}, E(w_{m,r}), E(w'_{m,r}) \mid m \geq 1, r \geq 1\}.$$

1. Prove that a morphism f is Sturmian if and only if f is acyclic and there exists a word $w \in \Omega$ such that $f(w)$ is a balanced word (in particular, an acyclic morphism f is Sturmian if and only if $f(w_{1,1})$ is a balanced word) (Berstel and Séébold 1994a).
2. Prove that no word of length less than or equal to 13 has the above property. (Richomme 1999b)

2.3.2 Let C be the set of morphic Sturmian characteristic words. Prove that, for any $c \in C$, the words $0c, 1c, 01c$ and $10c$ are morphic (Berstel and Séébold 1994a).

2.3.3 Prove that a morphism f is standard if and only if $f(0)$, $f(1)$ and $f(01)$ are standard words (De Luca 1997b).

2.3.4 Let $\alpha = [0, 1+d_1, d_2, \ldots]$ be the continued fraction of an irrational number α. Define an infinite word δ_α over $\{0, 1\}$ by

$$\delta_\alpha = 0^{d_1} 1^{d_2} 0^{d_3} 1^{d_4} \cdots.$$

Show that α is a Sturm number if and only if δ_α is purely periodic (Droubay, Justin, and Pirillo 2001).

Notes

The history of Sturmian words goes back to the astronomer Jean Bernoulli III (Bernoulli 1772). The book of Venkov 1970 describes early work by Christoffel 1875 and Markoff 1882. The first in-depth study is by Morse and Hedlund 1940. They also introduced the term "Sturmian", more precisely Sturmian trajectories, named after the mathematician Charles François Sturm (1803–55), born in Geneva, who taught at the École Polytechnique in Paris from 1840. He is famous for his rule to compute the roots of an algebraic equation. As described by Hedlund and Morse, Sturmian words are obtained in considering the zeros of solutions $u(x)$ of linear homogeneous second order differential equations

$$y'' + \phi(x)y = 0,$$

where $\phi(x)$ is continuous of period 1. If k_n is the number of zeros of u in the interval $[n, n+1)$, then the infinite word $01^{k_0} 0^{k_1} 0^{k_2} \cdots$ is Sturmian (or eventually periodic). The papers by Coven and Hedlund 1973 and Coven 1974 contain many combinatorial properties (in particular the description of two-sided infinite words of minimal complexity), and the paper by Stolarsky 1976 shows the relation with continued fractions, fixpoints, and Beatty sequences. The last twenty years have seen large developments, from the point of view of arithmetics, dynamical systems and combinatorics on words. There are surveys by T. C. Brown 1993, Berstel 1996, Ziccardi 1995, partly De Luca 1997a and for finite factors of Sturmian words Bender, Patashnik, and Rumsey 1994. Sturmian words are known under many other names. Each reflects the emphasis on a particular property. Thus, they are called two-distance sequences (see e.g. Lunnon and Pleasants 1992), Beatty sequences (de Bruijn 1989, 1981), characteristic sequences (Christoffel 1875), spectra (Boshernitzan and Fraenkel 1981, 1984; the *spectrum* of a number α is the multiset $\{\lfloor n\alpha \rfloor \mid n \geq 1\}$ in the book Graham, Knuth, and Patashnik 1989), digitized straight lines, cutting sequences and even musical sequences in a special case (Series 1985).

Sturmian words are of lowest possible complexity. For an overview on complexity of infinite words, see Allouche 1994. Two-sided infinite words of complexity $P(n) = n + 1$ include strictly mechanical words (Problem 2.1.1, Coven and Hedlund 1973). There is a large literature on infinite words with slightly more than minimal complexity (Coven 1974, Alessandri 1996, Cassaigne 1996, Ferenczi 1995, Rote 1994, Hubert 1995, 1996, Rauzy 1988). An extension to three letters has been initiated by Arnoux and Rauzy 1991, Arnoux, Mauduit, Shiokawa, and Tamura 1994, Castelli, Mignosi, and Restivo 1999 (the last paper relates Arnoux–Rauzy words to central words over three letters). Several properties have been extended to larger alphabets by Droubay et al. 2001. The property of balance and Theorem 2.1.5 are due to Morse and Hedlund 1940; our exposition benefits from Coven and Hedlund 1973. In particular, Proposition 2.1.3 is there. Theorem 2.1.13 is also from Morse and Hedlund 1940. The argument of the proof of Lemma 2.1.15 is from Tijdeman 1996. Christoffel words were investigated in Christoffel 1875. A systematic geometric study is in Borel and Laubie 1991, 1993. Several propositions of Subsection 2.1.3, Propositions 2.1.18, 2.1.19, 2.1.23, are from Mignosi 1989. He uses rotations (in a slightly different setting).

Mechanical words are also known as digitized straight lines. They have been considered for a long time in pattern recognition, where the problem is to compute the slope and the intercept of a finite Sturmian word as fast as possible, to test whether a word is a finite Sturmian word and, if not, to get the polygonal decomposition (see Bruckstein 1991, Dorst and Smeulders 1991 and the literature quoted there, also Berstel and Pocchiola 1996). Words generated by rotations are in fact more general than Sturmian words when the partition of $[0, 1)$ is defined independently of the angle of rotation (see Alessandri 1996, Gambaudo, Lanford, and Tresser 1984, Iwanik 1994, Rauzy 1988, Sidorov and Vershik 1993). Interval exchange is even more general, because the exchange functions are piecewise rotations (see e.g. Rauzy 1979, Didier 1997).

Standard pairs were introduced in a slightly different form in Rauzy 1985. His construction is known as *Rauzy's rules* (see also Problem 2.2.2).

Theorem 2.2.4 and its corollaries are from De Luca and Mignosi 1994. Theorem 2.2.11 is from De Luca and Mignosi 1994. It appears in a similar form in Coven and Hedlund 1973; see also Pedersen 1988.

Lemmas 2.2.17 and 2.2.18 are from Parvaix 1997. Proposition 2.2.24 was proved by Fraenkel, Mushkin, and Tassa 1978, see also Brown 1993. Theorem 2.2.31 is from Mignosi 1991, although the present proof is different. The proof of Theorem 2.2.36 given here is from De Luca and Mignosi 1994. There are several other proofs, in Mignosi 1991, Berstel and Pocchiola 1993. The formula also appeared in Koplowitz, Lindenbaum, and Bruckstein 1990.

The proof of Theorem 2.2.37 by the factor graphs is from Berthé 1996. The result is also known as the *three-distance theorem*. There is a large literature on this subject (see Berthé 1996 and the survey paper Alessandri and Berthé 1998).

Sturmian morphisms were investigated in Séébold 1991. The equivalence between (i) and (ii) of Theorem 2.3.7 is due to Mignosi and Séébold 1993, the proof of equivalence to the third condition is adapted from Berstel and Séébold 1994a. Proposition 2.3.11 is from Berstel and Séébold 1994b. Theorem 2.3.12 appears in De Luca 1997c. The results of Subsection 2.3.4 are from Séébold 1998. The relation to automorphisms of free groups is from Wen and Wen 1994a. The proof given here is simpler than the original one. For results on free groups and their automorphisms, see e.g. Magnus, Karrass, and Solitar 1966 or Lyndon and Schupp 1977. Theorem 2.3.25 is from Crisp, Moran, Pollington, and Shiue 1993. Several weak versions of this theorem were known earlier (see Brown 1993 for a discussion). Our proof is adapted from Berstel and Séébold 1994a. A self-contained proof exists by Komatsu and van der Poorten 1996. The characterization of Sturm numbers is from Allauzen 1998. Several generalizations to noncharacteristic Sturmian words were proposed (See e.g. Komatsu 1996, Arnoux, Ferenczi, and Hubert 2000).

CHAPTER 3

Unavoidable Patterns

3.0. Introduction

In Chapter 1, avoidable and unavoidable sets of words have been defined. The focus was then on the case of finite sets of words. In the present chapter, we turn to particular infinite sets of words, defined as *pattern languages*. A *pattern* is a word that contains special symbols called *variables*, and the associated pattern language is obtained by replacing the variables with arbitrary nonempty words, with the condition that two occurrences of the same variable have to be replaced with the same word.

The archetype of a pattern is the square, $\alpha\alpha$. The associated pattern language is $L = \{uu \mid u \in A^+\}$, and it is now a classical result that L is an avoidable set of words if A has at least three elements, whereas it is an unavoidable set of words if A has only one or two elements. Indeed, an infinite square-free word on three letters can be constructed, and it is easy to check that every binary word of length 4 contains a square. For short, we will say that the pattern $\alpha\alpha$ is 3-avoidable and 2-unavoidable.

General patterns can contain more than just one variable. For instance, $\alpha\beta\alpha$ represents words of the form uvu, with $u, v \in A^+$ (this pattern is unavoidable whatever the size of the alphabet; see Proposition 3.1.2). They could also be allowed to contain constant letters, which unlike variables are never replaced with arbitrary words, but this is not very useful in the context of avoidability, so we will consider here only "pure" patterns, constituted only of variables.

There are in fact two separate notions of avoidability for patterns. The difference is in how the alphabet is specified. This may seem a minor point, but it results in completely different problems. In Section 3.2, we study "absolute" avoidability, where a pattern is said to be avoidable when there exists one alphabet for which the corresponding pattern language is avoidable, and unavoidable if the language is unavoidable whatever the size of the alphabet. This part of the theory is well advanced, the main result being the existence of an algorithm deciding whether a given pattern is avoidable or not.

In Section 3.3, on the contrary, the alphabet is fixed. This gives a hierarchy of avoidability notions, depending on the size of this alphabet, and which is nicely expressed by associating an *avoidability index* to every pattern. Here, no general decidability theorem is known, but some bounds can be given, and an exhaustive classification is possible in certain simple cases.

3.1. Definitions and basic properties

3.1.1. Patterns and avoidability

Let us first fix notation and give some formal definitions. Throughout the chapter, we will mainly make use of two distinct alphabets. The first one, A, which is always assumed to be finite, is the usual alphabet on which ordinary words are constructed, and its elements, denoted by a, b, c, etc., are just called *letters*. The second alphabet, E, is used in patterns. Its elements are denoted by α, β, γ, etc. and are called *variables*, and words in E^* are called *patterns*. This distinction is meant to help the understanding of the roles of the different words used, but on some occasions it may be necessary to treat a pattern as an ordinary word, which amounts to taking $A = E$.

The *pattern language* associated to a pattern $p \in E^*$ is the language on A containing all the words $h(p)$, where h is a nonerasing morphism from E^* to A^* that substitutes an arbitrary nonempty word for every variable. It is denoted by $p(A^+)$. A word $w \in A^*$ is said to *encounter* the pattern p if it contains an element of the pattern language as a factor, i.e. if $F(w) \cap p(A^+) \neq \emptyset$. Equivalently, we say that p *occurs* or *appears* in w, otherwise w is said to *avoid* p. These definitions also apply to infinite words $w \in A^\omega$.

For example, consider the pattern $p = \alpha\alpha\beta\beta\alpha$. The pattern language associated to p on the alphabet A is $p(A^+) = \{uuvvu \mid u, v \in A^+\}$. The word 1011011000111 contains p (through $h: \alpha \mapsto 011, \beta \mapsto 0$), whereas the word 0000100010111 avoids p.

Given two patterns p and p', we can treat p' as a word and check whether it encounters p. If this is the case, we denote this by $p|p'$ (which can also be read as "p *divides* p'"). The relation on E^* defined in this way is clearly reflexive and transitive, so it is a preorder on E^*. When $p|p'$ and $p'|p$ hold together, the patterns p and p' are said to be *equivalent*, and this occurs if and only if they differ by a permutation of E.

3.1. Definitions and basic properties

A pattern p is *avoidable on A* if there are infinitely many words in A^* that avoid p, i.e. if $p(A^+)$ is an avoidable set of words in A^*. This is equivalent, by König's lemma (Proposition 1.2.3) to the existence of one infinite word in A^ω avoiding p. If on the contrary every long enough word in A^* encounters p, then p is *unavoidable on A*.

If the cardinality of A is k and p is avoidable on A, then p is said to be *k-avoidable*. Obviously, changing the names of the letters has no influence on the patterns that can be avoided (the situation would be different if we were considering more general patterns where constants are allowed), therefore a pattern is k-avoidable if and only if it is avoidable on any k-letter alphabet. A pattern which is not k-avoidable is *k-unavoidable*.

In the above example, an infinite word avoiding $p = \alpha\alpha\beta\beta\alpha$ can be constructed, as we shall see in Lemma 3.3.2. We then say that p is 2-avoidable. On the other hand, p is 1-unavoidable, as are all other patterns (every unary word longer than p trivially contains p).

Finally, a pattern which is avoidable on A for some A will simply be called *avoidable*, and a pattern which is unavoidable on A for every A will be called *unavoidable*.

On several occasions we will need to delete certain variables from a pattern. If V is a subset of E, we will denote by δ_V the morphism from E^* to $(E \setminus V)^*$ that maps a variable in V to the empty word, and a variable in $E \setminus V$ to itself. In general, there is no link between the avoidability of p and that of $\delta_V(p)$.

3.1.2. Powers

The simplest class of patterns is certainly the class of powers of a single variable, α^n. The first two, $\alpha^0 = \varepsilon$, $\alpha^1 = \alpha$, are trivially unavoidable as they are encountered by any nonempty word. The situation changes radically for $n \geq 2$, with $\alpha^2 = \alpha\alpha$ being 3-avoidable, and α^n for $n \geq 3$ being 2-avoidable, as shown by the next proposition.

Recall that the Thue–Morse infinite word $t = abbabaab\cdots$ is the fixed point of the binary morphism $\theta: a \mapsto ab, b \mapsto ba$ (see Example 1.2.9 in Chapter 1). Moreover, let $u = abcacbabcbac\cdots$ be the fixed point of the ternary morphism $\mu: a \mapsto abc, b \mapsto ac, c \mapsto b$.

PROPOSITION 3.1.1. (i) *The Thue–Morse infinite word t avoids the patterns $\alpha\alpha\alpha$ and $\alpha\beta\alpha\beta\alpha$.*

(ii) *The infinite word $u = abcacbabcbac\cdots$ avoids the pattern $\alpha\alpha$.*

Proof. Property (ii) can be reduced to (i) using the following observation. Let $\pi: \{a, b, c\}^* \to \{a, b\}^*$ be the morphism defined by $\pi(a) = abb$, $\pi(b) =$

ab, and $\pi(c) = a$. Then $\pi(u) = t$. Indeed, it is easy to check that $\pi \circ \mu = \theta \circ \pi$, hence $\theta(\pi(u)) = \pi(\mu(u)) = \pi(u)$. By Proposition 1.2.8, θ has a unique fixed point beginning with a, so that $\pi(u) = t$. Now, if a square is found in u, i.e. if vv occurs for some word $v \in \{a, b\}^+$, then $\pi(vv)$ occurs in t. Moreover, the first letter of $\pi(v)$ is a, as well as the letter following $\pi(vv)$ in t. But then we have found an occurrence of $\alpha\beta\alpha\beta\alpha$ or of $\alpha\alpha\alpha$ (if $\pi(v) = a$) in t.

Let us now prove (i). We proceed by contradiction, assuming that there is an occurrence of $\alpha\alpha\alpha$ or $\alpha\beta\alpha\beta\alpha$ in t. Consider the shortest such occurrence, $uvuvu$ with $u \in A^+$ and $v \in A^*$. It occurs for the first time in t at position n.

Since aaa and bbb are not factors of t, $|uvuvu| \geq 5$. All factors of length 5 of t contain either aa or bb as a factor, hence $uvuvu$ and therefore uvu contain aa or bb. These words occur only at odd positions in t, hence the positions of all occurrences of uvu, among which n and $n + |uv|$, have the same parity. Thus $|uv|$ is even.

Let now u' be the word formed by letters of u that have an even position in t, and v' similarly from v. Then $u'v'u'v'u'$ occurs in t, and $|u'v'u'v'u'| < |uvuvu|$. If $u' \neq \varepsilon$, this contradicts the minimality of $uvuvu$. If $u' = \varepsilon$, letters at odd positions should be considered instead. ∎

Since the infinite word u avoids squares, it is said to be *square-free*. Similarly, since two overlapping occurrences of the same word contain one of the patterns $\alpha\alpha\alpha$ or $\alpha\beta\alpha\beta\alpha$, and the Thue–Morse infinite word t avoids them, it is said to be *overlap-free*.

Many other patterns are avoided by the Thue–Morse infinite word. This can usually be proved with arguments similar to those used in the proof of Proposition 3.1.1. One of these arguments is the presence of *synchronizing words*, here aa and bb. It is fairly general and is used in most avoidability proofs (see Subsection 3.3.2), as well as the general structure of the proof (consider an occurrence with minimal length, then construct a shorter one to reach a contradiction). The other argument, the use of parity, is specific to the Thue–Morse word, or at least to infinite words generated by uniform morphisms, and has to be replaced for other kinds of infinite words.

3.1.3. Sesquipowers

The only unavoidable patterns we have seen for the moment are the empty pattern ε and the pattern α. To construct other unavoidable patterns, we need new variables.

With two variables α and β, we can construct the pattern $\alpha\beta$ which is obviously unavoidable (any word of length at least 2, regardless of the

3.1. Definitions and basic properties

alphabet, contains an occurrence of $\alpha\beta$). More interesting is the pattern $\alpha\beta\alpha$:

PROPOSITION 3.1.2. *The pattern $\alpha\beta\alpha$ is unavoidable. More precisely, if the cardinality of A is k, any word of length at least $2k + 1$ contains an occurrence of $\alpha\beta\alpha$, and this bound is tight.*

Proof. Let $w \in A^*$ be a word of length at least $2k + 1$. Then one of the letters in A, say a, occurs at least three times in w. Write $w = w_0 a w_1 a w_2 a w_3$, and let $h(\alpha) = a$ and $h(\beta) = w_1 a w_2$. Then $h: \{\alpha, \beta\}^* \to A^*$ is a nonerasing morphism and $h(\alpha\beta\alpha)$ is a factor of w. If $A = \{a_1, a_2, \ldots, a_k\}$, then $a_1 a_1 a_2 a_2 \cdots a_k a_k$ is a word of length $2k$ avoiding $\alpha\beta\alpha$. ∎

The above construction applies in fact to any pattern (although it is not so easy to get a tight bound then).

PROPOSITION 3.1.3. *Let p be a pattern unavoidable on A, and ζ a variable that does not occur in p. Then the pattern $p\zeta p$ is unavoidable on A.*

Proof. Let $k = \text{Card } A$. Since p is unavoidable on A, there is an integer l such that any word in A^l contains p. This is a finite set with k^l elements. Let now $N = k^l(l + 1) + l$, and consider any word $w \in A^N$. The word w can be viewed as the concatenation of $k^l + 1$ words of length l, separated by individual letters. Among these $k^l + 1$ factors of length l, at least two are equal, say $w = w_0 v w_1 v w_2$ with $|v| = l$ and $|w_1| \geq 1$. There is a nonerasing morphism $h:(\text{Alph } p)^* \to A^*$ such that $v = v_0 h(p) v_1$, and letting $h(\zeta) = v_1 w_1 v_0$, we find that $h(p\zeta p)$ is a factor of w. ∎

If we apply Proposition 3.1.3 recursively starting with the empty word, we can construct an infinite family of unavoidable patterns. Let α_n, for $n \in \mathbb{N}$, be different variables in E. Let $Z_0 = \varepsilon$, and for all $n \in \mathbb{N}$, $Z_{n+1} = Z_n \alpha_n Z_n$. The patterns Z_n are called *Zimin words*, or *sesquipowers*.

PROPOSITION 3.1.4. *The Zimin patterns Z_n are all unavoidable.*

Proof. Let A be a finite alphabet. We have seen that $Z_0 = \varepsilon$, $Z_1 = \alpha_0$, and $Z_2 = \alpha_0 \alpha_1 \alpha_0$ are unavoidable on A. If Z_n is unavoidable on A, then by Proposition 3.1.3, $Z_{n+1} = Z_n \alpha_n Z_n$ is also unavoidable on A (note that we are allowed to apply Proposition 3.1.3 since α_n does not occur in Z_n). Since all sesquipowers Z_n are unavoidable on any A, they are unavoidable. ∎

116   3. Unavoidable Patterns

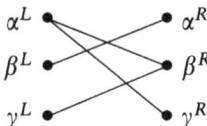

Figure 3.1. The adjacency graph of $\alpha\beta\alpha\gamma\beta\alpha$.

3.2. Deciding avoidability: the Zimin algorithm

3.2.1. Reduction of patterns

To show that avoidability is decidable, we shall prove that it is equivalent to a property of *irreducibility*, defined below, which can itself be checked through a recursive algorithm.

THEOREM 3.2.1. *A pattern is avoidable if and only if it is irreducible*

The proof will be carried out in the following two subsections. Let us first define the reduction process.

Let $p \in E^*$ be a pattern. The *adjacency graph* of p is the bipartite graph $AG(p)$ with two copies of E as vertices, denoted by E^L and E^R, and with an edge between ξ^L and η^R if and only if $\xi\eta$ is a factor of p. For instance, the adjacency graph of $\alpha\beta\alpha\gamma\beta\alpha$, shown on Figure 3.1, has six vertices and four edges.

A nonempty subset F of Alph p is called a *free set* (for p) if there exists no path in $AG(p)$ linking a left-side vertex ξ^L to a right-side vertex η^R with ξ and η in F. To find all the free sets, one should first determine the connected components of $AG(p)$. In the example above, the adjacency graph has two connected components, and two free sets, $\{\alpha\}$ and $\{\beta\}$.

Given a pattern p and a free set F for p, we say that p reduces in one step to q by the deletion of F if $q = \delta_F(p)$ is the pattern obtained by deleting from p all occurrences of letters in F. We shall denote this fact by $p \xrightarrow{F} q$. We say that p reduces to q if there is a sequence of one-step reductions leading from p to q, and denote this by $p \xrightarrow{*} q$. Finally, a pattern p is *reducible* if it reduces to the empty pattern, $p \xrightarrow{*} \varepsilon$, and p is *irreducible* otherwise.

In the above example, $p = \alpha\beta\alpha\gamma\beta\alpha$ reduces to $\beta\gamma\beta$ by deletion of the free set $\{\alpha\}$, and $\beta\gamma\beta$ itself reduces to γ, which in turns reduces to ε. The pattern p is therefore reducible. However, if we had begun with the deletion of the free set $\{\beta\}$, we would have obtained the irreducible pattern

$\alpha\alpha\gamma\alpha$. To prove that a pattern is irreducible, it is therefore necessary to recursively explore all possible sequences of one-step reductions and make sure that none of them leads to the empty pattern. (Since the exploration of a potentially large tree is needed, one should not expect this algorithm to be very efficient in practice.) This recursive algorithm will be called the *Zimin algorithm*.

It should be noted that it is sometimes necessary to delete free sets having more than one element, for instance for the pattern $p = \alpha\beta\alpha\gamma\alpha'\beta\alpha\delta\alpha\beta\alpha'\gamma\alpha'\beta\alpha'$ where all deletions of a free singleton lead to an irreducible pattern, while p can be reduced to the empty pattern starting with the deletion of the free set $\{\alpha, \alpha'\}$.

We shall need some additional notation concerning adjacency graphs. Given a pattern p and a set X of vertices of $AG(p)$, we denote by $C(X, p)$ the set of vertices of $AG(p)$ that are in the same connected component as an element of X, by $C_L(X, p)$ the set of variables $\xi \in E$ such that $\xi^L \in C(X, p)$, and by $C_R(X, p)$ the set of variables $\xi \in E$ such that $\xi^R \in C(X, p)$. If F is a free set and we apply these definitions to the set F^L of left-side vertices associated to F, then we obtain the two sets $C_L(F^L, p)$ and $C_R(F^L, p)$. The fact that F is a free set can then be expressed as $F \subset C_L(F^L, p) \setminus C_R(F^L, p)$.

3.2.2. Reducible patterns are unavoidable

Let us first prove the easier direction of Theorem 3.2.1: if a pattern is reducible, then it is unavoidable. This is a particular case of the following lemma, taking $q = \varepsilon$, the empty pattern being obviously unavoidable.

LEMMA 3.2.2. *If $p \stackrel{*}{\to} q$ and q is unavoidable, then p is also unavoidable.*

Proof. It suffices to prove Lemma 3.2.2 in the case where p reduces to q in one step by deletion of a free set F, the general case being then deduced by induction on the number of steps.

We shall prove that p is unavoidable on any alphabet A by induction on the size of A. For Card $A = 1$ the assertion is obviously true. Assume now that $A = A' \cup \{a\}$ and that p is unavoidable on A'. Let $L = A'^+ \setminus A'^* p(A'^+) A'^*$ be the set of nonempty words on A' avoiding p, which is finite by assumption, and $M = aA^* \setminus A^* p(A^+) A^*$ be the set of words on A avoiding p and starting with a. Each word of M which is not a power of a can be represented as a nonempty product of words in $N = \{a^i w a^j \mid w \in L, 0 < i < |p|, 0 \le j < |p|\}$, which is obviously finite. In other terms, if we view N as a new alphabet, $M \subset i(N^+) \cup a^+$, where i is the morphism from the free monoid N^* to A^* mapping an element of N (viewed as a letter) to itself (viewed as a word).

Let ζ be a variable that does not occur in p, and $E' = E \cup \{\zeta\}$. Then $q\zeta$ is an unavoidable pattern and every sufficiently long word on N contains an occurrence of this pattern. Consequently, for any $w \in N^*$ long enough, there is a nonerasing morphism f from E'^* to N^* such that $f(q\zeta)$ is a factor of w. For a variable ξ, $f(\xi) \in N^+$ hence $i(f(\xi)) \in aA^+$. We now define a new morphism g from E^* to A^* as follows:
  (i) $g(\xi) = i(f(\xi))$ if $\xi \in E \setminus (C_L(F^L, p) \cup C_R(F^L, p))$,
  (ii) $g(\xi) = a^{-1}i(f(\xi))$ if $\xi \in C_R(F^L, p) \setminus C_L(F^L, p)$,
  (iii) $g(\xi) = i(f(\xi))a$ if $\xi \in C_L(F^L, p) \setminus (C_R(F^L, p) \cup F)$,
  (iv) $g(\xi) = a^{-1}i(f(\xi))a$ if $\xi \in C_L(F^L, p) \cap C_R(F^L, p)$,
  (v) $g(\xi) = a$ if $\xi \in F$.

Note that the five cases are indeed exclusive of each other, and that this defines a nonerasing morphism. Moreover, as we shall see below, $g(p)$ is a factor of $i(f(q\zeta))$. Consequently, $i(w)$ encounters p and cannot be in M, which means that M is finite and that p is unavoidable on A.

It remains to show that $g(p)$ is a factor of $i(f(q\zeta))$. We shall prove by induction on k, $1 \le k \le |p|$, that if p_k is the prefix of length k of p and $p_k \xrightarrow{F} q_k$, where q_k is a prefix of q, then $rg(p_k)$ is equal to $i(f(q_k))s_k$, where r is a or ε (depending on whether the first letter of p is in $C_R(F^L, p)$ or not) and s_k is a or ε (depending on whether the last letter of p_k is in $C_L(F^L, p)$ or not). For $k = 1$, this is obvious from the definition of g. Assume that $rg(p_k) = i(f(q_k))s_k$, and let $p_{k+1} = p_k\eta$. The last letter of p_k is denoted by ξ, so that there is an edge from ξ^L to η^R in $AG(p)$. We have to show that $rg(p_{k+1}) = i(f(q_{k+1}))s_{k+1}$, with $s_{k+1} = a$ if $\eta \in C_L(F^L, p)$, $s_{k+1} = \varepsilon$ otherwise. Writing $rg(p_{k+1}) = rg(p_k)g(\eta) = i(f(q_k))s_kg(\eta)$, it reduces to $s_kg(\eta) = i(f(\eta))s_{k+1}$ if $\eta \notin F$, or to $s_kg(\eta) = s_{k+1}$ if $\eta \in F$. This is again obvious from the definition of g, observing that $s_k = a$ occurs if and only if $\eta \in C_R(F^L, p)$, since this is equivalent to $\xi \in C_L(F^L, p)$. ∎

3.2.3. Irreducible patterns are avoidable

We now turn to the other direction of Theorem 3.2.1: if a pattern is unavoidable, then it is reducible. This part of the proof relies on several lemmas.

LEMMA 3.2.3. *Suppose that $f(q)$ is a factor of p for some nonerasing morphism f of E^* (so that q divides p), and that F is a free set for p. Let F' be the set of variables ξ such that $f(\xi) \in F^+$. Then F' is a free set for q. Moreover, if $p \xrightarrow{F} p'$ and $q \xrightarrow{F'} q'$, then $f'(q')$ is a factor of p', where*

3.2. Deciding avoidability: the Zimin algorithm

$f' = \delta_F \circ f|_{E\setminus F'}$ is the nonerasing morphism from $(E \setminus F')^*$ to $(E \setminus F)^*$ mapping a variable ξ to the pattern obtained by deleting the elements of F from $f(\xi)$.

Proof. Given a variable $\xi \in E$, we denote by $f_1(\xi)$ the first letter of $f(\xi)$ and by $f_2(\xi)$ the last letter of $f(\xi)$. If $\xi\eta$ occurs in q, then $f_2(\xi)f_1(\eta)$ occurs in p. Mapping ξ^L to $f_2(\xi)^L$ and ξ^R to $f_1(\xi)^R$, we see that $AG(q)$ is mapped to a subgraph of $AG(p)$. If there were a path from ξ^L to η^R in $AG(q)$, with ξ and η in F', then there would be a path from $f_2(\xi)^L$ to $f_1(\eta)^R$ in $AG(p)$, which is not the case since $f_2(\xi)$ and $f_1(\eta)$ are elements of the free set F. Therefore F' is a free set for q, and the rest of the lemma is obvious. ∎

Observe that if $\xi \in F'$ occurs in q, then $f(\xi)$ must have length 1, since two variables of F cannot occur consecutively in p.

LEMMA 3.2.4. *Suppose that p, p', q are patterns such that $p \xrightarrow{*} p'$ and $f(q)$ is a factor of p for some nonerasing morphism f. Then there exist a pattern q' and a nonerasing morphism f' such that $q \xrightarrow{*} q'$ and $f'(q')$ is a factor of p', with the additional condition that $f(\mathrm{Alph}\, q \setminus \mathrm{Alph}\, q') \subset (\mathrm{Alph}\, p \setminus \mathrm{Alph}\, p')^*$ (if a variable ξ is deleted from q, then $f(\xi)$ contains only variables deleted from p).*

Proof. This is just the iteration of Lemma 3.2.3, and can be proved by induction on the number of reduction steps from p to p'. If $p = p'$, then the result is obvious, with $q' = q$ and $f' = f$. Assume now that $p \xrightarrow{*} p'' \xrightarrow{F} p'$ and that we have constructed q'' and f'' such that $q \xrightarrow{*} q''$, $f''(q'')$ is a factor of p'', and $f(\mathrm{Alph}\, q \setminus \mathrm{Alph}\, q'') \subset (\mathrm{Alph}\, p \setminus \mathrm{Alph}\, p'')^*$. Then apply Lemma 3.2.3 to p'' and q'', to construct a free set F' for q'', a pattern q' such that $q'' \xrightarrow{F'} q'$ and a nonerasing morphism f' such that $f'(q')$ is a factor of p' and $f''(F') \subset F^+$.

$$\begin{array}{ccccc} p & \xrightarrow{*} & p'' & \xrightarrow{F} & p' \\ f \uparrow & & f'' \uparrow & & f' \uparrow \\ q & \xrightarrow{*} & q'' & \xrightarrow{F'} & q' \end{array}$$

The additional condition also holds, since $\mathrm{Alph}\, q \setminus \mathrm{Alph}\, q' = (\mathrm{Alph}\, q \setminus \mathrm{Alph}\, q'') \cup F'$, with $f(\mathrm{Alph}\, q \setminus \mathrm{Alph}\, q'') \subset (\mathrm{Alph}\, p \setminus \mathrm{Alph}\, p'')^* \subset (\mathrm{Alph}\, p \setminus \mathrm{Alph}\, p')^*$ and $\delta_F(f''(F')) \subset \delta_F(F^+) \subset \{\varepsilon\}$, hence $f(F') \subset (\mathrm{Alph}\, p \setminus \mathrm{Alph}\, p')^*$. ∎

LEMMA 3.2.5. *Let $q = \delta_V(p)$ be a pattern obtained from another pattern p by deleting the variables in a set V (not necessarily a free set), Suppose that there are a pattern r and a nonerasing morphism f such that $r \stackrel{\bullet}{\to} q$ and $f(p)$ is a factor of r, and that $\xi \in V$ if and only if $f(\xi) \in (\text{Alph} \, r \setminus \text{Alph} \, q)^*$. Then $p \stackrel{\bullet}{\to} q$.*

Proof. Apply Lemma 3.2.4, with r and q playing respectively the roles of p and p'. There exist a pattern q' and a nonerasing morphism f' such that $p \stackrel{\bullet}{\to} q'$ and $f'(q')$ is a factor of q, with $f(\text{Alph} \, p \setminus \text{Alph} \, q') \subset (\text{Alph} \, r \setminus \text{Alph} \, q)^*$. Then $\text{Alph} \, p \setminus \text{Alph} \, q' \subset V$, and $\delta_V(q') = \delta_V(p) = q$, hence $|q'| \geq |q|$. On the other hand, since $f'(q')$ is a factor of q, $|q'| \leq |q|$. We therefore have $|q| = |q'|$, and this is only possible if $q = q'$. We thus have $p \stackrel{\bullet}{\to} q$. ∎

Let us define, for any positive integer k, a morphism φ_k on the $4k$-letter alphabet $A_k = \{a_0, a_1, \ldots, a_{2k-1}, b_0, b_1, \ldots, b_{2k-1}\}$. For $0 \leq i \leq 2k - 1$, set $\varphi_k(a_i) = a_0 b_i a_1 b_{i+1} \ldots a_{k-1} b_{i+k-1}$ and $\varphi_k(b_i) = a_k b_i a_{k+1} b_{i+1} \ldots a_{2k-1} b_{i+k-1}$, where indices are taken modulo $2k$. The morphism φ_k is uniform with length $2k$. Using the morphism φ_k, we can construct an infinite word $w^{(k)} = \varphi_k^\omega(a_0)$, the fixed point of φ_k. We shall now prove that any irreducible pattern is avoided by $w^{(k)}$ for some k.

LEMMA 3.2.6. *Let v be a factor of length at least 2 of $w^{(k)}$. Then there are an integer i, $0 \leq i \leq 4k - 1$, and a letter $x \in A_k$ such that, whenever v occurs at position $n \geq 0$ in $w^{(k)}$, one has*
  (i) $n \equiv i \pmod{4k}$,
  (ii) *the letter at position $n' = \lfloor \frac{n}{2k} \rfloor$ in $w^{(k)}$ is $w_{n'}^{(k)} = x$.*

Proof. Let us assume that $|v| = 2$; the general case follows trivially.

Observe that the a's and b's alternate in the images under φ_k, hence in $w^{(k)}$. Consequently, the letter following an occurrence of $\varphi_k(a_i)$ is a_k, and the letter following an occurrence of $\varphi_k(b_i)$ is a_0, so that the letters at even positions in $w^{(k)}$ cycle periodically in the set $\{a_0, a_1, \ldots, a_{2k-1}\}$, i.e. $w_{2i}^{(k)} = a_i$, where the index in a_i is taken modulo $2k$. In other terms, the letters a_i are *synchronizing letters* that indicate the position in $w^{(k)}$ modulo $4k$. Since any factor of length 2 contains at least one letter a_i, the position of an occurrence of such a factor is unique modulo $4k$.

If v occurs at position n, then it starts within the image of $x = w_{n'}^{(k)}$. It is either a factor of $\varphi_k(x)$, or formed with the last letter of $\varphi_k(x)$ followed by the first letter of the image of the next letter, that is a_k or a_0. There are exactly $8k^2$ such words of length 2, all different. Therefore x is uniquely

3.2. Deciding avoidability: the Zimin algorithm

defined by v, and can be computed using the following rules, where all indices are modulo $2k$:
- if $v = a_i b_j$ with $0 \le i \le k-1$, then $x = a_{j-i}$,
- if $v = b_j a_{i+1}$ with $0 \le i \le k-1$, then $x = a_{j-i}$,
- if $v = a_i b_j$ with $k \le i \le 2k-1$, then $x = b_{j+k-i}$,
- if $v = b_j a_{i+1}$ with $k \le i \le 2k-1$, then $x = b_{j+k-i}$.

In other terms, the letters b_j are *recognizing letters* that, together with a neighboring a_i, allow us to reconstruct the word one particular factor of $w^{(k)}$ comes from under the action of φ_k. ∎

LEMMA 3.2.7. *Let p be a pattern, k an integer such that $2k >$ Card Alph p, and v a factor of $w^{(k)}$ such that $\varphi_k(v)$ encounters p. Then there is a pattern q such that $p \xrightarrow{*} q$ and v encounters q.*

Proof. Let h be a nonerasing morphism such that $h(p)$ is a factor of $\varphi_k(v)$. According to Lemma 3.2.6, to each letter $x \in A$ are associated $2k$ words of length 2 that recognize $\varphi_k(x)$. Since $2k >$ Card Alph p and these $2k$ words end with different letters, at least one of these words ends with a letter which is not the first letter of any $h(\xi)$ with ξ occurring in p. Let us choose one such word d_x and call it the *decisive word* for x. By construction, whenever d_x occurs in $h(p)$ it occurs within some $h(\xi)$.

Let V be the set of variables ξ such that $h(\xi)$ contains no decisive word, and $q = \delta_V(p)$. We define a nonerasing morphism h' from $(E \setminus V)^*$ to A^* by $h'(\xi) = x_1 x_2 \ldots x_m$, where $d_{x_1}, d_{x_2}, \ldots, d_{x_m}$ are the decisive words occurring in $h(\xi)$, in the order in which they occur. Since each decisive word corresponds to exactly one letter in v, and consecutive decisive words correspond to consecutive letters, the word $h'(q)$ is a factor of v, i.e. v encounters q.

Note that V is not necessarily a free set, so we have not yet proved that $p \xrightarrow{*} q$. For this we shall define a morphism f such that $f(p) \xrightarrow{*} q$ and apply Lemma 3.2.5. The morphism f, from E^* to $(E \cup A)^*$ (elements of A being treated as additional variables), is defined as follows:
  (i) if $\xi \in V$, then we set $f(\xi) = h(\xi)$,
  (ii) if $h(\xi)$ does not contain a_0 or a_k, but contains a decisive word d_x (only one decisive word may occur in this case), then $\varphi_k(x) = a_i v_1 h(\xi) v_2$, with $i \in \{0, k\}$ and $v_1, v_2 \in A^*$, and we set $f(\xi) = h(\xi) v_2 \xi v_1 h(\xi)$,
  (iii) if $\xi \notin V$ and $h(\xi) = v_1 \varphi_k(w) a_j v_2$ with $j \in \{0, k\}$ and $v_1, v_2, w \in A^*$, with w of maximal length, then let a_i be the first letter of $\varphi_k(w) a_j$, x_1 be the first letter of $h'(\xi)$, and x_2 be the last letter of $h'(\xi)$. We set $f(\xi) = v_1 v_1' \xi v_2' v_2$, where $v_1' = \varepsilon$ if the first decisive word d_{x_1} of

$h(\xi)$ occurs in $v_1 a_i$, $v'_1 = \varphi_k(x_1)$ otherwise, and similarly $v'_2 = \varepsilon$ if the last decisive word d_{x_2} of $h(\xi)$ occurs in $a_j v_2$, $v'_2 = (a_{j+k})^{-1} \varphi_k(x_2) a_j$ otherwise.

Deleting the elements of A from the pattern $f(p)$, one obtains exactly q, since $f(\xi) \in A^*$ when $\xi \in V$ and $f(\xi) \in A^* \xi A^*$ otherwise: $\delta_A(f(p)) = \delta_V(p) = q$. Moreover, two variables ξ_1 and ξ_2 consecutive in q are separated in $f(p)$ by the word $v_1 a_i v_2 \in A^*$, where $a_{i+k} v_1$ is the image by φ_k of the last letter in $h'(\xi_1)$ and $a_i v_2$ is the image by φ_k of the first letter in $h'(\xi_2)$. This allows us to reduce $f(p)$ as follows. First, let $F_0 = \{b_0, b_1, \ldots, b_{2k-1}\}$. This is a free set for $f(p)$, since an element of F_0 can only be followed by an element of $E \cup \{a_0, a_1, \ldots, a_{2k-1}\}$, which in turn can only be preceded by an element of F_0. Thus $f(p) \xrightarrow{F_0} p_0$, where the pattern p_0 contains only variables in E and letters a_i, occurring in sequences $a_1 a_2 \ldots a_{k-1} a_k a_{k+1} \ldots a_{2k-1}$ or $a_{k+1} a_{k+2} \ldots a_{2k-1} a_0 a_1 \ldots a_{k-1}$ between two variables in E. Then $F_1 = \{a_1, a_{k+1}\}$ is a free set for p_0, allowing the reduction $p_0 \xrightarrow{F_1} p_1$. We continue deleting $F_i = \{a_i, a_{k+i}\}$ for i from 2 to $k-1$, until we get a pattern $p_{k-1} \in (E \cup \{a_0, a_k\})^*$, in which the elements of E and $F_k = \{a_0, a_k\}$ alternate, so that F_k is again a free set and $p_{k-1} \xrightarrow{F_k} q$.

Consequently, $f(p) \xrightarrow{*} q$ and the pattern $r = f(p)$ satisfies the hypotheses of Lemma 3.2.5, so that $p \xrightarrow{*} q$. ∎

LEMMA 3.2.8. *If the infinite word $w^{(k)}$ encounters a pattern p containing less than $2k$ distinct variables, then p is reducible.*

Proof. There is a positive integer m such that $\varphi_k^m(a_0)$ encounters p. Applying Lemma 3.2.7, we obtain a pattern p_1 such that $p \xrightarrow{*} p_1$ and $\varphi_k^{m-1}(a_0)$ encounters p_1. This process can be repeated since the condition $2k > \operatorname{Card} \operatorname{Alph} p_1$ still holds (the number of variables involved cannot increase), yielding patterns p_i for $1 \leq i \leq m$ such that $p \xrightarrow{*} p_i$ and $\varphi_k^{m-i}(a_0)$ encounters p_i. But then a_0 encounters p_m, which means that p_m is either a single variable or the empty pattern, hence p is reducible. ∎

Proof of Theorem 3.2.1. By Lemma 3.2.2, if p is reducible, then p is unavoidable. Conversely, assume that p is unavoidable. Then for all k, in particular for $k = \left\lceil \frac{\operatorname{Card} \operatorname{Alph} p + 1}{2} \right\rceil$, $w^{(k)}$ encounters p. By Lemma 3.2.8, p is reducible. ∎

COROLLARY 3.2.9. *The infinite word $w^{(k)}$ avoids all avoidable patterns with at most $2k - 1$ variables.*

3.3. Avoidability on a fixed alphabet

Proof. If $w^{(k)}$ encounters a pattern p with at most $2k-1$ variables, then by Lemma 3.2.8 p is reducible, hence by Lemma 3.2.2 p is unavoidable. ∎

COROLLARY 3.2.10. *If all variables that occur in a pattern p occur at least twice, then p is avoidable.*

Proof. If p is unavoidable, then it reduces to a single variable α. But then α is a variable that occurs only once in p. ∎

COROLLARY 3.2.11. *Let p be a pattern with n variables. If $|p| \geq 2^n$, then p is avoidable.*

Proof. We prove by induction on n that if p is unavoidable, then $|p| < 2^n$. For $n = 1$, the result holds according to Proposition 3.1.1. Assume that it holds for n, and consider an unavoidable pattern p with $n + 1$ variables. According to Corollary 3.2.10, there is a variable α that occurs only once in p. Then p can be written as $p = p_1 \alpha p_2$, where p_1 and p_2 are patterns with at most n variables, which are both unavoidable since they divide p. By the induction hypothesis, p_1 and p_2 have length at most $2^n - 1$, hence p has length at most $2^{n+1} - 1$. ∎

3.3. Avoidability on a fixed alphabet

3.3.1. The avoidability index

As shown by the case of the square $\alpha\alpha$, an avoidable pattern need not be avoidable on a two-letter alphabet, in other terms the same pattern can be 2-unavoidable but k-avoidable for some larger k. This leads us to define the *avoidability index* of a pattern $p \in E^*$, $\mu(p)$. It is the smallest integer k such that p is k-avoidable, or ∞ if p is unavoidable. Clearly, $2 \leq \mu(p) \leq \infty$ since no pattern is 1-avoidable. In some sense, the avoidability index of a pattern measures how easy it is to avoid this pattern. With the preorder on E^* defined by divisibility, the function $\mu: E^* \to \mathbb{N} \cup \{\infty\}$ is nonincreasing: if $p|q$, then $\mu(p) \geq \mu(q)$.

Contrary to the situation of Section 3.2 where the size of the alphabet did not matter, there is no known algorithm to determine the status of a given pattern, that is to compute its avoidability index. Even for very short patterns the value of $\mu(p)$ may be unknown. For instance, it is not known at the time of writing whether $\mu(\alpha\alpha\beta\beta\gamma\gamma)$ is equal to 2 or 3, although there is some experimental evidence that the index is 2.

Therefore, the results that we shall present here are only partial. They are of two kinds: computations in a simple case where patterns can be completely classified according to their avoidability index, and bounds in the general case.

3.3.2. The binary case

We will now restrict to a particular class of patterns, namely binary patterns, i.e. patterns with at most two different variables. In this subsection, we set $E = \{\alpha, \beta\}$.

Let us first review the unary case, studied in Subsection 3.1.2. The empty pattern and the pattern reduced to one variable are unavoidable: $\mu(\varepsilon) = \mu(\alpha) = \infty$. Squares are 3-avoidable but 2-unavoidable, hence $\mu(\alpha\alpha) = 3$. Finally, cubes and larger powers are 2-avoidable, hence $\mu(\alpha^k) = 2$ for $k \geq 3$.

The fact that $\alpha\alpha$ is a 3-avoidable 2-unavoidable pattern gives us readily some information about binary patterns. Only a finite number of binary patterns are 3-unavoidable. Indeed, a pattern which is divisible by $\alpha\alpha$ must be 3-avoidable, and since $\alpha\alpha$ is 2-unavoidable, there are only finitely many binary patterns which are not divisible by $\alpha\alpha$, namely ε, α, β, $\alpha\beta$, $\beta\alpha$, $\alpha\beta\alpha$, and $\beta\alpha\beta$. All of these are in fact unavoidable, which implies that the avoidability index of a binary pattern can only be 2, 3, or ∞.

The remaining question is to distinguish patterns that have avoidability index 3 from patterns that have avoidability index 2.

LEMMA 3.3.1. *The binary patterns $\alpha\alpha$, $\alpha\alpha\beta$, $\alpha\alpha\beta\alpha$, $\alpha\beta\beta\alpha$, $\alpha\alpha\beta\beta$, $\alpha\beta\alpha\beta$, $\alpha\alpha\beta\alpha\alpha$, and $\alpha\alpha\beta\alpha\beta$ have avoidability index 3.*

Proof. Since all these patterns are divisible by $\alpha\alpha$, they are 3-avoidable. A simple backtracking algorithm is sufficient to check that they are 2-unavoidable. The results are summarized in the following table, where w is an example of a binary word avoiding p of maximal length, and N is the total number of binary words avoiding p, including the empty word and unary words.

p	w	$|w|$	N
$\alpha\alpha$	aba	3	7
$\alpha\alpha\beta$	abab	4	13
$\alpha\alpha\beta\alpha$	abababaaa	9	91
$\alpha\beta\beta\alpha$	aabbbaaabb	10	93
$\alpha\alpha\beta\beta$	abaaabaaaba	11	147
$\alpha\beta\alpha\beta$	abaabbaaabbaabbab	18	477
$\alpha\alpha\beta\alpha\alpha$	abaaaabbbbabababab	18	1 699
$\alpha\alpha\beta\alpha\beta$	ababababaabbaaabbbaaaabbbbaaabbbaabbab	38	26 241

∎

3.3. Avoidability on a fixed alphabet

LEMMA 3.3.2. *The binary patterns $\alpha\alpha\alpha$, $\alpha\beta\alpha\beta\alpha$, $\alpha\beta\alpha\beta\beta\alpha$, $\alpha\alpha\beta\alpha\beta\beta$, $\alpha\beta\alpha\alpha\beta$, and $\alpha\alpha\beta\beta\alpha$ have avoidability index 2.*

Proof. According to Proposition 3.1.1, $\alpha\alpha\alpha$ and $\alpha\beta\alpha\beta\alpha$ are avoided by the Thue–Morse infinite word, and therefore 2-avoidable.

The two patterns $\alpha\beta\alpha\beta\beta\alpha$ and $\alpha\alpha\beta\alpha\beta\beta$ are avoided by the infinite word $u = v^\omega(a)$, where v is the uniform morphism that maps a to aab and b to bba. The proof is similar to that of Proposition 3.1.1, so we will only summarize it. Assume that the pattern p is not avoided by the infinite word u, and consider an element $h(p)$ of $F(u) \cap p(A^+)$ of minimal length and the position n of its first occurrence in u. Then discuss the value of n, $|h(\alpha)|$, and $|h(\beta)|$ modulo 3. In each of the 27 cases, either a contradiction is immediately reached, or an earlier or shorter occurrence of p in u can be constructed. Some preliminary observations on the structure of u, such as the fact that all squares that occur in u have length a multiple of 3, except aa and bb, can greatly reduce the number of cases that actually need to be considered.

The pattern $\alpha\beta\alpha\alpha\beta$ is avoided by the infinite word $v = \psi(\mu^\omega(a))$, where μ is the ternary morphism defined before Proposition 3.1.1 ($\mu(a) = abc$, $\mu(b) = ac$, $\mu(c) = b$) and ψ maps a to aaa, b to bbb, and c to $ababab$. We know by Proposition 3.1.1 that $\mu^\omega(a)$ is square-free. The first step is to prove a *synchronization lemma* for ψ applied on $\mu^\omega(a)$: if x is a factor of v of length 7 or more, then there exist a unique triple (s, y, p) and two letters (not necessarily unique) d and e such that dye is a factor of $\mu^\omega(a)$, s is a proper suffix of $\psi(d)$, p is a proper prefix of $\psi(e)$, and $x = s\psi(y)p$. Then this synchronization lemma can be used to prove that the only squares that occur in v are a^2, b^2, $(aa)^2$, $(ab)^2$, $(ba)^2$, $(bb)^2$, and $(baba)^2$. Now, assume that $h(\alpha\beta\alpha\alpha\beta)$ occurs in v (there is no need to consider a minimal occurrence here). Then $h(\alpha)$ is one of a, b, aa, ab, ba, bb, $baba$. If $h(\alpha\beta)$ has length 7 or more, then the synchronization lemma can be applied to it and it can be written as $s\psi(y)p$, leaving finitely many possibilities for s, p and $h(\alpha)$. Otherwise, there are finitely many possibilities for $h(\alpha)$ and $h(\beta)$. In each of the cases, either the word $h(\alpha\beta\alpha\alpha\beta)$ contains a small factor that obviously cannot occur in v, or applying ψ^{-1} yields a square in $\mu^\omega(a)$.

The pattern $\alpha\alpha\beta\beta\alpha$ is avoided by the infinite word $v = \chi(\mu^\omega(a))$, where μ is as above and χ maps a to aa, b to aba, and c to abb. The proof is similar. Here the synchronization lemma applies to words of length 5 or more, and the length of squares is not bounded. ∎

THEOREM 3.3.3. *Binary patterns fall in three categories:*

- *The 7 binary patterns ε, α, β, αβ, βα, αβα, and βαβ are unavoidable (their avoidability index is ∞).*

- *The 22 binary patterns αα, ββ, ααβ, αββ, βαα, ββα, ααβα, ααββ, αβαα, αβαβ, αββα, βααβ, βαβα, βαββ, ββαα, ββαβ, ααβαα, ααβαβ, αβαββ, βαβαα, ββαβα, and ββαββ have avoidability index 3.*

- *All other binary patterns, and in particular all binary patterns of length 6 or more, have avoidability index 2.*

Proof. All binary patterns of length 6 are divisible by a pattern mentioned in Lemma 3.3.2, or by its mirror image: they are therefore 2-avoidable, and have avoidability index 2. Consequently, all larger patterns also have avoidability index 2. There remains a list of 29 patterns, 7 of which are unavoidable. The other 22 patterns are the patterns mentioned in Lemma 3.3.1, or their mirror images, up to a renaming of the variables. They all have avoidability index 3. ∎

3.3.3. A bound on the avoidability index

In our proof of Theorem 3.2.1, we constructed an infinite word that avoids a given irreducible pattern p. The construction uses a number of different letters that depend only of the number of variables in p. Thus, we have a bound (probably far from optimal) on $\mu(p)$ as a function of Card Alph p.

THEOREM 3.3.4. *If p is an avoidable pattern, then $\mu(p) \leq 4 \left\lceil \frac{\text{Card Alph}\, p+1}{2} \right\rceil \leq 2\,\text{Card Alph}\, p + 4$.*

Proof. Let $k = \left\lceil \frac{\text{Card Alph}\, p+1}{2} \right\rceil$, so that $2k > \text{Card Alph}\, p$. By Corollary 3.2.9, the infinite word $w^{(k)}$ avoids p. Since $w^{(k)}$ is defined over an alphabet of $4k$ letters, p is $4k$-avoidable. ∎

The proof is so short because all of the work has already been done in Subsection 3.2.3.

3.3.4. A bound on the length of 2-unavoidable patterns

We have seen in Theorem 3.3.3 that there are only finitely many 2-unavoidable binary patterns, namely that all binary patterns of length 6 or more are 2-avoidable. We shall now try to generalize this to a bound for patterns with more variables.

3.3. Avoidability on a fixed alphabet

For $n, k \geq 1$, let us denote by ℓ_{nk} the smallest integer l (or ∞ if no such integer exists) such that every pattern of length l with n variables is k-avoidable. We extend the notation to $k = \infty$ with the convention that "∞-avoidable" means "avoidable". Clearly, $\ell_{nk} \leq \ell_{n'k'}$ if $n \leq n'$ and $k \geq k'$.

We know already that $\ell_{n1} = \infty$ since no pattern is 1-avoidable. Theorem 3.3.3 implies that $\ell_{22} = 6$ and $\ell_{2k} = 4$ for $k \geq 3$, and Proposition 3.1.1 implies that $\ell_{12} = 3$ and $\ell_{1k} = 2$ for $k \geq 3$. The value of $\ell_{n\infty}$ is given by Corollary 3.2.11: $\ell_{n\infty} = 2^n$.

If we are able to prove that all avoidable patterns on n variables are N_n-avoidable, then we can deduce that $\ell_{nk} = 2^n$ for all $k \geq N_n$. According to Theorem 3.3.4, this holds for $N_n = 4 \lceil \frac{n+1}{2} \rceil$.

We shall now prove that ℓ_{nk} is finite for all $k \geq 2$. Obviously, it is sufficient for this to bound ℓ_{n2} for all n.

THEOREM 3.3.5. *For all $n \geq 1$, $\ell_{n2} < 200 \cdot 5^n$.*

To prove Theorem 3.3.5, we have to construct a binary infinite word that avoids long enough patterns. We shall do so by starting with an infinite word $w_n \in A^\omega$ on an N_n-letter alphabet $A = \{a_0, a_1, \ldots, a_{N_n-1}\}$ avoiding all avoidable patterns on n variables, then recoding it on the binary alphabet $A' = \{a, b\}$ with a morphism $g : A^* \to A'^*$.

Such words w_n have been constructed in Subsection 3.2.3, where we called them $w^{(k)}$, with $k = \frac{N_n}{4} = \lceil \frac{n+1}{2} \rceil$. However, the construction of g is independent of the choice of w_n, so any other definition could be used instead for w_n.

Let $n \geq 1$, and $E = \{\alpha_1, \alpha_2, \ldots, \alpha_n\}$. Assume that $B_1, B_2, \ldots, B_{n-1}$ are $n-1$ fixed integers greater than 1. Let

$$M = T_n + 2B_{n-1} + \sum_{i=2}^{n} iB_{n+1-i},$$

where $T_n = \frac{1}{2}(N_n - 1)(n+1)(n+2) + n^2 + 2n + 4$, and let P be the set of patterns $p \in E^*$ of length at least M such that, for all i between 1 and $n-1$, every factor q of p of length B_i contains at least $i+1$ distinct variables.

Consider the morphism

$$g : A^* \longrightarrow A'^*$$
$$a_i \longmapsto a^{B_{n-1}+1} b^{n+i} z_1 b^i z_2 b^i \cdots z_{n-2} b^i z_{n-1} b^i a b^{n+nN_n+i} a$$

where $z_i = (ab^i)^{B_{n-i}-1} ab^{n+iN_n}$.

Assume that p is a pattern in P that is not avoided by the infinite word $g(w_n)$. Then there are a prefix y of w_n, a nonerasing morphism $\varphi: E^* \to A'^*$ and words $t_1, t_2 \in A'^*$ such that $g(y) = t_1 \varphi(p) t_2$.

Let E' be the set of variables $\xi \in E$ such that $\varphi(\xi)$ contains aa. The value of M has been chosen so that $M = \max\{|g(a_i)|\} + B_{n-1} + 1$. Since $|\varphi(p)| \geq |p| \geq M$, the word $\varphi(p)$ contains at least one occurrence of $a^{B_{n-1}+2}$ at the boundary between two images of letters under g. Then either the set E' is nonempty, or there are B_{n-1} consecutive variables in p that are mapped to a by φ. In the latter case, since any variable in E occurs at least once in any factor of length B_{n-1} of p, the morphism φ maps all variables to a and $\varphi(p) = a^{|p|}$, which is obviously impossible since a^M is not a factor of $g(w_n)$.

Let us now define a new function $h: E \to \{0, 1, \ldots, n+(n+1)N_n - 1\}^*$ in the following way. For any $\xi \in E$, $\varphi(\xi)$ can be written in the form $\varphi(\xi) = b^{j_0} a b^{j_1} a \ldots a b^{j_m}$, with $0 \leq j_i < n+(n+1)N_n$, and we set $h(\xi) = j_1 j_2 \ldots j_{m-1}$, leaving out j_0 and j_m. Note that $h(\xi)$ can be the empty word if $m \leq 1$, but that $h(\xi)$ is never the empty word when $\xi \in E'$.

A *cut* of g is a pair of words (u_1, u_2) in A'^* such that $u_1 u_2 = g(a_i)$ for some i and there exist $p_1, p_2 \in E^*$ and $s_1, s_2 \in A^*$ with $p_1 p_2 = p$, $t_1 \varphi(p_1) = g(s_1) u_1$ and $\varphi(p_2) t_2 = u_2 g(s_2)$. A cut is called an *initial cut* when $h(\xi) = \varepsilon$ for any variable ξ in p_1.

If u is a factor occurring only once in all $g(a_i)$, we say that u is cut if there exists a noninitial cut $(v_1 u_1, u_2 v_2)$ where u_1 and u_2 are nonempty words such that $u_1 u_2 = u$.

LEMMA 3.3.6. *There exists an integer r with $0 \leq r \leq n$ such that for all $0 \leq i < N_n$, $ab^{n+rN_n+i}a$ is not cut.*

Proof. Let r be the smallest nonnegative integer such that the word $h(\xi)$ does not end in r for any variable $\xi \in E$. Since E has n elements, $r \leq n$. We shall prove that $ab^{n+rN_n+i}a$ is never cut when $0 \leq i < N_n$.

Note first that the word $ab^{n+rN_n+i}a$ occurs exactly once in $g(a_i)$ and not in any $g(a_j)$ with $j \neq i$. Indeed, it occurs as a factor of $a^{B_{n-1}+1} b^{n+i} z_1$ if $r = 0$, of $z_r b^i z_{r+1}$ if $1 \leq r \leq n-1$, or as a suffix if $r = n$.

Suppose that for some i, $ab^{n+rN_n+i}a$ is cut. Then there is a noninitial cut $(v_1 u_1, u_2 v_2)$ with $u_1 = ab^j$, $u_2 = b^{n+rN_n+i-j}a$, $v_1 u_1 u_2 v_2 = g(a_i)$, $p_1 p_2 = p$, $t_1 \varphi(p_1) = g(s_1) v_1 u_1$ and $\varphi(p_2) t_2 = u_2 v_2 g(s_2)$. Since this cut is noninitial, there exist variables $\xi_0, \xi_1, \ldots, \xi_m$ such that p_1 ends in $\xi_0 \xi_1 \ldots \xi_m$, $h(\xi_0) \neq \varepsilon$ and $h(\xi_l) = \varepsilon$ for $1 \leq l \leq m$. In particular, for $1 \leq l \leq m$, $\varphi(\xi_l)$ contains at most one occurrence of a.

Assume first that $r = 0$. Then $\varphi(\xi_0 \xi_1 \ldots \xi_m)$ is a suffix of $t_1 \varphi(p_1) = g(s_1) a^{B_{n-1}+1} b^j$. If $m \geq B_{n-1}$, then the factor $\xi_1 \ldots \xi_m$ of p must contain n

3.3. Avoidability on a fixed alphabet

different variables, which contradicts the fact that E' is nonempty (and of course, $h(\xi) \neq \varepsilon$ when $\xi \in E'$). Therefore $m < B_{n-1}$, and $\varphi(\xi_1 \ldots \xi_m)$ can take at most $B_{n-1} - 1$ of the a's in $a^{B_{n-1}+1}$. Consequently $\varphi(\xi_0)$ ends in aa or $aab^{j'}$, and $h(\xi_0)$ ends in 0, a contradiction with the definition of r.

Assume now that $r = n$. Then for all ξ in E, $h(\xi)$ is nonempty and ends in an integer between 0 and $n - 1$. Necessarily $m = 0$ and $t_1\varphi(p_1)$ ends in $ab^{n+(n-1)N_n+i}ab^j$, hence $h(\xi_0)$ ends in $n + (n - 1)N_n + i$, which is not an integer between 0 and $n - 1$.

Finally, assume that $1 \leq r \leq n - 1$. Then $t_1\varphi(p_1) = g(s_1)v_1u_1$ ends in the word $ab^{n+(r-1)N_n}(ab^r)^{B_{n-r}-1}ab^j$. Let F be the set of variables ξ such that $h(\xi)$ either is empty or ends in a value larger than or equal to n. The cardinal of F is at most $n - r$, hence p does not contain more than $B_{n-r-1} - 1$ consecutive variables in F. In particular, $m < B_{n-r-1}$. If $m < B_{n-r-1} - 1$, one finds that $\varphi(\xi_0)$ ends in $ab^rab^{j'}$, hence $h(\xi_0)$ ends in r, a contradiction with the definition of r. If $m = B_{n-r-1} - 1$, then either $\varphi(\xi_0)$ ends in $ab^rab^{j'}$ as above, or it ends in $ab^{n+(r-1)N_n}ab^{j'}$ and $\xi_0 \in F$, so that we have B_{n-r-1} consecutive variables in F, which is again excluded. ∎

LEMMA 3.3.7. *The infinite word $g(w_n)$ avoids all patterns $p \in P$.*

Proof. Assume that p is a pattern in P that is not avoided by $g(w_n)$. Let r be as given by Lemma 3.3.6, and ψ be the morphism from E^* to A^* defined by $\psi = \pi \circ h$, where $\pi(n+rN_n+i) = a_i$ and $\pi(j) = \varepsilon$ otherwise, and h is naturally extended to a morphism. Then $\psi(p)$ is a factor of y (recall that $\varphi(p)$ is a factor of $g(y)$). Indeed, let first p' be the suffix of p starting with the first occurrence of a variable ξ such that $h(\xi) \neq \varepsilon$. Obviously $\psi(p') = \psi(p)$. Any occurrence of a letter a_i in y such that $g(a_i)$ falls inside $\varphi(p')$ is marked by the word $ab^{n+rN_n+i}a$, which is not cut (not even by an initial cut, thanks to the definition of p'). This word corresponds therefore to exactly one occurrence of a_i in $\psi(p')$. Conversely, each letter in $\psi(p')$ corresponds to one occurrence of a marker $ab^{n+rN_n+i}a$ in $\varphi(p')$, hence to a letter a_i in y.

Let $q = \delta_V(p)$, where V is the set of variables that are mapped to ε by ψ. The pattern q is not empty. Indeed, cut p into three approximately equal parts, $p = p_1p_2p_3$. Since $|p| \geq M \geq 4B_{n-1}$, one can take $|p_j| \geq B_{n-1}$ for each $j = 1, 2, 3$, i.e. p_j must contain at least one occurrence of each letter in E. In particular, there is a letter in E' in p_1, which implies that $|p'| > |p_2p_3|$. There is also a letter in E' in p_3. Since $|\varphi(p)| \geq |p| > B_{n-1}+2$, there is at least one variable ξ such that $\varphi(\xi)$ contains at least one b, and this variable occurs at least once in p_2. Consequently, p_1 and p_3

overlap two different blocks $a^{B_{n-1}+2}$, and between them there is some $g(a_i)$, which occurs completely within $\varphi(p')$, except maybe for some power of a at the beginning, and which contains one marker $ab^{n+rN_n+i}a$, so that $\psi(p') = \psi(q)$ is not empty.

Since $|p| \geq M \geq 4B_{n-1}$, p contains at least four occurrences of each variable in E, so that q contains at least four occurrences of each variable that occurs in it, and is therefore avoidable according to Corollary 3.2.10. The infinite word $g(w_n)$ contains $\psi(q)$ as a factor, in contradiction with the fact that $g(w_n)$ avoids all avoidable patterns on at most n variables. ∎

Proof of Theorem 3.3.5. Assume that we have proved that for $i < n$, every pattern of length S_i with i variables is 2-avoidable. Let $B_1 = S_1$, $B_2 = S_2$, ..., $B_{n-1} = S_{n-1}$ and apply Lemma 3.3.7. Then a pattern p of length M with n variables either is in P, in which case it is avoided by $g(w_n)$ according to Lemma 3.3.7, or for some $i < n$ contains a factor q of length B_i with at most i variables, which is itself 2-avoidable by the induction hypothesis.

We have just proved by induction that every pattern of length S_n with n variables is 2-avoidable, where (S_n) is the numeric sequence starting with $S_1 = 3$ and satisfying the recurrence relation

$$S_n = T_n + 2S_{n-1} + \sum_{i=2}^{n} iS_{n+1-i}.$$

Let $\lambda > \left(1 - 2^{-\frac{1}{3}}\right)^{-1}$, so that $K = \left(1 - \frac{1}{\lambda}\right)^{-2} + \frac{2}{\lambda} - 1 < 1$, and

$$C = \max\left\{\frac{3}{\lambda}, \frac{1}{1-K} \max\left\{\frac{T_n}{\lambda^n} \,\Big|\, n \geq 1\right\}\right\}.$$

Then $S_n \leq C\lambda^n$ for all $n \geq 1$, since $3 = S_1 \leq C\lambda$ and, if $S_i \leq C\lambda^i$ for $1 \leq i < n$, then

$$S_n \leq C\lambda^n \left(\frac{T_n}{C\lambda^n} + \frac{2}{\lambda} + \sum_{i=2}^{n} i\lambda^{1-i}\right)$$

with $\frac{2}{\lambda} + \sum_{i=2}^{n} i\lambda^{1-i} < K$ and $\frac{T_n}{C\lambda^n} \leq 1 - K$.

Since $\left(1 - 2^{-\frac{1}{3}}\right)^{-1} \approx 4.847$, we can take $\lambda = 5$. Then, with $N_n = 4\left\lceil\frac{n+1}{2}\right\rceil$, we find that $C < 200$. Note that the actual value of N_n influences only the constant C, as long as $N_n = o(n^{-2}\lambda^n)$. ∎

Problems

Section 3.1

3.1.1 A set of patterns $P \subset E^*$ is said to be avoidable on A^* if there exists an infinite word in A^ω that avoids all elements of P. P is said to be k-avoidable if it is avoidable on a k-letter alphabet, and avoidable if it is k-avoidable for some k. Show that if P is finite, then P is avoidable if and only if all its elements are avoidable. Does this equivalence still hold if P is infinite? What if avoidability is replaced with k-avoidability?

3.1.2 A *simple formula* is a finite set of patterns, and is denoted by

$$f = p_1 \cdot p_2 \cdot \cdots \cdot p_m ,$$

the order being unimportant. A word u is said to encounter the simple formula f if there exists a nonerasing morphism $h: E^* \to A^*$ such that all the words $h(p_1), h(p_2), \ldots, h(p_m)$ are factors of u. Avoidability and k-avoidability of simple formulas are then defined as for patterns. (Note that this is a different notion from the avoidability of sets in Problem 3.1.1.) Show that f is avoidable on A if and only if the pattern $p_1 \zeta_1 p_2 \zeta_2 \cdots p_{m-1} \zeta_{m-1} p_m$ is avoidable on A, where $\zeta_1, \ldots, \zeta_{m-1}$ are distinct variables that do not occur in f. How can this fact be used for practical checking of unavoidability? (Compare the number of binary words avoiding $\alpha\alpha\beta\gamma\beta\alpha\beta\alpha$ with the number of binary words avoiding the equivalent simple formula $\alpha\alpha\beta \cdot \beta\alpha\beta\alpha$.)

*3.1.3 Definitions of Problems 3.1.1 and 3.1.2 can be combined to construct the *algebra of formulas* $\mathscr{F}(\mathscr{E})$: a *formula* is a set of simple formulas, denoted by $f_1 + f_2 + \cdots + f_k$. $(\mathscr{F}(\mathscr{E}), +, \cdot)$ is a commutative algebra, the neutral elements being 0 (the formula containing no simple formula) and 1 (the simple formula containing no pattern). A word u is said to encounter the formula f if it encounters one of its elements f_i. All words encounter 1, and no word encounters 0. Design an equivalence relation $\sim$ on $\mathscr{F}(\mathscr{E})$ such that $f \sim f'$ implies that $\mu(f) = \mu(f')$, and that allows rewriting sequences such as

$$\alpha\alpha\beta\alpha\gamma\alpha\beta\delta\beta\alpha\alpha + \alpha\beta\beta\alpha\beta \sim \alpha\alpha\beta\alpha \cdot \alpha\beta \cdot \beta\alpha\alpha + \alpha\beta\beta\alpha\beta$$
$$\sim \alpha\alpha\beta\alpha \cdot \beta\alpha\alpha + \alpha\beta\beta\alpha\beta$$
$$\sim \alpha\alpha\beta\alpha \cdot \beta\alpha\alpha + \beta\alpha\alpha\beta\alpha$$
$$\sim \alpha\alpha\beta\alpha \cdot \beta\alpha\alpha$$

$$\sim \alpha\alpha\beta\alpha \cdot \beta\alpha\alpha + \alpha\alpha\beta\alpha\alpha$$
$$\sim \alpha\alpha\beta\alpha \cdot \beta\alpha\alpha + \alpha\alpha \cdot \alpha\alpha$$
$$\sim \alpha\alpha\beta\alpha \cdot \beta\alpha\alpha + \alpha\alpha$$
$$\sim \alpha\alpha \ .$$

Two patterns equivalent in the sense of Subsection 3.1.1 should also be equivalent for this relation.

3.1.4 A pattern p is said to be *D0L-avoidable* on A if there exist a morphism $f: A^* \to A^*$ and a letter $a \in A$ such that the morphic infinite word $f^\omega(a)$ is properly defined (see Proposition 1.2.8) and avoids p. Let here $A = \{a, b\}$.

1. Show that the patterns $\alpha\beta\beta\gamma\alpha\beta\beta\gamma$ and $\alpha\beta\alpha\gamma\alpha\beta\alpha\gamma$ are avoidable on A but D0L-unavoidable on A. (*Hint*: A binary morphic word contains arbitrarily long squares.)

2. Show that the pattern $\alpha\beta\alpha\alpha\beta\alpha$ is D0L-unavoidable on A. (*Hint*: A binary morphic word either is cube-free or contains arbitrarily long cubes.)

3. More generally, show that for all $n \geq 2$, the pattern $Z_n Z_n$, i.e. the square of the sesquipower of order n, is avoidable on A but is D0L-unavoidable on A.

**3.1.5 Prove or disprove that, if p is avoidable on A, there exist an alphabet B (possibly larger than A), two morphisms $f: B^* \to B^*$ and $g: B^* \to A^*$, and a letter $a \in B$ such that the infinite word $g(f^\omega(a))$ is well defined and avoids p.

Section 3.2

3.2.1 Show that a pattern is unavoidable if and only if it divides some sesquipower Z_n. (*Hint*: If p reduces to q in one step, and q divides Z_n, then p divides Z_{n+1}.)

*3.2.2 A pattern p such that $AG(p)$ is a connected graph is called a *locked pattern*. Show that all locked patterns are 4-avoidable. (*Hint*: Modify Lemma 3.2.7 to prove that $w^{(1)}$ avoids locked patterns, using the fact that a variable ξ of p such that $|h(\xi)|$ is odd would constitute a free set, and that consequently only two different letters can occur at the beginning of all $h(\xi)$, so that decisive words can be found.)

3.2.3 Given a pattern $p = \xi_1 \xi_2 \cdots \xi_m$ and $k \leq m$, the *k-chop* of p is the pattern

$$(\xi_1 \xi_2 \cdots \xi_k)\zeta_1(\xi_2 \xi_3 \cdots \xi_{k+1})\zeta_2 \cdots \zeta_{m-k}(\xi_{m-k+1}\xi_{m-k+2}\cdots \xi_m)$$

where $\zeta_1, \ldots, \zeta_{m-k}$ are new variables, or equivalently the simple formula (see Problem 3.1.2)

$$\xi_1\xi_2 \cdots \xi_k \cdot \xi_2\xi_3 \cdots \xi_{k+1} \cdots \cdots \xi_{m-k+1}\xi_{m-k+2} \cdots \xi_m$$

formed of all factors of length k of p. Show that the 2-chop of a pattern is either 4-avoidable or unavoidable. (*Hint*: If the 2-chop of p is irreducible, then the reduction algorithm stops on a locked 2-chop (see Problem 3.2.2), which is 4-avoidable and divides the 2-chop of p.) What prevents this proof from extending to k-chops with arbitrary k?

Section 3.3

3.3.1 Construct a family of 2-unavoidable patterns R_n on n variables, similar to sesquipowers, such that $|R_n| = 3 \cdot 2^{n-1} - 1$. Conclude that $\forall n \geq 1, \ell_{n2} \geq 3 \cdot 2^{n-1}$.

**3.3.2 Show that $\forall n \geq 1, \ell_{n2} = 3 \cdot 2^{n-1}$ and $\forall n \geq 1, \forall k \geq 3, \ell_{nk} = 2^n$.

3.3.3 Given a positive integer l, we say that a word u contains an *l-occurrence* of p if it contains a factor $h(p)$ where $|h(\xi)| \geq l$ for all $\xi \in E$. The pattern p is said to be *weakly k-avoidable* if there exist an integer l and an infinite word on k letters without l-occurrences of p. Show that every avoidable pattern is weakly 2-avoidable.

3.3.4 Show that the pattern $\alpha\beta\zeta_1\beta\gamma\zeta_2\gamma\alpha\zeta_3\beta\alpha\zeta_4\alpha\gamma$ has avoidability index 4. (*Hint*: This pattern is locked (see Problems 3.2.2 and 3.2.3) and the equivalent simple formula $\alpha\beta \cdot \beta\gamma \cdot \gamma\alpha \cdot \beta\alpha \cdot \alpha\gamma$ (see Problem 3.1.2) is 3-unavoidable.)

**3.3.5 Prove or disprove that all avoidable patterns have avoidability index at most 4.

Notes

Finding infinite words that avoid repetitions (mainly squares or cubes) is an old problem that can be traced back to Thue 1906, 1912, and was rediscovered or studied by many authors, including Adian 1979, Aršon 1937, Berstel 1979a, 1984, Dean 1965, Dekking 1976, Entringer, Jackson, and Schatz 1974, Evdokimov 1968, Hawkins and Mientka 1956, Istrail 1977, Leech 1957, Morse and Hedlund 1944, Pleasants 1970, Salomaa 1981, Shyr 1977, Zech 1958.

The present notion of pattern was introduced independently by Bean, Ehrenfeucht, and McNulty 1979 and Zimin 1979, 1982. We adopt here the vocabulary of Bean et al. Zimin calls *blocking term* what we call *unavoidable pattern*, and σ-*deletion of variables* what we call *deletion of a free set*. Theorem 3.2.1 and most lemmas in its proof are taken from Zimin (Lemmas 3.2.2, 3.2.3, 3.2.4, and 3.2.5 are Zimin's Lemmas 4, 6, 7 and 8). Zimin also introduced sesquipowers as a family of unavoidable patterns.

Baker, McNulty, and Taylor 1989 introduced the adjacency graph and free sets, which allow a nice presentation of pattern reduction. They defined locked patterns (see Problem 3.2.2) and gave the first example of an avoidable pattern which is not avoidable on a ternary alphabet (see Problem 3.3.4). They also gave a first linear bound on the avoidability index of patterns with a given number of variables (nonlinear bounds can be derived from the proofs of Zimin and Bean et al., but they did not make them explicit). The bound we give here in Theorem 3.3.4 was found by Mel'ničuk (an unpublished paper communicated by P. Goralčik). We adapted Zimin's proof to make use of Mel'ničuk's construction, slightly modified, with the help of notes from lectures of Goralčik given at LITP, Paris, in 1992 and Volkov in a talk given at Marquette University, Milwaukee, in 1991.

The classification of binary patterns presented in Subsection 3.3.2 was started by Schmidt 1986, 1989, who proved that binary patterns of length 13 are 2-avoidable. The bound was reduced to the optimal value 6 by Roth 1992 and the classification completed by Cassaigne 1993b, 1994b. Vaniček 1989 independently established the classification (see also Goralčik and Vaniček 1991). A similar classification for ternary patterns was started by Nilgens 1991 and continued (with semi-automatized proofs of avoidability) by Cassaigne 1994b, but is not yet complete.

The bound in Theorem 3.3.5 was found by Cassaigne and Roth (see Cassaigne 1994b).

Formulas (see Problems 3.1.2 and 3.1.3) were defined by Cassaigne 1993b. Their study has not been carried very far. For instance, there is no classification of binary formulas. Avoidability by D0L and HD0L words (see Problems 3.1.4 and 3.1.5) was also studied by Cassaigne 1993b, 1994a. The notion of l-occurrences of patterns (see Problem 3.3.3) was studied by Roth 1991.

A list of open problems on patterns was published by Currie 1993, who offers prizes for the resolution of some of them. None of them seems to have been solved yet. Problems 3.1.5 and 3.3.5 are in that list.

One question we did not include here is the study of the set of words (finite or infinite) avoiding a given pattern: its growth, its topological structure, etc. For references, see e.g. Cassaigne 1993a and Currie 1993.

CHAPTER 4

Sesquipowers

4.0. Introduction

In this chapter we shall be concerned with *sesquipowers*. Any nonempty word is a sesquipower of *order* 1. A word w is a sesquipower of order n if $w = uvu$, where u is a sesquipower of order $n - 1$. Sesquipowers have many interesting combinatorial properties which have applications in various domains. They can be defined by using *bi-ideal sequences*.

A finite or infinite sequence of words $f_1, \ldots, f_n, \ldots$ is called a bi-ideal sequence if for all $i > 0$, f_i is both a prefix and a suffix of f_{i+1} and, moreover, $2|f_i| \leq |f_{i+1}|$. A sesquipower of order n is then the nth term of a bi-ideal sequence. Bi-ideal sequences have been considered, with different names, by several authors in algebra and combinatorics (see Notes).

In Sections 4.2 and 4.3 we analyze some interesting combinatorial properties of bi-ideal sequences and the links existing between bi-ideal sequences, recurrence and n-divisions. From these results we will obtain in Section 4.4 an improvement (Theorem 4.4.5) of an important combinatorial theorem of Shirshov. We recall (see Lothaire 1983) that Shirshov's theorem states that for all positive integers p and n any sufficiently large word over a finite totally ordered alphabet will have a factor f which is a pth power or is n-*divided*, i.e., f can be factorized into nonempty blocks as $f = x_1 \cdots x_n$ with the property that all the words that one obtains by a nontrivial rearrangement of the blocks are lexicographically less than f.

In Theorem 4.4.5, we link bi-ideal sequences and the Shirshov property. Indeed, in this case the n-divided factor f is the nth term of a bi-ideal sequence whose canonical factorization $(x_1, \ldots, x_n)$ is an n-division of f. Moreover, $x_1, \ldots, x_n$ are Lyndon words such that $x_1 > x_2 > \cdots > x_n$.

In Section 4.5 some applications of bi-ideal sequences to finiteness conditions for finitely generated semigroups will be given. These conditions are based on different concepts such as permutation properties, iteration conditions, and minimal conditions on principal bi-ideals.

4.1. Bi-ideal sequences

A sequence $f_1, \ldots, f_n, \ldots$ of words of A^* is called a *bi-ideal sequence* if $f_1 \in A^+$ and for all $i > 0$

$$f_{i+1} \in f_i A^* f_i.$$

If the sequence is of finite length n, then $(f_1, \ldots, f_n)$ is called a bi-ideal sequence of *order* n. If $f_1, \ldots, f_n, \ldots$ is a bi-ideal sequence, then there exists a unique sequence of words $g_1, g_2, \ldots, g_n, \ldots$ such that for all $i > 0$

$$f_{i+1} = f_i g_i f_i.$$

Thus a bi-ideal sequence is any sequence of words $f_1, \ldots, f_n, \ldots$ satisfying the following requirements: for all $i > 0$
  (i) f_i is both a prefix and a suffix of f_{i+1},
  (ii) $2|f_i| \leq |f_{i+1}|$.

If $(f_1, \ldots, f_n)$ is a bi-ideal sequence of order n, then the last term $f = f_n$ will be called a *sesquipower of order* n. Obviously, any sesquipower of order n is also a sesquipower of order k for all $k = 1, \ldots, n-1$. Thus with any word $f \in A^+$ one can associate a positive integer, called the *degree* of f, defined as the maximal order of any bi-ideal sequence having f as last term. However, in general, for a given word $f \in A^+$ there can exist different bi-ideal sequences of order equal to the degree of f and having f as last term. For instance, the word $f = abababababa$ of degree 3 is the last term of the two bi-ideal sequences (a, aba, f) and $(a, ababa, f)$.

EXAMPLE 4.1.1. Recall that the *Fibonacci word* (see Chapter 1 and Chapter 8)

$$f = abaababaabaababaabab a abaababaabaab \cdots$$

is the limit of the sequence of words $(f_n)_{n \geq 0}$, inductively defined by $f_0 = b$, $f_1 = a$, $f_{n+1} = f_n f_{n-1}$, for all $n > 0$. The sequences $(f_{2k})_{k \geq 1}$ and $(f_{2k+1})_{k \geq 0}$ of the terms of even and odd index, respectively, are infinite bi-ideal sequences (converging to f). Indeed, one has

$$f_2 = ab \text{ and } f_{2k} = f_{2k-2} f_{2k-3} f_{2k-2}, \; k > 1,$$

and

$$f_1 = a \text{ and } f_{2k+1} = f_{2k-1} f_{2k-2} f_{2k-1}, \; k \geq 1.$$

Bi-ideal sequences are closely related to Zimin's words, as defined in Chapter 3. Let $X = \{x_1,\ldots,x_n,\ldots\}$ be a possibly infinite alphabet, called the *pattern alphabet*, and Z_n, $n > 0$, be the sequence of words inductively defined as

$$Z_1 = x_1, \quad Z_{n+1} = Z_n x_{n+1} Z_n \quad \text{for} \quad n > 0.$$

Thus one has

$$Z_2 = x_1 x_2 x_1, \quad Z_3 = x_1 x_2 x_1 x_3 x_1 x_2 x_1, \quad \ldots.$$

Let $\phi: X^* \to A^*$ be any nonerasing morphism from X^* to A^*. One easily verifies that the sequence $(\phi(Z_n))_{n>0}$ is a bi-ideal sequence. Conversely, if $f_1,\ldots,f_n,\ldots$ is a bi-ideal sequence with $f_{n+1} = f_n g_n f_n$, and $\phi: X^* \to A^*$ is the morphism defined as

$$\phi(x_1) = f_1, \quad \phi(x_{n+1}) = g_n, \quad \text{for } n > 0,$$

then

$$f_n = \phi(Z_n)$$

for all $n > 0$. The following theorem, which shows that for all $n > 0$ the pattern Z_n is unavoidable, is Proposition 3.1.4.

THEOREM 4.1.2. *Let A be a k-letter alphabet. For any $n > 0$, there exists a positive integer $M(k,n)$, such that any word of A^* of length at least $M(k,n)$ contains as a factor a sesquipower of order n.* ■

4.2. Canonical factorizations

In this section we investigate some interesting combinatorial properties of bi-ideal sequences which will be useful later in order to prove some extensions of a theorem of Shirshov.

Let n be a positive integer and $(w_1,\ldots,w_n)$ be a sequence of n words of A^+. The sequence $(w_1,\ldots,w_n)$ is called an *n-sequence* if for any $i = 1,\ldots,n-1$

$$w_i \in w_{i+1} \cdots w_n A^*.$$

The sequence $(w_1,\ldots,w_n)$ is called an *inverse n-sequence* if

$$w_{i+1} \in A^* w_1 \cdots w_i,$$

for any $i = 1,\ldots,n-1$.

We analyze now an important relationship between bi-ideal sequences of order n and n-sequences (inverse n-sequences). From this one derives two canonical factorizations of the last term of any bi-ideal sequence of order n.

Let $(f_i)_{i=1,\ldots,n}$ be a bi-ideal sequence of order n with $f_1 \in A^+$ and set $f_{i+1} = f_i g_i f_i$, with $g_i \in A^*$ for $i = 1,\ldots,n-1$. Let $w_n = f_1$ and

$$w_{n-i} = f_i g_i, \quad 1 \leq i \leq n-1.$$

One has $f_{i+1} = f_i g_i f_i = w_{n-i} f_i$ for $1 \leq i \leq n-1$, so that, by iteration, one has

$$f_{i+1} = w_{n-i} \cdots w_n, \quad 0 \leq i \leq n-1. \tag{4.2.1}$$

Moreover, since $w_i = f_{n-i} g_{n-i}$ for $1 \leq i \leq n-1$, Equation (4.2.1) implies

$$w_i = w_{i+1} \cdots w_n g_{n-i} \in w_{i+1} \cdots w_n A^*. \tag{4.2.2}$$

It follows from Equation (4.2.1) that $f_n = w_1 w_2 \cdots w_n$. The n-tuple $(w_1, w_2, \ldots, w_n)$ is called the *canonical factorization* of f_n.

One can also introduce the inverse canonical factorization of f_n by setting $w'_1 = f_1$ and

$$w'_{i+1} = g_i f_i, \quad 1 \leq i \leq n-1.$$

One easily derives that for all $i = 1, \ldots, n$

$$w'_1 \cdots w'_i = f_i, \tag{4.2.3}$$

where

$$w'_{i+1} = g_i w'_1 \cdots w'_i \in A^* w'_1 \cdots w'_i, \quad 1 \leq i \leq n-1. \tag{4.2.4}$$

It follows from Equation (4.2.3) that $f_n = w'_1 \cdots w'_n$. The n-tuple $(w'_1, \ldots, w'_n)$ is called the *inverse canonical factorization* of f_n.

By Equations (4.2.2) and (4.2.4), the canonical (inverse canonical) factorization of the last term of a bi-ideal sequence of order n is an n-sequence (inverse n-sequence).

Conversely, one easily verifies that if $(w_1, w_2, \ldots, w_n)$ is an n-sequence (inverse n-sequence), then the sequence of words $f_i = w_{n-i+1} \cdots w_n$ ($f_i = w_1 \cdots w_i$), $1 \leq i \leq n$, is a bi-ideal sequence of order n whose last term has a canonical (inverse canonical) factorization given by $(w_1, w_2, \ldots, w_n)$.

EXAMPLE 4.2.1. Consider the bi-ideal sequence of order 3, $f_1 = a$, $f_2 = aba$ and $f_3 = ababaaba$. In this case $g_1 = b$ and $g_2 = ba$. The canonical factorization of f_3 is the 3-sequence $(ababa, ab, a)$. The inverse canonical factorization is the inverse 3-sequence $(a, ba, baaba)$.

4.2. Canonical factorizations

Let us remark that although a sesquipower f of order n may have several canonical factorizations, these are uniquely determined by the bi-ideal sequences of order n having f as last term. As an example, the word $f = ababababab$ is the last term of the two bi-ideal sequences of order 3

$$(a, aba, f) \quad \text{and} \quad (a, ababa, f).$$

Thus f has the two canonical factorizations

$$(abababab, ab, a) \quad \text{and} \quad (ababab, abab, a)$$

which uniquely correspond to the preceding bi-ideal sequences.

The following proposition, called the *reciprocity law*, summarizes the links existing between the two canonical factorizations of the last term of a bi-ideal sequence of order n, expressed by Equations (4.2.1) and (4.2.3).

PROPOSITION 4.2.2. *Let $(w_1, \ldots, w_n)$ and $(w'_1, \ldots, w'_n)$ be the canonical factorizations of the nth term f_n of a bi-ideal sequence. For any $i = 0, \ldots, n-1$ one has*

$$w'_1 \cdots w'_{i+1} = f_{i+1} = w_{n-i} \cdots w_n.\quad\blacksquare$$

We denote by $\mathscr{S}_n$ the symmetric group on $\{1, \ldots, n\}$. Let $<$ be a total order in A^*. A sequence $(u_1, u_2, \ldots, u_n)$ of n words of A^+ is called an *n-division* (with respect to the order $<$) if for any nontrivial permutation σ of $\mathscr{S}_n$ one has

$$u_{\sigma(1)} u_{\sigma(2)} \cdots u_{\sigma(n)} < u_1 u_2 \cdots u_n.$$

A word u is called *n-divided* (with respect to the order $<$) if there exists an n-division $(u_1, u_2, \ldots, u_n)$ such that $u = u_1 u_2 \cdots u_n$.

An n-division $(u_1, u_2, \ldots, u_n)$ with respect to the relation $>$, which is the inverse of $<$, is also called an *inverse n-division* with respect to the order $<$. A word which is n-divided with respect to $>$ is also called *inversely n-divided* with respect to $<$.

In the following we shall consider n-divisions (inverse n-divisions) and n-divided (inversely n-divided) words only with respect to the lexicographic order, so that the above terms will be used without specifying the total order. We shall denote, when there are no ambiguities, the lexicographic order on A^* simply by $<_A$ or $<$.

EXAMPLE 4.2.3. Let the alphabet $A = \{a, b\}$ be ordered by $a < b$. The word $w = ababbaba$ is 3-divided by the sequence $(ababb, ab, a)$ and inversely 3-divided by $(a, ba, bbaba)$. Moreover, one can easily verify that $(ababb, ab, a)$ is the only 3-division of w and that w does not admit 4-divisions.

We recall the two following basic properties of the lexicographic order (Lothaire 1983).

PROPOSITION 4.2.4. *For all u, v, w, w' in A^**
  (i) $u < v \iff wu < wv$,
  (ii) *if $v \notin uA^*$, then $u < v \Rightarrow uw < vw'$.* ∎

The following proposition characterizes n-divisions.

PROPOSITION 4.2.5. *An n-sequence $(w_1, \ldots, w_n)$ is an n-division (inverse n-division) if and only if for all $i = 1, \ldots, n-1$, one has $w_{i+1}w_i < w_iw_{i+1}$ ($w_{i+1}w_i > w_iw_{i+1}$).*

Proof. ($\Rightarrow$) Suppose that the n-sequence $(w_1, \ldots, w_n)$ is an n-division and let us prove that $w_{i+1}w_i < w_iw_{i+1}$ for $1 \leq i \leq n-1$. Assume, by contradiction, that $w_{i+1}w_i \geq w_iw_{i+1}$ for some integer i, $1 \leq i \leq n-1$. This implies, by Proposition 4.2.4(1), that $w_1 \cdots w_{i-1}w_iw_{i+1} \leq w_1 \cdots w_{i-1}w_{i+1}w_i$. If $w_iw_{i+1} = w_{i+1}w_i$, then

$$w_1 \cdots w_{i-1}w_iw_{i+1}w_{i+2} \cdots w_n = w_1 \cdots w_{i-1}w_{i+1}w_iw_{i+2} \cdots w_n,$$

which is a contradiction. Let us then suppose $w_iw_{i+1} < w_{i+1}w_i$. Since $|w_iw_{i+1}| = |w_{i+1}w_i|$, it follows by the property 4.2.4(ii) of the lexicographic order that

$$w_1 \cdots w_{i-1}w_iw_{i+1}w_{i+2} \cdots w_n < w_1 \cdots w_{i-1}w_{i+1}w_iw_{i+2} \cdots w_n$$

which is again a contradiction.

($\Leftarrow$) We begin by proving that $w_jw_i < w_iw_j$ for any i, j with $1 \leq i < j \leq n$. If $i = j-1$ the result follows from the hypotheses. Then let us suppose that $i < j-1$. We can write

$$w_i \in w_{j-1}w_j \cdots w_n A^*$$

and

$$w_iw_j = w_{j-1}w_j\mu, \text{ for some } \mu \in A^*.$$

By assumption, one has

$$w_jw_{j-1} < w_{j-1}w_j.$$

Since $i \leq j-2$, we can write

$$w_i = w_{j-1}\lambda, \text{ for some } \lambda \in A^*.$$

4.3. Sesquipowers and recurrence

Since $|w_j w_{j-1}| = |w_{j-1} w_j|$ and $w_j w_{j-1} < w_{j-1} w_j$, in view of the property 4.2.4(ii) of the lexicographic ordering, one has

$$w_j w_i = w_j w_{j-1} \lambda < w_{j-1} w_j \mu = w_i w_j.$$

Any permutation can be obtained by a sequence of exchanges of adjacent elements. Thus, the above property shows that for any nontrivial permutation $\sigma \in \mathscr{S}_n$ one has $w_{\sigma(1)} w_{\sigma(2)} \cdots w_{\sigma(n)} < w_1 w_2 \cdots w_n$. The proof in the case of the inverse n-division is perfectly symmetric. ∎

By an argument similar to that of the preceding proposition, one can prove the following.

PROPOSITION 4.2.6. *An inverse n-sequence $(w_1, \ldots, w_n)$ is an n-division (inverse n-division) if and only if for all i, $1 \leq i \leq n-1$, $w_{i+1} w_i < w_i w_{i+1}$ ($w_{i+1} w_i > w_i w_{i+1}$).*

4.3. Sesquipowers and recurrence

Let $f_1, \ldots, f_n, \ldots$ be an infinite bi-ideal sequence, where $f_{i+1} = f_i g_i f_i$ for all $i > 0$. Since for all $i > 0$, f_i is a prefix of the next term f_{i+1} the sequence (f_n) converges to the infinite word

$$x = f_1 (g_1 f_1)(g_2 f_2) \cdots (g_n f_n) \cdots.$$

Let us observe that one can rewrite x as

$$x = w_1 w_2 \cdots w_n \cdots,$$

where $w_1 = f_1$, $w_{i+1} = g_i f_i$, $i > 0$. For all $n > 0$, $(w_1, \ldots, w_n)$ is the inverse canonical factorization of f_n.

A word $x \in A^\omega$ is a *sesquipower* if it is the limit of a bi-ideal sequence.

PROPOSITION 4.3.1. *A word $x \in A^\omega$ is recurrent if and only if it is a sesquipower.*

Proof. Let $x \in A^\omega$ be recurrent. We construct, inductively, a bi-ideal sequence $(f_n)_{n>0}$ such that f_n is a prefix of x, for any $n > 0$. We set $f_1 = x_1$. Suppose, by induction, that we have constructed the bi-ideal sequence up to the ith element f_i, with $i > 0$. Since f_i is a prefix of x and x is recurrent, f_i will occur in x infinitely many times, so that there will exist a word $g \in A^*$ such that $f_{i+1} = f_i g f_i$ is still a prefix of x. Thus there exists a bi-ideal sequence $(f_n)_{n>0}$ whose elements are prefixes of x. This implies that $x = \lim_{n \to \infty} f_n$, i.e., x is a sesquipower.

Conversely, let $x \in A^\omega$ be a sesquipower, i.e., $x = \lim_{n \to \infty} f_n$, where (f_n) is a bi-ideal sequence. Let $w \in F(x)$. There exists $\lambda \in A^*$ such that λw is a prefix of x. Thus $w \in F(f_k)$ for a suitable positive integer k. Now for any $p > 0$, f_k will occur at least 2^p times in f_{k+p}. This shows that the number of occurrences of w in x has no upper bound. ∎

Recall from Chapter 1, that a word $w \in A^\omega$ is eventually periodic if there exist words $u \in A^*$ and $v \in A^+$ such that $w = uv^\omega$. The following proposition is straightforward and is left as an exercise (see Problem 4.3.1).

PROPOSITION 4.3.2. *Let $w \in A^\omega$ be an eventually periodic word. If w is recurrent, then w is periodic.*

Uniformly recurrent words are sesquipowers which have a special interest since any factor of them occurs an infinite number of times but with bounded gaps (see Subsection 1.5.2). Moreover, as shown in Proposition 1.5.12, for any infinite set $L \subseteq A^*$ there exists a uniformly recurrent word $x \in A^\varsigma$ such that $F(x) \subseteq F(L)$.

For a uniformly recurrent word w, we denote by r_w the *recurrence index* of w. We recall that for any $n > 0$ each factor u of w of length n will occur in any factor of w of length $r_w(n)$.

When a word is uniformly recurrent one can 'localize' in any sufficiently large factor of it (whose length depends on the recurrence index) the occurrence of a sesquipower of any order.

PROPOSITION 4.3.3. *Let $t \in A^\omega$ be a uniformly recurrent word. For any $n > 0$ there exists a positive integer $D(n)$ such that for any $w \in A^*, a \in A$, with $wa \in F(t)$ and $|w| \geq D(n)$ one has that*

$$w = \lambda f_n,$$

where $\lambda \in A^*$, and f_n is the nth term of a bi-ideal sequence $f_{i+1} = f_i g_i f_i$, with $g_i \in aA^*$, $i = 1, \ldots, n-1$, $f_1 \in aA^*$ and $|f_i| \leq D(i)$ for $i = 1, \ldots, n$.

Proof. The proof is by induction on n. For $n = 1$ we set $D(1) = r_t(1)$, where r_t is the recurrence index of the word t. Let $w \in A^*$, $|w| \geq D(1)$ and $wa \in F(t)$. Then in w the letter a has to occur, so that we can factorize w as $w = xay$ with $x, y \in A^*$ and $|ay| \leq D(1)$. The statement follows if we set $f_1 = ay$. Now let $n > 1$. By induction we may suppose that there exists an integer $D(n-1)$ satisfying the statement for $n-1$. Then we set

$$D(n) = r_t(D(n-1) + 1) + D(n-1).$$

4.3. Sesquipowers and recurrence

Let $w \in A^*$, $a \in A$ be such that $|w| \geq D(n)$ and $wa \in F(t)$. We can write $w = xv$, with $|x| \geq r_t(D(n-1)+1)$ and $|v| = D(n-1)$. Since $va \in F(t)$, by the induction hypothesis one has

$$v = \lambda' f_{n-1},$$

with $\lambda' \in A^*$, and f_{n-1} is the $(n-1)$th term of a bi-ideal sequence $f_{i+1} = f_i g_i f_i$, with $f_1 \in aA^*$, $g_i \in aA^*$, $i \in \{1, \ldots, n-2\}$, and $|f_i| \leq D(i)$ for $i \in \{1, \ldots, n-1\}$. By the properties of the recurrence index r_t, one has that x contains va as a factor and then also $f_{n-1}a$. Hence one can write $x = \lambda f_{n-1} a \mu$, with $\lambda, \mu \in A^*$, so that

$$w = \lambda f_{n-1} a \mu \lambda' f_{n-1}.$$

Therefore, if we set $g_{n-1} = a\mu\lambda'$, one has $f_n = f_{n-1} g_{n-1} f_{n-1}$ with $g_{n-1} \in aA^*$. Since $|f_{n-1}a\mu| \leq r_t(D(n-1)+1)$ and $|\lambda' f_{n-1}| = D(n-1)$ it follows that $|f_n| \leq D(n)$. ∎

Let p be a positive integer. A finite or infinite word w is called *p-power-free* if it does not have a factor of the form u^p with $u \neq \varepsilon$. An infinite word w is called *ω-power-free* if for any $u \in F(w)$, $u \neq \epsilon$, there exists an integer p such that $u^p \notin F(w)$.

It is clear that a finite word which is 2-power-free (i.e., square-free) is primitive, whereas the converse is not, generally, true. If a word is p-power-free, then it is also ω-power-free. However, there exist infinite words which are ω-power-free even though for any $p > 1$ they have a factor which is a p-power (see Problem 4.3.2). The following lemma shows that a uniformly recurrent word $w \in A^\omega$ is ω-power-free if and only if it is not periodic.

LEMMA 4.3.4. *A uniformly recurrent word $w \in A^\omega$ is either periodic or ω-power-free.*

Proof. Suppose by contradiction that $w \in A^\omega$ is a uniformly recurrent word which is neither periodic nor ω-power-free. Hence, there exists a word $u \in A^+$, which one can always take to be primitive, such that for all $n > 0$, $u^n \in F(w)$. Let us now consider an occurrence of u in w. This is determined by a word $\lambda \in A^*$ such that λu is a prefix of w. Since, by Proposition 4.3.2, w is not eventually periodic, there exist $n > 0$ and $v \in A^+$ such that $|v| = |u|$, $v \neq u$ and $\lambda u^n v$ is still a prefix of w. Let $m > 0$ be such that $|u^m| > r_w((n+1)|u|)$, where r_w is the recurrence index of w. Hence, u^m has as a factor the word $u^n v$. Since u is primitive and $|u| = |v|$, one easily derives $u = v$ which is a contradiction. ∎

4.4. Extensions of a theorem of Shirshov

In this section we prove (Theorem 4.4.4) that an ω-power-free one-sided infinite word has a factor which is the last term of a bi-ideal sequence of any order n whose canonical factorization is an n-division. We need to recall some properties and prove some lemmas on Lyndon words.

The set of all Lyndon words on the alphabet A will be denoted by $\mathscr{L}_A$, or simply $\mathscr{L}$ when there is no confusion. For $A = \{a,b\}$ and $a < b$, a list of Lyndon words of increasing length is

$$a, b, ab, aab, abb, aaab, aabb, abbb, aaaab, aaabb, aabab, \ldots.$$

The following proposition (Lothaire 1983) gives an equivalent definition of Lyndon words.

PROPOSITION 4.4.1. *A nonempty word w is a Lyndon word if and only if it is strictly less than any of its proper nonempty suffixes.*

Let A be a finite totally ordered alphabet, let $a = \min A$ and denote by $<_A$ the lexicographic order on A^*. The set

$$Y = a^+(A \setminus \{a\})^+$$

is a code on the alphabet A (see Chapter 1). Let X be a finite subset of Y and B be an alphabet such that $\text{Card}(B) = \text{Card}(X)$. If δ is a bijection of B and X, then it can be extended to a injective morphism $\delta: B^* \to Y^*$. We can then totally order B by setting for $x, y \in B$

$$x <_B y \iff \delta(x) <_A \delta(y). \tag{4.4.1}$$

The total order of B can be extended to the lexicographic order $<_B$ of B^*.

LEMMA 4.4.2. *For $u, v \in B^*$, one has $u <_B v \Rightarrow \delta(u) <_A \delta(v)$.*

Proof. Suppose first that u is a left factor of v, i.e., $v = u\xi$ with $\xi \in B^*$. One has then $\delta(v) = \delta(u)\delta(\xi)$, i.e., $\delta(u)$ is a prefix of $\delta(v)$, so that $\delta(u) <_A \delta(v)$. Let us then suppose that

$$u = hx\xi, \quad v = hy\eta,$$

with $h, \xi, \eta \in B^*$, $x, y \in B$ and $x <_B y$. From Formula (4.4.1), $x <_B y \Rightarrow \delta(x) <_A \delta(y)$. We have to consider two cases:

Case 1: $\delta(x) = rbs$, $\delta(y) = rct$ with $r, s, t \in A^*$, $b, c \in A$ and $b < c$. One has then $\delta(u) = \delta(h)rbs\delta(\xi)$, $\delta(v) = \delta(h)rct\delta(\eta)$ and $\delta(u) <_A \delta(v)$.

4.4. Extensions of a theorem of Shirshov

Case 2: $\delta(x)$ is a proper prefix of $\delta(y)$, i.e., $\delta(y) = \delta(x)\zeta$ with $\zeta \in A^+$. Since $\delta(x), \delta(y) \in X$, one has $\delta(x) = a^h f$, $\delta(y) = a^k g$ with $h, k > 0$ and $f, g \in (A \setminus \{a\})^+$. It follows that ζ begins with a letter $b > a$. Hence, we can write $\zeta = b\zeta'$. One has then $\delta(u) = \delta(h)\delta(x)\delta(\xi)$ and $\delta(v) = \delta(h)\delta(x)b\zeta'\delta(\eta)$. Hence, if $\delta(\xi) = \epsilon$ then $\delta(u)$ is a prefix of $\delta(v)$ so that $\delta(u) <_A \delta(v)$. If, on the contrary, $\delta(\xi) \neq \epsilon$ then $\delta(\xi) \in aA^*$ and again $\delta(u) <_A \delta(v)$. ∎

LEMMA 4.4.3. *If* $w \in \mathscr{L}_B$, *then* $\delta(w) \in \mathscr{L}_A$.

Proof. Let $w = x_1 \cdots x_n \in \mathscr{L}_B$ with $x_i \in B$ $(i = 1, \ldots, n)$. From Proposition 4.4.1 one has

$$x_1 \cdots x_n <_B x_i \cdots x_n$$

for all $i \geq 2$. This implies by Lemma 4.4.2

$$\delta(x_1) \cdots \delta(x_n) <_A \delta(x_i) \cdots \delta(x_n). \quad (4.4.2)$$

To prove that $\delta(w) \in \mathscr{L}_B$ we have to show that for any factorization $\delta(w) = uv$ with $u \neq \epsilon$, $v \in A^+$ one has $\delta(w) <_A v$. This is the case by Formula (4.4.2) when $v = \delta(x_i) \cdots \delta(x_n)$ for a suitable $i \geq 2$. Let us then suppose that there exist an integer i and words $\delta' \in A^+$, $\delta'' \in A^*$ such that $0 < i \leq n$ and

$$\delta(x_i) = \delta'\delta'', \quad v = \delta''\delta(x_{i+1}) \cdots \delta(x_n)$$

(in the case $i = n$, $v = \delta''$). By definition $\delta(x_i) = \delta'\delta'' = a^k f$ with $k > 0$ and $f \in (A \setminus \{a\})^+$. If $|\delta'| \geq k$, then δ'' begins with a letter $b > a$, i.e., $\delta'' = b\zeta$. Thus

$$v = b\zeta\delta(x_{i+1}) \cdots \delta(x_n).$$

Since $\delta(w)$ begins with the letter a the result in this case trivially follows. Let us then suppose that $|\delta'| < k$ so that

$$v = a^{k-p}f\delta(x_{i+1}) \cdots \delta(x_n),$$

with $p > 0$. Now $a^k f <_A a^{k-p} f$, so that by Property 4.2.4(ii) one derives

$$\delta(x_i) \cdots \delta(x_n) = a^k f \delta(x_{i+1}) \cdots \delta(x_n) <_A a^{k-p} f \delta(x_{i+1}) \cdots \delta(x_n) = v.$$

Therefore,

$$\delta(x_1) \cdots \delta(x_n) <_A \delta(x_i) \cdots \delta(x_n) <_A v,$$

which concludes the proof. ∎

THEOREM 4.4.4. *Let x be an ω-power-free one-sided infinite word over a finite and totally ordered alphabet A. For any $n > 1$, x has a factor s which is the nth term of a bi-ideal sequence whose canonical factorization $(w_1, \ldots, w_n)$ is an n-division of s such that the words w_i are Lyndon words with*

$$w_1 > w_2 > \cdots > w_n.$$

Proof. We shall prove the theorem in the case of uniformly recurrent ω-power-free bi-infinite words. Indeed, by Proposition 1.5.12, one derives that for any one-sided infinite ω-power-free word x there exists a uniformly recurrent ω-power-free two-sided infinite word x' such that $F(x') \subseteq F(x)$. Thus if the assertion of the theorem holds for x', then it will hold also for x. We give now a procedure in order to find for any positive integer n a sequence of totally ordered finite alphabets Σ_i ($i = 1, \ldots, n+1$) and a sequence of bi-infinite uniformly recurrent and ω-power-free words $x_i \in \Sigma_i^\zeta$, with $\Sigma_i = \text{Alph } x_i$, such that $x_1 = x$ and for any i, $2 \leq i \leq n+1$, there exists a injective morphism

$$\delta_i : \Sigma_i^* \to \Sigma_{i-1}^*,$$

with the property

$$\delta_i(x_i) = x_{i-1}.$$

The construction is given inductively. We totally order $\Sigma_1 = \text{Alph } x \subseteq A$ and extend this order to the lexicographic order of Σ_1^*. Let us suppose that we have constructed the sequence until the ith step. Then let $x_i \in \Sigma_i^\zeta$, where $\Sigma_i = \text{Alph } x_i$ is supposed to be totally ordered. This order can be extended to the lexicographic order $<_{\Sigma_i}$ of Σ_i^*. Let $b_i = \min \Sigma_i$. One can then construct the sets

$$X_i = b_i^+(\Sigma_i \setminus \{b_i\})^+$$

and

$$\Lambda_{i+1} = F(x_i) \cap X_i.$$

By induction x_i is uniformly recurrent and ω-power-free. Hence, there are not enough long subwords in x_i which either are powers of b_i or do not contain b_i as a factor, so that Λ_{i+1} is finite.

The set Λ_{i+1} is a code having a *synchronization delay* equal to 1, i.e., for any pair $(y_1, y_2) \in \Lambda_{i+1} \times \Lambda_{i+1}$ and $\alpha, \beta \in \Sigma_i^*$ one has

$$\alpha y_1 y_2 \beta \in \Lambda_{i+1}^* \Longrightarrow \alpha y_1, y_2 \beta \in \Lambda_{i+1}^*.$$

One can then consider an alphabet Σ_{i+1} with $\text{Card}(\Sigma_{i+1}) = \text{Card}(\Lambda_{i+1})$ and a bijection $\gamma_{i+1} : \Sigma_{i+1} \to \Lambda_{i+1}$. Let $\delta_{i+1} : \Sigma_{i+1}^* \to \Sigma_i^*$ be the morphism

4.4. Extensions of a theorem of Shirshov

which extends γ_{i+1}. Since Λ_{i+1} is a code, δ_{i+1} is injective. Moreover, from the bounded synchronization delay (equal to 1) one has that x_i can be uniquely factorized in terms of the elements of Λ_{i+1}. Then there exists a bi-infinite word $x_{i+1} \in \Sigma_{i+1}^{\zeta}$ such that

$$\delta_{i+1}(x_{i+1}) = x_i.$$

Since, by the inductive hypothesis, x_i is uniformly recurrent and ω-power-free, one derives that so will be x_{i+1}. Indeed, it is trivial that x_{i+1} is not ω-power-free. Let us prove that x_{i+1} is recurrent. In fact let $u \in F(x_{i+1})$. One has that $\delta_{i+1}(u) \in F(x_i)$. Moreover, there exists a letter $x \in \Sigma_i \setminus \{b_i\}$ such that $x\delta_{i+1}(u)b_i \in F(x_i)$. The factor $x\delta_{i+1}(u)b_i$ is recurrent in the unique factorization of x_i in terms of the elements of Λ_{i+1}. This implies that u will be recurrent in x_{i+1}. Moreover, x_{i+1} is uniformly recurrent. In fact one easily derives that for any n an upper bound to $r_{x_{i+1}}(n)$ is given by $r_{x_i}((n+2)l_M)$ where $l_M = \max\{|y| \mid y \in \Lambda_{i+1}\}$.

Now we define in Σ_{i+1} a total ordering by setting for $x, y \in \Sigma_{i+1}$

$$x <_{\Sigma_{i+1}} y \iff \delta_{i+1}(x) <_{\Sigma_i} \delta_{i+1}(y).$$

This total order of Σ_{i+1} can be extended to the lexicographic order $<_{\Sigma_{i+1}}$ of Σ_{i+1}^*. Moreover, for any $u, v \in \Sigma_{i+1}^*$ one has by Lemma 4.4.2

$$u <_{\Sigma_{i+1}} v \implies \delta_{i+1}(u) <_{\Sigma_i} \delta_{i+1}(v).$$

Let $w_0, w_1, w_2, \ldots, w_n \in \Sigma_1^+ \subseteq A^*$ be defined by

$$w_n = b_1 = \min \Sigma_1,$$

and for $i = 2, \ldots, n+1$

$$w_{n-i+1} = \delta_2(\delta_3(\cdots \delta_i(b_i) \cdots))$$

with $b_i = \min \Sigma_i$.

We prove that $s = w_1 w_2 \cdots w_n \in F(x)$. Moreover, s is the nth term of a bi-ideal sequence whose canonical factorization is $(w_1, w_2, \ldots, w_n)$. By construction for any $i > 1$ and for any $b \in \Sigma_i$ one has

$$\delta_i(b) = b_{i-1} u_1, \quad \text{with} \quad u_1 \in \Sigma_{i-1}^+,$$

thus, applying the same relation to the first letter of u_1, we obtain

$$\delta_{i-1}(\delta_i(b)) = \delta_{i-1}(b_{i-1})b_{i-2}u_2, \quad \text{with} \quad u_2 \in \Sigma_{i-2}^+.$$

Iterating this procedure one has

$$\delta_2(\cdots \delta_{i-1}(\delta_i(b)) \cdots) = \delta_2(\cdots \delta_{i-1}(b_{i-1}) \cdots) \cdots \delta_2(b_2)b_1 u_{i-1},$$

with $u_{i-1} \in \Sigma_1^+ \subseteq A^+$. Thus, for $b = b_i$, we obtain for any i, $2 \le i \le n+1$,

$$w_{n-i+1} = w_{n-i+2} \cdots w_n u_{i-1}. \tag{4.4.3}$$

Thus, in particular, one has

$$w_0 = w_1 \cdots w_n u_n.$$

Since $w_0 \in F(x)$ then $s = w_1 \cdots w_n \in F(x)$. Moreover, Equation (4.4.3) implies that $(w_1, w_2, \ldots, w_n)$ is an n-sequence, so that $s = w_1 \cdots w_n$ is the last term of a bi-ideal sequence of order n whose canonical factorization is $(w_1, \ldots, w_n)$.

Let us denote by $\mathscr{L}_i$ the set of the Lyndon words on the alphabet Σ_i. For each $i = 2, \ldots, n$ the word $b_i = \min \Sigma_i$ is a Lyndon word on the alphabet Σ_i. Moreover, by Lemma 4.4.3 the injective morphism δ_i preserves Lyndon words. Thus, $\delta_i(b_i) \in \mathscr{L}_{i-1}$ and

$$w_{n-i+1} = \delta_2(\delta_3(\cdots \delta_i(b_i) \cdots)) \in \mathscr{L}_1.$$

All the words w_i, $(i = 1, \ldots, n)$ are then Lyndon on the alphabet A. Moreover, since from Equation (4.4.3) w_{n-i+2} is a proper prefix of w_{n-i+1}, one has

$$w_1 > w_2 > \cdots > w_n.$$

From this one easily derives (see Problem 4.4.1) that $(w_1, \ldots, w_n)$ is an n-division and this concludes the proof. ∎

As a consequence of the preceding proposition we give the following theorem.

THEOREM 4.4.5. *For all k, p, n positive integers there exists a positive integer $N(k, p, n)$ such that for any totally ordered alphabet A of cardinality k any word $w \in A^*$ whose length is at least $N(k, p, n)$ is such that*
  (i) *there exists $u \ne \epsilon$ such that $u^p \in F(w)$ or*
  (ii) *there exists $s \in F(w)$ which is the nth term of a bi-ideal sequence whose canonical factorization $(w_1, \ldots, w_n)$ is an n-division of s. Moreover, the words w_i, $i = 1, \ldots, n$, are Lyndon words such that*

$$w_1 > w_2 > \cdots > w_n.$$

Proof. Let A be a totally ordered alphabet of cardinality k. The set of all words of A^* which satisfy either (i) or (ii) is a two-sided ideal $J_{k,n,p}$, or simply J, of A^*. Let $C = A^* \setminus J$. The set C is closed by factors, so that if we suppose that C is infinite, then by König's lemma (Proposition 1.2.3) there exists a one-sided infinite word $x \in A^\omega$ such that $F(x) \subseteq C$. Now either x has a factor which is a p-power and then $F(x) \cap J \neq \emptyset$, or x is p-power-free. In this case it follows from Theorem 4.4.4 that again $F(x) \cap J \neq \emptyset$. Hence, in both the cases we reach a contradiction. Thus C is finite and this proves the assertion. ∎

4.5. Finiteness conditions for semigroups

Let S be a semigroup. One can naturally embed S in a monoid S^1 as follows. If S is a monoid, then $S^1 = S$. If S has no identity, then S^1 is obtained from S by adjoining an extra element 1 satisfying the property $s1 = 1s = s$ for all $s \in S^1$.

If $s, t \in S$ we say that s is a *factor* of t if $t \in S^1 s S^1$. If $t \in sS^1$ ($t \in S^1 s$) then s is called a *left factor* (*right factor*) of t. For any $t \in S$ we denote by $F(t)$ the set of the factors of t. For any subset X of S, $F(X) = \bigcup_{t \in X} F(t)$. One says that X is *factorial* or *closed by factors* if $F(X) = X$.

A semigroup (group) S is *finitely generated* if there exists a finite subset X of S such that the subsemigroup (subgroup) $\langle X \rangle$ generated by X is S. A semigroup (group) S is called *locally finite* if every finitely generated subsemigroup (subgroup) of S is finite.

When a semigroup S is finitely generated by a set X, one can introduce an alphabet A having the same cardinality of X. As is well known any bijection $\delta : A \to X$ can be extended to a unique surjective morphism

$$\phi : A^+ \to S.$$

When δ is the identity map, ϕ is usually called the *canonical epimorphism*. Moreover, one has $S \cong A^+ / \phi \phi^{-1}$.

Let us suppose that A is totally ordered. Recall from Chapter 1 that the *radix order* $<_a$ on A^+ is defined for u, v in A^+ by

$$u <_a v \iff (|u| < |v|) \text{ or } (|u| = |v| \text{ and } u < v),$$

where $<$ is the lexicographic order. From the definition it follows that $<_a$ is a well-order.

A word w is said to be *reducible* with respect to the morphism ϕ and to the order $<_a$, or simply *reducible*, if there exists $u \in A^+$ such that

$$u <_a w \text{ and } \phi(u) = \phi(w).$$

A word which is not reducible will be called *irreducible*. Let $s \in S$. In the set $\phi^{-1}(s)$ there is a unique minimal element with respect to $<_a$ usually called the *canonical representative* of s. Hence, the set of all canonical representatives of the elements of S is the set of all irreducible elements of S. For any set $T \subseteq S$, we denote by C_T the set of the canonical representatives of the elements of T.

An infinite (bi-infinite) word x is *reducible*, relative to the morphism $\phi: A^+ \to S$ and to the order $<_a$, if there exists $w \in F(x)$ which is reducible. An infinite word x which is not reducible, i.e., any $w \in F(x)$ is irreducible, is called *irreducible*.

We recall now some lemmas and propositions on canonical representatives.

LEMMA 4.5.1. *Let S be a finitely generated semigroup and T be any subset of S closed by factors. Then the set C_T is closed by factors.*

Proof. Let $x \in C_T$ and u be a factor of x, i.e., $x = \lambda u \mu$, with $\lambda, \mu \in A^*$. Since $\phi(x) = \phi(\lambda)\phi(u)\phi(\mu)$ and T is closed by factors one has $\phi(u) \in T$ and then $u \in \phi^{-1}(T)$. Suppose now that $u' \in A^+$ exists such that $u' <_a u$ and $\phi(u') = \phi(u)$. If $|u'| < |u|$, then $x' = \lambda u' \mu$ is such that $|x'| < |x|$ and $\phi(x') = \phi(x)$ which is a contradiction. Let us then suppose $|u'| = |u|$ and $u' < u$. Thus $x' = \lambda u' \mu < \lambda u \mu$ and $\phi(x') = \phi(x)$ which is again a contradiction. Hence, $u \in C_T$ which concludes the proof. ∎

LEMMA 4.5.2. *Let S be a finitely generated semigroup. If T is an infinite subset of S closed by factors, then there exists a uniformly recurrent irreducible word x such that $F(x) \subseteq C_T$.*

Proof. Since C_T is closed by factors then by Proposition 1.5.12 there exists a uniformly recurrent word $x \in A^\omega$ such that $F(x) \subseteq C_T$, so that x is irreducible. ∎

As in the case of free monoids, one can introduce for any semigroup S the notions of bi-ideal sequence and n-sequence. A sequence

$$s_1, \ldots, s_n, \ldots$$

of elements of a semigroup S is a *bi-ideal sequence* if for any $i > 0$

$$s_{i+1} \in s_i S^1 s_i.$$

When the sequence is finite and of length n, then $(s_1, \ldots, s_n)$ is called a bi-ideal sequence of *order n*.

From Lemma 4.5.2 one easily derives the following.

4.5. Finiteness conditions for semigroups

PROPOSITION 4.5.3. *Let S be a finitely generated semigroup. If T is an infinite subset of S closed by factors, then there exists a bi-ideal sequence $(s_n)_{n>0}$ such that for all $n > 0$, $s_n \in T$ and for all positive integers i, j, $i \neq j$, $s_i \neq s_j$.*

Proof. By Lemma 4.5.2 there exists an irreducible uniformly recurrent word $x \in A^\omega$ such that $F(x) \subseteq C_T$. By Proposition 4.3.1, the word x is a sesquipower, so that there exists a bi-ideal sequence $(f_n)_{n>0}$ such that $x = \lim_{n \to \infty} f_n$. Since for every $n > 0$, $f_n \in F(x) \subseteq C_T$ we have $\phi(f_n) \in T$, where ϕ is the canonical epimorphism. Moreover, since x is irreducible it follows that for all positive integers $i, j, i \neq j$, $\phi(f_i) \neq \phi(f_j)$. The image by ϕ of the bi-ideal sequence $(f_n)_{n>0}$ is then a bi-ideal sequence $(s_n)_{n>0}$, with $s_n = \phi(f_n) \in T$ for all $n > 0$, such that $s_i \neq s_j$ for $i \neq j$. ∎

A sequence $t_1, \ldots, t_n$ of n elements of a semigroup S is called an *n-sequence* if for all $i = 1, \ldots, n-1$,

$$t_i \in t_{i+1} \cdots t_n S^1. \tag{4.5.1}$$

As in the case of free monoids the notions of bi-ideal sequence of order n and of n-sequence are related. In fact, let $s_1, \ldots, s_n$ be a bi-ideal sequence of S where

$$s_{i+1} = s_i g_i s_i$$

with $g_i \in S^1, i = 1, \ldots, n-1$. As one easily verifies setting $t_n = s_1$ and $t_{n-i} = s_i g_i$, $i = 1, \ldots, n-1$, the sequence $(t_1, \ldots, t_n)$ is an n-sequence. Conversely, if $(t_1, \ldots, t_n)$ is an n-sequence, then the sequence $s_i = t_{n-i+1} \cdots t_n, i = 1, \ldots, n$, is a bi-ideal sequence of order n.

Let us, finally, observe that in the case of a group G any sequence $g_1, \ldots, g_n$ of n elements of G is an n-sequence.

4.5.1. Permutation property

Let S be a finitely generated semigroup and $\phi: A^+ \to S$ be the canonical epimorphism. For any $s \in S$ the *order* of s is the order (cardinality) of the subsemigroup $\langle s \rangle$ generated by $\{s\}$. The order of s is finite if and only if there exist positive integers i and j, $i < j$, depending on the element s, such that

$$s^i = s^j. \tag{4.5.2}$$

Let j be the minimal integer for which the above relation is satisfied. The integer i, which is unique, is called the *index* and $p = j - i$ the *period* of s. The order of $\langle s \rangle$ is then given by $i + p - 1$. In the case of a group, due to cancellativity, condition (4.5.2) simply becomes $s^p = 1$, where 1 denotes the identity of the group and the period p depends on the element s.

A semigroup (group) S is *periodic* (or *torsion*) if any element $s \in S$ generates a subsemigroup (subgroup) of finite order.

The problem of whether a finitely generated and periodic group is finite was posed by W. Burnside in 1902 and, subsequently, extended to the case of semigroups. However, the condition of a finite generation and the periodicity are not sufficient to assure the finiteness of a semigroup or a group (see Notes).

A finitely generated, torsion and commutative semigroup is obviously finite. A. Restivo and C. Reutenauer introduced in 1984 a property of semigroups, called the *permutation property*, which generalizes commutativity and is such that a finitely generated and torsion semigroup is finite if and only if it is permutable. Let us give the following.

Let S be a semigroup and n be an integer > 1. A sequence $s_1, \ldots, s_n$ of n elements of S is called *permutable* if the product $s_1 \cdots s_n$ remains invariant under some nontrivial permutation of its factors, i.e., there exists a permutation $\sigma \in \mathscr{S}_n$, different from the identity, such that

$$s_1 s_2 \cdots s_n = s_{\sigma(1)} s_{\sigma(2)} \cdots s_{\sigma(n)}.$$

We say that a semigroup S is *n-permutable* if any sequence of n elements of S is permutable. Obviously, 2-permutability is equivalent to commutativity. We say that S is *permutable* if there exists an integer $n > 1$ such that S is n-permutable.

THEOREM 4.5.4. *Let S be a finitely generated and periodic semigroup. S is finite if and only if it is permutable.*

The original proof of the theorem is based on the theorem of Shirshov. We prove here a slight more general version of Theorem 4.5.4 by considering a weak notion of permutability of a semigroup S. One requires that for some integer $n > 1$, not all the sequences of n elements are permutable, but only the n-sequences of S.

THEOREM 4.5.5. *Let S be a finitely generated and periodic semigroup. S is finite if and only if there exists an integer $n \geq 2$ such that any n-sequence of S is permutable.*

Proof. The "only if" part is trivial, so we prove the "if" part. Let $n \geq 2$ be an integer such that any n-sequence of S is permutable. Let $\phi: A^+ \to S$ be the canonical epimorphism and suppose by contradiction that S is infinite. By Lemma 4.5.2 there exists an irreducible and uniformly

4.5. Finiteness conditions for semigroups

recurrent word t. Since S is periodic, t is ω-power-free. Indeed, otherwise, by Lemma 4.3.4, t is periodic so that it contains a factor u^p such that

$$\phi(u^p) = \phi(u)^p = \phi(u)^q = \phi(u^q),$$

with $1 \leq q < p$, and this contradicts the irreducibility of t. By Theorem 4.4.4, t contains a factor x which is the nth term of a bi-ideal sequence whose canonical factorization is an n-division. We can write $x = w_1 w_2 \cdots w_n$, where $(w_1, w_2, \ldots, w_n)$ is the canonical factorization of x which is an n-sequence. Let us set $s_i = \phi(w_i)$, for $i = 1, \ldots, n$. Since $(s_1, s_2, \ldots, s_n)$ is an n-sequence of S, it is permutable. Then, for a nontrivial permutation $\sigma \in \mathscr{S}_n$ one has

$$\phi(x) = s_1 s_2 \cdots s_n = s_{\sigma(1)} s_{\sigma(2)} \cdots s_{\sigma(n)}.$$

On the other hand one has $x > w_{\sigma(1)} w_{\sigma(2)} \cdots w_{\sigma(n)}$ and this contradicts the irreducibility of t. ∎

4.5.2. Iteration property

In this subsection we consider some finiteness conditions for semigroups based on *iteration properties*. These properties are very important in formal language theory, since they are related to the 'pumping properties' of regular languages.

Let S be a semigroup and m and n two integers such that $m > 0$ and $n \geq 0$. We say that the sequence $s_1, s_2, \ldots, s_m$ of m elements of S is *n-iterable* if there exist i, j such that $1 \leq i \leq j \leq m$ and

$$s_1 \cdots s_m = s_1 \cdots s_{i-1} (s_i \cdots s_j)^n s_{j+1} \cdots s_m. \tag{4.5.3}$$

We say that S is (m, n)-*iterable*, or satisfies the property $C(n, m)$, if all sequences of m elements of S are n-iterable.

Let us observe that property $C(1, m)$ is always trivially true. Moreover, the property $C(0, m)$ is actually a cancellation property (the *m-cancellation property*) which obviously implies the finiteness of any finitely generated semigroup satisfying it, so that a semigroup satisfies properly the iteration property $C(n, m)$ only if $n > 1$.

PROPOSITION 4.5.6. *If S is a finite semigroup, then S satisfies $C(n, m)$ with $m = \mathrm{Card}(S) + 1$ and any $n \geq 0$. If S satisfies $C(n, m)$, with $n \neq 1$, then S is periodic.*

Proof. Let $s_1, \ldots, s_m$ be a sequence of m elements of S, with $m = \operatorname{Card}(S) + 1$. We consider then the sequence

$$s_1, s_1 s_2, \ldots, s_1 s_2 \cdots s_m.$$

Since $m > \operatorname{Card}(S)$ there exist integers i, j such that $1 \le i < j \le m$ and

$$s_1 \cdots s_i = s_1 \cdots s_i(s_{i+1} \cdots s_j) = s_1 \cdots s_i(s_{i+1} \cdots s_j)^n,$$

for all $n \ge 0$. Thus

$$s_1 \cdots s_i s_{j+1} \cdots s_m = s_1 \cdots s_i(s_{i+1} \cdots s_j)^n s_{j+1} \cdots s_m,$$

so that $C(n, m)$ holds for all $n \ge 0$.

If S satisfies $C(n, m)$, then for any $s \in S$, consider the sequence $s_1 = s_2 = \cdots = s_m = s$. One has that there exist integers i, j such that $1 \le i \le j \le m$ and

$$s^m = s^{m+(n-1)(j-i+1)},$$

so that, since $n \ne 1$, S is periodic. ∎

Let us consider in a semigroup S the equivalence relations $\mathscr{R}$ and $\mathscr{L}$ defined, for $s, t \in S$, by

$$s \mathrel{\mathscr{R}} t \iff sS^1 = tS^1, \quad s \mathrel{\mathscr{L}} t \iff S^1 s = S^1 t.$$

Moreover we set $\mathscr{H} = \mathscr{R} \cap \mathscr{L}$ and $\mathscr{D} = \mathscr{R} \vee \mathscr{L}$, where $\mathscr{R} \vee \mathscr{L}$ denotes the smallest equivalence relation on S containing $\mathscr{R}$ and $\mathscr{L}$. From the definition one has that $\mathscr{R}$ and $\mathscr{L}$ are left invariant and right invariant, respectively.

LEMMA 4.5.7. *Let S be a finitely generated semigroup satisfying $C(n,m)$, with $n > 1$, and $\phi: A^+ \to S$ be the canonical epimorphism. For any uniformly recurrent word w there exists a positive integer M such that for any $u, v \in F(w)$ with $|u| > M$*

$$uv \in F(w) \implies \phi(u) \mathrel{\mathscr{R}} \phi(uv).$$

Proof. It is sufficient to prove that there exists a positive integer M such that for each $u \in F(w)$ and $a \in A$ if $ua \in F(w)$, then $\phi(u) \mathrel{\mathscr{R}} \phi(ua)$. By Proposition 4.3.3 there exists a positive integer $M = D(m)$ such that for any $u \in A^*$, $a \in A$, with $ua \in F(w)$ and $|u| \ge M$ one has that

$$u = \lambda f_m,$$

4.5. Finiteness conditions for semigroups

where $\lambda \in A^*$, and f_m is the mth term of a bi-ideal sequence $f_{i+1} = f_i g_i f_i$, with $f_1 \in aA^*$ and $g_i \in aA^*$, $i = 1, \ldots, m-1$. Let us write f_m as

$$f_m = w_1 \cdots w_m$$

where $(w_1, \ldots, w_m)$ is the canonical factorization of f_m. As S satisfies $C(n, m)$ there exist integers i, j such that $1 \leq i \leq j \leq m$ and

$$w_1 \cdots w_m \equiv w_1 \cdots w_{i-1}(w_i \cdots w_j)^n w_{j+1} \cdots w_m, \quad (4.5.4)$$

where $\equiv$ denotes the congruence relation $\phi \phi^{-1}$. One can rewrite the preceding formula as

$$w_1 \cdots w_m \equiv w_1 \cdots w_i \cdots w_j w_i v, \quad (4.5.5)$$

with $v \in A^*$. Let us first suppose $j < m$. By Equation (4.2.2) one derives, by iteration,

$$w_i = w_{j+1} \cdots w_m g_{m-j} u,$$

with $u \in A^*$. Hence, since $g_{m-j} \in aA^*$ one has

$$w_1 \cdots w_m \equiv w_1 \cdots w_j w_{j+1} \cdots w_m g_{m-j} \zeta = w_1 \cdots w_j w_{j+1} \cdots w_m a \xi$$

with $\zeta, \xi \in A^*$. This implies

$$\phi(f_m) \, \mathcal{R} \, \phi(f_m a).$$

Since the relation $\mathcal{R}$ is left invariant it follows that $\phi(u) \, \mathcal{R} \, \phi(ua)$. In the case $j = m$ since $w_i \in aA^*$ for all $i = 1, \ldots, m$, from Formula (4.5.5) one has again $\phi(f_m) \, \mathcal{R} \, \phi(f_m a)$ which implies $\phi(u) \, \mathcal{R} \, \phi(ua)$. ∎

THEOREM 4.5.8. *Let S be a finitely generated semigroup. S is finite if and only if it satisfies $C(2, m)$ for a suitable $m > 0$.*

Proof. The 'only if' part follows from Proposition 4.5.6. Let us then prove the 'if' part. Let S be a finitely generated semigroup satisfying $C(2, m)$ for a suitable $m > 0$ and suppose, by contradiction, that S is infinite. From Lemma 4.5.2 there exists a uniformly recurrent word w which is irreducible with respect to the canonical epimorphism $\phi: A^+ \to S$. By Lemma 4.5.7 there exists a positive integer M such that for any $u, v \in F(w)$ with $|u| > M$, $uv \in F(w) \Rightarrow \phi(u) \, \mathcal{R} \, \phi(uv)$. Let now u be any factor of w such that $|u| > M$. Since w is uniformly recurrent we can consider $m+1$ consecutive nonoverlapping occurrences of u and then the factor v of w

$$v = u x_1 u x_2 \cdots u x_m u,$$

with $x_i \in A^*$ for $(i = 1, \ldots, m)$. From condition $C(2, m)$ there exist i, j such that $1 \leq i \leq j \leq m$ and

$$v \equiv ux_1 \cdots ux_{i-1}(ux_i \cdots ux_j)^2 ux_{j+1} \cdots ux_m u. \tag{4.5.6}$$

Moreover, from Lemma 4.5.7 one has that

$$\phi(u) \mathcal{R} \phi(ux_i \cdots ux_m u),$$

for all $i = 1, \ldots, m$. This implies that for any $i = 1, \ldots, m$, there exists a word $t \in A^*$ depending on i, such that

$$u \equiv ux_i \cdots ux_m ut.$$

One has then

$$v = ux_1 ux_2 \cdots ux_m u \equiv ux_1 \cdots ux_{i-1}(ux_i \cdots ux_j)^2 ux_{j+1} \cdots ux_m ut\zeta,$$

having set $\zeta = x_{j+1} \cdots ux_m u$. By Formula (4.5.6), it follows that

$$v \equiv ux_1 \cdots ux_{i-1} ux_i \cdots ux_m ut\zeta$$
$$\equiv ux_1 \cdots ux_{i-1} ux_{j+1} \cdots ux_m u.$$

Hence, v is reducible; that is a contradiction. ∎

A special form of iteration property is the *iteration on the right*. A semigroup S satisfies the condition $D(n, m)$, $m > 0, n \geq 0$, if for any sequence $s_1, s_2, \ldots, s_m$ of m elements of S there exist integers i, j such that $1 \leq i \leq j \leq m$ and

$$s_1 \cdots s_j = s_1 \cdots s_{i-1}(s_i \cdots s_j)^n.$$

It is clear that if a semigroup satisfies $D(n, m)$, then it satisfies $C(n, m)$. Thus from Theorem 4.5.8, $D(2, m)$ is a finiteness condition for finitely generated semigroups. Moreover, it is straightforward to derive that any finite semigroup S satisfies $D(n, m)$ for a suitable m, depending on the cardinality of S, and for all $n \geq 0$.

Another important property, strictly related to the iteration property, is the *strong periodicity*. Let S be a semigroup. We denote by $E(S)$ the set of its idempotent elements.

Let m be a positive integer. A semigroup S is *strongly m-periodic* if for any sequence $s_1, \ldots, s_m$ of m elements of S there exist integers i and j such that $1 \leq i \leq j \leq m$ and $s_i \cdots s_j \in E(S)$.

A semigroup S is *strongly periodic* if there exists a positive integer m such that S is strongly m-periodic. The origin of the term strongly

4.5. Finiteness conditions for semigroups

m-periodic is due to the fact that if S is strongly m-periodic, then S is certainly periodic and, moreover, the index and the period of any element are less than or equal to m. It is clear from the definition that if a semigroup S is strongly m-periodic, then S satisfies $C(2, m)$ and $D(2, m)$.

The following interesting theorem holds.

THEOREM 4.5.9. *Let S be a finitely generated semigroup. The following conditions are equivalent.*
  (i) *S is finite.*
  (ii) *$S\backslash E(S)$ is finite.*
  (iii) *S is strongly periodic.*

Proof. The implication (i)⇒(ii) is trivial. For the implication (ii)⇒(iii) one uses the theorem of Ramsey (Lothaire 1983, Chapter 4). Let $F = S\backslash E(S)$ and $p = \text{Card}(F)$. We prove that S is strongly m-periodic with $m = R(2, 3, p+1) - 1$ where R denotes the function of Ramsey's theorem. Let $s_1, \ldots, s_m$ be a sequence of m elements of S. We define

$$B_0 = \{\{i, j\} \mid 1 \leq i < j \leq m+1 \text{ and } s_i s_{i+1} \cdots s_{j-1} \in E(S)\},$$

and for any $f \in F$,

$$B_f = \{\{i, j\} \mid 1 \leq i < j \leq m+1 \text{ and } s_i s_{i+1} \cdots s_{j-1} = f\}.$$

By Ramsey's theorem there exist $1 \leq i_1 < i_2 < i_3 \leq m+1$ such that $\{i_1, i_2\}, \{i_2, i_3\}$ and $\{i_1, i_3\}$ are all in the same class. This class is certainly B_0 because, otherwise, there will exist $f \in F$ such that

$$f = s_{i_1} \cdots s_{i_3-1} = (s_{i_1} \cdots s_{i_2-1})(s_{i_2} \cdots s_{i_3-1}) = f^2$$

which is a contradiction.

(iii) ⇒ (i) If a semigroup S is strongly m-periodic, then it satisfies the condition $C(2, m)$. Then, by Theorem 4.5.8, S is finite. ∎

4.5.3. Minimal conditions on principal bi-ideals

We recall that a *bi-ideal* B of S is a subsemigroup of S such that

$$BSB \subseteq B.$$

A bi-ideal is called *principal* if it is of the form $sS^1 s$, where s is any element of S.

A semigroup S satisfies the *minimal condition on principal bi-ideals* if any strictly descending chain

$$s_1 S^1 s_1 \supset s_2 S^1 s_2 \supset \cdots \supset s_n S^1 s_n \supset \cdots,$$

with $s_1, s_2, \ldots, s_n, \ldots \in S$, has a finite length.

THEOREM 4.5.10. *Let S be a finitely generated semigroup. If S satisfies the minimal condition on principal bi-ideals and if all subgroups of S are finite, then S is finite.*

Proof. Let S be a semigroup satisfying the hypotheses of the statement and let $\phi: A^+ \to S$ be the canonical epimorphism. Suppose by contradiction that S is infinite. Then, by Proposition 4.5.3, there exists a bi-ideal sequence $(s_n)_{n>0}$ such that for all positive integers i, j, with $i \neq j$, one has $s_i \neq s_j$. Since $s_{n+1} \in s_n S^1 s_n$, for $n \geq 1$, it follows that

$$s_n S^1 s_n \supseteq s_{n+1} S^1 s_{n+1}.$$

Thus we have a descending chain

$$s_1 S^1 s_1 \supseteq s_2 S^1 s_2 \supseteq \cdots \supseteq s_n S^1 s_n \supseteq \cdots.$$

By the minimal condition on principal bi-ideals, there exists an integer k such that $s_k S^1 s_k = s_n S^1 s_n$, for any $n \geq k$. Let n be any integer $\geq k$. One has $s_n S^1 s_n = s_{n+1} S^1 s_{n+1} = s_{n+2} S^1 s_{n+2}$, and, moreover, $s_{n+1} \in s_n S^1 s_n$. Thus we have

$$s_{n+1} = s_n t s_n = s_{n+1} h s_{n+1} = s_{n+2} r s_{n+2}, \qquad (4.5.7)$$

for some $t, h, r \in S^1$. Moreover, since $s_{n+2} \in s_{n+1} S^1 s_{n+1}$, one has

$$s_{n+2} = s_{n+1} z s_{n+1}, \qquad (4.5.8)$$

for some $z \in S^1$. From Equations (4.5.7) and (4.5.8) one derives that for any $n \geq k$ $s_{n+1} \mathcal{R} s_{n+2}$, $s_{n+1} \mathcal{L} s_{n+2}$ and, therefore, $s_{n+1} \mathcal{H} s_{n+2}$. In conclusion, for any $n > k$ all the elements s_n lie in the same $\mathcal{H}$-class H. From Equation (4.5.7) one has that s_{n+1} is a regular element (see Problem 4.5.5) for any $n \geq k$. The $\mathcal{H}$-class H is in a regular $\mathcal{D}$-class D, hence it is finite since it has the same cardinality as a maximal subgroup contained in D (see Problem 4.5.5), and, by hypothesis, all subgroups of S are finite. Then there exist two integers i, j, with $k < i < j$ such that $s_i = s_j$ which is a contradiction. ∎

Problems

Section 4.1

4.1.1 A sequence $f_1, f_2, \ldots, f_n$ of n words on the alphabet A is called a *quasi-ideal* sequence of order n if $f_1 \in A^+$ and for all $i = 1, 2, \ldots, n-1$ one has

$$f_{i+1} \in f_i A^* \cap A^* f_i.$$

Thus a bi-ideal sequence of order n is also a quasi-ideal sequence of order n. Give an example of a quasi-ideal sequence which is not a bi-ideal sequence.

4.1.2 Any word $w \in A^+$ is a *quasi-power* of order 1. For any $n > 0$ a word w is called a *quasi-power* of order $n+1$ if there exists a quasi-power u of order n such that

$$w \in uA^+ \cap A^+ u.$$

The *quasi-power degree* of w is the maximal order of w as a quasi-power. Show that if w is a quasi-power of degree n, then there exists a *unique* quasi-ideal sequence $(f_1, \ldots, f_n)$ such that $f_n = w$.

4.1.3 A word w has a *bord* $u \in A^+$, if $w \in uA^+ \cap A^+ u$. As is well known a word w has the proper period p ($0 < p < |w|$) if and only if there exists a bord u such that $p = |w| - |u|$. Show that if w is a quasi-power of degree n and $(f_1, \ldots, f_n)$ is the unique quasi-ideal sequence such that $f_n = w$, then $(f_1, \ldots, f_{n-1})$ is the sequence of all the bords of w. Thus $n-1$ is the number of all proper periods of w.

Section 4.2

4.2.1 Let $(w_1, \ldots, w_n)$ and $(w'_1, \ldots, w'_n)$ be the canonical factorizations of the nth term of a bi-ideal sequence. Prove that

- for each i, $1 \leq i \leq n-1$, one has $w'_i w'_{i+1} < w'_{i+1} w'_i$ if and only if $w_{n-i+1} w_{n-i} < w_{n-i} w_{n-i+1}$,
- $(w_1, \ldots, w_n)$ is an n-division (inverse n-division) if and only if $(w'_1, \ldots, w'_n)$ is an inverse n-division (n-division).

4.2.2 Let $(w_1, \ldots, w_m)$ be a sequence of words. We say that $(u_1, \ldots, u_n)$ is a *derived sequence* of $(w_1, \ldots, w_m)$ if there exist $n+1$ integers

$j_1, j_2, \ldots, j_{n+1}$ such that $1 \leq j_1 < j_2 < \cdots < j_{n+1} \leq m+1$, and

$$u_1 = w_{j_1} \cdots w_{j_2-1}, \ldots, u_n = w_{j_n} \cdots w_{j_{n+1}-1}.$$

Prove that a derived sequence $(u_1, \ldots, u_n)$ of an m-sequence $(w_1, \ldots, w_m)$ (inverse m-sequence) is an n-sequence (inverse n-sequence).

Section 4.3

4.3.1   Prove that if an eventually periodic word $w \in A^\omega$ is recurrent, then w is periodic.

4.3.2   Let $A = \{a, b, c, d\}$ and m be the Thue–Morse word on the alphabet $\{a, b, c\}$ which can be generated iterating on the letter a the morphism ϕ defined by $\phi(a) = abc$, $\phi(b) = ac$, $\phi(c) = b$ (see Lothaire 1983). Let us denote by p_i the prefix of m of length i and construct the word

$$w = dp_1(dp_2)^2(dp_3)^3 \cdots (dp_n)^n \cdots.$$

Show that for any $p > 1$, w has a factor which is a p-power. However, w is ω-power-free.

Section 4.4

4.4.1   Let $l_1, l_2, \ldots, l_n$ be n Lyndon words with $l_1 > l_2 > \cdots > l_n$. Let $w_i = l_i^{k_i}$, with $k_i \geq 1$, $1 \leq i \leq n$. Prove that the word $w = w_1 w_2 \cdots w_n$ is n-divided and $(w_1, w_2, \ldots, w_n)$ is an n-division. (*Hint*: Use the property (see Lothaire 1983) that if x and y are Lyndon words and $x < y$, then xy is a Lyndon word and $x < xy < y$.)

Section 4.5

4.5.1   The notion of finite, as well as infinite, irreducible words can be given with respect to any partial order $\leq$ in A^+. A partial order in A^+ is a *well-partial-order*, if any subset X of A^+ has at least one and at most a finite number of minimal elements in X. In such a case if $\phi: A^+ \to S$ is a morphism of A^+ onto the semigroup S, then for any $s \in S$ the set $\phi^{-1}(s)$ of the representatives of s has a finite > 0 number of representatives.

Prove that if a well-partial-order $\leq$ in A^+ is monotone (i.e., invariant with respect to concatenation), then the set C_T of all irreducible representatives of any factorial set $T \subseteq S$ is closed by factors.

4.5.2 Prove that if H is an abelian subgroup of a group G such that the index m of H in G is finite, then G is n-permutable with $n = 2m$.

4.5.3 A semigroup S is called *weakly permutable*, if there exists an integer $n > 1$ such that for any sequence $s_1, s_2, \ldots, s_n$ of n elements of S there exist two permutations $\sigma, \tau \in \mathcal{S}_n$, $\sigma \neq \tau$ such that
$$s_{\sigma(1)} s_{\sigma(2)} \cdots s_{\sigma(n)} = s_{\tau(1)} s_{\tau(2)} \cdots s_{\tau(n)}.$$
It is obvious that if a semigroup S is permutable, then it is weakly permutable. Show that the converse is not, in general, true.

4.5.4 A semigroup S is called a *band* if all its elements are idempotents, i.e., for any $s \in S$, $s = s^2$. Show that a finitely generated band is finite.

4.5.5 Let S be a semigroup. The relations $\mathcal{R}$, $\mathcal{L}$ and $\mathcal{D}$ satisfy the following properties.

  (i) The relations $\mathcal{R}$ and $\mathcal{L}$ commute, and so $\mathcal{D} = \mathcal{RL} = \mathcal{LR}$.

  (ii) Any two $\mathcal{H}$-classes in the same $\mathcal{D}$-class have the same cardinality.

  (iii) An $\mathcal{H}$-class H_e containing an idempotent e is equal to the maximal subgroup of S having e as identity.

An element s of a semigroup S is called *regular* if there exists $x \in S$ such that $s = sxs$. A $\mathcal{D}$-class D of a semigroup is called *regular* if all its elements are regular. Let D be a $\mathcal{D}$-class of a semigroup S. Show that the following hold.

  1. D is regular if and only if it contains a regular element.
  2. D is regular if and only if it contains an idempotent.
  3. Any two maximal subgroups in D are isomorphic.

(See Clifford and Preston 1961, Chapter 2.)

4.5.6 Define in a semigroup S the quasi-order relation $\leq_B$ by: for $s, t \in S$
$$s \leq_B t \iff s = t \text{ or } s \in tS^1 t.$$
A semigroup satisfies the condition $\min_B$ if any strictly descending chain $s_1 >_B s_2 >_B \cdots >_B s_n >_B \cdots$ of elements of S has a finite length.
Prove that if a semigroup S satisfies the minimal condition on principal bi-ideals, then it satisfies $\min_B$.

4.5.7 A nonempty subset Q of a semigroup S is called a quasi-ideal of S if
$$QS \cap SQ \subseteq Q.$$
Show that
- every quasi-ideal of S is a bi-ideal of S,
- a subset of a semigroup S is a quasi-ideal if and only if it is the intersection of a right ideal of S and a left ideal of S,
- a semigroup S is a group if and only if it contains no proper quasi-ideal (bi-ideal).

Notes

The name of bi-ideal sequence appears in Coudrain and Schützenberger 1966 who introduced these sequences in the frame of semigroup theory. Actually, these sequences were considered ten years earlier by Jacobson 1964 in his book on ring theory. A bi-ideal sequence of order n was called by Jacobson an *n-sequence*. Zimin's words Z_n were introduced by Zimin 1982. A word which is the nth term of a bi-ideal sequence was also called a *sesquipower of order n* by Simon 1988 and a *quasi-power of order n* by Berstel and Reutenauer 1988. Theorem 4.1.2 was first proved in Coudrain and Schützenberger 1966.

Shirshov's theorem appears in Shirshov 1957 (see also Lothaire 1983, Chapter 7). A different proof of Shirshov's theorem based on an unavoidable regularity related to Lyndon words was given by Reutenauer 1986. A proof which uses the uniform recurrence is given by Justin and Pirillo 1991. An improvement of Shirshov's theorem in which the n-divided factor is the nth term of a bi-ideal sequence was given in De Luca and Varricchio 1991a. Theorem 4.4.5, whose proof is in De Luca and Varricchio 1999, is a further generalization since the n-division is a strictly decreasing sequence of Lyndon words.

The problem of whether a finitely generated and periodic group is finite was posed by W. Burnside in 1902 and, subsequently, extended to the case of semigroups. A negative answer to the Burnside problem was given by Golod 1964. This author by means of a technique of proof discovered with I. R. Shafarevich, based on the nonfiniteness of a dimension of a suitable algebra associated with a field, was able to show the existence of an infinite 3-generated p-group.

The permutation property of semigroups was introduced in Restivo and Reutenauer 1984, where the proof of Theorem 4.5.4, based on Shirshov's theorem, was given. A characterization of permutable groups is given in Curzio et al. 1985. Curzio, Longobardi, and Maj 1983 give an algebraic proof that a finitely generated and torsion group is finite if and only if it is permutable. An extension of Theorem 4.5.4 based on the weaker notion of ω-permutability appears in De Luca and Varricchio 1990.

The notion of strong periodicity and Theorem 4.5.9 are due to Simon 1980. The proof of Simon makes use of a finiteness condition due to Hotzel 1979. The proof that condition $D(n,m)$ is a finiteness condition for finitely generated semigroups appears in De Luca and Restivo 1984 for $n = 2$ and in De Luca and Varricchio 1991b for $n = 3$. A more constructive proof of the result in the case of $D(2,m)$, as well as an upper bound to the cardinality of the semigroup, was given by Hashiguchi 1986.

The proof that condition $C(n,m)$ is a finiteness condition for finitely generated semigroups in the cases $n = 2$ and $n = 3$ appears in De Luca and Varricchio 1991a. The proof makes use of a deep structure theorem on finitely generated semigroups (the J-depth decomposition theorem) (De Luca and Varricchio 1999).

Theorem 4.5.10 is due to Coudrain and Schützenberger 1966. An extension of this result under the weaker hypothesis that the subgroups of the given finitely generated semigroup are locally finite appears in De Luca and Varricchio 1994.

CHAPTER 5

The Plactic Monoid

5.0. Introduction

Young tableaux have had a long history since their introduction by A. Young a century ago. It is only in the 1960s that there came to the fore a monoid structure on them, a structure taking into account most of their combinatorial properties, and having applications to the different fields in which Young tableaux were used.

Summarizing what had been his motivation to spend so much time on the plactic monoid, M.P. Schützenberger drew out three reasons: (1) it allows us to embed the ring of symmetric polynomials into a noncommutative ring; (2) it is the syntactic monoid of a function on words generalizing the maximal length of a nonincreasing subword; (3) it is a natural generalization to alphabets with more than two letters of the monoid of parentheses.

The starting point of the theory is an algorithm, due to C. Schensted, for the determination of the maximal length of a nondecreasing subword of a given word. The output of this algorithm is a tableau, and if one decides to identify the words leading to the same tableau, one arrives at the plactic monoid, whose defining relations were determined by D. Knuth.

The first significant application of the plactic monoid was to provide a complete proof of the Littlewood–Richardson rule, a combinatorial algorithm for multiplying Schur functions (or equivalently, to decompose tensor products of representations of unitary groups, a fundamental issue in many applications, e.g., in particle physics), which had been in use for almost 50 years before being fully understood. In fact, as will be shown in Section 5.4, the algebra of Schur functions can be lifted to the plactic algebra, and even to the free associative algebra. Once this crucial step is realized, all the proofs become straightforward.

Subsequent applications, also connected with group theory, physics and geometry, include a combinatorial description of the Kostka–Foulkes polynomials, which arise as entries of the character table of the finite linear groups $GL_n(\mathbf{F}_q)$, as Poincaré polynomials of certain algebraic varieties, or in the solution of certain lattice models in statistical mechanics. One can also mention a noncommutative version of the Demazure character formula, and the construction of keys, leading to a better understanding of the standard bases of Lakshmibai and Seshadri, and to a combinatorial description of the Schubert polynomials.

Quite recently, the combinatorics of Young tableaux has been illuminated by the theory of quantum groups, and especially by Kashiwara's theory of crystal bases. Roughly speaking, quantum groups are deformations depending on a parameter q of certain algebras classically associated with a Lie group G, which give back the classical object for $q = 1$. With some care, it is possible to take the limit $q \to 0$ in certain formulas, and to recover in this way classical bijections such as the Robinson–Schensted correspondence.

From a group-theoretic point of view, the combinatorics of Young tableaux is associated with root systems of type A. By means of quantum groups, it is now possible to define plactic monoids for other root systems, and to use them for describing the corresponding Littlewood–Richardson rules. There is also a similar construction taking into account the combinatorics of quasi-symmetric functions (the hypoplactic monoid).

Conventions. In this chapter, A will denote a totally ordered alphabet of n letters $a_1 < a_2 < \cdots < a_n$. In the examples, we shall usually take $A = \{1, 2, \ldots, n\}$.

5.1. Schensted's algorithm

Consider the following problem: given a word $w \in A^*$ on the totally ordered alphabet A, find the length of the longest nondecreasing subwords of w.

C. Schensted has given an elegant algorithmic solution, which does not require the actual determination of a maximal nondecreasing subword. His method relies on the notion of a *Young tableau*, a combinatorial structure issuing from group theory.

A nondecreasing word $v \in A^*$ is called a *row*. Let $u = x_1 \cdots x_r$ and $v = y_1 \cdots y_s$ be two rows ($x_i, y_j \in A$). We say that u *dominates* v ($u \triangleright v$) if $r \leq s$ and for $i = 1, \ldots, r$, $x_i > y_i$. Clearly, every word w has a unique factorization $w = u_1 \cdots u_k$ as a product of rows of maximal length. A *tableau* is a word w such that $u_1 \triangleright u_2 \triangleright \cdots \triangleright u_k$. It is customary to think of tableaux as planar objects and to represent w as the left justified superposition of its rows. For instance, taking $A = \{1 < 2 < \cdots\}$,

$$t = 68\ 4556\ 223357\ 1112444$$

is a tableau whose planar representation is

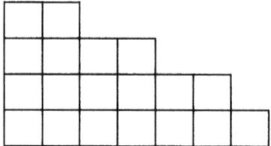

Similarly, a strictly decreasing word is called a *column*. Reading from bottom to top the lengths of the rows of a tableau t, one obtains a nonincreasing sequence $\lambda = (\lambda_1 \geq \lambda_2 \geq \cdots \geq \lambda_k)$ which is called the *shape* of t. Such a sequence is called a *partition* of the integer $|\lambda| = \lambda_1 + \cdots + \lambda_k$. On our example, $\lambda = (7, 6, 4, 2)$. The graphical representation of a partition by a planar diagram of boxes is called its *Ferrers* (or *Young*) *diagram*. Thus, the Ferrers diagram of $(7, 6, 4, 2)$ is

The *conjugate partition* λ' of λ is obtained by reading the heights of the columns of the diagram of λ. For example, the conjugate partition of $(7, 6, 4, 2)$ is $(4, 4, 3, 3, 2, 2, 1)$.

Schensted's algorithm associates to each $w \in A^*$ a tableau $t = P(w)$. The elementary step of the algorithm consists in the insertion of a letter into a row. Given a row $v = y_1 \cdots y_s$ and a letter x, the insertion of x into v is $P(vx) = vx$ if vx is a row, and $P(vx) = y_i v'$ otherwise, where y_i is the leftmost letter of v which is strictly greater than x, and v' is obtained from v through replacing y_i by x. To insert a letter x into a tableau $t = v_1 \cdots v_k$, one first inserts x into the bottom row v_k. Then, if $v_k x$ is not a row, $P(v_k x) = y v'_k$ and one inserts y into v_{k-1}, and so on. The process terminates when one reaches the top row v_1, or when a letter has been inserted at the right end of a row. For example, the insertion of 3

5.2. Greene's invariants and the plactic monoid

in the tableau t above goes through the following steps:

$$P(1112444 \cdot 3) = 4 \cdot 1112344,$$
$$P(223357 \cdot 4) = 5 \cdot 223347,$$
$$P(4556 \cdot 5) = 6 \cdot 4555,$$
$$P(68 \cdot 6) = 8 \cdot 66,$$

and the result is

$$P(t \cdot 3) = 8 \cdot 66 \cdot 4555 \cdot 223347 \cdot 1112344.$$

In a more formal way, the map P is defined recursively by

$$P(tx) = \begin{cases} tx & \text{if } v_k x \text{ is a row,} \\ P(v_1 \cdots v_{k-1} y) v'_k & \text{if } P(v_k x) = y v'_k \end{cases}$$

for a tableau t with row decomposition $t = v_1 \cdots v_k$, and for an arbitrary word $w \in A^*$, $P(wx) = P(P(w)x)$.

As an example of the general case, the successive steps of the calculation of $P(132541)$ are

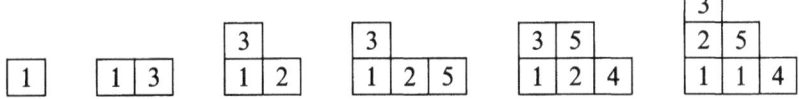

THEOREM 5.1.1. *The maximal length of a nondecreasing subword of w is equal to the length of the bottom row of $P(w)$.*

Similarly, the maximal length of a decreasing subword of w is equal to the height of the first column of $P(w)$.

For example, the maximal nondecreasing subwords of $w = 132541$ are 125, 124, 135 and 134. Note that 114, the bottom row of $P(w)$, is not a subword of w.

Schensted's theorem will be proved in the next section. Actually, we will prove a more general result due to C. Greene, which gives an interpretation of the lengths of all rows and the heights of all columns of $P(w)$.

5.2. Greene's invariants and the plactic monoid

For $w \in A^*$, let $l_k(w)$ be the maximum of the sum of the lengths of k disjoint nondecreasing subwords of w. Similarly, let $l'_k(w)$ be the maximum of the sum of the lengths of k decreasing subwords of w.

Let $\lambda = (\lambda_1, \ldots, \lambda_r)$ be the shape of $P(w)$, and let $\lambda' = (\lambda'_1, \ldots, \lambda'_s)$ be the conjugate partition.

THEOREM 5.2.1. *For $k = 1, \ldots, r$, $\lambda_k = l_k(w) - l_{k-1}(w)$, and for $k = 1, \ldots, s$, $\lambda'_k = l'_k(w) - l'_{k-1}(w)$ (where $l_0(w) = l'_0(w) = 0$).*

To prove this theorem, it is natural to investigate the relationship between two words having the same Schensted tableau. Therefore, we introduce an equivalence relation $\sim$ on A^* defined by

$$u \sim v \iff P(u) = P(v).$$

For words of length ≤ 2, one has $u \sim v \Leftrightarrow u = v$, since each such word is either a row or a column. The first nontrivial relations occur in length 3, and come from the tableaux of shape $(2, 1)$. With three letters $x < y < z$ we have four nonmonotonic words whose P-symbols are

$$P(xzy) = P(zxy) = \begin{array}{|c|c|}\hline z & \\\hline x & y \\\hline\end{array}, \quad P(yzx) = P(yxz) = \begin{array}{|c|c|}\hline y & \\\hline x & z \\\hline\end{array}, \qquad (5.2.1)$$

and similarly, with two distinct letters $x < y$,

$$P(xyx) = P(yxx) = \begin{array}{|c|c|}\hline y & \\\hline x & x \\\hline\end{array}, \quad P(yxy) = P(yyx) = \begin{array}{|c|c|}\hline y & \\\hline x & y \\\hline\end{array}. \qquad (5.2.2)$$

We will prove in what follows that $\sim$ is in fact the congruence on A^* generated by the relations implied by (5.2.1), (5.2.2). It is the quotient of the free monoid by these relations that will be the main object of this chapter.

DEFINITION 5.2.2. The *plactic monoid* on the alphabet A is the quotient $\mathrm{Pl}(A) = A^* / \equiv$, where $\equiv$ is the congruence generated by the *Knuth relations*

$$xzy \equiv zxy \quad (x \leq y < z), \qquad (5.2.3)$$
$$yxz \equiv yzx \quad (x < y \leq z). \qquad (5.2.4)$$

The first step in proving Greene's theorem is

PROPOSITION 5.2.3. *Every word is congruent to its Schensted tableau, that is,*

$$w \equiv P(w).$$

5.2. Greene's invariants and the plactic monoid

Proof. By definition of $\equiv$, the proposition is true for $|w| \leq 3$. We proceed by induction on $|w|$. Assume that for a word w we have $P(w) \equiv w$, and let x be a letter. We have to show that $P(wx) \equiv wx$, or equivalently $P(wx) \equiv P(w) \cdot x$. The definition of the map P allows us to reduce this verification to the case where w is a row. Assuming this, if wx is a row then $P(wx) = wx$, and otherwise, $P(wx) = yw'$ where y is the leftmost letter in w which is $> x$, and w' is obtained from w by replacing y by x. Then, writing $w = uyv$, we have $wx \equiv uyxv$ by a sequence of applications of (5.2.4), and $uyxv \equiv yuxv$ by a sequence of applications of (5.2.3). ∎

Next, we show

PROPOSITION 5.2.4. *If $w \equiv w'$, then $l_k(w) = l_k(w')$ for all k.*

Proof. We can assume that w' is obtained from w by a single Knuth transformation. Let us write, for instance,

$$w = uxzyv, \qquad w' = uzxyv \qquad (x \leq y < z).$$

Clearly, all nondecreasing subwords of w' are also subwords of w. Hence, $l_k(w) \geq l_k(w')$. Conversely, let $(w_1, \ldots, w_k)$ be a k-tuple of disjoint nondecreasing subwords of w. Then, w_i is also a subword of w', unless $w_i = u'xzv'$, where u' and v' are subwords of u and v. If y does not occur in any of the remaining w_j, then w_i can be replaced by $w'_i = u'xyv'$, which is a nondecreasing subword of w'. Otherwise, if some $w_j = u''yv''$, then one replaces the pair (w_i, w_j) by $w'_i = u'xyv''$ and $w'_j = u''zv'$. The case of a Knuth transformation of type (5.2.4) is similar. Therefore, we have $l_k(w) \leq l_k(w')$. ∎

Thus the integers $l_k(w)$ are not modified by Knuth's transformations (5.2.3) (5.2.4). They are called *Greene's plactic invariants*. Two other important plactic invariants, the charge and cocharge, will be studied in Section 5.6.

Proof of Theorem 5.2.1. Using Propositions 5.2.3 and 5.2.4, the only thing to prove is that for a tableau t of shape λ, $l_k(t) = \lambda_1 + \cdots + \lambda_k$. Taking for $w_1, \ldots, w_k$ the k longest rows of t, we see that $l_k(t) \geq \lambda_1 + \cdots + \lambda_k$. Conversely, a nondecreasing subword w of t uses at most one letter from each column of the planar representation of t, therefore k disjoint nondecreasing subwords can use at most $\lambda_1 + \cdots + \lambda_k$ letters of t. ∎

We are now in a position to prove the cross-section theorem:

THEOREM 5.2.5. *The equivalence $\sim$ coincides with the plactic congruence. In particular, each plactic class contains exactly one tableau.*

Proof. Let us assume that $w \sim w'$. Then, by Proposition 5.2.3,

$$w \equiv P(w) = P(w') \equiv w'.$$

Conversely, suppose that $w \equiv w'$. Then, from Proposition 5.2.4 and Theorem 5.2.1 we see that $P(w)$ and $P(w')$ have the same shape. Now, let z be the greatest letter of w and w', and write $w = uzv$, $w' = u'zv'$, where z does not occur either in v or in v'. Then, we claim that $uv \equiv u'v'$. Indeed, we can assume that w and w' differ by a single Knuth transformation. If z is not involved in this transformation, then either $u \equiv u'$ and $v = v'$, or $u = u'$ and $v \equiv v'$. And if z is involved, erasing z in (5.2.3) or (5.2.4) leaves us with $xy = xy$ or $yx = yx$, so that $uv = u'v'$.

By induction on the length of w, we can assume that $P(uv) = P(u'v')$. From the description of Schensted's algorithm, since z is the greatest letter, it is clear that after erasing z in $P(uzv)$, one is left with $P(uv)$. Therefore, $P(w)$ is obtained from $P(uv)$ by adding a box z at a place imposed by the shape of $P(w)$, and since the same is true for w', we conclude that $P(w) = P(w')$. ∎

5.3. The Robinson–Schensted–Knuth correspondence

We have seen in the preceding section that the set $\text{Tab}(A)$ of all tableaux over the alphabet A is a cross-section of the canonical projection $\pi: A^* \to \text{Pl}(A) = A^*/\equiv$. It is now a natural question to investigate the structure of the plactic classes $\pi^{-1}(t)$, $t \in \text{Tab}(A)$. As we will see, the elements of $\pi^{-1}(t)$ are also parameterized by certain tableaux.

Let us say that a tableau is *standard* if its entries are the integers $1, 2, \ldots, n$, each of them occurring exactly once. The set of standard tableaux is denoted by STab. For a partition λ, we denote by $\text{Tab}(\lambda, A)$ (resp. $\text{STab}(\lambda)$) the set of tableaux over A (resp. of standard tableaux) of shape λ.

By keeping track of the successive steps of the insertion algorithm, one can define a map $Q: A^* \to \text{STab}$ such that $w \mapsto (P(w), Q(w))$ is one-to-one. More precisely, let $w = y_1 \cdots y_m$. Observe that a standard tableau t is nothing but a chain of partitions $\lambda^{(1)} \subset \lambda^{(2)} \subset \cdots \subset \lambda^{(m)}$ such that the diagram of $\lambda^{(i+1)}$ is obtained from that of $\lambda^{(i)}$ by adding one box, which is the one labeled $i+1$ in t. Now, $Q(w)$ is by definition the standard tableau encoding the chain of shapes of $P(y_1), P(y_1 y_2), \ldots, P(w)$. For example,

5.3. The Robinson–Schensted–Knuth correspondence

the chain of insertions seen above gives

$$Q(132541) = \begin{array}{|c|c|c|} \hline 6 & & \\ \hline 3 & 5 & \\ \hline 1 & 2 & 4 \\ \hline \end{array}.$$

Clearly, $Q(w)$ has the same shape as $P(w)$.

THEOREM 5.3.1. *The map*

$$\rho : A^* \longrightarrow \coprod_\lambda \mathrm{Tab}(\lambda, A) \times \mathrm{STab}(\lambda)$$
$$w \longmapsto (P(w), Q(w))$$

is a bijection, called the Robinson–Schensted correspondence.

Proof. The inverse map ρ^{-1} can be explicitly constructed. The idea is that, given a row v and a letter y, there exist a unique row v' and letter x such that $yv \equiv v'x$. This shows that the insertion process described in Section 5.1 can be reversed, provided that one specifies the box to be erased. Given a pair $(t, t') \in \mathrm{Tab}(\lambda, A) \times \mathrm{STab}(\lambda)$, one constructs $w = \rho^{-1}(t, t')$ by deleting successively in t the boxes labeled $n, n - 1, \ldots, 1$ in t'. ∎

COROLLARY 5.3.2. *Q induces a bijection between the plactic class of each tableau t and $\mathrm{STab}(\lambda)$, where λ is the shape of t. In particular, the cardinality of the class of t is equal to*

$$f_\lambda := \mathrm{Card}\,\mathrm{STab}(\lambda).$$

Restricting ρ to the set of standard words on $A = \{1, 2, \ldots, n\}$, which can be identified with the symmetric group $\mathfrak{S}_n$, one obtains a bijection

$$\mathfrak{S}_n \longleftrightarrow \coprod_\lambda \mathrm{STab}(\lambda) \times \mathrm{STab}(\lambda). \tag{5.3.1}$$

It provides in particular a bijective proof of an identity of Frobenius:

$$n! = \sum_{|\lambda|=n} f_\lambda^2,$$

a special case of the fact that the cardinality of a finite group is equal to the sum of the squares of the dimensions of its irreducible representations (over $\mathbb{C}$).

As shown by the next theorem, there is some compatibility between the Robinson–Schensted map and the group structure of $\mathfrak{S}_n$.

THEOREM 5.3.3. *For $\sigma \in \mathfrak{S}_n$, $Q(\sigma) = P(\sigma^{-1})$.*

The original proof of Schützenberger proceeded by induction on n. We give below a simple derivation based on Greene's theorem.

To this end, it will be convenient to represent a permutation σ by a *biword* (or word in *biletters*, that is, pairs of letters $(a, b) \in A \times B$ in the product of two alphabets, denoted here for convenience by $\begin{bmatrix} a \\ b \end{bmatrix}$)

$$\sigma \leftrightarrow \begin{bmatrix} i_1 \cdots i_n \\ j_1 \cdots j_n \end{bmatrix}$$

where each $j_k = \sigma(i_k)$. Among the biwords representing σ, we have two distinguished ones $\begin{bmatrix} \text{id} \\ \sigma \end{bmatrix}$ and $\begin{bmatrix} \sigma^{-1} \\ \text{id} \end{bmatrix}$, which are obtained by sorting one of them using the lexicographic order on biletters with priority on the top or bottom row.

More generally, for a biword $\begin{bmatrix} u \\ v \end{bmatrix}$ where $u, v \in A^*$ are not necessarily standard, we denote by $\begin{bmatrix} u' \\ v' \end{bmatrix}$ the nondecreasing rearrangement of $\begin{bmatrix} u \\ v \end{bmatrix}$ for the lexicographic order with priority on the top row, and by $\begin{bmatrix} u'' \\ v'' \end{bmatrix}$ the nondecreasing rearrangement for the lexicographic order with priority on the bottom row. Thus, for

$$\begin{bmatrix} u \\ v \end{bmatrix} = \begin{bmatrix} 21335424 \\ 13652414 \end{bmatrix},$$

we have

$$\begin{bmatrix} u' \\ v' \end{bmatrix} = \begin{bmatrix} 12233445 \\ 31156442 \end{bmatrix} \text{ and } \begin{bmatrix} u'' \\ v'' \end{bmatrix} = \begin{bmatrix} 22514433 \\ 11234456 \end{bmatrix}.$$

The crucial property is the following.

LEMMA 5.3.4. *For any biword $\begin{bmatrix} u \\ v \end{bmatrix}$, the tableaux $P(v')$ and $P(u'')$ have the same shape.*

Proof. Let $\begin{bmatrix} u \\ v \end{bmatrix} = \begin{bmatrix} u_1 \cdots u_m \\ v_1 \cdots v_m \end{bmatrix}$ and consider a nondecreasing subword $\beta = v_{i_1} \cdots v_{i_r}$ of v'. Then, by definition of $\begin{bmatrix} u' \\ v' \end{bmatrix}$, $\alpha = u_{i_1} \cdots u_{i_r}$ is also

5.3. The Robinson–Schensted–Knuth correspondence

nondecreasing, and
$$\begin{bmatrix} u_{i_1} \\ v_{i_1} \end{bmatrix} \leq \cdots \leq \begin{bmatrix} u_{i_r} \\ v_{i_r} \end{bmatrix}$$
for *both* lexicographic orders. Therefore, α is also a nondecreasing subword of u''. From this remark, we see that there is a bijection between the k-tuples of disjoint nondecreasing subwords of v' and those of u''. By Theorem 5.2.1 the conclusion follows. ∎

Proof of Theorem 5.3.3. Let $\sigma \in \mathfrak{S}_n$ and $\begin{bmatrix} u' \\ v' \end{bmatrix} = \begin{bmatrix} \text{id} \\ \sigma \end{bmatrix}$, $\begin{bmatrix} u'' \\ v'' \end{bmatrix} = \begin{bmatrix} \sigma^{-1} \\ \text{id} \end{bmatrix}$.
The left factors of σ are encoded by the biwords
$$\begin{bmatrix} u(k)' \\ v(k)' \end{bmatrix} = \begin{bmatrix} 1 & 2 & \cdots & k \\ \sigma_1 & \sigma_2 & \cdots & \sigma_k \end{bmatrix}$$
for which we have
$$\begin{bmatrix} u(k)'' \\ v(k)'' \end{bmatrix} = \begin{bmatrix} \sigma^{-1}|_{[1,k]} \\ (\sigma_1 \cdots \sigma_k)\uparrow \end{bmatrix}$$
where $(\sigma_1 \cdots \sigma_k)\uparrow$ is the increasing rearrangement of the left factor $\sigma_1 \cdots \sigma_k$, and for a word $w \in A^*$ and a subset B of A, $w|_B$ denotes the subword of w obtained by erasing the letters which are not in B. From Lemma 5.3.4, at each step of the insertion algorithm, we have that $P(\sigma_1 \cdots \sigma_k)$ and $P(\sigma^{-1}|_{[1,k]})$ have the same shape. So at the end, $P(\sigma^{-1}) = Q(\sigma)$. ∎

In fact, Theorem 5.3.3 can be readily generalized to give a similar result for the insertion tableau $Q(w)$ of an arbitrary word $w \in A^*$. To do this, we need the notion of *standardization*.

Let $x_1 < x_2 < \cdots < x_r$ be the letters occurring in w, with respective multiplicities $m_1, \ldots, m_r$. By labeling from 1 to m_1 the occurrences of x_1, reading from left to right, then from $m_1 + 1$ to $m_1 + m_2$ the occurrences of x_2, and so on, we get a standard word, denoted by std(w). For example

$$\text{std}(31156442) = 41278563.$$

This defines in particular the standardization of a tableau. It is immediate to check from Knuth's relations that

LEMMA 5.3.5. *If $w \equiv w'$, then $\text{std}(w) \equiv \text{std}(w')$. In particular, $P(\text{std}(w)) = \text{std}(P(w))$.* ∎

It is also clear from the description of the Robinson–Schensted algorithm that the following holds.

LEMMA 5.3.6. $Q(w) = Q(\text{std}(w))$. ∎

We can now state

COROLLARY 5.3.7. *For any $w \in A^*$, $Q(w) = P(\text{std}(w)^{-1})$.*

Proof. By Theorem 5.3.3, $P(\text{std}(w)^{-1}) = Q(\text{std}(w))$, which is equal to $Q(w)$ by Lemma 5.3.6. ∎

In the Robinson–Schensted correspondence for nonstandard words, there is a dissymmetry between the left tableau $P(w)$ and the right tableau $Q(w)$. Lemma 5.3.4 shows the way to restore the symmetry, by extending the correspondence to commutative classes of biwords, i.e. monomials in commutative biletters $\binom{x}{y}$. Given two words $u = u_1 \cdots u_m$ and $v = v_1 \cdots v_m$ of the same length, we denote by $\binom{u}{v} = \binom{u_1}{v_1} \cdots \binom{u_m}{v_m}$ the associated monomial in commutative biletters (not to be confused with the biword $\begin{bmatrix} u \\ v \end{bmatrix}$).

DEFINITION 5.3.8. Let $\binom{u}{v}$ be a monomial, and $\begin{bmatrix} u' \\ v' \end{bmatrix}$, $\begin{bmatrix} u'' \\ v'' \end{bmatrix}$ be the two biwords associated as above to the biword $\begin{bmatrix} u \\ v \end{bmatrix}$. The *Knuth correspondence* κ is defined by

$$\kappa \binom{u}{v} = (P(v'), P(u'')).$$

By Corollary 5.3.7, we recover the Robinson–Schensted correspondence by encoding $w = y_1 \cdots y_m$ as the monomial $\binom{1}{y_1} \cdots \binom{m}{y_m}$. By Lemma 5.3.4, we know that $P(v')$ and $P(u'')$ have the same shape. It will follow from the alternative description given below that κ is a bijection between monomials in biletters and pairs of tableaux of the same shape. Recall that the *evaluation* of a word is the vector $\text{ev}(w) = (|w|_{a_1}, |w|_{a_2}, \ldots, |w|_{a_n})$, where $A = \{a_1, \ldots, a_n\}$.

PROPOSITION 5.3.9. *$P(u'')$ is the unique tableau of evaluation $\text{ev}(u'')$ such that $\text{std}(P(u'')) = Q(v')$.*

5.3. The Robinson–Schensted–Knuth correspondence

Proof. By lexicographic sorting of $\begin{bmatrix} \mathrm{std}(u) \\ \mathrm{std}(v) \end{bmatrix}$ we have $(\mathrm{std}(v)')^{-1} = \mathrm{std}(u)''$. Since lexicographic sorting obviously commutes with standardization, it follows that $(\mathrm{std}(v'))^{-1} = \mathrm{std}(u'')$. Hence,

$$\begin{aligned} Q(v') &= P((\mathrm{std}(v')^{-1})) \quad \text{(Corollary 5.3.7)} \\ &= P(\mathrm{std}(u'')) \\ &= \mathrm{std}(P(u'')) \quad \text{(Lemma 5.3.5)}. \end{aligned}$$ ∎

Therefore, to compute the inverse image of a pair of tableaux (t, t') under the Knuth correspondence, we can apply the inverse Robinson–Schensted map to $(t, \mathrm{std}(t'))$ to get $v' = \rho^{-1}(t, \mathrm{std}(t'))$. Then, $\kappa^{-1}(t, t') = \begin{pmatrix} t' \uparrow \\ v' \end{pmatrix}$.

Note that the symmetry

$$\kappa \begin{pmatrix} u \\ v \end{pmatrix} = (t, t') \iff \kappa \begin{pmatrix} v \\ u \end{pmatrix} = (t', t),$$

which generalizes Theorem 5.3.3, is incorporated in the definition of κ. In particular, taking $t' = t$, κ establishes a bijection between $\mathrm{Tab}(A)$ and the set of *symmetric* monomials in biletters, i.e. those such that $\begin{pmatrix} u \\ v \end{pmatrix} = \begin{pmatrix} v \\ u \end{pmatrix}$ (which amounts to saying that for any $x, y \in A$, $\begin{pmatrix} x \\ y \end{pmatrix}$ and $\begin{pmatrix} y \\ x \end{pmatrix}$ occur with the same multiplicity). As an immediate consequence of this observation, we can compute the generating series of the numbers

$$d_\alpha := |\{t \in \mathrm{Tab}(A) \mid \mathrm{ev}(t) = \alpha\}| \quad (\alpha \in \mathbb{N}^A)$$

which are the cardinalities of the multihomogeneous components of the plactic monoid.

Theorem 5.3.10. *Let $\xi_1, \xi_2, \ldots$ be commuting indeterminates. Then,*

$$\sum_{\alpha \in \mathbb{N}^A} d_\alpha \xi^\alpha = \prod_i \frac{1}{1 - \xi_i} \prod_{i<j} \frac{1}{1 - \xi_i \xi_j}.$$

Proof. The commutative image $\underline{t}$ of a tableau t under $a_i \mapsto \xi_i$ is obtained from $\begin{pmatrix} u \\ v \end{pmatrix} = \kappa^{-1}(t, t)$ by mapping each biletter $\begin{pmatrix} i \\ j \end{pmatrix}$ to $(\xi_i \xi_j)^{1/2}$. Now, the generating series of all symmetric monomials in biletters is clearly

$$\prod_i \frac{1}{1 - \begin{pmatrix} i \\ i \end{pmatrix}} \prod_{i<j} \frac{1}{1 - \begin{pmatrix} i \\ j \end{pmatrix}\begin{pmatrix} j \\ i \end{pmatrix}}.$$ ∎

COROLLARY 5.3.11. *For Card $A = n$, the cardinality of the homogeneous component of degree k of $\mathrm{Pl}(A)$ is equal to the coefficient of z^k in*

$$\frac{1}{(1-z)^n} \cdot \frac{1}{(1-z^2)^{n(n-1)/2}}.$$ ∎

5.4. Schur functions and the Littlewood–Richardson rule

Let $\xi_1, \xi_2, \ldots, \xi_n$ be commuting indeterminates as in the preceding section, and retain the notation $w \mapsto \underline{w}$ for the commutative image $a_i \mapsto \xi_i$ of a word $w \in A^*$.

DEFINITION 5.4.1. Let λ be a partition. The generating function

$$s_\lambda(\xi_1, \ldots, \xi_n) = \sum_{t \in \mathrm{Tab}(\lambda, A)} \underline{t}$$

is called a *Schur function*.

Although it is not obvious from this definition, s_λ is a symmetric polynomial in $\xi_1, \ldots, \xi_n$ (this will be proved in Section 5.6). Most of the combinatorial constructions of Section 5.3 imply interesting and classical Schur function identities. For example, Schur's identity 5.3.10 can be rewritten as

$$\sum_\lambda s_\lambda(\xi_1, \ldots, \xi_n) = \prod_i \frac{1}{1 - \xi_i} \prod_{i<j} \frac{1}{1 - \xi_i \xi_j}.$$

From Theorem 5.3.1 we get

$$\frac{1}{1 - (\xi_1 + \cdots + \xi_n)} = \sum_\lambda f_\lambda s_\lambda(\xi_1, \ldots, \xi_n).$$

Indeed, the left-hand side is clearly the generating function of A^*.

Finally, from the bijectivity of Knuth's correspondence, we obtain a classical and fundamental identity which can be traced back to Cauchy. To state it, we need a second set $\eta_1, \ldots, \eta_n$ of commuting variables. Sending the biletter $\begin{pmatrix} a_i \\ a_j \end{pmatrix}$ onto $\xi_i \eta_j$ and the pair (t, t') to the product of the commutative image of t in the variables ξ and of t' in the variables η, we get

THEOREM 5.4.2.
$$\prod_{i,j} \frac{1}{1 - \xi_i \eta_j} = \sum_\lambda s_\lambda(\xi) s_\lambda(\eta).$$

5.4. Schur functions and the Littlewood–Richardson rule

Group-theoretical arguments show that a product of Schur functions is equal to a positive sum of Schur functions:

$$s_\lambda(\xi)s_\mu(\xi) = \sum_\nu c^\nu_{\lambda\mu} s_\nu(\xi) \qquad (5.4.1)$$

where $c^\nu_{\lambda\mu} \in \mathbb{N}$. The calculation of the coefficients $c^\nu_{\lambda\mu}$ is of interest in many fields. A combinatorial interpretation of these numbers implying an efficient algorithm for their computation was given without proof by Littlewood and Richardson.

The most illuminating proof of this rule proceeds by lifting the calculus of Schur functions to the algebra $\mathbb{Z}[\mathrm{Pl}(A)]$ of the plactic monoid, introducing the *plactic Schur function*

$$S_\lambda(A) = \sum_{t \in \mathrm{Tab}(\lambda, A)} t,$$

where tableaux are evaluated in the plactic monoid. This plactic Schur function can be seen as the projection in $\mathbb{Z}[\mathrm{Pl}(A)]$ of any one of the *free Schur functions*

$$\mathbf{S}_t(A) = \sum_{Q(w)=t} w \ \in \mathbb{Z}\langle A \rangle$$

indexed by $t \in \mathrm{STab}(\lambda)$. In fact the Littlewood–Richardson rule will be deduced from a statement in the free algebra $\mathbb{Z}\langle A \rangle$.

Recall that the shuffle product is the linear operation on $\mathbb{Z}\langle A \rangle$ defined by

$$ua \sqcup\!\sqcup vb = (u \sqcup\!\sqcup vb)a + (ua \sqcup\!\sqcup v)b$$

for all words u, v and letters a, b; with the empty word as neutral element.

THEOREM 5.4.3. *Let A' and A'' be two subalphabets such that $a' < a''$, for all $a \in A'$, $a'' \in A''$. For $t' \in \mathrm{Tab}(A')$ and $t'' \in \mathrm{Tab}(A'')$ we have*

$$\left(\sum_{P(w')=t'} w' \right) \sqcup\!\sqcup \left(\sum_{P(w'')=t''} w'' \right) = \sum_{t \in \mathrm{Sh}(t',t'')} \sum_{P(w)=t} w$$

where $\mathrm{Sh}(t', t'')$ is the set of all tableaux t such that $t|_{A'} = t'$ and $P(t|_{A''}) = t''$, that is, of all tableaux t occurring in the shuffle product of t' and a word in the plactic class of t''.

Thus the shuffle of a plactic class of A' and a plactic class of A'' is a union of plactic classes of A (identifying a class and the sum of its elements). It is in fact a direct consequence of the following.

LEMMA 5.4.4. *Let I be an interval of A. Then*

$$w \equiv w' \Rightarrow w|_I \equiv w'|_I .$$

Proof. It is enough to check the lemma in the case when w' differs from w by a single Knuth transformation, and this amounts to the observation that erasing x or z in (5.2.3) or (5.2.4), we are left with $xy = xy$ or $yz = yz$. ∎

Proof of Theorem 5.4.3. The words occurring in the shuffle are exactly those w such that $w|_{A'} \equiv t'$ and $w|_{A''} \equiv t''$. By Lemma 5.4.4, this set of words is saturated with respect to the plactic congruence, hence is a union of plactic classes. ∎

We can now state the plactic version of the Littlewood–Richardson rule.

THEOREM 5.4.5. *The plactic Schur functions span a commutative subalgebra of* $\mathbb{Z}[\text{Pl}(A)]$ *and we have*

$$S_\lambda(A)S_\mu(A) = \sum_\nu c^\nu_{\lambda\mu} S_\nu(A) ,$$

where the $c^\nu_{\lambda\mu}$ are the same as in (5.4.1). *In particular $c^\nu_{\lambda\mu}$ is equal to the number of factorizations in* $\text{Pl}(A)$ *of any tableau* $t \in \text{Tab}(\nu, A)$ *as a product $t't''$ with $t' \in \text{Tab}(\lambda, A)$ and $t'' \in \text{Tab}(\mu, A)$.*

Proof. We first work in the free associative algebra $\mathbb{Z}\langle A\rangle$ and consider a product $\mathbf{S}_{t'}(A)\mathbf{S}_{t''}(A)$ where t', t'' are arbitrary standard tableaux of respective shapes λ and μ, with $p = |\lambda|$, $q = |\mu|$. We identify as above a word w' of length p with a monomial in commutative biletters:

$$w' = \begin{pmatrix} 1 \cdots p \\ w' \end{pmatrix} .$$

Then, by reordering biletters, we can write in view of Proposition 5.3.9

$$\mathbf{S}_{t'} = \sum_{Q(w')=t'} \begin{pmatrix} 1 \cdots p \\ w' \end{pmatrix} = \sum_{P(u)=t'} \overrightarrow{\begin{pmatrix} u \\ r' \end{pmatrix}} ,$$

where the notation means that the second sum is over all words u and r' such that the biword $\begin{bmatrix} u \\ r' \end{bmatrix}$ is increasing for the lexicographic order with bottom priority, and that $P(u) = t'$. Similarly, using for w'' of length q the identification

$$w'' = \begin{pmatrix} (p+1) \cdots (p+q) \\ w'' \end{pmatrix}$$

5.4. Schur functions and the Littlewood–Richardson rule

we can express $S_{t''}$ as

$$S_{t''} = \sum_{P(v)=t''[p]} \overrightarrow{\binom{v}{r''}},$$

where $t''[p]$ denotes the tableau obtained from t'' by adding p to all its entries. Now sorting lexicographically (with bottom priority) any of the biwords $\begin{bmatrix} u \\ r' \end{bmatrix}\begin{bmatrix} v \\ r'' \end{bmatrix}$, one gets a biword $\begin{bmatrix} w \\ r \end{bmatrix}$ such that w occurs in $u \sqcup\!\sqcup v$. Conversely, all increasing biwords $\begin{bmatrix} w \\ r \end{bmatrix}$ such that w occurs in $u \sqcup\!\sqcup v$ arise in this way from the sorting of a unique product $\begin{bmatrix} u \\ r' \end{bmatrix}\begin{bmatrix} v \\ r'' \end{bmatrix}$ of increasing biwords. Thus, by Theorem 5.4.5,

$$S_{t'}S_{t''} = \sum_t \sum_{P(w)=t} \overrightarrow{\binom{w}{r}},$$

where the outer sum is over all standard tableaux t which occur in the shuffle of t' and of a word congruent to $t''[p]$. Hence

$$S_{t'}S_{t''} = \sum_t S_t, \qquad (5.4.2)$$

sum over the same tableaux t, and taking the plactic image we obtain

$$S_\lambda S_\mu = \sum_v c_{\lambda\mu}^v S_v \qquad (5.4.3)$$

where $c_{\lambda\mu}^v$ is the number of standard tableaux of shape v which occur in the shuffle of t' and of a word in the class of $t''[p]$. Taking the commutative image of (5.4.3), we see that the $c_{\lambda\mu}^v$ are the same as in (5.4.1), which implies that the plactic Schur functions span a subalgebra of $\mathbb{Z}[\mathrm{Pl}(A)]$ isomorphic to the commutative algebra spanned by the ordinary Schur functions. Finally the interpretation of $c_{\lambda\mu}^v$ in terms of factorizations in $\mathrm{Pl}(A)$ follows directly from the definition of plactic Schur functions. ∎

As an illustration of (5.4.2), one can check that for

$$t' = t'' = \begin{array}{|c|c|} \hline 3 & \\ \hline 1 & 2 \\ \hline \end{array}$$

the product $S_{t'}S_{t''}$ is equal to $\sum_t S_t$ where t ranges over the following tableaux:

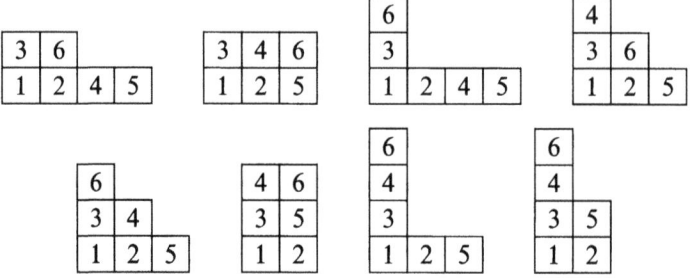

COROLLARY 5.4.6. *Let $R(\lambda, k)$ (resp. $C(\lambda, k)$) be the set of partitions whose diagrams are obtained by adding k boxes to the diagram of λ, no two of them being added in the same column (resp. in the same row). Then,*

$$S_\lambda S_{(k)} = \sum_{v \in R(\lambda,k)} S_v,$$

$$S_\lambda S_{(1^k)} = \sum_{v \in C(\lambda,k)} S_v.$$

Proof. Let $m = |\lambda|$. To calculate $S_t \cdot S_{12\cdots k}$, we have to look for the standard tableaux in the shuffle of the plactic class of t with the one-element class

$$(m+1)(m+2)\cdots(m+k).$$

Clearly, these tableaux can only be obtained by dispatching at the periphery of t the letters $(m+1), \ldots, (m+k)$ from left to right and in this order, and the resulting shapes are exactly those of $R(\lambda, k)$. The second formula is proved similarly. ∎

To recover the classical formulation of Littlewood and Richardson, we need the notion of a *Yamanouchi word*. We say that w is a Yamanouchi word on $A = \{1, 2, \ldots, n\}$ if any right factor v of w satisfies $|v|_1 \geq |v|_2 \geq \cdots \geq |v|_n$.

LEMMA 5.4.7. *The Yamanouchi words of a given evaluation $\mu = (\mu_1, \ldots, \mu_n)$ form a single plactic class whose representative tableau is the Yamanouchi tableau*

```
...
2 2 ... 2
1 1 ... ... 1
```

– that is, the unique tableau with shape and evaluation μ.

5.4. Schur functions and the Littlewood–Richardson rule

Proof. It is immediate to check that if w is a Yamanouchi word, and if w' is obtained from w by a single Knuth transformation, then w' is also a Yamanouchi word. Therefore, a plactic class which contains a Yamanouchi word contains only Yamanouchi words. Now, a tableau is a Yamanouchi word if and only if its bottom row contains only 1's, the next row contains only 2's, and so on. Hence there is a unique Yamanouchi tableau, namely, the unique tableau of shape μ and evaluation μ, and the lemma follows from Theorem 5.2.5. ∎

We can now see that the classical version of the Littlewood–Richardson rule is a direct consequence of (5.4.2). Indeed, to calculate $c_{\lambda\mu}^\nu$, we can choose for t' and t'' the standard tableaux of respective shapes λ and μ in which each row consists of consecutive integers. These tableaux are the standardized tableaux of the Yamanouchi tableaux of the same shapes, so that the words w'' in the plactic class of $t''[p]$ are precisely the shifted standardized words of the Yamanouchi words y'' of evaluation μ. Hence, if one erases in the tableaux t the entries of t', which are irrelevant, and replaces the word w'' by the unique Yamanouchi word y'' of which it is the standardized form, one obtains the classical Litewood-Richardson tableaux, i.e., the skew Yamanouchi tableaux of shape ν/λ and evaluation μ. Continuing the preceding example, one would obtain

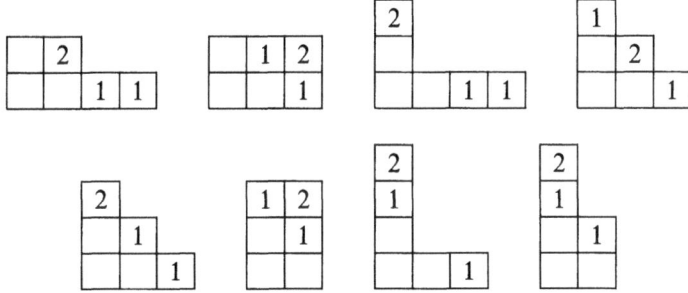

Another useful formulation of the rule is the following.

COROLLARY 5.4.8. *Let y_μ denote the unique Yamanouchi tableau of shape μ. Then $c_{\lambda\mu}^\nu$ is equal to the number of tableaux t of shape λ such that $t \cdot y_\mu$ is a Yamanouchi word of evaluation ν.*

Proof. By Theorem 5.4.5, $c_{\lambda\mu}^\nu$ is the number of factorizations $y_\nu = t \cdot t'$ in Pl(A), with $t \in \text{Tab}(\lambda, A)$ and $t' \in \text{Tab}(\mu, A)$. Equivalently, by Lemma 5.4.7, $c_{\lambda\mu}^\nu$ is the number of Yamanouchi words w of weight ν such that $w = t \cdot t'$ in A^*, for some $t \in \text{Tab}(\lambda, A)$ and $t' \in \text{Tab}(\mu, A)$. Then the right factor t' must be a Yamanouchi tableau, that is $t' = y_\mu$. ∎

For example, the coefficient $c_{(3,2),(2,1)}^{(4,3,1)}$ is equal to 2, corresponding to the following two tableaux t:

2	3	
1	1	2

2	2	
1	1	3

5.5. Coplactic operations

The set of words w having a given insertion tableau $t = Q(w)$ is called a *coplactic* class. In the preceding section we have seen that the sum $\mathbf{S}_t$ of the elements of a coplactic class is a pertinent lifting of a Schur function to the free algebra $\mathbb{Z}\langle A \rangle$. In this section, we show that coplactic classes can be endowed with a structure of colored graph.

We introduce linear operators e_i, f_i, σ_i, $i = 1, \ldots, n-1$, acting on $\mathbb{Z}\langle A \rangle$ in the following way. Consider first the case of the two-letter subalphabet $A_i = \{a_i, a_{i+1}\}$. Let $w = x_1 \cdots x_m$ be a word on A_i. Bracket every factor $a_{i+1} a_i$ of w. The letters which are not bracketed constitute a subword w_1 of w. Then bracket every factor $a_{i+1} a_i$ of w_1. There remains a subword w_2. Continue this procedure until it stops, giving a word w_k of type $w_k = a_i^r a_{i+1}^s = x_{j_1} \cdots x_{j_{r+s}}$. The image of w_k under e_i, f_i or σ_i is given by

$$e_i(a_i^r a_{i+1}^s) = \begin{cases} a_i^{r+1} a_{i+1}^{s-1} & (s \geq 1), \\ 0 & (s = 0), \end{cases}$$

$$f_i(a_i^r a_{i+1}^s) = \begin{cases} a_i^{r-1} a_{i+1}^{s+1} & (r \geq 1), \\ 0 & (r = 0), \end{cases}$$

$$\sigma_i(a_i^r a_{i+1}^s) = a_i^s a_{i+1}^s.$$

Let $w_k' = x_{j_1}' \cdots x_{j_{r+s}}'$ denote the image of w_k. The image of the initial word w is then $w' = y_1 \cdots y_m$, where $y_i = x_i'$ if $i \in \{j_1, \ldots, j_{r+s}\}$ and $y_i = x_i$ otherwise.

For example, if $w = (a_2 a_1) a_1 a_1 a_2 (a_2 a_1) a_1 a_1 a_1 a_2$, we have

$$w_1 = a_1 a_1 (a_2 a_1) a_1 a_1 a_2 \quad \text{and} \quad w_2 = a_1 a_1 a_1 a_1 a_2.$$

Thus,

$$e_1(w) = a_2 a_1 \; \underline{a_1 a_1} \; a_2 a_2 a_1 a_1 \; \underline{a_1 a_1 a_1},$$
$$f_1(w) = a_2 a_1 \; \underline{a_1 a_1} \; a_2 a_2 a_1 a_1 \; \underline{a_1 a_2 a_2},$$
$$\sigma_1(w) = a_2 a_1 \; \underline{a_1 a_2} \; a_2 a_2 a_1 a_1 \; \underline{a_2 a_2 a_2},$$

5.5. Coplactic operations

where the underlined letters are those of the subword w'_2. Finally, the general action of the operators e_i, f_i, σ_i on w is defined by the previous rules applied to the subword $w|_{A'_i}$, the other letters remaining unchanged.

THEOREM 5.5.1. *Let h be anyone of the operators e_i, f_i, σ_i.*
  (i) *Let $w \in A^*$ and suppose that $h(w) \neq 0$. Then $Q(h(w)) = Q(w)$.*
  (ii) *Let w' be congruent to w. Then $h(w) \equiv h(w')$.*

Proof. (i) Suppose first that $A = \{a_1, a_2\}$, and let us give the proof in the case $h = f_1$. Let $w \in A^*$ be such that $f_1 w \neq 0$. This means that $w = u a_1 v$ where $u \equiv (a_2 a_1)^k a_1^{r-1}$ $(r \geq 1)$, $v \equiv a_2^s (a_2 a_1)^l$ and that we have $f_1(w) = u a_2 v$. Clearly, $Q(u a_2) = Q(u a_1)$. Next, the insertion of v into $P(u a_2)$ will produce the same sequence of shapes as the insertion of v into $P(u a_1)$. Indeed, write $v = v_1 \cdots v_k$ and assume by induction that $P(u a_1 v_1 \cdots v_{r-1})$ and $P(u a_2 v_1 \cdots v_{r-1})$ have the same shape. If $v_r = a_2$, then clearly $P(u a_1 v_1 \cdots v_r)$ and $P(u a_2 v_1 \cdots v_r)$ will also have the same shape. If $v_r = a_1$, then since $v \equiv a_2^s (a_2 a_1)^l$, we see that $r \geq 2$ and that the tableau $P(u a_1 v_1 \cdots v_{r-1})$ has at least one a_2 in its bottom row. Thus the insertion of a_1 in both tableaux will produce again two tableaux of the same shape.

The proof is similar in the case $h = e_1$, and this also implies the case $h = \sigma_1$ since $\sigma_1 w$ is of the form either $f_1^p w$ or $e_1^q w$.

Consider now the general case $A = \{a_1, \ldots, a_n\}$, and suppose that $h = f_i, e_i$ or σ_i. By Corollary 5.3.7, we have to prove that $P(\operatorname{std}(h(w))^{-1}) = P(\operatorname{std}(w)^{-1})$. Recall that $\operatorname{std}(w)^{-1}$ is the word u'' obtained from the representation of w as the biword $\begin{bmatrix} u \\ v \end{bmatrix} = \begin{bmatrix} \operatorname{id} \\ w \end{bmatrix}$ (see Section 5.3). Set $w_1 = h(w)$ and $\begin{bmatrix} u_1 \\ v_1 \end{bmatrix} = \begin{bmatrix} \operatorname{id} \\ w_1 \end{bmatrix}$. Then, we can write $v'' = \alpha a_i^r a_{i+1}^s \beta$ where a_i and a_{i+1} do not occur in α and β, $v''_1 = \alpha a_i^{r'} a_{i+1}^{s'} \beta$ $(r+s = r'+s')$, $u'' = \gamma \varepsilon \delta$ where $|\alpha| = |\gamma|$ and $|\beta| = |\delta|$, and finally $u''_1 = \gamma \varepsilon_1 \delta$. By the above proof for a two-letter alphabet, $\varepsilon_1 \equiv \varepsilon$. Therefore, $u''_1 \equiv u''$ as required.

(ii) Suppose that w' differs from w by a single Knuth transformation, and let us take for example $h = f_i$. Write $w = \alpha xzy\beta$ and $w' = \alpha zxy\beta$, where we assume that $x < y < z$. Let a (resp. a') be the letter a_i of w which is changed into a_{i+1} by f_i. We claim that if a is a letter of α (resp. β), then a' is the letter occupying the same position in w'. This is clear because the transformation $xzy \to zxy$ does not modify the relative positions of consecutive letters a_i and a_{i+1}. Therefore, $f_i(w) \equiv f_i(w')$ trivially if a is a letter of α or of β. Otherwise, a is one of the letters x, y, z of w and a' is the same letter in w'. Hence, according as $a = x, y$

or z, we have

$$f_i(w) = \begin{cases} \alpha a_{i+1} zy\beta \\ \alpha xz a_{i+1}\beta \\ \alpha x a_{i+1} y\beta \end{cases} \equiv f_i(w') = \begin{cases} \alpha z a_{i+1} y\beta \\ \alpha zx a_{i+1}\beta \\ \alpha a_{i+1} xy\beta \end{cases}.$$

Note that in the case $a = y$, we must have $z \geq a_{i+2}$, because if $z = a_{i+1}$, $y = a_i$, then zy would be put between brackets. In the case $w = \alpha xyx\beta$ and $w' = \alpha yxx\beta$, the reasoning given above remains unchanged, except when $x = a_i$, $y = a_{i+1}$, and a does not belong to α or β. In this case, we have

$$f_i(w) = f_i(\alpha a_i a_{i+1} a_i \beta) = \alpha a_{i+1} a_{i+1} a_i \beta,$$

and

$$f_i(w') = f_i(\alpha a_{i+1} a_i a_i \beta) = \alpha a_{i+1} a_i a_{i+1} \beta \equiv f_i(w).$$

The case of a Knuth transformation $yxz \equiv yzx$ ($x < u \leq z$) is treated similarly. ∎

We shall now make use of the operators e_i, f_i to define a graph Γ on A^*. The vertices of this graph are all the words $w \in A^*$, and we put labeled arrows between words according to the following rule:

$$(w \xrightarrow{i} w') \iff (f_i w = w').$$

Note that if $f_i w = w' \neq 0$, then $e_i w' = w$, hence at each vertex w there is at most one incident arrow of color i (and also, by definition, at most one outgoing arrow of color i). Hence the subgraph obtained by erasing all arrows of color $j \neq i$ is extremely simple: it is just a collection of disjoint i-strings

$$w_1 \xrightarrow{i} w_2 \longrightarrow \cdots \xrightarrow{i} w_k$$

of various lengths $k \geq 0$. However, when all the colors are considered simultaneously, a rich combinatorial structure emerges. Let us call "connected components of Γ" the connected components of the underlying nonoriented unlabeled graph.

PROPOSITION 5.5.2. (i) *The connected components of Γ are the coplactic classes.*

(ii) *Two coplactic classes are isomorphic as subgraphs of Γ if and only if they are indexed by two standard tableaux of the same shape.*

Proof. (i) By Theorem 5.5.1(i), any connected component of Γ is contained in a coplactic class. Conversely, let w be a non-Yamanouchi word. Then there exists an index i such that $e_i w \neq 0$. If $w' = e_i w$ is not a Yamanouchi

5.5. Coplactic operations

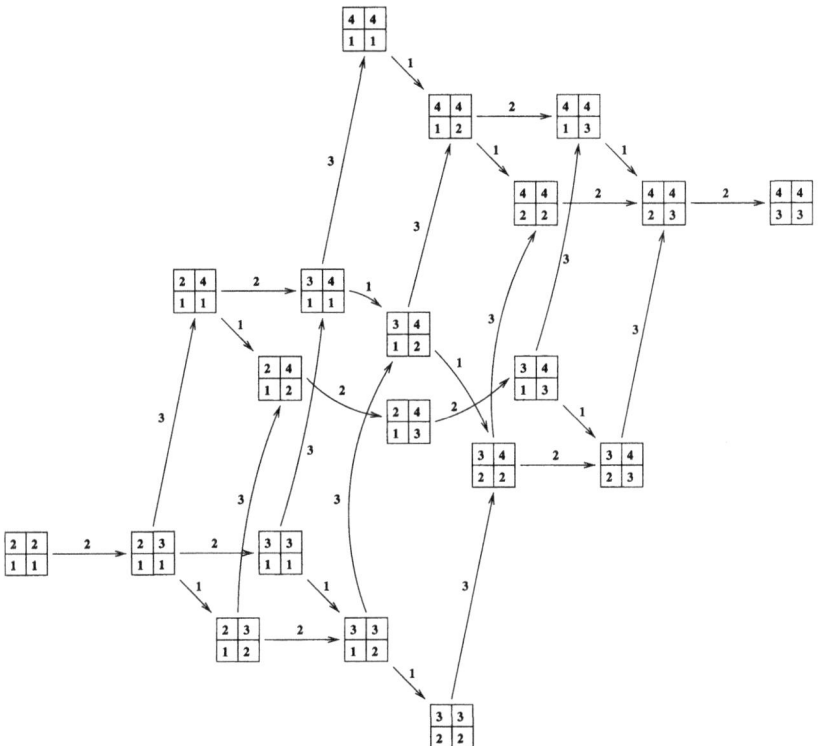

Figure 5.1. The graph structure of the coplactic class of $t = 2211$.

word, we can again find j such that $e_j w' = w'' \neq 0$. Iterating this procedure, we construct a chain of arrows connecting w to the unique Yamanouchi word in its coplactic class. Hence any two words of the same coplactic class are connected by a sequence of arrows going through the same Yamanouchi word.

(ii) It follows from Theorem 5.5.1(ii) that two coplactic classes indexed by standard tableaux of the same shape are isomorphic as subgraphs. Conversely, if two coplactic classes C, C' correspond to two standard tableaux t, t' of respective shapes $\lambda \neq \lambda'$, then the Yamanouchi words of these classes have evaluations λ and λ'. It is easy to check from the definition of f_i that for a Yamanouchi word of evaluations $\lambda = (\lambda_1, \ldots, \lambda_k)$, one has

$$\max\{p \mid f_i^p y \neq 0\} = \lambda_i - \lambda_{i+1} .$$

Hence the unique vertices of C and C' with no incident arrows have outgoing strings of different lengths, and C and C' are not isomorphic. ∎

As an illustration Figure 5.1 shows the graph structure of the coplactic class of $t = 2211$ for $A = \{1, 2, 3, 4\}$. These graphs are examples of *crystal graphs* in the sense of Kashiwara.

5.6. Cyclage and canonical embeddings

In this section we investigate the behavior of the previous constructions under circular permutations on words. We denote by ζ the bijection on A^* defined by $\zeta(x_1 x_2 \cdots x_n) = x_2 \cdots x_n x_1$ ($x_i \in A$).

PROPOSITION 5.6.1. *The cyclic shift ζ commutes with the maps σ_i.*

Proof. We have to prove that $\zeta \sigma_i(w) = \sigma_i \zeta(w)$, $w \in A^*$. If the first letter x_1 of w is different from a_i and a_{i+1} there is nothing to prove. Otherwise we distinguish four cases. Let us say that a letter x_k of w is *free* if it does not occur inside a pair of corresponding brackets at the end of the bracketing procedure described in Section 5.5. We then have the following cases: (i) $x_1 = a_i$ and no a_{i+1} is free; (ii) $x_1 = a_i$ and at least one a_{i+1} is free; (iii) $x_1 = a_{i+1}$ is free; (iv) $x_1 = a_{i+1}$ is not free. In each case, the verification is immediate. ∎

LEMMA 5.6.2. *Let $t \in A^*$ be a tableau and σ be any product of σ_i. Then the following conditions are equivalent:*
  (i) $\sigma(t) = t$;
  (ii) $\sigma(P(\zeta(t))) = P(\zeta(t))$.

Proof. Since ζ is bijective,
$$\sigma(t) = t \Leftrightarrow \zeta(\sigma(t)) = \zeta(t).$$

By Proposition 5.6.1, $\zeta(\sigma(t)) = \sigma(\zeta(t))$, which has the same Q-symbol as $\zeta(t)$ by Theorem 5.5.1(i). Thus
$$\sigma(t) = t \Leftrightarrow P(\sigma(\zeta(t))) = P(\zeta(t))$$

because of Theorem 5.3.1. Now, again by Theorem 5.5.1, $P(\sigma(w)) = \sigma(P(w))$ for any $w \in A^*$ and the statement follows. ∎

THEOREM 5.6.3. *The operators σ_i satisfy the Moore–Coxeter relations*

$$\sigma_i^2 = 1, \tag{5.6.1}$$
$$\sigma_i \sigma_j = \sigma_j \sigma_i \quad (|i - j| > 1), \tag{5.6.2}$$
$$\sigma_i \sigma_{i+1} \sigma_i = \sigma_{i+1} \sigma_i \sigma_{i+1}. \tag{5.6.3}$$

5.6. Cyclage and canonical embeddings

In other words, the map ρ sending the elementary transposition $(i, i+1)$ onto σ_i is a linear representation of the symmetric group $\mathfrak{S}_n$ in $\mathbb{Z}\langle A \rangle$.

Proof. Relations (5.6.1) and (5.6.2) are obviously satisfied. To prove (5.6.3), we have to show that $(\sigma_i \sigma_{i+1})^3(w) = w$ for any $w \in A^*$. From Theorem 5.5.1, it is enough to check this when $w = t$ is a tableau. Let $t = uv$ where v is the bottom row of t. By Lemma 5.6.2, it is equivalent to show that $(\sigma_i \sigma_{i+1})^3 P(uv) = P(vu)$. Now, in the tableau $t' = P(vu)$ all the letters a_1, a_2 lie in the bottom row. Writing $t' = u'v'$ and $t'' = P(v'u')$, and iterating, we construct a sequence $t^{(k)}$ of tableaux such that all the letters $a_1, \ldots, a_{k+1}$ of $t^{(k)}$ are in its first row, and such that

$$(\sigma_i \sigma_{i+1})^3(t) = t \iff (\sigma_i \sigma_{i+1})^3(t^{(k)}) = t^{(k)}.$$

But $t^{(n-1)}$ is a row, and $(\sigma_i \sigma_{i+1})^3(t^{(n-1)})$ has to be a row with the same evaluation, hence $(\sigma_i \sigma_{i+1})^3(t^{(n-1)}) = t^{(n-1)}$. ∎

COROLLARY 5.6.4. *The free Schur functions S_t are invariant under the above action of $\mathfrak{S}_n$. As a consequence, the commutative Schur functions $s_\lambda(\xi)$ are symmetric in the usual sense.*

We next investigate which transformations on tableaux arise when the map P is applied to circular permutations of words. Let Row(A) denote the subset of Tab(A) consisting of rows.

DEFINITION 5.6.5. *Let t be a tableau which is not a row. We put*

$$\mathscr{C}(t) = P(\zeta(t)).$$

The map $\mathscr{C}$: Tab$(A) \setminus$ Row$(A) \to$ Tab(A) is called cyclage.

To describe properties of the cyclage map, we need to use a plactic invariant on words called *cocharge*. Let w be a word. Let σ be any permutation such that $v = \sigma(w)$ has a dominant evaluation, that is

$$|v|_{a_1} \geq |v|_{a_2} \geq \cdots \geq |v|_{a_n}.$$

Write v on a circle, adding a "point at infinity" $*$ (see Figure 5.2). Then label each letter of v according to the following algorithm, reading the word clockwise.

1. Start at $*$ and label the first unlabeled a_1 with 0.

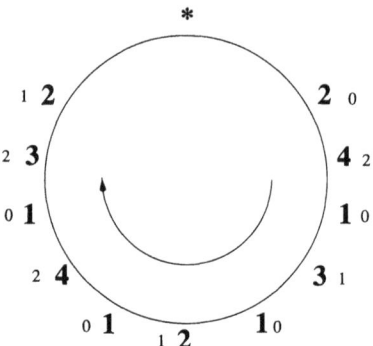

Figure 5.2. The calculation of the cocharge of $w = 23141213142$ (labels are written in small type).

2. After labeling an a_i with the number c, label the first unlabeled a_{i+1} with $c+1$ if it is obtained without crossing $*$, and with c otherwise. If there is no unlabeled a_{i+1}, go to the first step again, while there are still unlabeled letters.

The sum of all labels is called the *cocharge* of w, and is denoted by $\text{coch}(w)$. The complementary statistic $\text{ch}(w) = \max\{\text{coch}(v) \mid \text{ev}(v) = \text{ev}(w)\} - \text{coch}(w)$ is called the *charge* of w. For example, the cocharge of $w = 23141213142$ (whose evaluation is dominant) is equal to 9, as shown in Figure 5.2.

LEMMA 5.6.6. (i) *If* $\mathscr{C}(t) = t'$, *then for any* $\sigma \in \mathfrak{S}(A)$, $\mathscr{C}(\sigma(t)) = \sigma(t')$.
(ii) *If* $w \equiv w'$ *then* $\text{coch}(w) = \text{coch}(w')$.
(iii) *For* $t \in \text{Tab}(A) \setminus \text{Row}(A)$, *we have* $\text{coch}(\mathscr{C}(t)) = \text{coch}(t) - 1$.
(iv) *If* $\mathscr{C}(t) = \mathscr{C}(t')$ *and* $t \neq t'$, *then* t *and* t' *must have different shapes.*

Proof. Part (i) results clearly from Theorem 5.5.1 and Proposition 5.6.1.

As to (ii), we note that by definition $\text{coch}(\sigma(w)) = \text{coch}(w)$ for $\sigma \in \mathfrak{S}(A)$, hence using Theorem 5.5.1 (ii) we can assume that w and w' have a dominant evaluation. For such words, the above calculation of the charge proceeds by extracting from w a sequence of standard subwords $w^{(i)}$ such that
$$\text{coch}(w) = \sum_i \text{coch}(w_i).$$
Now, it is clear that replacing a factor $a_i a_j$ by $a_j a_i$ when $|i - j| \neq 1$ does not change these subwords, and thus does not change the cocharge.

5.6. Cyclage and canonical embeddings

Similarly, one checks that replacing a factor $a_{i+1}a_ia_i$ (resp. $a_{i+1}a_{i+1}a_i$) by $a_ia_{i+1}a_i$ (resp. $a_{i+1}a_ia_{i+1}$) does not modify these standard subwords. Hence, cocharge is invariant under plactic relations.

Let now $t = xw$, $x \in A$, be a tableau of dominant evaluation which is not a row. Then $x \neq a_1$, and the order in which letters are labeled in the word xw is the same as in wx. Thus, all labels are preserved except the label of x which is decreased by 1, and

$$\mathrm{coch}(P(wx)) = \mathrm{coch}(wx) = \mathrm{coch}(xw) - 1$$

which proves (iii).

To prove (iv), assume that t and t' are two different tableaux of the same shape, and write $t = xw$, $t' = x'w'$ with $x, x' \in A$. Then w and w' also are two tableaux of the same shape, say λ. By Corollary 5.4.6, $S_\lambda S_{(1)}$ is a multiplicity-free sum of tableaux in $\mathbb{Z}[\mathrm{Pl}(A)]$, hence $wx \neq w'x'$, that is, $\mathscr{C}(t) \neq \mathscr{C}(t')$. ∎

We shall now use the map $\mathscr{C}$ to define a graph structure on the set $\mathrm{Tab}(A)$. Specifically, consider the oriented graph with set of vertices $\mathrm{Tab}(A)$ and edges defined by

$$t \longrightarrow t' \quad \Longleftrightarrow \quad \mathscr{C}(t) = t'.$$

Since the cyclage map does not change the evaluation of tableaux this graph decomposes into the disjoint union of the subgraphs with sets of vertices $\mathrm{Tab}(\cdot, \mu)$ for all evaluations μ. The following theorem describes these subgraphs and shows how they can all be naturally embedded into the subgraph of standard tableaux.

THEOREM 5.6.7. (i) *The subgraph* $\mathrm{Tab}(\cdot, \mu)$ *is a rooted-tree with root the unique row-tableau of evaluation μ. Two evaluations which differ by a permutation give rise to isomorphic trees.*

(ii) *Let μ and v be two evaluations such that*

$$\mu_k = v_k \quad \text{for } k \neq i, j,$$
$$\mu_i > \mu_j,$$
$$v_i = \mu_i - 1,$$
$$v_j = \mu_j + 1.$$

Then there exists a unique embedding $\mathscr{I}_{\mu v}$ of $\mathrm{Tab}(\cdot, \mu)$ into $\mathrm{Tab}(\cdot, v)$ commuting with $\mathscr{C}$ and such that $\mathscr{I}_{\mu v}(t)$ has the same shape as t for all t.

(iii) *Similarly, for any evaluation μ there exists a unique embedding $\mathscr{I}_\mu$ of $\mathrm{Tab}(\cdot, \mu)$ into STab preserving shapes and commuting with $\mathscr{C}$.*

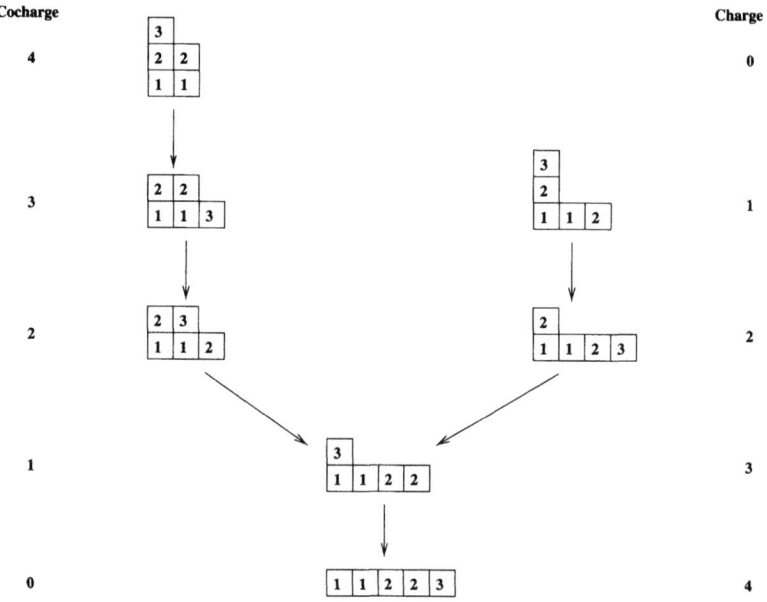

Figure 5.3. The tree structure of $\mathrm{Tab}(\cdot,(2,2,1))$.

Proof By Lemma 5.6.6(iii), the map $\mathscr{C}$ decreases cocharge by 1. Hence, the cyclage graph has no cycle and is a union of trees. It is clear from the definition of cocharge that row-tableaux are the only words with cocharge 0. Therefore, the subgraph $\mathrm{Tab}(\cdot,\mu)$ is a rooted-tree with root the unique row of evaluation μ. If $v = \sigma(\mu)$ for some $\sigma \in \mathfrak{S}(A)$, then, by Lemma 5.6.6(i), $\mathrm{Tab}(\cdot,\mu)$ and $\mathrm{Tab}(\cdot,v)$ are isomorphic as trees, which proves (i).

Let $\sigma \in \mathfrak{S}(A)$ be any permutation such that $\sigma(a_i) = a_1$ and $\sigma(a_j) = a_2$. Let $\mu' = \sigma(\mu)$ and $v' = \sigma(v)$. Given $t = xw$ in $\mathrm{Tab}(\cdot,\mu')$ its image under f_1 is nonzero and is the tableau in $\mathrm{Tab}(\cdot,v')$ obtained by changing the rightmost a_1 into a_2. This operation clearly commutes with $\mathscr{C}$, since the letter x which is cycled does not interfere, in the computation of $P(wx)$, with the subtableau of w consisting of the occurrences of a_1 and a_2. Therefore, the image of $\mathrm{Tab}(\cdot,\mu')$ under f_1 is a subtree of $\mathrm{Tab}(\cdot,v')$. Moreover, if two tableaux of the same shape have the same image under cyclage, then they are identical according to Lemma 5.6.6(iv). Hence there can be only one map from $\mathrm{Tab}(\cdot,\mu')$ to $\mathrm{Tab}(\cdot,v')$ preserving shape and commuting with $\mathscr{C}$. Finally, using σ^{-1}, one obtains from this embedding of $\mathrm{Tab}(\cdot,\mu')$ in $\mathrm{Tab}(\cdot,v')$ an embedding of $\mathrm{Tab}(\cdot,\mu)$ in $\mathrm{Tab}(\cdot,v)$ with the same properties, and (ii) is proved.

5.6. Cyclage and canonical embeddings

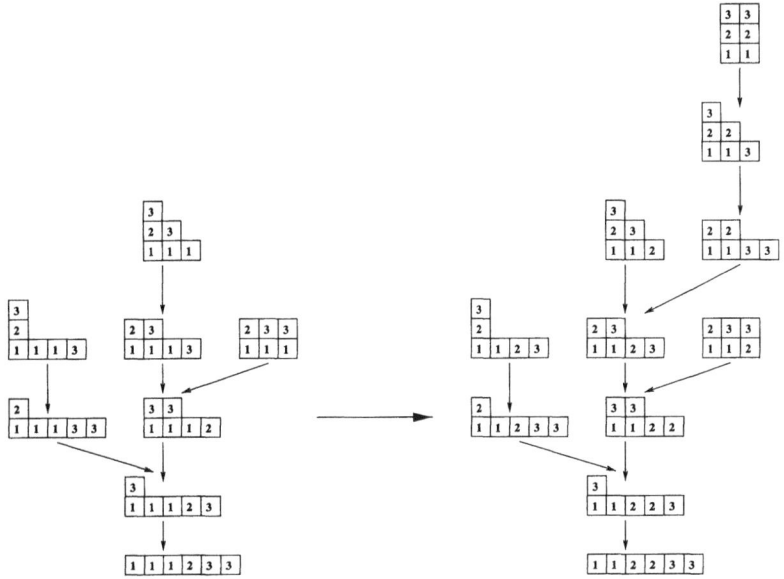

Figure 5.4. The embedding of $\text{Tab}(\cdot, (3, 1, 2))$ in $\text{Tab}(\cdot, (2, 2, 2))$.

Composing the preceding embeddings, one obtains for each evaluation μ at least one embedding of $\text{Tab}(\cdot, \mu)$ into $\text{Tab}(\cdot, (1, \ldots, 1))$ preserving shapes and commuting with $\mathscr{C}$. The uniqueness of such an embedding is again ensured by Lemma 5.6.6(iv). ∎

Figure 5.3 and Figure 5.4 illustrate Theorem 5.6.7 by displaying the tree structure of $\text{Tab}(\cdot, (2, 2, 1))$ and the canonical embedding of $\text{Tab}(\cdot, (3, 1, 2))$ in $\text{Tab}(\cdot, (2, 2, 2))$.

The main motivation for studying cyclage and the related plastic invariants given by charge and cocharge is to develop a combinatorial approach to the *Kostka–Foulkes polynomials* $K_{\lambda\mu}(q)$ which arise in many contexts, ranging from the character theory of the finite linear groups $GL_n(\mathbf{F}_q)$ to the geometry of flag varieties or the solution of certain models in statistical mechanics. Actually, one has the following important result.

THEOREM 5.6.8. *The Kostka polynomial is equal to the generating function of the charge on the set* $\text{Tab}(\lambda, \mu)$ *of tableaux of shape λ and weight μ:*

$$\sum_{t \in \text{Tab}(\lambda, \mu)} q^{\text{ch}(t)} = K_{\lambda\mu}(q).$$

The proof of this theorem is outside the scope of this chapter.

Problems

Section 5.1

5.1.1 (The Erdös–Szekeres theorem) Prove that any permutation of $n^2 + 1$ elements contains a monotonic sub-sequence of length $n+1$. Show that there exist permutations of n^2 elements with no monotonic sub-sequence with length greater than n.

Section 5.2

5.2.1 Let $\bar{w}$ denote the mirror image of a word w. Let w be a standard word, and $t = P(w)$. Show that $P(\bar{w}) = t^T$, the transposed tableau of t.

5.2.2 Let w be a standard word. Show that the sequence w^n stabilizes in $\mathrm{Pl}(A)$, in the following sense: for n sufficiently large, $w^{n+1} \equiv c \cdot w^n$, where c is the column such that $\mathrm{ev}(c) = \mathrm{ev}(w)$.

5.2.3 Let w be a standard word. Let $V(w)$ be the set of words v such that $wv \equiv vr$, where r is a row. Show that the set of words of minimal length in $V(w)$ is a plactic class.

5.2.4 The *column reading* $C(t)$ of a tableau t is the word obtained by reading the planar representation of t column-wise, from left to right and from top to bottom. Show that for any tableau, $C(t) \equiv t$.

5.2.5 (Plactic monoid and quantum matrices) Let $\mathscr{A}$ be the associative unital $\mathbb{Q}[q, q^{-1}]$-algebra generated by elements $x_{11}, x_{12}, x_{21}, x_{22}$ subject to the relations

$$x_{12}x_{11} = qx_{11}x_{12},$$
$$x_{21}x_{11} = qx_{11}x_{21},$$
$$x_{22}x_{21} = qx_{21}x_{22},$$
$$x_{22}x_{12} = qx_{12}x_{22},$$
$$x_{12}x_{21} = x_{21}x_{12},$$
$$x_{22}x_{11} = x_{11}x_{22} + (q - q^{-1})x_{12}x_{21}.$$

1. Show that $D = x_{11}x_{22} - q^{-1}x_{12}x_{21}$ commutes with the x_{ij}, hence is central in $\mathscr{A}$.
2. Introduce the $\mathbb{Z}[q]$-lattice $\mathscr{L}$ in $\mathscr{A}$ spanned by the elements $D^k x_{11}^l x_{22}^m$ ($k, l, m \in \mathbb{N}$).
   a. Show that every diagonal monomial $x_{i_1 i_1} \cdots x_{i_k i_k}$ ($i, j \in \{1, 2\}$) belongs to $\mathscr{L}$. (*Hint*: Prove that $x_{22}x_{11} = (1 - q^2)D + q^2 x_{11}x_{22}$.)

b. Let $w = i_1 \cdots i_k$, $w' = j_1 \cdots j_k \in \{1,2\}^*$. Prove that
$$w \equiv w' \iff x_{i_1 i_1} \cdots x_{i_k i_k} \equiv x_{j_1 j_1} \cdots x_{j_k j_k} \pmod{q\mathscr{L}}.$$

Section 5.3

5.3.1 Show that the number a_n of involutions in $\mathfrak{S}_n$ is equal to the number of standard tableaux of weight n. Show that
$$\sum_{n \geq 0} a_n \frac{z^n}{n!} = e^{z + \frac{z^2}{2}}.$$

Section 5.4

5.4.1 Show that if $\lambda = (k^l)$ and $\mu = (r^s)$ are partitions of rectangular shapes, all the coefficients $c_{\lambda\mu}^\nu$ are 0 or 1, and give a simple graphical description of the partitions ν such that $c_{\lambda\mu}^\nu = 1$.

5.4.2 For an integer k, let $h_k = s_{(k)}$ be the Schur function indexed by the one-part partition (k), and for a partition $\mu = (\mu_1, \ldots, \mu_r)$, set $h_\mu = h_{\mu_1} h_{\mu_2} \cdots h_{\mu_r}$. The *Kostka numbers* $K_{\lambda\mu}$ are defined as the coefficients of the expansion $h_\mu = \sum_\lambda K_{\lambda\mu} s_\lambda$. Show that $K_{\lambda\mu}$ is equal to the number of tableaux of shape λ and evaluation μ.

5.4.3 Let $X = \{x_1, x_2, \ldots, x_n\}$ be a set of commuting indeterminates, and let $E(t) = \prod_i (1 + tx_i) = \sum_k e_k t^k$, $H(t) = \prod_i (1 - tx_i)^{-1} = \sum_k h_k t^k$ be the generating functions of the elementary and complete symmetric functions of X. Let $p_k = \sum_i x_i^k$ be the power sum symmetric functions.

1. Show that $\sum_{k \geq 1} p_k t^{k-1} = H'(t) E(-t)$.

2. Deduce from 1. that $p_m = \sum_{k=0}^{m-1} (-1)^k s_{(m-k, 1^k)}$.

3. The *character table* of the symmetric group $\mathfrak{S}_n$ is a square matrix χ_μ^λ indexed by pairs of partitions of n, in which χ_μ^λ is equal to the coefficient of s_λ in the product of power sums $p_\mu = p_{\mu_1} p_{\mu_2} \cdots p_{\mu_r}$. Using 2. and the Littlewood–Richardson rule, compute the character tables of the groups $\mathfrak{S}_n$ for $n \leq 6$.

Section 5.5

5.5.1 Let $w = x_1 \cdots x_m \in A^*$. One says that the integer $i < m$ is a *descent* of w if $x_i > x_{i+1}$. The *major index* maj(w) of w is the sum of its descents. We denote by Des(w) the descent set of w.

A *recoil* of a standard tableau t is an entry i of t such that $i+1$ occurs in a higher row. Let $\operatorname{Rec}(t)$ be the set of recoils of t. The *index* of a tableau is $\operatorname{ind}(t) = \sum_{i \in \operatorname{Rec}(t)} i$.
It is customary to encode a subset $E = \{e_1, \ldots, e_{r-1}\} \subseteq \{1, 2, \ldots, m-1\}$ by a *composition* of m, i.e. a vector $I = (i_1, \ldots, i_r)$ of positive integers with sum $|I| = m$. The encoding $I = C(E)$ of E is specified by $e_k = i_1 + i_2 + \cdots + i_k$. The composition $I = C(\operatorname{Des}(w))$ is called the *descent composition* of w. Conversely, the set E defined in this way from a composition I is called the descent set of I and denoted by $\operatorname{Des}(I)$. As above, one sets $\operatorname{maj}(I) = \sum_k e_k$.

1. Show that for any word, $\operatorname{Des}(w) = \operatorname{Rec} Q(w)$.
2. For a composition I, define the *noncommutative ribbon Schur function* $R_I \in \mathbb{Z}\langle A \rangle$ by

$$R_I = \sum_{\operatorname{Des}(w) = \operatorname{Des}(I)} w.$$

a. Show that $R_I = \sum_{\operatorname{Rec}(t) = \operatorname{Des}(I)} S_t$.
b. Show that $w \mapsto Q(w)$ defines a bijection between the set of Yamanouchi words of evaluation λ and $\operatorname{STab}(\lambda)$.
c. Let r_I be the commutative image of R_I, and $r_I = \sum_\lambda c_\lambda^I s_\lambda$ its expansion in the Schur basis. Show that r_I is equal to the number of Yamanouchi words of evaluation λ with descent composition I.
3. Prove the identity between formal series

$$\overrightarrow{\prod_{k \geq 0} \prod_{i \geq 1}} (1 - q^k a_i)^{-1} = \sum_{m \geq 0} \frac{1}{(q)_m} \sum_{|w|=m} q^{\operatorname{maj}(w)} w,$$

where $(q)_m = (1-q)(1-q^2) \cdots (1-q^m)$.
4. By taking the commutative image of the above identity, and applying Cauchy's identity to the alphabets $Q = \{1, q, q^2, \ldots\}$ and X, show that $\sum_{|I|=m} c_\lambda^I q^{\operatorname{maj}(I)} = (q)_m s_\lambda(Q)$ and obtain the generating function of the major index on the set of standard tableaux of a given shape:

$$\sum_{t \in \operatorname{STab}(\lambda)} q^{\operatorname{maj}(t)} = (q)_m s_\lambda(Q).$$

This is equal to the Kostka polynomial $K_{\lambda, 1^m}(q)$.

Section 5.6

5.6.1 (Catabolism) Let $k:$ Tab $\to$ Tab be the map $t = t'v \mapsto vt'$ where v is the bottom row of t. Let $\varphi(t)$ be the sequence of shapes of $t, k(t), k^2(t), \ldots$.
1. Show that the restriction of φ to STab is one-to-one.
2. Show that φ is invariant under the action of $\mathfrak{S}(A)$ (i.e., $\varphi(\sigma(t)) = \varphi(t)$).
3. Show that φ is invariant under the canonical embeddings Tab$(\lambda) \hookrightarrow$ Tab$(1^n) =$ STab.

Notes

The name *plactic monoid* was coined by Schützenberger with reference to the *tectonique des plaques*. The basic theory of the plactic monoid was systematically developed in Lascoux and Schützenberger 1981.

Schensted's algorithm appeared in Schensted 1961. It was realized later that Robinson, in an attempt to prove the Littlewood–Richardson rule, had already formulated in Robinson 1938 the correspondence (5.3.1), which is essentially equivalent to Schensted's result (Theorem 5.3.1).

Theorem 5.2.5 is due to Knuth 1970. Greene's invariants were introduced in Greene 1974. Theorem 5.3.3 appears in Schützenberger 1963. It was already stated, without proof, in Robinson 1938.

The left-hand side of Theorem 5.3.10 can be interpreted as the sum of the characters of all irreducible polynomial representations of $GL_n(\mathbb{C})$. Using this interpretation, Theorem 5.3.10 is a classical identity of Schur (see Littlewood 1950).

For an account of the theory of symmetric functions see Littlewood 1950 or Macdonald 1995. The proof of the Littlewood–Richardson rule given in Section 5.4 first appeared in Schützenberger 1977. Corollary 5.4.6 is known by geometers as the *Pieri rule*.

Lascoux and Schützenberger 1988 is the basic reference for the material of Section 5.5, with emphasis on the operators σ_i. Our exposition here, which stresses the role played by the operators e_i and f_i, is strongly influenced by Kashiwara's theory of crystal bases (see Kashiwara 1991, Kashiwara 1994, Lascoux, Leclerc, and Thibon 1995, Leclerc and Thibon 1996). The connection between the Robinson–Schensted correspondence and quantum groups was first observed in Date, Jimbo, and Miwa 1990.

Concerning the statistics charge and cocharge, the cyclage, and their applications to Kostka–Foulkes polynomials, see Schützenberger 1978, Lascoux and Schützenberger 1980, Lascoux 1991. Another combinatorial description of the Kostka–Foulkes polynomials in terms of the geometry of crystal graphs was given in Lascoux et al. 1995.

The Littlewood–Richardson rule and the plactic monoid have been generalized to other root systems by Littelmann (see Littelmann 1994, Littelmann 1996). A monoid associated in a similar way to Gessel's quasi-symmetric functions has been introduced in Krob and Thibon 1997.

Problem 5.1.1 is a classical result that appears for instance in Knuth 1973. Problem 5.2.5 is from Leclerc and Thibon 1996. More on character tables (Problem 5.4.3) can be found in Macdonald 1995. Problem 5.5.1 is from Gelfand et al. 1995.

CHAPTER 6

Codes

6.0. Introduction

The theory of codes provides some jewels of combinatorics on words that we want to describe in this chapter.

A basic result is the defect theorem (Theorem 6.2.1), which states that if a set X of n words satisfies a nontrivial relation, then these words can be expressed simultaneously as products of at most $n-1$ words. It is the starting point of the chapter. In Chapters 9 and 13, other defect properties are studied in different contexts.

A nontrivial relation is simply a finite word w which ambiguously factorizes over X. This means that X is not a code. The defect effect still holds if X is not an ω-code, i.e., if the nontrivial relation is an infinite, instead of a finite, word (Theorem 6.2.4).

The defect theorem implies several well-known properties on words that are recalled in this chapter. For instance, the fact that two words which commute are powers of the same word is a consequence. Another consequence is that a two-element code or more generally an elementary set is an ω-code. The latter property appears to be a crucial step in one of the proofs of the D0L equivalence problem.

A remarkable phenomenon appears when, for a finite code X, neither the set X nor its reversal $\tilde{X}$ is an ω-code. In this case the defect property is stronger: the n elements of X can be expressed as products of at most $n-2$ words (Theorem 6.3.4). It follows that for codes X with three elements, either X or $\tilde{X}$ is an ω-code. The proof of this property is rather long. It uses in a very elegant and subtle way techniques of combinatorics on words.

In this chapter, we also present a deep result by Schützenberger about finite maximal codes. It states that if a finite maximal code X is an ω-code, or equivalently if X has bounded decoding delay, then X is a prefix

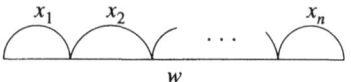

Figure 6.1. An X-factorization of the word w.

code (Theorem 6.4.1). The original proof is complex. The proof given here is short and elementary.

6.1. X-factorizations

6.1.1. Codes

Let $X \subseteq A^+$. A sequence $(x_1, x_2, \ldots, x_n)$ of n words of X is an X-*factorization* of a word $w \in A^*$ if $w = x_1 x_2 \cdots x_n$. See Figure 6.1.

A set $X \subseteq A^+$ is a *code* if any word $w \in A^*$ has at most one X-factorization. This definition is equivalent to the one given in Chapter 1.

The simplest codes are *prefix codes* $X \subseteq A^+$. Recall that they are sets such that no word of X is a proper prefix of another word of X. *Suffix codes* are defined symmetrically as sets such that no word of X is a proper suffix of another word of X.

EXAMPLE 6.1.1. Let $A = \{a, b\}$. The set $X = \{a, ab, ba\}$ is not a code because $w = aba$ has two distinct X-factorizations, namely (a, ba), (ab, a).

EXAMPLE 6.1.2. The set $X = \{a, ab, bb\}$ over the alphabet $\{a, b\}$ is a suffix code.

The name of "code" is motivated by the next proposition. Roughly speaking, if the letters of a source alphabet B are put in one-to-one correspondence with the words of a code X over a target alphabet A, then a source message $r \in B^*$ is encoded into a coded message $w \in A^*$ by replacing any letter of r by the corresponding word of X. Unique decipherability is insured by the fact that w has exactly one X-factorization.

PROPOSITION 6.1.3. *A set $X \subseteq A^+$ is a code if and only if any morphism $\varphi: B^* \to A^*$ induced by a bijection from B onto X is injective.*

With the notation of the proposition, we say that φ is a *coding morphism for X*.

6.1. X-factorizations

Proof of proposition. Let $\varphi: B^* \to A^*$ be a morphism induced by a bijection from B onto X. Let $r, s \in B^*$ be such that $\varphi(r) = \varphi(s)$. Let us prove that $r = s$. Set $r = \alpha_1 \cdots \alpha_n$, $s = \beta_1 \cdots \beta_m$ with $\alpha_i, \beta_j \in B$ and $n, m \geq 0$. Since $\varphi(r) = \varphi(s)$, this word has the two X-factorizations $(\varphi(\alpha_1), \ldots, \varphi(\alpha_n))$ and $(\varphi(\beta_1), \ldots, \varphi(\beta_m))$. But X is a code, hence $n = m$ and $\varphi(\alpha_i) = \varphi(\beta_i)$ for all i. As φ is injective on B, one has $\alpha_i = \beta_i$ for all i, and $r = s$.

For the converse, let X be a subset of A^+ and $\varphi: B^* \to A^*$ be an injective morphism induced by a bijection from B onto X. Let $w \in A^*$ be a word with X-factorizations $(x_1, \ldots, x_n)$, $(y_1, \ldots y_m)$. Through the bijection φ, let $x_i = \varphi(\alpha_i)$, $y_j = \varphi(\beta_j)$ for letters α_i, β_j. Thus $w = \varphi(\alpha_1 \cdots \alpha_n) = \varphi(\beta_1 \cdots \beta_m)$. As φ is injective, one has $\alpha_1 \cdots \alpha_n = \beta_1 \cdots \beta_m$, that is, $n = m$ and $\alpha_i = \beta_i$ for all i. It follows that the two X-factorizations are equal. Thus X is a code. ∎

EXAMPLE 6.1.1 (*continued*). Let $\varphi: B^* = \{\alpha, \beta, \gamma\}^* \to A^*$ be defined by $\varphi(\alpha) = a$, $\varphi(\beta) = ab$ and $\varphi(\gamma) = ba$. It is not injective since $\varphi(\alpha\gamma) = \varphi(\beta\alpha)$.

PROPOSITION 6.1.4. *Let $\varphi: B^* \to A^*$ be an injective morphism. If $Z \subseteq B^+$ is a code, then $\varphi(Z)$ is a code. If $X \subseteq A^+$ is a code, then $\varphi^{-1}(X)$ is a code.*

Proof. Suppose that Z is a code. Consider a word $w \in A^*$ with the $\varphi(Z)$-factorizations $(x_1, \ldots, x_n)$, $(y_1, \ldots, y_m)$ such that $x_i = \varphi(z_i)$, $y_j = \varphi(t_j)$ and $z_i, t_j \in Z$. Then $\varphi(z_1 \cdots z_n) = \varphi(t_1 \cdots t_m)$ and $z_1 \cdots z_n = t_1 \cdots t_m$ since φ is injective. As Z is a code, $n = m$ and $z_i = t_i$ for all i. It follows that $\varphi(z_i) = \varphi(t_i)$ for any i, showing that $\varphi(Z)$ is a code. A similar argument shows that $\varphi^{-1}(X)$ is a code if X is a code. ∎

EXAMPLE 6.1.5. The set $Z = \{\alpha\alpha, \alpha\beta, \alpha\gamma, \beta, \gamma\}$ is a code over the alphabet $B = \{\alpha, \beta, \gamma\}$. Let $\varphi: B^* \to A^*$ be the morphism induced by $\varphi(\alpha) = a$, $\varphi(\beta) = ab$ and $\varphi(\gamma) = bb$. The set $\varphi(Z) = \{aa, aab, abb, ab, bb\}$ is a code.

We end this subsection with a characterization of codes by a property on the monoids that they generate.

We recall that a submonoid M of A^* has a unique minimal generating set $(M - \varepsilon) - (M - \varepsilon)^2$ (see Chapter 1). For convenience, we call it the *base* of M.

Let $M = X^*$ be the submonoid of A^* generated by a set $X \subseteq A^+$. If X is a code, then X is necessarily the base of M. Otherwise, X contains a word w which belongs to $(M - \varepsilon)^2$. This word has thus two X-factorizations: (w) itself and $(x_1, \ldots, x_n)$, with $n \geq 2$, a contradiction.

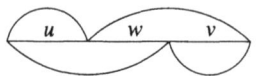

Figure 6.2. The stability condition.

However, it may happen that the base of a submonoid M is not a code. We have seen such a base in Example 6.1.1: the monoid $M = \{a, ab, ba\}^*$ is generated by the set $X = \{a, ab, ba\}$ which is also the base of M, but X is not a code.

When the base X of a submonoid M is a code, we say that M is *free*. A submonoid M of A^* is *stable* if for any $u, w, v \in A^*$,

$$u, wv, uw, v \in M \Rightarrow w \in M.$$

Figure 6.2 gives a pictorial representation of the stability condition.

PROPOSITION 6.1.6. *Let M be a submonoid of A^*. Then M is free if and only if M is stable.*

Proof. Suppose that M is free, i.e., the base X of M is a code. Take u, wv, uw, $v \in M = X^*$. Consider the X-factorizations

$$(x_1, \ldots, x_k), (x_{k+1}, \ldots, x_n), (y_1, \ldots, y_\ell), (y_{\ell+1}, \ldots, y_m)$$

of u, wv, uw, v respectively. Since X is a code, the X-factorizations

$$(x_1, \ldots, x_k, x_{k+1}, \ldots x_n), (y_1, \ldots, y_\ell, y_{\ell+1}, \ldots y_m)$$

of $u(wv) = (uw)v$ are equal. Moreover, $\ell \geq k$ since $|uw| \geq |u|$, showing that

$$uw = x_1 \cdots x_k x_{k+1} \cdots x_\ell = u x_{k+1} \cdots x_\ell.$$

Hence, $w = x_{k+1} \cdots x_\ell \in M$, and M is stable.

For the converse, assume that M is stable but its base X is not a code. There exists a word $z \in M$ with X-factorizations $(x_1, \ldots, x_n)$, $(y_1, \ldots, y_m)$ such that $x_1 \neq y_1$. We can suppose that $y_1 = x_1 w$ with w a nonempty word. Hence

$$u = x_1, wv = x_2 \cdots x_n, uw = y_1, v = y_2 \cdots y_m \in M.$$

But M is stable, thus $w \in M$. Consequently, $y_1 = x_1 w \in X \cap (M - \varepsilon)^2$, showing that X is not the base of M. This leads to the contradiction. ∎

6.1. X-factorizations

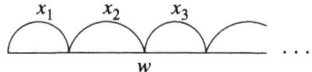

Figure 6.3. An X-factorization of the infinite word w.

COROLLARY 6.1.7. *The intersection of an arbitrary family of free submonoids of A^* is a free submonoid of A^*.*

Proof. Take a family of free submonoids M_i of A^*, indexed by a set I. Denote by M the set $\bigcap_{i \in I} M_i$. It is a submonoid of A^*. Let $u, w, v \in A^*$ be such that $u, wv, uw, v \in M$. As any M_i is free and then stable, the word w belongs to $M_i \subseteq M$. It follows that M is stable and thus free. ∎

6.1.2. ω-Codes

In this subsection, infinite instead of finite X-factorizations are studied.

Let $X \subseteq A^+$. An *X-factorization* of a word $w \in A^\omega$ is an infinite sequence $(x_1, x_2, \ldots, x_n, \ldots)$ of elements of X such that $w = x_1 x_2 \cdots x_n \cdots$ (see Figure 6.3).

A set $X \subseteq A^+$ is an *ω-code* if any word of A^ω has at most one X-factorization. Any ω-code is a code as two distinct X-factorizations $(x_1, \ldots, x_n)$, $(y_1, \ldots, y_m)$ of a word $w \in A^+$ lead to two distinct X-factorizations of the word $w^\omega \in A^\omega$: $(x_1, \ldots, x_n, x_1, \ldots, x_n, \ldots)$, $(y_1, \ldots, y_m, y_1, \ldots, y_m, \ldots)$. The converse is false, as shown in Example 6.1.2.

EXAMPLE 6.1.2 *(continued)*. The code $X = \{a, ab, bb\}$ is not an ω-code because $(a, bb, bb, \ldots)$ and $(ab, bb, bb, \ldots)$ are two distinct X-factorizations of the word ab^ω.

Let $u, v \in A^*$ be two words. We write $u \leq v$ when u is prefix of v.

A code $X \subseteq A^+$ has a *bounded decoding delay* if there is an integer $d \geq 0$ such that for any $x, x', y_1, \ldots, y_d \in X$ and $z \in X^*$,

$$xy_1 \cdots y_d \leq x'z \Rightarrow x = x'.$$

In other words, the knowledge of a prefix $xy_1 \cdots y_d \in X^{d+1}$ of a word $x'z \in x'X^*$ does not allow the situation depicted in Figure 6.4. The smallest integer d in the preceding definition is called the *decoding delay* of the code.

This notion extends in a natural way the concept of prefix code, since prefix codes have a decoding delay $d = 0$.

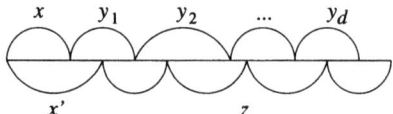

Figure 6.4. An impossible situation for a decoding delay d.

EXAMPLE 6.1.8. The code $X = \{a, ab\}$ is not a prefix code, but it has a decoding delay $d = 1$. More generally, the code $X = \{a, a^d b\}$ has a decoding delay d.

EXAMPLE 6.1.2 (*continued*). The suffix code $X = \{a, ab, bb\}$ does not have a bounded decoding delay, because for any $d \geq 0$, $a(bb)^d$ is a prefix of $ab(bb)^d$.

The next proposition shows the relationship between ω-codes and codes with bounded decoding delay (see also Problem 6.1.3).

PROPOSITION 6.1.9. *Any code with bounded decoding delay is an ω-code. Conversely, any finite ω-code has a bounded decoding delay.*

Proof. Let $X \subseteq A^+$ be a code with decoding delay $d \geq 0$. Assume that X is not an ω-code. Then, there exists $w \in A^\omega$ with two X-factorizations $(x, y_1, y_2, \ldots, y_m, \ldots)$, $(x', x_1, x_2, \ldots, x_n, \ldots)$ such that $x \neq x'$. This is impossible since $xy_1 \cdots y_d \leq x' x_1 \cdots x_n$ for some $n \geq 0$.

For the converse, assume that X does not have a bounded decoding delay. Then for any $d \geq 0$, there exist $n \geq 0$ and $x, x', y_1, \ldots, y_d, x_1, \ldots, x_n \in X$ such that

$$xy_1 \cdots y_d \leq x' x_1 \cdots x_n \quad \text{and} \quad x \neq x'.$$

As X is finite, for a large enough d, there exists a suffix w of a word of X which is repeated as follows (see Figure 6.5):

$$xu = x'vw, \quad xuu' = x'vv'w$$

with $u = y_1 \cdots y_c$, $u' = y_{c+1} \cdots y_d$, $v = x_1 \cdots x_\ell$, $v' = x_{\ell+1} \cdots x_m$, $\ell < m \leq n$. It follows that the word $xuu'^\omega = x'vv'^\omega$ has two distinct X-factorizations, showing that X is not an ω-code. ∎

The concepts of coding morphism, free and stable submonoid carry over to ω-codes as follows. The related propositions remain true, with similar proofs.

6.1. X-factorizations

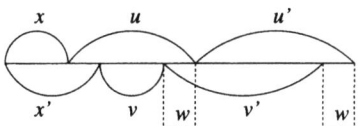

Figure 6.5. The suffix w is repeated.

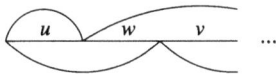

Figure 6.6. The stability condition for infinite words.

PROPOSITION 6.1.10. *A set $X \subseteq A^+$ is an ω-code if and only if any morphism $\varphi: B^\infty \to A^\infty$ induced by a bijection from B onto X is injective on B^ω.* ∎

We say that φ is a *coding morphism for X*.

PROPOSITION 6.1.11. *Let $\varphi: B^\infty \to A^\infty$ be a morphism injective on B^ω. If $Z \subseteq B^+$ is an ω-code, then $\varphi(Z)$ is an ω-code. If $X \subseteq A^+$ is an ω-code, then $\varphi^{-1}(X)$ is an ω-code.* ∎

We recall that a binoid $M \subseteq A^\infty$ is *finitary* if its minimal generating set X is a subset of A^+. In particular $M = X^\infty$ (see Chapter 1). For convenience we call X the *base* of M.

If the base X of a finitary binoid $M \subseteq A^\infty$ is an ω-code, we say that M is *free*.

Freeness for a submonoid M of A^* means that its base X does not satisfy any *nontrivial relation* over finite words. Freeness for a finitary binoid of A^∞ means that its base does not satisfy any *nontrivial relation* over infinite words. A nontrivial relation on finite words implies a nontrivial relation on infinite words because any ω-code is a code.

A binoid M of A^∞ is *stable* if for any $u, w \in A^*$ and $v \in A^\omega$,

$$u, wv, uw, v \in M \Rightarrow w \in M$$

(see Figure 6.6).

PROPOSITION 6.1.12. *Let M be a finitary binoid of A^∞. Then M is free if and only if M is stable.* ∎

EXAMPLE 6.1.2 (*continued*). Let $M = X^\infty$ be the finitary binoid generated by $X = \{a, ab, bb\}$. It is not free since its base X is not an ω-code. It is not stable because the words $u = a$, $wv = b(bb)^\omega$, $uw = ab$, $v = (bb)^\omega$ belong to M, but $w = b$ does not belong to M. Nevertheless the submonoid X^* generated by X is free because X is a code. It is stable by Proposition 6.1.6.

6.2. Defect

This section is devoted to the defect theorem which is a basic result on sets of words. It states that if a set X of n words satisfies a nontrivial relation, then these words can be expressed simultaneously as products of at most $n - 1$ words. Originally the defect theorem relies on a nontrivial relation over A^*. It still works with a nontrivial relation over A^∞. We also show that the $n - 1$ words can be taken in a code associated to X, which is "minimal" in a certain sense.

6.2.1. Defect theorem

Given a set $X \subseteq A^+$, we consider the family $\mathscr{F}$ of all the free submonoids of A^* containing X. It is not empty since A^* belongs to $\mathscr{F}$. It is closed for arbitrary intersection by Corollary 6.1.7. It follows that the *smallest* free submonoid of A^* containing X exists: it is equal to $\bigcap_{M \in \mathscr{F}} M$. This set is called the *free hull* of X. In particular, the base Y of the free hull of X is a code and $X \subseteq Y^*$.

Let $X \subseteq Y^*$, with Y a code over the alphabet A. As any word of X has a unique Y-factorization, Y can be viewed as an alphabet for X. We denote by $\text{Alph}_Y(X)$ the set of words of Y appearing in the Y-factorization of the elements of X. We define $\text{First}_Y(X)$ as the set of all y_1 such that there exists an Y-factorization $(y_1, \ldots, y_n)$ of some $x \in X$. We define $\text{Last}_Y(X)$ similarly. This notation is much used below in Subsection 6.3.2. We simply write $\text{Alph}(X)$, $\text{First}(X)$ and $\text{Last}(X)$ when $Y = A$.

THEOREM 6.2.1 (Defect theorem). *Let $Y \subseteq A^+$ be the base of the free hull of a finite set $X \subseteq A^+$. If X is not a code, then*

$$\text{Card}(Y) \leq \text{Card}(X) - 1.$$

Proof. Suppose that Y is the base of the free hull of X. Then Y is a code such that $X \subseteq Y^*$. Any element x of X has thus a unique Y-factorization $(y_1, \ldots, y_n)$. Hence the function $\alpha: X \to Y$ such that

6.2. Defect

$\alpha(x) = y_1 = \text{First}_Y(x)$ is well defined. It is not injective. Indeed, there exists a word $w \in A^*$ with two X-factorizations $(x_1, \ldots, x_k)$, $(x'_1, \ldots, x'_\ell)$ such that $x_1 \neq x'_1$, because X is not a code. But w has only one Y-factorization over the code Y. This implies that $\alpha(x_1) = \alpha(x'_1)$. If α is not surjective, then there is a word y in $Y - \alpha(X)$. Consider the set $Z = (Y-y)y^*$. Clearly $X \subseteq Z^* \subsetneq Y^*$. Moreover it is not difficult to check that Z is a code (see also Example 6.2.9 below and Subsection 6.2.4). This is impossible because Y^* is the smallest free submonoid which contains X. Consequently, $\alpha: X \to Y$ is surjective, not injective, and $\text{Card}(Y) \leq \text{Card}(X) - 1$. ∎

The proof of Theorem 6.2.1 is based on the following property of the free hull.

PROPOSITION 6.2.2. *Let $X \subseteq A^+$ and Y be the base of the free hull of X. Then*
$$\text{First}_Y(X) = Y.$$
∎

We end this subsection with a nice link between the free hull of X and its dependency graph. The *dependency graph* G_X of a finite set $X \subseteq A^+$ is an undirected graph with X as set of vertices and edges $(x, x') \in X \times X$ whenever there exist $z, z' \in X^*$ such that $xz = x'z'$.

PROPOSITION 6.2.3. *Let $Y \subseteq A^+$ be the base of the free hull of a finite set $X \subseteq A^+$. If X is not a code, then*
$$\text{Card}(Y) \leq c(X) \leq \text{Card}(X) - 1$$
where $c(X)$ is the number of connected components of the dependency graph G_X of X.

Proof. The inequality $c(X) \leq \text{Card}(X) - 1$ holds because X is not a code. To show the first inequality, we define a function α from the set of connected components of G_X into Y as follows. Let C be a connected component and $x \in X$ which belongs to C. Then $\alpha(C) = y$ such that $\text{First}_Y(x) = y$. This function is well defined because if (x, x') is an edge of G_X, then $xz = x'z'$ for some $z, z' \in X^*$. Hence $xz = x'z'$ has a unique Y-factorization beginning with y. By Proposition 6.2.2, α is surjective, showing that $\text{Card}(Y) \leq c(X)$. ∎

EXAMPLE 6.1.1 (*continued*). The set $X = \{a, ab, ba\}$ is not a code. The smallest free submonoid M containing X is equal to $\{a, b\}^*$. Indeed M is stable by Proposition 6.1.6. Since $u = a, wv = ba, uw = ab, v = a \in M$, we have $b \in M$. This shows that $a, b \in M$. Hence the base of the free hull of X is $Y = \{a, b\}$ and $\text{First}_Y(X) = Y$. The dependency graph G_X of X has two connected components, namely $\{a, ab\}$ and $\{ba\}$.

6.2.2. Infinite words

All the arguments of Subsection 6.2.1 can be repeated in the context of infinite words.

Given a set $X \subseteq A^+$, the smallest free finitary binoid of A^∞ containing X is called the *ω-free hull* of X. This binoid always exists. Indeed let M be the intersection of all free finitary binoids of A^∞ containing X. Then M is a stable binoid because the intersection of stable binoids is again a stable binoid. If M is not finitary, then $M' = (M \cap A^*)^\infty$ is a stable (and thus free) finitary binoid which is strictly included in M and contains X. This is impossible. Note that the base Y of the ω-free hull of X is an ω-code and $X \subseteq Y^*$.

The defect theorem remains true. The proof is similar.

THEOREM 6.2.4 (Defect theorem). *Let $Y \subseteq A^+$ be the base of the ω-free hull of a set $X \subseteq A^+$. Then $\text{First}_Y(X) = Y$. If X is a finite set which is not an ω-code, then*

$$\text{Card}(Y) \leq \text{Card}(X) - 1.$$

Proof. Clearly $\text{First}_Y(X) \subseteq Y$. As in the proof of Theorem 6.2.1, we define $\alpha: X \to Y$ such that $\alpha(x) = \text{First}_Y(x)$, that is, $\alpha(x)$ is the first element y_1 in the Y-factorization $(y_1, \ldots, y_n)$ of x. This function is surjective. Otherwise, considering $y \in Y - \alpha(X)$, we get a set $Z = (Y - y)y^*$ which is an ω-code (see Example 6.2.9 and Proposition 6.2.10) and such that $X \subseteq Z^\infty \subsetneq Y^\infty$. This is impossible. It follows that $\text{First}_Y(X) = Y$. If X is a finite set which is not an ω-code, then α is not injective and $\text{Card}(Y) \leq \text{Card}(X) - 1$. ∎

EXAMPLE 6.1.1 (*continued*). Recall that the set $X = \{a, ab, bb\}$ is a code but not an ω-code. The base of its free hull is X itself because X^* is a free submonoid. The ω-free hull M of X is the free finitary binoid $M = \{a, b\}^\infty$. Indeed M is stable by Proposition 6.1.12. Since $u = a, wv = b(bb)^\omega, uw = ab, v = (bb)^\omega \in X^\infty \subseteq M$, we have $b \in M$.

Proposition 6.2.3 also holds in the context of infinite words. The definition of the graph G_X is slightly different. There is an edge $(x, x') \in X \times X$ if and only if there exist $z, z' \in X^\omega$ (instead of X^*) with $xz = x'z'$.

6.2.3. Consequences

We state three corollaries of the defect theorem.

COROLLARY 6.2.5. *If two words commute, or more generally satisfy a nontrivial relation on finite or infinite words, then they are powers of the same word.*

Proof. Let $x, y \in A^+$ satisfying a nontrivial relation. Then $X = \{x, y\}$ is either not a code or not an ω-code, depending on whether the relation is on finite or infinite words. By the defect theorem (Theorem 6.2.1 or 6.2.4), the base of the free or ω-free hull of X is a one-element set $\{z\} \subseteq A^+$ such that $\{x, y\} \subseteq z^*$. ∎

Recall that any ω-code is a code. The converse is true for codes with two elements as shown in the next corollary. This property is false for larger codes (see Example 6.1.2).

COROLLARY 6.2.6. *Any two-element code is an ω-code.*

Proof. Let $X = \{x, y\}$ be a code. If X is not an ω-code, then $X \subseteq z^*$ where $\{z\}$ is the base of the ω-free hull of X (Theorem 6.2.4). This is impossible because X is a code. ∎

The final corollary deals with elementary sets. A finite set X is *simplifiable* if $X \subseteq Y^*$ with $\text{Card}(Y) \leq \text{Card}(X) - 1$. Otherwise it is *elementary*.

COROLLARY 6.2.7. *Any elementary set is an ω-code.*

Proof. By the defect theorem, if X is not an ω-code, then $X \subseteq Y^*$ with Y an ω-code such that $\text{Card}(Y) \leq \text{Card}(X) - 1$. Hence X is not an elementary set. ∎

The converse of this corollary is false, as shown by the next example.

EXAMPLE 6.2.8. The set $X = \{a, abc, abcbc\}$ is an ω-code. However, X is simplifiable because $X \subseteq \{a, bc\}^*$.

6.2.4. Composition of codes

This subsection is devoted to the operation of composition of codes. It clarifies the construction of the code $(Y - y)y^*$ in the proof of Theorems 6.2.1 and 6.2.4.

Take two sets $Y \subseteq A^+, Z \subseteq B^+$ with

$$B = \text{Alph}(Z).$$

We say that Y, Z are *composable* if there exists a bijection φ from B onto Y. The set
$$X = \varphi(Z) \subseteq Y^*$$
is obtained by replacing the letters of Z by the corresponding (through φ) words of Y. The set X resulting of the *composition* of Y and Z is denoted by
$$X = Y \circ_\varphi Z,$$
or more simply by
$$X = Y \circ Z.$$

EXAMPLE 6.1.5 (*continued*). The sets $Y = \{a, ab, bb\}$, $Z = \{\alpha\alpha, \alpha\beta, \alpha\gamma, \beta, \gamma\}$ are composable, thanks to the bijection φ such that $\varphi(\alpha) = a$, $\varphi(\beta) = ab$ and $\varphi(\gamma) = bb$. The set $X = Y \circ_\varphi Z$ is equal to $\{aa, aab, abb, ab, bb\}$.

EXAMPLE 6.2.9. Let Y be a set over A and $Z = (B-b)b^*$ over B, with b a particular letter of B. Let φ be a bijection from B onto Y. Denote by y the word of Y equal to $\varphi(b)$. Then $X = Y \circ_\varphi Z$ is the set $(Y - y)y^*$.

The operation of composition conserves the code or the ω-code property.

PROPOSITION 6.2.10. *Let $Y \subseteq A^+$, $Z \subseteq B^+$ be two composable sets. If Y, Z are codes (resp. ω-codes), then $X = Y \circ Z$ is a code (resp. ω-code).*

Proof. Let $\varphi: B = \mathrm{Alph}(Z) \to Y$ be the function used for the composition of Y and Z. Suppose that Y, Z are codes. The function φ extends into a morphism injective on B^* by Proposition 6.1.3. It follows by Proposition 6.1.4 that $X = \varphi(Z)$ is a code. The case of ω-codes is solved similarly thanks to Propositions 6.1.10 and 6.1.11. ∎

Example 6.1.5 illustrates this proposition for codes.

EXAMPLE 6.2.9 (*continued*). The $Z = (B-b)b^*$ is a code and an ω-code. If Y is a code (resp. ω-code), then X is also a code (resp. ω-code). This property is used in the proofs of Theorems 6.2.1 and 6.2.4.

Given a finite set X, the length $\sum_{x \in X} |x|$ of X is denoted by $\mathrm{Lg}(X)$.

The composition of codes is an associative operation, that is
$$X \circ (Y \circ Z) = (X \circ Y) \circ Z. \tag{6.2.1}$$

Note also that if $X = Y \circ Z$, then
$$\mathrm{Card}(Z) = \mathrm{Card}(X) \text{ and } \mathrm{Lg}(Z) \leq \mathrm{Lg}(X) \tag{6.2.2}$$
with $\mathrm{Lg}(Z) = \mathrm{Lg}(X)$ if and only if $Y = \mathrm{Alph}(X)$.

6.3. More defect

Figure 6.7. The two forbidden situations for ξ-codes.

The next proposition gives conditions such that a set $X \subseteq A^+$ is written as $Y \circ Z$ for a given code $Y \subseteq A^+$.

PROPOSITION 6.2.11. *Let $X, Y \subseteq A^+$ such that Y is a code. If*

$$X \subseteq Y^* \text{ and } \mathrm{Alph}_Y(X) = Y,$$

then $X = Y \circ Z$ for some $Z \subseteq B^+$. Moreover, if X is a code, then Z is a code.

Informally, as Y is a code such that $X \subseteq Y^*$ and $\mathrm{Alph}_Y(X) = Y$, the set X can be viewed as written over the alphabet Y, instead of the alphabet A. This is the definition of the set Z.

Proof of proposition. Set $Z = \varphi^{-1}(X)$ where $\varphi: B^\infty \to A^\infty$ is a coding morphism for Y. By definition, φ induces a bijection from B onto Y. As $\mathrm{Alph}_Y(X) = Y$, we get $\mathrm{Alph}(Z) = B$. Thus, $X = Y \circ_\varphi Z$. If X is a code, then $Z = \varphi^{-1}(X)$ is also a code by Proposition 6.1.4. ∎

6.3. More defect

6.3.1. ξ-Codes

In this subsection, we study sets $X \subseteq A^+$ such that both X and its reversal $\widetilde{X}$ are ω-codes. These sets are called ξ-*codes*. Roughly speaking, no right-infinite word, and no left-infinite word, is "ambiguously" factorized by words of X (see Figure 6.7).

Given a set $X \subseteq A^+$, in the same way we have defined the free hull of X and its ω-free hull, we can define the ξ-*free hull* of X. It is the smallest finitary binoid M of A^∞ containing X such that M and $\widetilde{M}$ are free. This binoid always exists. Its base Y is a ξ-code such that $X \subseteq Y^*$.

We get the next two properties as a consequence of Theorem 6.2.4.

PROPOSITION 6.3.1. *Let $X \subseteq A^+$ and Y be the base of the ξ-free hull of X. Then*
$$\text{First}_Y(X) = Y \text{ and } \text{Last}_Y(X) = Y.$$ ∎

THEOREM 6.3.2. *Let $Y \subseteq A^+$ be the base of the ξ-free hull of a finite subset X of A^+. If X is not a ξ-code, then*
$$\text{Card}(Y) \leq \text{Card}(X) - 1.$$ ∎

EXAMPLE 6.3.3. The set $X = \{ab, ababbb, bab\}$ is a ξ-code. Thus it is equal to the base of its ξ-free hull. The set $Y = \{ab, ababbb, bbbb\}$ is not a ξ-code since $(ab, ab, bbbb, bbbb, \ldots)$ and $(ababbb, bbbb, bbbb, \ldots)$ are two distinct X-factorizations of the word $abab(bb)^\omega$. The base of its ξ-free hull is the set $\{ab, bb\}$ which is also the base of its ω-free hull.

Theorem 6.3.2 leads to a refinement of Corollaries 6.2.6 and 6.2.7, namely: any two-element code is a ξ-code; any elementary set is a ξ-code.

If in Theorem 6.3.2, the hypothesis that "X is not a ξ-code" is replaced by "X is a code, but neither X nor $\tilde{X}$ is an ω-code", we get a stronger defect. This defect effect is noteworthy.

THEOREM 6.3.4. *Let $Y \subseteq A^+$ be the base of the ξ-free hull of a finite set $X \subseteq A^+$. If X is a code, but neither X nor $\tilde{X}$ is an ω-code, then*
$$\text{Card}(Y) \leq \text{Card}(X) - 2.$$

EXAMPLE 6.3.5. The set $X = \{a, ab, bbab, bbbb\}$ is a code, but neither X nor $\tilde{X}$ is an ω-code. Indeed, $(a, bbbb, bbbb, \ldots), (ab, bbbb, bbbb, \ldots)$ are X-factorization of the word ab^ω, and $(ba, bbbb, bbbb, \ldots), (babb, bbbb, bbbb, \ldots)$ are $\tilde{X}$-factorizations of the word bab^ω. The base $Y = \{a, b\}$ of the ξ-free hull of X has cardinality $\text{Card}(Y) = \text{Card}(X) - 2$.

The proof of Theorem 6.3.4 is based on Proposition 6.3.6. The proof of this proposition needs several steps. Subsection 6.3.2 below is completely dedicated to it.

PROPOSITION 6.3.6. *If X is a finite code satisfying*
$$\text{Card}(\text{Alph}(X)) = \text{Card}(X) - 1$$
and
$$\text{Alph}(X) = \text{First}(X) = \text{Last}(X),$$
then either X or $\tilde{X}$ is an ω-code.

6.3. More defect

EXAMPLE 6.1.2 (*continued*). The code $X = \{a, ab, bb\}$ satisfies the condition of Proposition 6.3.6 since $\text{Alph}(X) = \{a, b\} = \text{First}(X) = \text{Last}(X)$. It is not an ω-code, but its reversal $\widetilde{X} = \{a, ba, bb\}$ is an ω-code.

Proof of Theorem 6.3.4. Assume the contrary: let $X \subseteq A^+$ be a code with a minimal $\text{Lg}(X)$ such that $X, \widetilde{X}$ are not ω-codes and the base Y of the ξ-free hull of X has cardinality

$$\text{Card}(Y) \geq \text{Card}(X) - 1.$$

We will show that X is a code satisfying the hypotheses of Proposition 6.3.6 but not the thesis. The main tool is the operation of composition of codes, in particular Propositions 6.2.10, 6.2.11 and Properties (6.2.1), (6.2.2).

By Theorem 6.3.2, $\text{Card}(Y) \leq \text{Card}(X) - 1$. Hence

$$\text{Card}(Y) = \text{Card}(X) - 1. \tag{6.3.1}$$

Let us prove that

$$Y = \text{Alph}(X). \tag{6.3.2}$$

As Y^∞ is the ξ-free hull of X, one has $X \subseteq Y^*$. By Proposition 6.3.1, $\text{Alph}_Y(X) = Y$. It follows by Proposition 6.2.11 that

$$X = Y \circ X' \tag{6.3.3}$$

for some finite code X'. Neither X' nor $\widetilde{X'}$ is an ω-code (as X and $\widetilde{X}$ are not; see Proposition 6.2.10). By (6.3.3), $\text{Lg}(X') \leq \text{Lg}(X)$. Suppose that $\text{Lg}(X') < \text{Lg}(X)$. Thus by definition of X, the base Y' of the ξ-free hull of X' satisfies

$$\text{Card}(Y') \leq \text{Card}(X') - 2. \tag{6.3.4}$$

As for X, write X' as the composition

$$X' = Y' \circ Z'.$$

Considering $X = Y \circ X' = Y \circ (Y' \circ Z') = (Y \circ Y') \circ Z'$, we get $X \subseteq (Y \circ Y')^* \subseteq Y^*$ and then

$$(Y \circ Y')^\infty \subseteq Y^\infty.$$

Being the composition of two ξ-codes, $Y \circ Y'$ is also a ξ-code. Hence, since Y^∞ is the ξ-free hull of X, we have $Y^\infty = (Y \circ Y')^\infty$, and then

$$Y = Y \circ Y'. \tag{6.3.5}$$

So

$$\text{Card}(X)-1 \stackrel{(6.3.1)}{=} \text{Card}(Y) \stackrel{(6.3.5)}{=} \text{Card}(Y') \stackrel{(6.3.4)}{\leq} \text{Card}(X')-2 \stackrel{(6.3.3)}{=} \text{Card}(X)-2$$

which is impossible. It follows that $\text{Lg}(X) = \text{Lg}(X')$ and by (6.3.3) $Y = \text{Alph}(X)$.

We end the proof. By (6.3.1) and (6.3.2), $\text{Card}(X)-1 = \text{Card}(\text{Alph}(X))$. By Proposition 6.3.1, $\text{First}(X) = \text{Alph}(X) = \text{Last}(X)$. Consequently, either X or $\tilde{X}$ is an ω-code by Proposition 6.3.6. This brings the contradiction. ∎

6.3.2. A particular class of codes

The proof of Proposition 6.3.6 is rather long. It uses in an elegant way techniques of combinatorics on words.

Let $u, v \in A^\infty$. We use the notation $u \leq v$ when u is prefix of v, and $u < v$ when u is a proper prefix of v. The longest common prefix of u and v is denoted by $u \wedge v$. The words u, v are called *incomparable* if neither $u \leq v$ nor $v \leq u$. The set of prefixes of words in X is denoted by $\text{Pref}(X)$. For suffixes, we use the notation $\text{Suff}(X)$.

In a first step, we study the following particular class of codes $X \subseteq A^+$.

HYPOTHESIS 6.3.7. *$X \subseteq A^+$ is a finite code, it is not an ω-code and*

$$\text{Card}(\text{First}(X)) = \text{Card}(X) - 1.$$

In this hypothesis, since X is not an ω-code, there is an infinite word w with two distinct X-factorizations. Hence all words of X begin with distinct letters, except for two words $x, y \in X$ such that

$$x < y.$$

We denote by Amb_X the set of words *ambiguously covered*, i.e.,

$$\text{Amb}_X = \text{Pref}(xX^\omega) \cap \text{Pref}(yX^\omega).$$

Note that Amb_X contains w and all its prefixes.

We now prove several preliminary results (6.3.8–6.3.11) under Hypothesis 6.3.7. We begin with a technical one.

LEMMA 6.3.8. *Let $u, v \in A^+$ be a pair of incomparable words.*

(i) If $u, v \in \text{Pref}(xX^\omega)$ (resp. $u, v \in \text{Pref}(yX^\omega)$), then $u \wedge v \in xX^*$ (resp. $u \wedge v \in yX^*$).

(ii) If $u \in \text{Pref}(xX^\omega)$ and $v \in \text{Pref}(yX^\omega)$, or vice versa, then $u \wedge v \in X^*$.

6.3. More defect

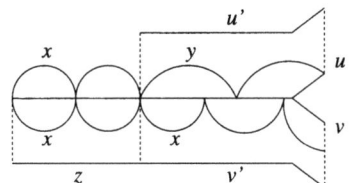

Figure 6.8. $(u, v) \in \mathcal{A}$ and $(u', v') \in \mathcal{B}$.

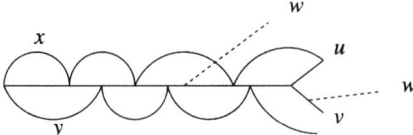

Figure 6.9. The two possible cases for w.

Proof. Define $\mathcal{A}$ as the set of pairs (u, v) of incomparable words such that either $u, v \in \text{Pref}(xX^\omega)$ and $u \wedge v \notin xX^*$, or $u, v \in \text{Pref}(yX^\omega)$ and $u \wedge v \notin yX^*$. Define also $\mathcal{B}$ as the set of pairs (u, v) of incomparable words such that either $u \in \text{Pref}(xX^\omega)$, $v \in \text{Pref}(yX^\omega)$ and $u \wedge v \notin X^*$, or $u \in \text{Pref}(yX^\omega)$, $v \in \text{Pref}(xX^\omega)$ and $u \wedge v \notin X^*$. Let us prove that both $\mathcal{A}$ and $\mathcal{B}$ are empty.

(1) Let us show that if $(u, v) \in \mathcal{A}$, then there exists $(u', v') \in \mathcal{B}$ with $|u' \wedge v'| < |u \wedge v|$. Suppose that $u, v \in \text{Pref}(xX^\omega)$ and $u \wedge v \notin xX^*$. There exist $x_1, \ldots, x_n, y_1, \ldots, y_m \in X$, $n, m \geq 1$, such that

$$xx_1 \cdots x_{n-1} < u \leq xx_1 \cdots x_n,$$
$$xy_1 \cdots y_{m-1} < v \leq xy_1 \cdots y_m.$$

Let i be maximum such that $x_j = y_j$, for all $j \in \{1, \ldots, i\}$. Let $z = xx_1 \cdots x_i$. Then $z \leq u \wedge v$. As $u \wedge v \notin xX^*$, z is a proper prefix of $u \wedge v$. Consider $x_{i+1} \neq y_{i+1}$. By Hypothesis 6.3.7, $\{x_{i+1}, y_{i+1}\} = \{x, y\}$. Hence the thesis holds for the pair (u', v') such that $u' = z^{-1}u$ and $v' = z^{-1}v$. See Figure 6.8.

(2) Let us show that if $(u, v) \in \mathcal{B}$, then there exists $(u', v') \in \mathcal{A}$ with $|u' \wedge v'| \leq |u \wedge v|$. By Hypothesis 6.3.7, there exists an infinite word $w \in \text{Amb}_X$. Either $w \wedge (u \wedge v) < u \wedge v$ or $u \wedge v \leq w$ (see Figure 6.9).

In the first case, let $u' \leq u \wedge v$ and $v' < w$ such that u', v' are incomparable. Then $u', v' \in \text{Amb}_X$. The word $u' \wedge v'$ cannot belong to $xX^* \cap yX^*$ because X is a code. One then checks that the pair (u', v') belongs to $\mathcal{A}$.

In the second case, as u, v are incomparable, $u \wedge v = u \wedge w$ or $u \wedge v = w \wedge v$ (see Figure 6.9). We only consider the first case; the second one is handled similarly. Define $u' = u$ and $v' < w$ such that u', v' are incomparable. Since $v' \in \text{Amb}_X$, it follows that $(u', v') \in \mathcal{A}$.

(3) We conclude the proof. If $\mathcal{A}$ is not empty, there exists a pair $(u, v) \in \mathcal{A}$ with minimal $|u \wedge v|$. Apply (a) and then (b) to this pair. We get a new pair $(u', v') \in \mathcal{A}$ such that $|u' \wedge v'| < |u \wedge v|$. This is impossible. A similar argument shows that $\mathcal{B}$ is empty. ∎

As X is not an ω-code, we know that there exists an infinite word in Amb_X. The proposition below states that this word is unique. It constitutes a nice combinatorial property of finite codes X which are not ω-codes and such that $\text{Card}(\text{First}(X)) = \text{Card}(X) - 1$.

PROPOSITION 6.3.9. *There exists a unique word $\sigma_X \in A^\omega$ such that*

$$\text{Amb}_X = \text{Pref}(\sigma_X).$$

Proof. Assume the contrary: there exist two incomparable words $u, v \in \text{Amb}_X$. These words both belong to $\text{Pref}(xX^\omega)$ and $\text{Pref}(yX^\omega)$. Since X is a code, $u \wedge v \notin xX^*$ or $u \wedge v \notin yX^*$. In both cases, we get a contradiction with Lemma 6.3.8(i). ∎

The next result is also interesting.

LEMMA 6.3.10. *If $u, v \in X^*$, then $u \wedge v \in X^*$.*

Proof. Suppose the contrary and take two words $u, v \in X^*$ such that $u \wedge v \notin X^*$ and $|u \wedge v|$ is minimal. It follows that u, v are incomparable words, and by minimality of $|u \wedge v|$, $u \in \text{Pref}(xX^\omega)$, $v \in \text{Pref}(yX^\omega)$, or the contrary. This is impossible in view of Lemma 6.3.8, part 2. ∎

The following lemma shows that σ_X has two distinct X-factorizations which are eventually periodic.

LEMMA 6.3.11. *There exist $r, s, r', s' \in X^*$ such that*

$$\sigma_X = rs^\omega = r's'^\omega$$

and $\text{First}_X(rs) \ne \text{First}_X(r's')$. Moreover, for any $t \in X^*$,

$$\sigma_X \ne t^\omega.$$

6.3. More defect

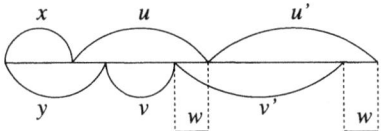

Figure 6.10. The suffix w appears twice.

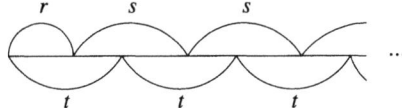

Figure 6.11. $\sigma_X = rs^\omega = t^\omega$.

Proof. By hypothesis, σ_X has two X-factorizations, one beginning with x and the other with y.

$$(x, x_1, x_2, \ldots), \quad (y, y_1, y_2, \ldots).$$

As X is a finite set, there exists a word $w \in \text{Suff}(X)$ which is repeated in the following way, for some $u = x_1 \cdots x_k \in X^*$, $u' = x_{k+1} \cdots x_n \in X^+$, $v = y_1 \cdots y_\ell \in X^*$, $v' = y_{\ell+1} \cdots y_m \in X^+$:

$$xu = yvw, \quad xuu' = yvv'w$$

(see Figure 6.10). It follows that $xuu'^\omega = yvv'^\omega = \sigma_X$ by uniqueness of σ_X (Proposition 6.3.9). The first part of the lemma is proved.

Assume now that $\sigma_X = t^\omega$ for some $t \in X^+$. Either $\text{First}_X(t) \neq \text{First}_X(rs)$ or $\text{First}_X(t) \neq \text{First}_X(r's')$. Suppose that we are in the first case. By replacing t and s by adequate powers, say $t^i, s^j \in X^+$, we can make the assumption that $|t| = |s|$ and $|t| > |r|$. Looking at Figure 6.11, one sees that $rs = tr$ with $\text{First}_X(rs) \neq \text{First}_X(tr)$. But X is a code: a contradiction. ∎

The next lemma is the last step before proving Proposition 6.3.6. There is no contradiction between this lemma and the previous one: Lemma 6.3.12 states that $\sigma_X = p^\omega$ for some $p \in A^+$; Lemma 6.3.11 states that $p \notin X^+$.

LEMMA 6.3.12. *If X and $\tilde{X}$ both satisfy Hypothesis 6.3.7, then σ_X and $\sigma_{\tilde{X}}$ are periodic.*

Proof. By Proposition 6.3.9,

$$\text{Amb}_X = \text{Pref}(\sigma_X).$$

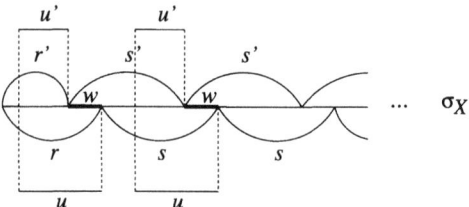

Figure 6.12. The case $u'w = u$ with $u, u' \in X^+$, $w \notin X^*$.

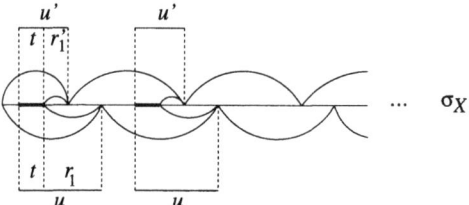

Figure 6.13. σ_X is a proper suffix of itself.

By Lemma 6.3.11,

$$\sigma_X = rs^\omega = r's'^\omega \tag{6.3.6}$$

where $r, s, r', s' \in X^+$ and $\text{First}_X(r) \neq \text{First}_X(r')$. With adequate powers of s, s', one can suppose that $|s| = |s'|$ and $|r|, |r'| \leq |s|$.

Since X is a code and $\text{First}_X(r) \neq \text{First}_X(r')$, one gets $r \neq r'$; let us say that $r'w = r$ for some $w \neq \varepsilon$. Again because X is a code, $w \notin X^*$. Consider

$$\tilde{u} = \tilde{r} \wedge \tilde{s}.$$

Hence w is suffix of u (see Figure 6.12). By Lemma 6.3.10 applied to $\tilde{X}$, $u \in X^+$ showing that $w \neq u$. Let

$$\tilde{u}' = \tilde{r}' \wedge \tilde{s}'.$$

We have $u'w = u$ and $u' \in X^+$ by Lemma 6.3.10 again (see Figure 6.12).

Let us factorize u, u' as $u = tr_1$, $u' = tr'_1$ with $t, r_1, r'_1 \in X^*$ and t of maximal length. As $w \notin X^*$, $r'_1 \neq \varepsilon$. It follows that $\text{First}_X(r_1) \neq \text{First}_X(r'_1)$. Therefore $r_1 s^\omega = r'_1 s'^\omega \in \text{Amb}_X$ (see Figure 6.13). This word is equal to σ_X by Proposition 6.3.9. Thus observe on Figure 6.13 that σ_X is a proper suffix of σ_X. It follows that σ_X is periodic.

6.3. More defect

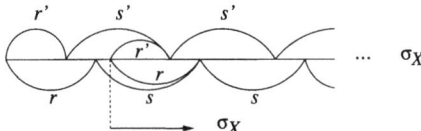

Figure 6.14. The words r, r' are chosen with minimal length.

The same argument applied to $\widetilde{X}$ instead of X shows that $\sigma_{\widetilde{X}}$ is periodic. ∎

REMARK 6.3.13. In the preceding proof, if r, r' are chosen with minimal length in (6.3.6), then $r = r_1$, $r' = r'_1$ and r is suffix of s, r' is suffix of s' (see Figure 6.14).

We are now ready to prove Proposition 6.3.6. The proof is *ad absurdum*. In particular, X, $\widetilde{X}$ both satisfy Hypothesis 6.3.7. The key results to get the contradiction are Proposition 6.3.9 and Lemmas 6.3.10-6.3.12. The proof goes back and forth between X and $\widetilde{X}$. To avoid any confusion, we systematically use the reversal notation, e.g. $\widetilde{v} \in \widetilde{X}^*$, whenever we work with $\widetilde{X}$.

Proof of Proposition 6.3.6. Assume that X and $\widetilde{X}$ are not ω-codes. As in the beginning of the previous proof, take

$$\sigma_X = rs^\omega = r's'^\omega$$

with $r, s, r', s' \in X^+$ and $|s| = |s'| \geq |r|, |r'|$. Recall that $r'w = r$ with

$$w \notin X^*.$$

We also choose r, r' with minimal length as in Remark 6.3.13. Consequently, $\mathrm{Last}_X(r) \neq \mathrm{Last}_X(s)$ and $\mathrm{Last}_X(r') \neq \mathrm{Last}_X(s')$. By Remark 6.3.13, $\widetilde{r} \leq \widetilde{s}$ and $\widetilde{r}' \leq \widetilde{s}'$.

Define $\widetilde{v} \in \mathrm{Pref}(\widetilde{X}^\omega)$ of maximal length such that (see Figure 6.15)

$$\widetilde{u} = \widetilde{r}\,\widetilde{v} \leq \widetilde{s}^\omega, \quad \widetilde{u}' = \widetilde{r}'\widetilde{v} \leq \widetilde{s}'^\omega. \tag{6.3.7}$$

As $\mathrm{Last}_X(r) \neq \mathrm{Last}_X(s)$, $\mathrm{Last}_X(r') \neq \mathrm{Last}_X(s')$, it follows that $\widetilde{u}, \widetilde{u}' \in \mathrm{Amb}_{\widetilde{X}}$, hence

$$\widetilde{u}, \widetilde{u}' \in \mathrm{Pref}(\sigma_{\widetilde{X}}).$$

If $\widetilde{u}$ is an infinite word, then $\widetilde{u} = \sigma_{\widetilde{X}} = \widetilde{s}^\omega$, with $\widetilde{s} \in \widetilde{X}^+$, a contradiction with Lemma 6.3.11. Thus $\widetilde{u}$ is a finite word. In the same way, $\widetilde{u}'$ is a finite word. This situation is summarized in Figure 6.15.

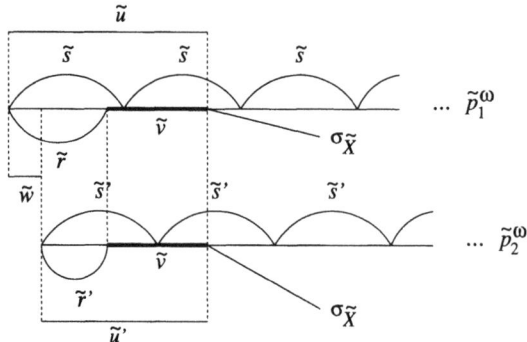

Figure 6.15. $\tilde{u}, \tilde{u}' \in \tilde{X}^+$ but $\tilde{v} \notin \tilde{X}^*$.

Let us go further. By Lemma 6.3.10, as $\tilde{u} = \tilde{s}^\omega \wedge \sigma_{\tilde{X}}$ and $\tilde{u}' = \tilde{s}'^\omega \wedge \sigma_{\tilde{X}}$, we have

$$\tilde{u}, \tilde{u}' \in \tilde{X}^+.$$

However

$$\tilde{v} \notin \tilde{X}^*.$$

Otherwise, let $a \in \text{Alph}(\tilde{X})$ be such that $\tilde{r}\tilde{v}a < \tilde{s}^\omega$ (see Figure 6.15). As $\text{First}(\tilde{X}) = \text{Alph}(\tilde{X})$, there exists $\tilde{z} \in \tilde{X}$ with $\text{First}(\tilde{z}) = a$. Therefore $\tilde{v}a$ is a word longer than $\tilde{v}$ such that $\tilde{v}a \in \text{Pref}(\tilde{X}^\omega)$ and $\tilde{r}\tilde{v}a < \tilde{s}^\omega$, a contradiction with the definition (6.3.7) of $\tilde{v}$.

Now, by Lemma 6.3.12, there exist primitive words $p, \tilde{q} \in A^+$ such that $\sigma_X = p^\omega$, $\sigma_{\tilde{X}} = \tilde{q}^\omega$. We can suppose that $|p| \geq |\tilde{q}|$. By Fine and Wilf's theorem (Proposition 1.2.1), as $\sigma_X = p^\omega = rs^\omega = r's'^\omega$, we have

$$s \in p_1^+, s' \in p_2^+$$

where p_1, p_2 and p are conjugate words.

Observe that the word $\tilde{u}$ cannot be too long:

$$|\tilde{u}| < |p| + |q|. \tag{6.3.8}$$

Indeed, if $|\tilde{u}| \geq |p|+|q|$, as $\tilde{u} \in \text{Pref}(\tilde{s}^\omega) \cap \text{Pref}(\sigma_{\tilde{X}}) = \text{Pref}(\tilde{p}_1^\omega) \cap \text{Pref}(\tilde{q}^\omega)$, we get $\tilde{p}_1 = \tilde{q}$ by Fine and Wilf's theorem. In particular, $\tilde{s} \in \tilde{q}^+$. This implies that $\sigma_{\tilde{X}} = \tilde{s}^\omega$ with $\tilde{s} \in \tilde{X}^+$. This is impossible by Lemma 6.3.11.

To get the final contradiction, we show that Inequality (6.3.8) never holds. For this, we come back to σ_X in the following way. Decompose $u, u' \in X^+$ as

$$u = tr_1, u' = tr_1'$$

6.3. More defect

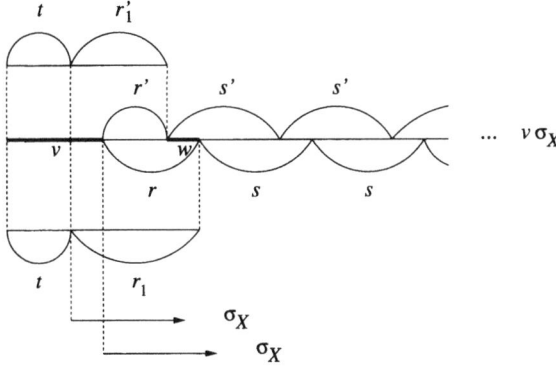

Figure 6.16. Back to σ_X.

where $t, r_1, r'_1 \in X^*$ and t is of maximal length (see Figure 6.16). We have $r'_1 \neq \varepsilon$ because $w \notin X^*$, and $r'_1 \neq r'$ because $v \notin X^*$. By definition of t, $\text{First}_X(r_1) \neq \text{First}_X(r'_1)$ and then $r_1 s^\omega = r'_1 s'^\omega = \sigma_X$ (Proposition 6.3.9). We have $|r'| < |r'_1|$ by minimality of $|r'|$ and because $r' \neq r'_1$. So r' is a proper suffix of r'_1 as depicted in Figure 6.16. Hence σ_X is a proper suffix of σ_X, that is $\sigma_X = t_1 \sigma_X$ with $v = t t_1$. Thus $\sigma_X = p^\omega = t_1^\omega$. As p is primitive, it follows that
$$|v| \geq |p|$$
(see Figure 6.16). Now let us look again at Figure 6.15. The words $\tilde{u}'$ and $\tilde{u} = \tilde{w}\tilde{u}'$ are prefixes of $\tilde{q}^\omega$ because $\sigma_{\tilde{x}} = \tilde{q}^\omega$. Since $|\tilde{u}'|, |\tilde{u}| \geq |v| \geq |p| \geq |q|$ and q is primitive, we deduce that $|\tilde{w}| \geq |\tilde{q}|$. Hence $|\tilde{u}| \geq |\tilde{w}| + |\tilde{v}| \geq |\tilde{q}| + |\tilde{p}|$. This ends the proof. ∎

6.3.3. Three-element codes

Theorem 6.3.4 leads to the following nice property of codes with three elements. This property is no longer true for larger codes (see Example 6.3.5).

COROLLARY 6.3.14. *Let $X = \{x, y, z\}$ be a code. Then X or $\tilde{X}$ is an ω-code.*

Proof. If not, by Theorem 6.3.4, $X \subseteq t^*$ where t^∞ is the ξ-free hull of X. This is impossible since X is a code. ∎

Another property which is characteristic of three-element codes, but does not hold for arbitrary ones, is given in the next proposition.

PROPOSITION 6.3.15. *For a three-element code $X = \{x, y, z\}$, there exists at most one infinite word σ with two X-factorizations $(x_1, x_2, \ldots)$, $(y_1, y_2, \ldots)$ such that $x_1 \neq y_1$.*

Proof. If X is an ω-code, such an infinite word σ cannot exist. Suppose that X is not an ω-code. If $\text{Card}(\text{First}(X)) = \text{Card}(X) - 1$, we have done by Proposition 6.3.9. Otherwise let

$$u = x^\omega \wedge y^\omega \wedge z^\omega \neq \varepsilon.$$

By Fine and Wilf's theorem (Proposition 1.2.1)

$$|u| < \min\{|x| + |y|, |y| + |z|, |z| + |x|\}.$$

Indeed, if $|u| \geq |x| + |y|$ for instance, then $x^i = y^j$ with $i, j \geq 1$. This is impossible because X is a code.

We show that for any $v \in \text{Pref}(X^*)$

$$|u| \leq |v| \Rightarrow u \leq v. \qquad (6.3.9)$$

Let v' be the prefix of v such that $|u| = |v'|$. If $v' \in \text{Pref}(x^\omega)$, then $u = v'$ by definition of u. If $v' = x^n y'$ with $y' \in \text{Pref}(y) - \varepsilon$ and $n \geq 1$, then $v' = u$ because $y' < u < x^\omega$. The other cases are proved similarly.

By (6.3.9), u is a prefix of each word xu, yu, zu. We define a new three-element code $X' = \{x', y', z'\}$ such that $x' = u^{-1}xu$, $y' = u^{-1}yu$, $z' = u^{-1}zu$. By (6.3.9), $w \in A^\omega$ has an X-factorization $(x_1, x_2, \ldots)$ if and only if $u^{-1}w$ has an X'-factorization $(u^{-1}x_1u, u^{-1}x_2u, \ldots)$. So X' is not an ω-code. Moreover the condition $\text{Card}(\text{First}(X')) = \text{Card}(X') - 1$ is now satisfied. Consequently, there exists exactly one infinite word $\sigma_{X'}$ with two distinct X'-factorizations (Proposition 6.3.9). The conclusion follows for X with $\sigma = u\sigma_{X'}$. ∎

EXAMPLE 6.3.16. The code $X = \{ba, bab, bb\}$ is such that $\text{Card}(\text{First}(X)) < \text{Card}(X) - 1$. As in the preceding proof, we construct the new code $X' = b^{-1}Xb = \{ab, abb, bb\}$, with $\sigma_{X'} = ab^\omega$. Thus there is a unique infinite word $\sigma = bab^\omega$ with two X-factorizations $(ba, bb, bb, \ldots)$ and $(bab, bb, bb, \ldots)$ that begin with distinct words of X.

The situation is much more complex for larger codes.

EXAMPLE 6.3.17. We associate with the code $X = \{ba, bab, abaa, aabaab\}$ the graph of Figure 6.17. An edge $u \to v$ means that there exists $x \in X$ such that $u = xv$ or $uv = x$. The infinite paths beginning with ba thus

6.4. A theorem of Schützenberger

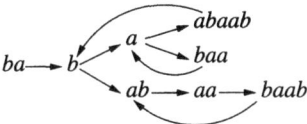

Figure 6.17. Graph of infinite words with two X-factorizations.

describe words of A^ω with two distinct X-factorizations. For instance the infinite path

$$ba \to b \to a \to baa \to a \to abaab \to b \to a \to baa \to a \to \cdots$$

describes the overlappings of Figure 6.18.

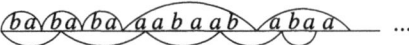

Figure 6.18. From a path to the associated pair of Y-factorizations.

6.4. A theorem of Schützenberger

In this last section, we want to describe a noteworthy property of finite maximal codes. A code $X \subseteq A^+$ is *maximal* if it cannot be strictly included in a code Y over the same alphabet A. The next theorem states two extremal behaviors of finite codes $X \subseteq A^+$ which are maximal: either X is not an ω-code or X is a prefix code. If one recalls that finite ω-codes are exactly finite codes with bounded decoding delay (Proposition 6.1.9), another interpretation of this result is: the decoding delay of a finite maximal code is either null or not bounded.

THEOREM 6.4.1 (Schützenberger's theorem). *Let $X \subseteq A^+$ be a finite maximal code. If X is an ω-code, then X is a prefix code.*

EXAMPLE 6.4.2. The set $X = \{a, ab, bb\}$ is a maximal code for the following reasons. One can easily prove (by induction on $|w|$) that any $w \in \{a, b\}^*$ belongs to $X^* \cup bX^*$. Assume that $Y = X \cup \{w\}$ is a code for some word $w \notin X$. Then $aw \in X^*$. This shows that aw has two distinct Y-factorizations: one X-factorization and the Y-factorization (a, w). Contradiction. Hence X is a maximal code. The set X is not a prefix code, and thus not an ω-code by Theorem 6.4.1. The reversal code $\widetilde{X}$ is prefix.

The next two examples show that the hypotheses of Theorem 6.4.1 are necessary.

EXAMPLE 6.4.3. The infinite code $X = ab^*$ is a maximal code. It is an ω-code which is not a prefix code.

EXAMPLE 6.4.4. The set $X = \{a, aba\}$ is a code which is not maximal. For instance, $X \cup \{bb\}$ is still a code. Both X and $\widetilde{X}$ are ω-codes without being prefix codes.

Proof of Theorem 6.4.1. (1) We first show that if $X \subseteq A^+$ is a maximal code, then X is *complete*, i.e.,

$$\forall w \in A^*, X^* w A^* \cap X^* \neq \emptyset \tag{6.4.1}$$

(see also Problem 6.4.1). This is trivially true for any alphabet $A = \{a\}$. Suppose that $\operatorname{Card}(A) \geq 2$ and some $w \in A^*$ satisfies

$$X^* w A^* \cap X^* = \emptyset. \tag{6.4.2}$$

In this equality, we can suppose that w is *unbordered*, i.e., if $w \in uA^+ \cap A^+u$, then $u = \varepsilon$. Indeed, if $\operatorname{First}(w) = a$, replace w by the unbordered word $wab^{|w|}$ such that $b \in A$, $b \neq a$.

The set $Y = X \cup \{w\}$ is not a code by hypothesis. So there exists a word $z \in Y^*$ with two Y-factorizations $(x_1, \ldots, x_n)$, $(y_1, \ldots, y_m)$ such that $x_1 \neq y_1$. As X is a code and by (6.4.2), w must appear among the x_i's *and* the y_j's. Consider the first occurrences of w

$$x_1, \ldots, x_{i-1} \in X, \quad x_i = w,$$
$$y_1, \ldots, y_{j-1} \in X, \quad y_j = w.$$

Again by (6.4.2), x_i and y_j must overlap and thus coincide since w is unbordered. Therefore, as $x_1 \neq y_1$, we have $i, j \geq 2$ and the word $x_1 \cdots x_{i-1}$ has two distinct X-factorizations $(x_1, \ldots, x_{i-1})$, $(y_1, \ldots, y_{j-1})$. This is impossible because X is a code.

(2) We now make the assumption that X is an ω-code which is not a prefix code. By Proposition 6.1.9, X has a decoding delay $d > 0$. Hence, define $t \in \operatorname{Pref}(X^*)$ of maximal length such that

$$xtA^* \cap yX^* \neq \emptyset, \text{ with } x, y \in X, x \neq y \tag{6.4.3}$$

(see Figure 6.19). Note that t is well defined since $0 < d < \infty$ and X is finite.

6.4. A theorem of Schützenberger

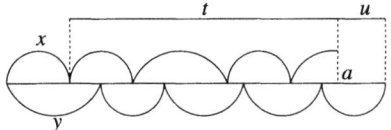

Figure 6.19. X is an ω-code which is not a prefix code.

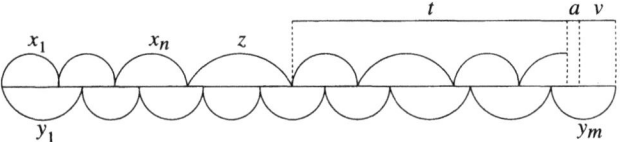

Figure 6.20. X is a maximal code.

Let $u \in A^*$ such that $xtu \in yX^*$. We can suppose that $u \neq \varepsilon$. Indeed, if $u = \varepsilon$, replace u by any element of X. We denote by a the first letter of u.

Let $w = zta$ such that z is a word of X of maximal length (recall that X is finite). By (6.4.1), there exist $x_1, \ldots, x_n, y_1, \ldots, y_m \in X$ and $v \in A^*$ such that

$$x_1 \cdots x_n wv = y_1 \cdots y_m$$

(see Figure 6.20). As $|t|$ is maximal for Property (6.4.3), $x_k = y_k$, for all $k \in \{1, \ldots, n\}$. By definition of z, we have $y_{n+1} \leq z$. Assume that $y_{n+1} < z$. Define $t' \in \text{Pref}(y_{n+2} \cdots y_m)$ such that $y_{n+1} t' = zt$ (see Figure 6.21). It follows that t' satisfies (6.4.3) with $|t'| > |t|$. This is impossible. Hence $y_{n+1} = z$ showing that ta belongs to $\text{Pref}(y_{n+2} \cdots y_m) \subseteq \text{Pref}(X^*)$. We have again a contradiction (see Figure 6.19): ta satisfies (6.4.3) since a is the first letter of u and $|ta|$ is longer than $|t|$. The conclusion is that X is necessarily a prefix code. ∎

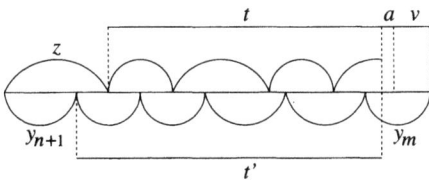

Figure 6.21. The case where y_{n+1} is a proper prefix of z.

Problems

Section 6.1

6.1.1 Show that a subset X of A^+ is a code if and only if for any $x \in X^+$, x^ω has only one X-factorization.

6.1.2 Show that a submonoid M of A^* is stable if and only if for any $u, v \in M$
$$uv, u, vu \in M \Rightarrow v \in M.$$

6.1.3 1. Show that for a rational code $X \subseteq A^+$, X has a bounded decoding delay if and only if X is an ω-code and $X^\omega \cap X^* \operatorname{Adh}(X) = \emptyset$, where
$$\operatorname{Adh}(X) = \{w \in A^\omega \mid \operatorname{Pref}(w) \subseteq \operatorname{Pref}(X)\}$$
(see Proposition 6.1.9).
2. Give an example of a rational ω-code which does not have a bounded decoding delay.

6.1.4 Find an example of a family $(X_n)_{n \in \mathbb{N}}$ of codes with bounded decoding delay such that the base of $\bigcap_{n \in \mathbb{N}} X_n^*$ is a code which does not have a bounded decoding delay. Compare with Corollary 6.1.7.

6.1.5 Let $A^{\mathbb{Z}}$ be the set of two-sided infinite words over A and A^ς the set of the equivalence classes on $A^{\mathbb{Z}}$ under the shift (observe the difference between A^ς and A^ζ as defined in Chapter 1). We denote by X^ς the subset of A^ς equal to
$$X^\varsigma = \{\cdots x_{-1} x_0 x_1 x_2 \cdots \mid x_n \in X, n \in \mathbb{Z}\}.$$

Given $X \subseteq A^+$, an X-decomposition of $w \in A^{\mathbb{Z}}$ is a strictly increasing sequence $(i_n)_{n \in \mathbb{Z}}$ of integers such that $i_0 \geq 0$, $i_{-1} < 0$ and $w_{[i_n, i_{n+1})} \in X$. An X-factorization of $w \in A^\varsigma$ is a sequence $(x_n)_{n \in \mathbb{Z}}$ of words of X such that
$$w = \cdots x_{-1} x_0 x_1 x_2 \cdots.$$

A set $X \subseteq A^+$ is called a $\mathbb{Z}$-code (resp. ς-code if any element of $A^{\mathbb{Z}}$ (resp. A^ς) has at most one X-decomposition (resp. X-factorization).

1. Show that any $\mathbb{Z}$-code is a ς-code. Give an example of a ς-code which is not a $\mathbb{Z}$-code.

2. Let φ be a bijection between B and $X \subseteq A^+$. Prove that φ is injective on B^ς if and only if X is a ς-code. Prove that φ is injective on B^ς and for any $r \in B^+$

$$r \text{ primitive} \Rightarrow \varphi(r) \text{ primitive}$$

if and only if X is a $\mathbb{Z}$-code.

Section 6.2

*6.2.1 Let X be a subset of A^+. Define a sequence $(M_n)_{n \in \mathbb{N}}$ of submonoids of A^* by

$$M_0 = X^*, \quad M_{n+1} = \left(M_n^{-1} M_n \cap M_n M_n^{-1}\right)^*.$$

Set $M = \bigcup_{n \in \mathbb{N}} M_n$.
1. Show that M is the free hull of X.
2. Prove that M is rational if X is rational.

6.2.2 Find a method for constructing the ω-free hull of a subset X of A^+.

6.2.3 1. Find a definition for the $\mathbb{Z}$-free hull of a finite set X.
2. If Y is the base of the $\mathbb{Z}$-free hull of X, prove that

$$\text{Card}(Y) \leq \text{Card}(X)$$

and that no better upper bound exists.
3. Prove that

$$\text{Card}(Y) \leq \text{Card}(X) - 1$$

if there exists a nonperiodic word $w \in A^{\mathbb{Z}}$ with two X-decompositions.
(See Problem 6.1.5.)

6.2.4 A submonoid M of A^* is *right-unitary* if

$$u, uw \in M \Rightarrow w \in M.$$

1. Show that $X \subseteq A^+$ is a prefix code if and only if it is the base of a right-unitary submonoid M of A^*.
2. Given a finite subset X of A^+, let Y be the base of the smallest right-unitary submonoid which contains X. Prove that Y exists and that

$$\text{Card}(Y) \leq \text{Card}(X),$$

showing that there is no defect.

3. Give an example where the preceding inequality becomes an equality.

6.2.5 Given a finite set $X \subseteq A^+$, we define four different ranks: the *free rank* $r_f(X)$ equal to Card(Y) with Y the base of the free hull of X, similarly the *ω-free rank* $r_\omega(X)$, the *right-unitary rank* $r_u(X)$, and finally the *combinatorial rank* $r(X)$ equal to $\min\{\text{Card}(Y) \mid X \subseteq Y^*\}$.
1. Show that $r(X) \leq r_u(X) \leq r_\omega(X) \leq r_f(X) \leq \text{Card}(X)$.
2. Find an example with strict inequalities.

6.2.6
1. Give a direct proof that any two-element code is an ω-code, i.e., without using the defect theorem.
2. Show that the constant d of the decoding delay can be arbitrarily large.

6.2.7 A morphism $\varphi : B^\infty \to A^\infty$ is *simplifiable* if $\varphi = \psi \circ \phi$ with $\phi : B^\infty \to C^\infty$, $\psi : C^\infty \to A^\infty$ two morphisms such that Card(C) $\leq$ Card(B) -1. If no such alphabet C exists, φ is called *elementary*. Show that a finite set $X \subseteq A^+$ is elementary if and only if there exists an elementary morphism $\varphi : B^\infty \to A^\infty$ such that $X = \varphi(B)$.

*6.2.8 Let $X \subseteq A^+$ be an elementary set. Show that the decoding delay of the ω-code X is bounded by

$$\sum_{x \in X} |x| - \text{Card}(X)$$

(use Problem 6.2.9).

6.2.9 Show that if $X = Y \circ Z$ with Y and Z two codes with decoding delay d_Y, d_Z respectively, then X is a code with decoding delay $d_X \leq d_Y + d_Z$.

*6.2.10 A code X is called *prefix–suffix composed* if

$$X = X_1 \circ X_2 \circ \cdots \circ X_k$$

with each X_n being a prefix or a suffix code.
1. Show that any two-element code is prefix–suffix composed.
2. Verify that the three-element code $X = \{a, aba, babaab\}$ is not prefix–suffix composed.
3. Find a finite maximal code which contains X (see Section 6.4 for the definition of a maximal code).

**6.2.11
1. Prove that every prefix–suffix composed code with n elements, $n \geq 3$, uses at most $2n - 3$ prefix and suffix codes.
2. Show that this upper bound is tight.

Section 6.3

6.3.1 Show that any ç-code is a ξ-code (see Problem 6.1.5 for the definition of a ç-code).

6.3.2 Let X be a subset of A^+ and Y be the base of its ξ-free hull.
1. Find an example of a set X which is not a code and such that $\text{Card}(Y) \leq \text{Card}(X) - 3$.
2. Find another example such that $\text{Card}(Y) = \text{Card}(X) - 2$.
3. Compare with Theorem 6.3.4 and Example 6.3.3.

6.3.3 Lemma 6.3.11 states that there are two pairs $(r, s), (r', s')$ of words $r, s, r', s' \in X^*$ such that $\text{First}_X(rs) \neq \text{First}_X(r's')$ and $\sigma_X = rs^\omega = r's'^\omega$. Take the words r, s, r', s' with minimal length. Show that there exists no other such pair.

*6.3.4 A variant of Theorem 6.3.4 is: If X is a code, but X and $\widetilde{X}$ are not ω-codes, then there exists Y such that $X \subseteq Y^*$ and $\text{Card}(Y) \leq \text{Card}(X) - 2$ (Y is not assumed to be the base of the ξ-free hull of X). Give a proof of this result, using Proposition 6.3.6 and elementary morphisms (see Problem 6.2.7 for the definition of an elementary morphism).

6.3.5 Prove that if $X = \{x, y\}$ is a code with two elements, then $\text{Amb}_X = \text{Pref}(xy \wedge yx)$.

*6.3.6 Let X be a three-element code which is not an ω-code. Prove that there exists no ω-code $Y \subseteq X^*$ such that $X^\omega = Y^\omega$.

Section 6.4

**6.4.1 Prove that a rational code is maximal if and only if it is complete.

*6.4.2 An elementary set $X \subseteq A^+$ is *maximal elementary* if X is not strictly included in any elementary set over A. Prove that an elementary set $X \subseteq A^+$ is maximal elementary if and only if $\text{Card}(X) = \text{Card}(A)$.

Notes

The theory of codes is a well-developed branch of theoretical computer science. We refer the reader to the books of Berstel and Perrin 1985, Shyr 1991, and to Jürgensen and Konstantinidis 1997.

Codes were investigated in depth for the first time in Schützenberger 1956, which contains Proposition 6.1.6. The notion of ω-code appears in Staiger 1986 under the name of ifl-codes. Codes for two-sided infinite words are introduced in Devolder and Timmerman 1992 (see Problem 6.1.5). It is also possible to consider codes composed with finite and infinite words (see Do Long Van 1982). This approach is not considered here.

The defect theorem stating that if a set X with n elements is not a code, then there exists Y with at most $n-1$ words such that $X \subseteq Y^*$, is folklore. It has been proved under various forms (Skordev and Sendov 1961, Lentin 1972, Makanin 1976, Ehrenfeucht and Rozenberg 1978). A defect theorem for sets X which are not ω-codes appears in Linna 1977. The proof given here for Theorems 6.2.1 and 6.2.4 is from Berstel et al. 1979. Proposition 6.2.3 appears in Harju and Karhumäki 1986. See the chapter 'Combinatorics on words' in the *Handbook of Formal Languages* for a presentation of the defect theorem and related material, Choffrut and Karhumäki 1997. Defect properties for other structures like two-sided infinite words or trees are studied in Karhumäki, Maňuch, and Plandowski 1998b, Mantaci and Restivo 1999 and Mantaci and Karhumäki 1999.

Elementary sets are related to elementary morphisms, a notion used in an elegant way for one of the proofs of the D0L equivalence problem given in Ehrenfeucht and Rozenberg 1978 (see also Rozenberg and Salomaa 1980).

A measure of the defect can be evaluated thanks to the combinatorial rank of a finite set X defined by $r(X) = \min\{\text{Card}(Y) \mid X \subseteq Y^*\}$. Other notions of rank can be defined such as the cardinality of the base of the free hull of X. See Harju and Karhumäki 1986 for a comparison of different kinds of ranks (see also Problem 6.2.5).

Algorithmic questions related to the defect effect are treated in several papers. The computation of the free hull of X is given in Spehner 1975 when X is finite and in Berstel et al. 1979 when X is rational (see Problem 6.2.1). The complexity results for the rank $r(X)$ are due to Néraud 1990a, Néraud 1993. It is there proved that deciding, for a given finite set X and a given number k, whether $r(X) \leq k$ is an NP-complete problem. The choice $k = 2$ makes the problem computationally easy as it can be solved in time $O(n \log^2 m)$ with $n = \text{Lg}(X)$ and $m = \max\{|x| \mid x \in X\}$.

Theorem 6.3.4 is a variant of a result by Honkala 1988 (where Y is not necessarily equal to the base of the ξ-free hull of X; see Problem 6.3.4). The proofs given in Section 6.3 to get Theorem 6.3.4 are strongly based on the material developed in Karhumäki 1985a, Karhumäki 1985b. It would be nice to have a shorter proof of this remarkable result. Corollary 6.3.14 and Proposition 6.3.15 about three-element codes are due to Karhumäki 1985a, Karhumäki 1985b.

Theorem 6.4.1 is due to Schützenberger 1966, solving a conjecture of E. N. Gilbert and Moore 1959. The original proof is tricky and in a certain way magic. The proof given in this chapter comes from Bruyère 1992, where another simple proof is also given which is based on properties of automata. The fact that for rational sets, complete codes are equivalent to maximal codes is a basic result in the theory of codes (see Schützenberger 1956, Berstel and Perrin 1985).

Let us mention some open problems.

Well-known algorithms exist to decide whether a given rational submonoid of A^* is generated by a code (see Berstel and Perrin 1985). Deciding whether a rational subset X of A^ω is equal to Y^ω with Y a (finite) ω-code is still an open question. This problem is solved in Litovsky 1991 with the condition that Y is an ω-code replaced by the condition that Y is a prefix code. Other partial interesting results can be found in Devolder 1999. See also Problem 6.3.6.

Some questions on three-element codes still remain open. One is the existence of a code $X = \{x, y, z\}$ which cannot be included in any finite maximal code. It is conjectured in Restivo, Salemi, and Sportelli 1989 that any three-element code is prefix–suffix composed, which implies that such an example X does not exist. In Derencourt 1996 a family of counterexamples to the conjecture of Restivo et al. 1989 is given, among which is the not prefix–suffix composed code $\{a, aba, babaab\}$ of Problem 6.2.10. However, all these examples can be included in a finite maximal code.

Another conjecture on three-element codes is proposed in Devolder 1993: a code $X = \{x, y, z\}$ such that

$$w^n \in X^* \Rightarrow w \in X^*$$

is a π-code, that is, any periodic word has at most one X-factorization.

Problem 6.1.1 is from Devolder et al. 1994. Problem 6.1.3 is from Devolder et al. 1994. Problems 6.1.4 and 6.2.1 appear in Berstel et al. 1979. Problem 6.2.3.3 is from Karhumäki et al. 1998b. Problem 6.2.5 is solved in Harju and Karhumäki 1986. Problem 6.2.8 comes from Rozenberg and Salomaa 1980. Problem 6.2.11 is solved in Derencourt 1996. Problem 6.3.3 comes from Karhumäki 1985a. Problem 6.3.4 is solved in Honkala 1988. Problem 6.3.6 appears in Julia 1996. Problem 6.4.2 is from Néraud 1990b.

CHAPTER 7

Numeration Systems

7.0. Introduction

This chapter deals with positional numeration systems. Numbers are seen as finite or infinite words over an alphabet of digits. A *numeration system* is defined by a pair composed of a base or a sequence of numbers, and of an alphabet of digits. In this chapter we study the representation of natural numbers, of real numbers and of complex numbers. We will present several generalizations of the usual notion of numeration system, which lead to interesting problems.

Properties of words representing numbers are well studied in number theory: the concepts of period, digit frequency, normality give rise to important results. Cantor sets can be defined by digital expansions.

In computer arithmetic, it is recognized that algorithmic possibilities depend on the representation of numbers. For instance, addition of two integers represented in the usual binary system, with digits 0 and 1, takes a time proportional to the size of the data. But if these numbers are represented with signed digits 0, 1, and -1, then addition can be realized in parallel in a time independent of the size of the data.

Since numbers are words, finite state automata are relevant tools to describe sets of number representations, and also to characterize the complexity of arithmetic operations. For instance, addition in the usual binary system is a function computable by a finite automaton, but multiplication is not.

The usual numeration systems, such as the binary and the decimal ones, are described in the first section. In fact, these systems are a particular case of all the various generalizations that will be presented in the next sections.

The second section is devoted to the study of the so-called beta-expansions, introduced by Rényi; see Notes. They consist in taking for base a real number $\beta > 1$. When β is actually an integer, we get the standard representation. When β is not an integer, a number may have several different β-representations. A particular β-representation, playing an important role, is obtained by a greedy algorithm, and is called the β-expansion; it is the greatest in the lexicographic order. The set of β-expansions of numbers of $[0, 1)$ is shift-invariant, and its closure, called the β-shift, is a symbolic dynamical system. We give several results on these topics. We do not cover the whole field, which is very lively and still growing. It has interesting connections with number theory and symbolic dynamics.

In the third section we consider the representation of integers with respect to a sequence of integers, which can be seen as a generalization of the notion of base. The most popular example is that of Fibonacci numbers. Every positive integer can be represented in such a system with digits 0 and 1. This field is closely related to the theory of beta-expansions.

The last section is devoted to complex numbers. Representing complex numbers as strings of digits allows us to handle them without separating real and imaginary parts. We show that every complex number has a representation in base $-n \pm i$, where n is an integer ≥ 1, with digits in $\{0, \ldots, n^2\}$. This numeration system enjoys properties similar to those of the standard beta-ary system.

For notations concerning automata and words the reader may want to consult Chapter 1.

7.1. Standard representation of numbers

In this section we will study standard numeration systems, where the base is a natural number. We will represent first the natural numbers, and then the nonnegative real numbers. The notation introduced in this section will be used in the other sections.

7.1.1. Representation of integers

Let $\beta \geq 2$ be an integer called the *base*. The (usual) *β-ary representation* of an integer $N \geq 0$ is a finite word $d_k \cdots d_0$ over the digit alphabet $A = \{0, \ldots, \beta - 1\}$, and such that

$$N = \sum_{i=0}^{k} d_i \beta^i.$$

Such a representation is unique, with the condition that $d_k \neq 0$. This representation is called *normal*, and is denoted by

$$\langle N \rangle_\beta = d_k \cdots d_0,$$

most significant digit first.

The set of all the representations of the positive integers is equal to A^*.

Let us consider the addition of two integers represented in the β-ary system. Let $d_k \cdots d_0$ and $c_k \cdots c_0$ be two β-ary representations of respectively N and M. It is not a restriction to suppose that the two representations have the same length, since the shorter one can be padded to the left by enough zeros. Let us form a new word $a_k \cdots a_0$, with $a_i = d_i + c_i$ for $0 \le i \le k$. Obviously, $\sum_{i=0}^{k} a_i \beta^i = N + M$, but the a_i's belong to the set $\{0, \ldots, 2(\beta - 1)\}$. So the word $a_k \cdots a_0$ has to be transformed into an equivalent one (i.e. having the same numerical value) belonging to A^*.

More generally, let C be a finite alphabet of integers, which can be positive or negative. The *numerical value* in base β on C^* is the function

$$\pi_\beta : C^* \longrightarrow \mathbb{Z}$$

which maps a word $w = c_n \cdots c_0$ of C^* onto $\sum_{i=0}^{n} c_i \beta^i$. The *normalization* on C^* is the partial function

$$\nu_C : C^* \longrightarrow A^*$$

that maps a word $w = c_n \cdots c_0$ of C^* such that $N = \pi_\beta(w)$ is non-negative onto its normal representation $\langle N \rangle_\beta$. Our aim is to prove that the normalization is computable by a finite transducer. We first prove a lemma.

LEMMA 7.1.1. *Let C be an alphabet containing A. There exists a right subsequential transducer that maps a word w of C^* such that $N = \pi_\beta(w) \ge 0$ onto a word v belonging to A^* and such that $\pi_\beta(v) = N$.*

Proof. Let $m = \max\{|c - a| \mid c \in C, a \in A\}$, and let $\gamma = m/(\beta - 1)$. First observe that, for $s \in \mathbb{Z}$ and $c \in C$, by the Euclidean division algorithm there exist unique $a \in A$ and $s' \in \mathbb{Z}$ such that $s + c = \beta s' + a$. Furthermore, if $|s| < \gamma$, then $|s'| \le (|s| + |c - a|)/\beta < (\gamma + m)/\beta = \gamma$.

Consider the subsequential finite transducer $(\mathcal{A}, \omega)$ over $C^* \times A^*$, where $\mathcal{A} = (Q, E, 0)$ is defined as follows. The set $Q = \{s \in \mathbb{Z} \mid |s| < \gamma\}$ is the set of possible carries, the set of edges is

$$E = \{s \xrightarrow{c/a} s' \mid s + c = \beta s' + a\}.$$

7.1. Standard representation of numbers

Observe that the edges are "letter to letter". The terminal function is defined by $\omega(s) = \langle s \rangle_\beta$ for $s \in Q$ such that $\pi_\beta(s) \geq 0$.

Now let $w = c_n \cdots c_0 \in C^*$ and $N = \sum_{i=0}^{n} c_i \beta^i$. Setting $s_0 = 0$, there is a unique path

$$s_0 \xrightarrow{c_0/a_0} s_1 \xrightarrow{c_1/a_1} s_2 \xrightarrow{c_2/a_2} \cdots \xrightarrow{c_{n-1}/a_{n-1}} s_n \xrightarrow{c_n/a_n} s_{n+1}.$$

By construction $N = a_0 + a_1\beta + \cdots + a_n\beta^n + s_{n+1}\beta^{n+1}$, hence the word $v = \omega(s_{n+1})a_n \cdots a_0$ has the same numerical value in base β as w.

Note that v is equal to the normal representation of N if and only if it does not begin with zeros. ∎

EXAMPLE 7.1.2. Figure 7.1 gives the right subsequential transducer realizing the conversion in base 2 from the alphabet $\{-1, 0, 1\}$ onto $\{0, 1\}$. The signed digit (-1) is denoted by $\bar{1}$.

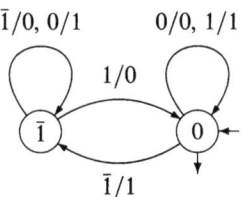

Figure 7.1. Right subsequential transducer realizing the conversion in base 2 from $\{\bar{1}, 0, 1\}$ onto $\{0, 1\}$.

The two following results are a direct consequence of Lemma 7.1.1.

PROPOSITION 7.1.3. *In base β, for every alphabet C of positive integers containing A, the normalization restricted to the domain $C^* \setminus 0C^*$ is a right subsequential function.*

Removing the zeros at the beginning of a word can be realized by a left sequential transducer, so the following property holds true for any alphabet.

PROPOSITION 7.1.4. *In base β, for every alphabet C containing A, the normalization on C^* is computable by a finite transducer.*

COROLLARY 7.1.5. *In base β, addition and subtraction (with possibly zeros ahead) are right subsequential functions.*

Proof. Take in Lemma 7.1.1 $C = \{0,\ldots,2(\beta-1)\}$ for addition, and $C = \{-(\beta-1),\ldots,\beta-1\}$ for subtraction. ∎

One proves easily that multiplication by a fixed integer is a right subsequential function, and that division by a fixed integer is a left subsequential function; see the Problems section. On the other hand, the following result shows that the power of functions computable by finite transducers is quite reduced.

PROPOSITION 7.1.6. *In base β, multiplication is not computable by a finite transducer.*

Proof. It is enough to show that the squaring function $\psi: A^* \longrightarrow A^*$ which maps $\langle N \rangle_\beta$ onto $\langle N^2 \rangle_\beta$ is not computable by a finite transducer. Take for instance $\beta = 2$, and consider $\langle 2^n - 1 \rangle_2 = 1^n$. Then $\psi(1^n) = \langle 2^{2n} - 2^{n+1} + 1 \rangle_2 = 1^{n-1}0^n 1$. Thus the image by ψ of the set $\{1^n \mid n \geq 1\}$ which is recognizable by a finite automaton is the set $\{1^{n-1}0^n 1 \mid n \geq 1\}$ which is not recognizable, thus ψ cannot be computed by a finite transducer. ∎

7.1.2. Representation of real numbers

Let $\beta \geq 2$ be an integer and set $A = \{0,\ldots,\beta-1\}$. A *β-ary representation* of a nonnegative real number x is an infinite sequence $(x_i)_{i \leq k}$ of $A^\mathbb{N}$ such that

$$x = \sum_{i \leq k} x_i \beta^i.$$

This representation is unique, and said to be *normal* if it does not end with $(\beta - 1)^\omega$, and if $x_k \neq 0$ when $x \geq 1$. It is traditionally denoted by

$$\langle x \rangle_\beta = x_k \cdots x_0 . x_{-1} x_{-2} \cdots .$$

If $x < 1$, then there exists some $i \geq 0$ such that $x < 1/\beta^i$. We then put $x_{-1}, \ldots, x_{-i+1} = 0$. The set of β-ary expansions of numbers ≥ 1 is equal to $(A \setminus 0)(A^\mathbb{N} \setminus A^*(\beta-1)^\omega)$, that of numbers of $[0, 1]$ is $A^\mathbb{N} \setminus A^*(\beta-1)^\omega$. The set $A^\mathbb{N}$ is the set of all β-ary representations (not necessarily normal).

The word $x_k \cdots x_0$ is the *integer part* of x and the infinite word $x_{-1} x_{-2} \cdots$ is the *fractional part* of x. Note that the natural numbers are exactly those having a zero fractional part (compare with the representation of complex numbers in Subsection 7.4.1).

7.1. Standard representation of numbers

If $\langle x \rangle_\beta = x_k \cdots x_0.x_{-1}x_{-2}\cdots$, then $x/\beta^{k+1} < 1$, and by shifting we obtain that

$$\langle x/\beta^{k+1} \rangle_\beta = .x_k \cdots x_0 x_{-1} x_{-2} \cdots$$

thus from now on we consider only numbers from the interval $[0,1]$. When $x \in [0,1]$, we will change our notation for indices and denote $\langle x \rangle_\beta = (x_i)_{i \geq 1}$.

Let C be a finite alphabet of integers, which can be positive or negative. The *numerical value* in base β on $C^\mathbb{N}$ is the function

$$\pi_\beta : C^\mathbb{N} \longrightarrow \mathbb{R}$$

which maps a word $w = (c_i)_{i \geq 1}$ of $C^\mathbb{N}$ onto $\sum_{i \geq 1} c_i \beta^{-i}$. The *normalization* on $C^\mathbb{N}$ is the partial function

$$\nu_C : C^\mathbb{N} \longrightarrow A^\mathbb{N}$$

that maps a word $w = (c_i)_{i \geq 1}$ such that $x = \pi_\beta(w)$ belongs to $[0,1]$ onto its β-ary expansion $\langle x \rangle_\beta \in A^\mathbb{N} \setminus A^*(\beta - 1)^\omega$.

PROPOSITION 7.1.7. *For every alphabet C containing A, the normalization on $C^\mathbb{N}$ is computable by a finite transducer.*

Proof. First we construct a finite transducer $\mathscr{B}$ where edges are the reverse of the edges of the transducer $\mathscr{A}$ defined in the proof of Lemma 7.1.1. Let $\mathscr{B} = (Q, F, 0, Q)$ with set of edges

$$F = \{t \xrightarrow{c/a} s \mid s \xrightarrow{c/a} t \in E\}.$$

Every state is terminal.

Let

$$s_0 \xrightarrow{c_1/a_1} s_1 \xrightarrow{c_2/a_2} s_2 \xrightarrow{c_3/a_3} \cdots \xrightarrow{c_n/a_n} s_n$$

be a path in $\mathscr{B}$ starting in $s_0 = 0$. Then

$$\frac{c_1}{\beta} + \cdots + \frac{c_n}{\beta^n} = \frac{a_1}{\beta} + \cdots + \frac{a_n}{\beta^n} - \frac{s_n}{\beta^n}.$$

Since $\mathscr{A}$ is sequential, the automaton $\mathscr{B}$ is unambiguous, that is, given an input word $(c_i)_{i \geq 1} \in C^\mathbb{N}$, there is a unique infinite path in $\mathscr{B}$ starting in 0 and labeled by $(c_i, a_i)_{i \geq 1}$ in $(C \times A)^\mathbb{N}$, and such that $\sum_{i \geq 1} c_i \beta^i = \sum_{i \geq 1} a_i \beta^i$, because for each n, $|s_n| < \gamma$.

To end the proof it remains to show that the function which, given a word in $A^\mathbb{N}$, transforms it into an equivalent word not ending with $(\beta - 1)^\omega$ is computable by a finite transducer, and this is clear from the fact that $A^\mathbb{N} \times (A^\mathbb{N} \setminus A^*(\beta - 1)^\omega)$ is a rational subset of $A^\mathbb{N} \times A^\mathbb{N}$ (see Chapter 1). ∎

COROLLARY 7.1.8. *Addition/subtraction, multiplication/division by a fixed integer of real numbers in base β are computable by a finite transducer.*

EXAMPLE 7.1.9. Figure 7.2 gives the finite transducer realizing nonnormalized addition (meaning that the result can end with the improper suffix 1^ω) of real numbers on the interval $[0, 1]$ in base 2.

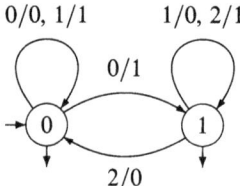

Figure 7.2. Finite transducer realizing nonnormalized addition of real numbers in base 2.

7.2. Beta-expansions

We now consider numeration systems where the base is a real number $\beta > 1$. Representations of real numbers in such systems were introduced by Rényi under the name of *beta-expansions*. They arise from the orbits of a piecewise-monotone transformation of the unit interval $T_\beta : x \mapsto \beta x \pmod 1$; see below. Such transformations were extensively studied in ergodic theory and symbolic dynamics.

7.2.1. Definitions

Let the base $\beta > 1$ be a real number. Let x be a real number in the interval $[0, 1]$. A *representation in base β* (or a β-representation) of x is an infinite word $(x_i)_{i \geq 1}$ such that

$$x = \sum_{i \geq 1} x_i \beta^{-i}.$$

A particular β-representation – called the β-*expansion* – can be computed by the "greedy algorithm" : denote by $\lfloor y \rfloor$ and $\{y\}$ the integer part and the fractional part of a number y. Set $r_0 = x$ and let for $i \geq 1$, $x_i = \lfloor \beta r_{i-1} \rfloor$, $r_i = \{\beta r_{i-1}\}$. Then $x = \sum_{i \geq 1} x_i \beta^{-i}$.

7.2. Beta-expansions

The β-expansion of x will be denoted by $d_\beta(x)$.

An equivalent definition is obtained by using the β-*transformation* of the unit interval which is the mapping

$$T_\beta : x \mapsto \beta x \,(\text{mod } 1).$$

Then $d_\beta(x) = (x_i)_{i \geq 1}$ if and only if $x_i = \lfloor \beta T_\beta^{i-1}(x) \rfloor$.

Let x be any real number greater than 1. There exists $k \in \mathbb{N}$ such that $\beta^k \leq x < \beta^{k+1}$. Hence $0 \leq x/\beta^{k+1} < 1$, thus it is enough to represent numbers from the interval $[0, 1]$, since by shifting we will get the representation of any positive real number.

EXAMPLE 7.2.1. Let $\beta = (1+\sqrt{5})/2$ be the golden ratio. For $x = 3 - \sqrt{5}$ we have $d_\beta(x) = 10010^\omega$.

If β is not an integer, the digits x_i obtained by the greedy algorithm are elements of the alphabet $A = \{0, \ldots, \lfloor \beta \rfloor\}$, called the *canonical alphabet*.

When β is an integer, the β-expansion of a number x of $[0, 1)$ is exactly the standard β-ary expansion, i.e. $d_\beta(x) = \langle x \rangle_\beta$, and the digits x_i belong to $\{0, \ldots, \beta-1\}$. However, for $x = 1$ there is a difference: $\langle 1 \rangle_\beta = 1$. but $d_\beta(1) = .\beta$. As we shall see later, the beta-expansion of 1 plays a key role in this theory.

Another characterization of a beta-expansion is the following one.

LEMMA 7.2.2. *An infinite sequence of nonnegative integers $(x_i)_{i \geq 1}$ is the β-expansion of a real number x in $[0, 1)$ (resp. of 1) if and only if for every $i \geq 1$ (resp. $i \geq 2$), $x_i \beta^{-i} + x_{i+1} \beta^{-i-1} + \cdots < \beta^{-i+1}$.*

Proof. Let $0 \leq x < 1$ and let $d_\beta(x) = (x_i)_{i \geq 1}$. By construction, for $i \geq 1$, $r_{i-1} = x_i/\beta + x_{i-1}/\beta^2 + \cdots < 1$, thus the result follows. ∎

A real number may have several β-representations. However, the β-expansion, obtained by the greedy algorithm, is characterized by the following property.

PROPOSITION 7.2.3. *The β-expansion of a real number x of $[0, 1]$ is the greatest of all the β-representations of x with respect to the lexicographic order.*

Proof. Let $d_\beta(x) = (x_i)_{i \geq 1}$ and let $(s_i)_{i \geq 1}$ be another β-representation of x. Suppose that $(x_i)_{i \geq 1} < (s_i)_{i \geq 1}$; then there exists $k \geq 1$ such that $x_k < s_k$ and $x_1 \cdots x_{k-1} = s_1 \cdots s_{k-1}$. From $\sum_{i \geq k} x_i \beta^{-i} = \sum_{i \geq k} s_i \beta^{-i}$ one gets $\sum_{i \geq k+1} x_i \beta^{-i} \geq \beta^{-k} + \sum_{i \geq k+1} s_i \beta^{-i}$, which is impossible since by Lemma 7.2.2 $\sum_{i \geq k+1} x_i \beta^{-i} < \beta^{-k}$. ∎

EXAMPLE 7.2.1 *(continued)*. Let β be the golden ratio. The β-expansion of $x = 3 - \sqrt{5}$ is equal to 10010^ω. Different β-representations of x are 01110^ω and $100(01)^\omega$ for instance.

As in the usual numeration systems, the order between real numbers is given by the lexicographic order on β-expansions.

PROPOSITION 7.2.4. *Let x and y be two real numbers from $[0, 1]$. Then $x < y$ if and only if $d_\beta(x) < d_\beta(y)$.*

Proof. Let $d_\beta(x) = (x_i)_{i \geq 1}$ and let $d_\beta(y) = (y_i)_{i \geq 1}$, and suppose that $d_\beta(x) < d_\beta(y)$. There exists $k \geq 1$ such that $x_k < y_k$ and $x_1 \cdots x_{k-1} = y_1 \cdots y_{k-1}$. Hence $x \leq y_1 \beta^{-1} + \cdots + y_{k-1} \beta^{-k+1} + (y_k - 1)\beta^{-k} + x_{k+1}\beta^{-k-1} + x_{k+2}\beta^{-k-2} + \cdots < y$ since $x_{k+1}\beta^{-k-1} + x_{k+2}\beta^{-k-2} + \cdots < \beta^{-k}$. The converse is immediate. ∎

If a representation ends in infinitely many zeros, like $v0^\omega$, the ending zeros are omitted and the representation is said to be *finite*. Note that the β-expansion of $x \in [0, 1]$ is finite if and only if $T_\beta^i(x) = 0$ for some i, and it is eventually periodic if and only if the set $\{T_\beta^i(x) \mid i \geq 1\}$ is finite. Numbers β such that $d_\beta(1)$ is eventually periodic are called *beta-numbers* and those such that $d_\beta(1)$ is finite are called *simple* beta-numbers.

REMARK 7.2.5. *The beta-expansion of 1 is never purely periodic.*

Indeed, suppose that $d_\beta(1)$ is purely periodic, $d_\beta(1) = (a_1 \cdots a_n)^\omega$, with n minimal, $a_i \in A$. Then $1 = a_1 \beta^{-1} + \cdots + a_n \beta^{-n} + \beta^{-n}$, which means that $a_1 \cdots a_{n-1}(a_n + 1)$ is a β-representation of 1, and $a_1 \cdots a_{n-1}(a_n + 1) > d_\beta(1)$, which is impossible.

EXAMPLE 7.2.6. (i) Let β be the golden ratio $(1 + \sqrt{5})/2$. The expansion of 1 is finite, equal to $d_\beta(1) = 11$.

(ii) Let $\beta = (3 + \sqrt{5})/2$. The expansion of 1 is eventually periodic, equal to $d_\beta(1) = 21^\omega$.

(iii) Let $\beta = 3/2$. Then $d_\beta(1) = 101000001 \cdots$. We shall see later that it is aperiodic.

7.2.2. The beta-shift

Recall that the set $A^{\mathbb{N}}$ is endowed with the lexicographic order, the product topology, and the (one-sided) shift σ, defined by $\sigma((x_i)_{i\geq 1}) = (x_{i+1})_{i\geq 1}$. Denote by D_β the set of β-expansions of numbers of $[0, 1)$. It is a shift-invariant subset of $A^{\mathbb{N}}$. The *β-shift* S_β is the closure of D_β and it is a subshift of $A^{\mathbb{N}}$. When β is an integer, S_β is the full β-shift $A^{\mathbb{N}}$.

The greedy algorithm computing the β-expansion can be rephrased as follows.

LEMMA 7.2.7. *The identity*

$$d_\beta \circ T_\beta = \sigma \circ d_\beta$$

holds on the interval $[0, 1]$.

Proof. Let $x \in [0, 1]$, and let $d_\beta(x) = (x_i)_{i\geq 1}$. Then $T_\beta(x) = \sum_{i\geq 1} x_i \beta^{-i}$, and the result follows. ∎

In the case where the β-expansion of 1 is finite, there is a special representation playing an important role. Let us introduce the following notation. Let $d_\beta(1) = (t_i)_{i\geq 1}$ and set $d_\beta^*(1) = d_\beta(1)$ if $d_\beta(1)$ is infinite and $d_\beta^*(1) = (t_1 \cdots t_{m-1}(t_m - 1))^\omega$ if $d_\beta(1) = t_1 \cdots t_{m-1} t_m$ is finite.

When β is an integer, β-representations ending with the infinite word $d_\beta^*(1)$ are the "improper" representations.

EXAMPLE 7.2.8. Let $\beta = 2$, then $d_\beta(1) = 2$ and $d_\beta^*(1) = 1^\omega$.

For $\beta = (1 + \sqrt{5})/2$, $d_\beta(1) = 11$ and $d_\beta^*(1) = (10)^\omega$.

The set D_β is characterized by the expansion of 1, as shown by the result below. Notice that the sets of finite factors of D_β and of S_β are the same, and that $d_\beta^*(1)$ is the supremum of S_β, but that, in case $d_\beta(1)$ is finite, $d_\beta(1)$ is not an element of S_β.

THEOREM 7.2.9. *Let $\beta > 1$ be a real number, and let s be an infinite sequence of nonnegative integers. The sequence s belongs to D_β if and only if for all $p \geq 0$*

$$\sigma^p(s) < d_\beta^*(1)$$

and s belongs to S_β if and only if for all $p \geq 0$

$$\sigma^p(s) \leq d_\beta^*(1).$$

Proof. First suppose that $s = (s_i)_{i \geq 1}$ belongs to D_β, then there exists x in $[0,1)$ such that $s = d_\beta(x)$. By Lemma 7.2.7, for every $p \geq 0$, $\sigma^p \circ d_\beta(x) = d_\beta \circ T_\beta^p(x)$. Since $T_\beta^p(x) < 1$ and d_β is a strictly increasing function (Proposition 7.2.4), $\sigma^p \circ d_\beta(x) = \sigma^p(s) < d_\beta(1)$.

In the case where $d_\beta(1) = t_1 \cdots t_m$ is finite, suppose there exists a $p \geq 0$ such that $\sigma^p(s) \geq d_\beta^*(1)$. Since $\sigma^p(s) < d_\beta(1)$, we get $s_{p+1} = t_1$, ..., $s_{p+m-1} = t_{m-1}$, $s_{p+m} = t_m - 1$. Iterating this process, we see that $\sigma^p(s) = d_\beta^*(1)$, which does not belong to D_β, a contradiction.

Conversely, let $d_\beta^*(1) = (d_i)_{i \geq 1}$ and suppose that for all $p \geq 0$, $\sigma^p(s) < d_\beta^*(1)$. By induction, let us show that for all $r \geq 1$, for all $i \geq 0$,

$$s_{p+1} \cdots s_{p+r} < d_{i+1} \cdots d_{i+r} \Rightarrow \frac{s_{p+1}}{\beta} + \cdots + \frac{s_{p+r}}{\beta^r} < \frac{d_{i+1}}{\beta} + \cdots + \frac{d_{i+r}}{\beta^r}.$$

This is obviously satisfied for $r = 1$. Suppose that $s_{p+1} \cdots s_{p+r+1} < d_{i+1} \cdots d_{i+r+1}$. First assume that $s_{p+1} = d_{i+1}$; then $s_{p+2} \cdots s_{p+r+1} < d_{i+2} \cdots d_{i+r+1}$. By induction hypothesis,

$$\frac{s_{p+2}}{\beta^2} + \cdots + \frac{s_{p+r+1}}{\beta^{r+1}} < \frac{d_{i+2}}{\beta^2} + \cdots + \frac{d_{i+r+1}}{\beta^{r+1}}$$

and the result follows. Next, suppose that $s_{p+1} < d_{i+1}$. Since for all $p \geq 0$, $\sigma^p(s) < d_\beta^*(1)$ we have $s_{p+2} \cdots s_{p+r+1} \leq d_1 \cdots d_r$, thus

$$\frac{s_{p+1}}{\beta} + \cdots + \frac{s_{p+r+1}}{\beta^{r+1}} \leq \frac{d_{i+1} - 1}{\beta} + \frac{d_1}{\beta^2} + \cdots + \frac{d_r}{\beta^{r+1}} < \frac{d_{i+1}}{\beta}$$

since $d_1/\beta^2 + \cdots + d_r/\beta^{r+1} < 1/\beta$.

Thus for all $p \geq 0$, for all $i \geq 0$,

$$\sum_{r \geq 1} s_{p+r} \beta^{-r} \leq \sum_{r \geq 1} d_{i+r} \beta^{-r}.$$

In particular for $i = 1$, $\sum_{r \geq 1} s_{p+r} \beta^{-r} \leq \sum_{r \geq 1} d_{r+1} \beta^{-r} < 1$ if β is not an integer, and the result follows by Lemma 7.2.2.

If β is an integer then $d_\beta^*(1) = (\beta - 1)^\omega$. If for all $p \geq 0$, $\sigma^p(s) < d_\beta^*(1)$, then every letter of s is smaller than or equal to $\beta - 1$ and s does not end with $(\beta - 1)^\omega$, therefore s belongs to D_β.

For the β-shift, we have the following situation. A sequence s belongs to $\overline{D_\beta}$ if and only if for each $n \geq 1$ there exists a word $v^{(n)}$ of D_β such that $s_1 \cdots s_n$ is a prefix of $v^{(n)}$. Hence, s belongs to S_β if and only if for every $p \geq 0$, for every $n \geq 1$, $\sigma^p(s_1 \cdots s_n 0^\omega) < d_\beta^*(1)$, or equivalently if $\sigma^p(s) \leq d_\beta^*(1)$. ∎

7.2. Beta-expansions

From this result follows the following characterization: a sequence is the β-expansion of 1 for a certain number β if and only if it is greater than all its shifted sequences.

COROLLARY 7.2.10. *Let $s = (s_i)_{i \geq 1}$ be a sequence of nonnegative integers with $s_1 \geq 1$ and for $i \geq 2$, $s_i \leq s_1$, and which is different from 10^ω. Then there exists a unique real number $\beta > 0$ such that $\sum_{i \geq 1} s_i \beta^{-i} = 1$. Furthermore, s is the β-expansion of 1 if and only if for every $n \geq 1$, $\sigma^n(s) < s$.*

Proof. Let f be the formal series defined by $f(z) = \sum_{i \geq 1} s_i z^i$, and denote by ρ its radius of convergence. Since $0 \leq s_i \leq s_1$, we get $\rho \geq 1/(s_1 + 1)$. Since for $0 < z < \rho$ the function f is continuous and increasing, and since $f(0) = 0$ and $f(z) > 1$ for z sufficient close to ρ, it follows that the equation $f(z) = 1$ has a unique solution. If $\beta > 1$ exists such that $f(1/\beta) = 1$, we get that $s_1/\beta \leq f(1/\beta) \leq s_1/(\beta - 1)$, thus β must be between s_1 and $s_1 + 1$. On the other hand, $f(1/(s_1 + 1)) \leq s_1/s_1 = 1$. If $s_1 \geq 2$, $f(1/s_1) \geq 1$. If $s_1 = 1$ and if the s_i's are eventually 0, then $f(1/s_1) \geq 1$, otherwise $\lim_{z \to 1} f(z) = +\infty$. Thus in any case there exists a real $\beta \in [s_1, s_1 + 1]$ such that $f(1/\beta) = 1$.

Now we make the following hypothesis (H) : for all $n \geq 1$, $\sigma^n(s) < s$. Suppose that the β-expansion of 1 is $d_\beta(1) = t \neq s$. Since s is a β-representation of 1, $s < t$. Hence, for each $n \geq 1$, $\sigma^n(s) < s < d_\beta(1)$. If $d_\beta(1)$ is infinite, by Theorem 7.2.9, s belongs to D_β, a contradiction.

If $d_\beta(1)$ is finite, say $d_\beta(1) = t_1 \cdots t_m$, either $s < d_\beta^*(1)$, and as above we get that s is in D_β, or $d_\beta^*(1) \leq s < d_\beta(1)$. In fact, s cannot be purely periodic because of hypothesis (H), thus it is different from $d_\beta^*(1)$. Thus s is necessarily of the form $(t_1 \cdots t_{m-1}(t_m - 1))^k t_1 \cdots t_m$ for some $k \geq 1$. So $s_{km+1} = t_1, \ldots, s_{km+m} = t_m$, and $\sigma^{km}(s) > s$ because $s_m = t_m - 1$, contradicting hypothesis (H). Hence the β-expansion of 1 is s.

Conversely, suppose that $s = d_\beta(1)$ for some $\beta > 1$. From Theorem 7.2.9, for every $n \geq 1$, $\sigma^n(s) < d_\beta^*(1)$. If $d_\beta(1)$ is infinite, $d_\beta(1) = d_\beta^*(1)$. If $d_\beta(1)$ is finite, $d_\beta^*(1) < d_\beta(1)$. ∎

Let us recall some definitions on symbolic dynamical systems or subshifts (see Section 1.5). Let $S \subseteq A^\mathbb{N}$ be a subshift, and let $I(S) = A^+ \setminus F(S)$ be the set of factors avoided by S. Denote by $X(S)$ the set of words of $I(S)$ which have no proper factor in $I(S)$. The subshift S is of *finite type* iff the set $X(S)$ is finite. The subshift S is *sofic* iff $X(S)$ is a rational set. It is equivalent to say that $F(S)$ is recognized by a finite automaton. The subshift S is said to be *coded* if there exists a prefix code $Y \subset A^*$ such that $F(S) = F(Y^*)$, or equivalently if S is the closure of Y^ω.

To the β-shift a prefix code $Y = Y_\beta$ is associated as follows. It is the set of words which, for each length, are strictly smaller than the prefix of $d_\beta(1)$ of the same length. More precisely: if $d_\beta(1) = (t_i)_{i \geq 1}$ is infinite, set $Y = \{t_1 \cdots t_{n-1} a \mid 0 \leq a < t_n, n \geq 1\}$, with the convention that if $n = 1$, $t_1 \cdots t_{n-1} = \varepsilon$; if $d_\beta(1) = t_1 \cdots t_m$, let $Y = \{t_1 \cdots t_{n-1} a \mid 0 \leq a < t_n, 1 \leq n \leq m\}$.

PROPOSITION 7.2.11. *The β-shift S_β is coded by the code Y.*

Proof. First, if $d_\beta(1) = (t_i)_{i \geq 1}$ is infinite, let us show that $D_\beta = Y^\omega$. Let $s \in D_\beta$. By Theorem 7.2.9, $s < d_\beta(1)$, thus can be written as $s = t_1 \cdots t_{n_1-1} a_{n_1} v_1$, with $a_{n_1} < t_{n_1}$ and $v_1 < d_\beta(1)$. Iterating this process, we see that $s \in Y^\omega$. Conversely, let $s = u_1 u_2 \cdots \in Y^\omega$, with $u_i = t_1 \cdots t_{n_i-1} a_{n_i}$, $a_{n_i} < t_{n_i}$. Then $s < d_\beta(1)$. For each $p \geq 0$, $\sigma^p(s)$ begins with a word of the form $t_{j_p} t_{j_p+1} \cdots t_{j_p+r-1} b_{j_p+r}$ with $b_{j_p+r} < t_{j_p+r}$, thus $\sigma^p(s) < \sigma^{j_p-1}(d_\beta(1)) < d_\beta(1)$.

Next, if $d_\beta(1) = t_1 \cdots t_m$ is finite, we claim that $Y^\omega = S_\beta$. First, let $s \in S_\beta$. By Theorem 7.2.9, $s \leq d_\beta^*(1)$, thus $s = t_1 \cdots t_{n_1-1} a_{n_1} v_1$, with $n_1 \leq m$, $a_{n_1} < t_{n_1}$ and $v_1 \leq d_\beta^*(1)$. Iterating the process we get $s \in S_\beta$. Conversely, let $s \in Y^\omega$, $s = u_1 u_2 \cdots$ with $u_i = t_1 \cdots t_{n_i-1} a_{n_i}$, $n_i \leq m$. As above, one gets that, for each $p \geq 0$, $\sigma^p(s) < d_\beta^*(1)$. ∎

We now compute the topological entropy of the β-shift

$$h(S_\beta) = -\log(\rho_{F(S_\beta)})$$

(see Subsection 1.5.3 for definitions and notations). In the case where the β-shift is sofic, by Theorem 1.5.14 the entropy $h(S_\beta)$ can be shown to be equal to $\log \beta$. We show below that the same result holds true for any kind of β-shift.

PROPOSITION 7.2.12. *The topological entropy of the β-shift is equal to $\log \beta$.*

Proof. For $n \geq 1$, the number of words of length n of Y is clearly equal to t_n, thus the generating series of Y is equal to

$$f_Y(z) = \sum_{n \geq 1} t_n z^n.$$

By Corollary 7.2.10, β^{-1} is the unique positive solution of $f_Y(z) = 1$. Since Y is a code, by Lemma 1.4.4 $\rho_{Y^*} = \beta^{-1}$. It is thus enough to show that $\rho_{Y^*} = \rho_{F(S_\beta)}$.

7.2. Beta-expansions

Let p_n be the number of factors of length n of the elements of S_β and let
$$f_{F(S_\beta)} = \sum_{n \geq 0} p_n z^n.$$

Let c_n be the number of words of length n of Y^*, and let
$$f_{F(Y^*)} = \sum_{n \geq 0} c_n z^n.$$

Since any word of Y^* is in $F(S_\beta)$, we have $c_n \leq p_n$. On the other hand, let w be a word of length n in $F(S_\beta)$. By Proposition 7.2.11, w can be uniquely written as $w = u_i t_1 \cdots t_i$, where $u_i \in Y^*$, $|u_i| = n - i$, and $0 \leq i \leq n$. Thus $p_n = c_n + \cdots + c_0$. Hence the series $f_{F(S_\beta)}$ and f_{Y^*} have the same radius of convergence, and the result is proved. ∎

We now show that the nature of the subshift as a symbolic dynamical system is entirely determined by the β-expansion of 1.

THEOREM 7.2.13. *The β-shift S_β is sofic if and only if $d_\beta(1)$ is eventually periodic.*

Proof. Suppose that $d_\beta(1)$ is infinite and eventually periodic:
$$d_\beta(1) = t_1 \cdots t_N (t_{N+1} \cdots t_{N+p})^\omega$$

with N and p minimal. We use the classical construction of minimal finite automata by right congruence classes (see Chapter 1). Let $F(D_\beta)$ be the set of finite factors of D_β. We construct an automaton $\mathcal{A}_\beta$ with $N + p$ states $q_1, \ldots, q_{N+p}$, where q_i, $i \geq 2$, represents the right class $[t_1 \cdots t_{i-1}]_{F(D_\beta)}$ and q_1 stands for $[\varepsilon]_{F(D_\beta)}$. For each i, $1 \leq i < N + p$, there is an edge labeled t_i from q_i to q_{i+1}. There is an edge labeled t_{N+p} from q_{N+p} to q_{N+1}. For $1 \leq i \leq N+p$, there are edges labeled by $0, 1, \ldots, t_i - 1$ from q_i to q_1. Let q_1 be the only initial state, and all states be terminal. That $F(D_\beta)$ is precisely the set recognized by the automaton $\mathcal{A}_\beta$ follows from Theorem 7.2.9. Note that, when the β-expansion of 1 happens to be finite, say $d_\beta(1) = t_1 \cdots t_m$, the same construction applies with $N = m$, $p = 0$ and all edges from q_m (labeled by $0, 1, \ldots, t_m - 1$) leading to q_1.

Suppose now that $d_\beta(1) = (t_i)_{i \geq 1}$ is not eventually periodic nor finite. There exists an infinite sequence of indices $i_1 < i_2 < i_3 < \cdots$ such that the sequences $t_{i_k} t_{i_k+1} t_{i_k+2} \cdots$ are all different for all $k \geq 1$. Thus for all pairs (i_j, i_ℓ), $j, \ell \geq 1$, there exists $p \geq 0$ such that, for instance, $t_{i_j+p} < t_{i_\ell+p}$ and $t_{i_j} \cdots t_{i_j+p-1} = t_{i_\ell} \cdots t_{i_\ell+p-1} = w$ (with the convention that, when $p = 0$,

$w = \varepsilon$). We have that $t_1 \cdots t_{i_j-1} w t_{i_j+p} \in F(D_\beta)$, $t_1 \cdots t_{i_\ell-1} w t_{i_\ell+p} \in F(D_\beta)$, $t_1 \cdots t_{i_\ell-1} w t_{i_\ell+p} \in F(D_\beta)$, but $t_1 \cdots t_{i_j-1} w t_{i_\ell+p}$ does not belong to $F(D_\beta)$. Hence $t_1 \cdots t_{i_j}$ and $t_1 \cdots t_{i_\ell}$ are not right congruent modulo $F(D_\beta)$. The number of right congruence classes is thus infinite, and $F(D_\beta)$ is not recognizable by a finite automaton. ∎

EXAMPLE 7.2.14. For $\beta = (3 + \sqrt{5})/2$, $d_\beta(1) = 21^\omega$, and the β-shift is sofic.

We have a similar result when the β-expansion of 1 is finite.

THEOREM 7.2.15. *The β-shift S_β is of finite type if and only if $d_\beta(1)$ is finite.*

Proof. Let us suppose that $d_\beta(1) = t_1 \cdots t_m$ is finite and let

$$Z = \bigcup_{2 \leq i \leq m-1} \{u \in A^i \mid u > t_1 \cdots t_i\} \cup \{u \in A^m \mid u \geq t_1 \cdots t_m\}.$$

Clearly $Z \subseteq A^+ \setminus F(S_\beta)$. The set $X(S_\beta)$ of words forbidden in S_β which are minimal for the factor order is a subset of Z. Since Z is finite, $X(S_\beta)$ is finite, and thus S_β is of finite type.

Conversely, suppose that the β-shift is of finite type. It is thus sofic, and by Theorem 7.2.13, $d_\beta(1)$ is eventually periodic. Suppose that $d_\beta(1)$ is not finite, $d_\beta(1) = t_1 \cdots t_N(t_{N+1} \cdots t_{N+p})^\omega$ with $N \geq 1$ and $p \geq 1$ minimal, and $t_{N+1} \cdots t_{N+p} \neq 0^p$. Let

$$Z = \{t_1 \cdots t_{j-1}(t_j + h_j) \mid 2 \leq j \leq N, \ 1 \leq h_j \leq t_1 - t_j\}$$
$$\cup \{t_1 \cdots t_N(t_{N+1} \cdots t_{N+p})^k t_{N+1} \cdots t_{N+j-1}(t_{N+j} + h_{N+j}) \mid$$
$$k \geq 0, \ 1 \leq j \leq p, \ 1 \leq h_{N+j} \leq t_1 - t_{N+j}\}.$$

Clearly $Z \subseteq A^+ \setminus F(S_\beta)$.

Case 1. Suppose there exists $1 \leq j \leq p$ such that $t_j > t_{N+j}$ and $t_1 = t_{N+1}, \ldots, t_{j-1} = t_{N+j-1}$. For $k \geq 0$ fixed, let $w^{(k)} = t_1 \cdots t_N(t_{N+1} \cdots t_{N+p})^k t_1 \cdots t_j \in Z$. We have $t_1 \cdots t_N(t_{N+1} \cdots t_{N+p})^k t_{N+1} \cdots t_{N+j-1} \in F(S_\beta)$. On the other hand, for $n \geq 2$, $t_n \cdots t_N(t_{N+1} \cdots t_{N+p})^k$ is strictly smaller in the lexicographic order than the prefix of $d_\beta(1)$ of the same length (the inequality is strict, since the t_i's are not all equal for $1 \leq i \leq N + p$), thus $t_n \cdots t_N(t_{N+1} \cdots t_{N+p})^k t_1 \cdots t_j \in F(S_\beta)$. Hence any strict factor of $w^{(k)}$ is in $F(S_\beta)$. Therefore for any $k \geq 0$, $w^{(k)} \in X(S_\beta)$, and $X(S_\beta)$ is thus infinite: the β-shift is not of finite type.

Case 2. No such j exists; then $d_\beta(1) = (t_1 \cdots t_N)^\omega$, which is impossible by Remark 7.2.5. ∎

7.2. Beta-expansions

EXAMPLE 7.2.16. For $\beta = (1 + \sqrt{5})/2$, the β-shift is of finite type, it is the golden mean shift described in Example 1.5.2.

7.2.3. Classes of numbers

Recall that an *algebraic integer* is a root of a monic polynomial with integral coefficients. An algebraic integer $\beta > 1$ is called a *Pisot number* if all its Galois conjugates have modulus less than 1. It is a *Salem number* if all its conjugates have modulus ≤ 1 and at least one conjugate has modulus 1. It is a *Perron number* if all its conjugates have modulus less than β.

EXAMPLE 7.2.17. (i) Every integer is a Pisot number. The golden ratio $(1 + \sqrt{5})/2$ and its square $(3 + \sqrt{5})/2$ are Pisot numbers, with minimal polynomials respectively $X^2 - X - 1$ and $X^2 - 3X + 1$.

(ii) A rational number which is not an integer is never an algebraic integer.

(iii) The number $(5 + \sqrt{5})/2$ is a Perron number which is neither Pisot nor Salem.

The most important result linking beta-shifts and numbers is the following one.

THEOREM 7.2.18. *If β is a Pisot number then the β-shift S_β is sofic.*

This result is a consequence of a more general result on β-expansions of numbers of the field $\mathbb{Q}(\beta)$ when β is a Pisot number. It is a partial generalization of the well-known fact that, when β is an integer, numbers having an eventually periodic β-expansion are the rational numbers of $[0, 1]$ (see Problems section).

PROPOSITION 7.2.19. *If β is a Pisot number then every number of $\mathbb{Q}(\beta) \cap [0, 1]$ has an eventually periodic β-expansion.*

Proof. Let $P(X) = X^d - a_1 X^{d-1} - \cdots - a_d$ be the minimal polynomial of $\beta = \beta_1$ and denote by $\beta_2, \ldots, \beta_d$ the conjugates of β. Let x be arbitrarily fixed in $\mathbb{Q}(\beta) \cap [0, 1]$. It can be expressed as

$$x = q^{-1} \sum_{i=0}^{d-1} p_i \beta^i$$

with q and p_i in $\mathbb{Z}$, $q > 0$ as small as possible in order to have uniqueness.

Let $(x_k)_{k\geq 1}$ be the β-expansion of x, and write

$$r_n = r_n^{(1)} = r_n(x) = \frac{x_{n+1}}{\beta} + \frac{x_{n+2}}{\beta^2} + \cdots = \beta^n\left(x - \sum_{k=1}^{n} x_k\beta^{-k}\right) = T_\beta^n(x) < 1.$$

For $2 \leq j \leq d$, let

$$r_n^{(j)} = r_n^{(j)}(x) = \beta_j^n\left(q^{-1}\sum_{i=0}^{d-1} p_i\beta_j^i - \sum_{k=1}^{n} x_k\beta_j^{-k}\right).$$

Let $\eta = \max\{|\beta_j| \mid 2 \leq j \leq d\} < 1$ since β is a Pisot number. Since $x_k \leq \lfloor \beta \rfloor$ we get

$$|r_n^{(j)}| \leq q^{-1}\sum_{i=0}^{d-1}|p_i|\eta^{n+i} + \lfloor\beta\rfloor\sum_{k=0}^{n-1}\eta^k$$

and, since $\eta < 1$, $\max_{1\leq j\leq d}\{\sup_n\{|r_n^{(j)}|\}\} < +\infty$.

We need a technical result. Set $R_n = (r_n^{(1)}, \ldots, r_n^{(d)})$ and let B be the matrix $B = (\beta_j^{-i})_{1\leq i,j\leq d}$.

LEMMA 7.2.20. *Let $x = q^{-1}\sum_{i=0}^{d-1} p_i\beta^i$. For every $n \geq 0$, there exists a unique d-uple $Z_n = (z_n^{(1)}, \ldots, z_n^{(d)})$ in $\mathbb{Z}^d$ such that $R_n = q^{-1}Z_n B$.*

Proof. By induction on n. First, $r_1 = r_1^{(1)} = \beta x - x_1$, thus

$$r_1 = q^{-1}\left(\sum_{i=0}^{d-1} p_i\beta^{i+1} - qx_1\right) = q^{-1}\left(\frac{z_1^{(1)}}{\beta} + \cdots + \frac{z_1^{(d)}}{\beta^d}\right)$$

using the fact that $\beta^d = a_1\beta^{d-1} + \cdots + a_d$, $a_j \in \mathbb{Z}$. Now, $r_{n+1} = r_{n+1}^{(1)} = \beta r_n - x_{n+1}$, hence

$$r_{n+1} = q^{-1}\left(z_n^{(1)} + \frac{z_n^{(2)}}{\beta} + \cdots + \frac{z_n^{(d)}}{\beta^{d-1}} - qx_{n+1}\right) = q^{-1}\left(\frac{z_{n+1}^{(1)}}{\beta} + \cdots + \frac{z_{n+1}^{(d)}}{\beta^d}\right)$$

since $z_n^{(1)} - qx_{n+1} \in \mathbb{Z}$.

Thus

$$r_n = r_n^{(1)} = \beta^n\left(q^{-1}\sum_{i=0}^{d-1} p_i\beta^i - \sum_{k=1}^{n} x_k\beta^{-k}\right) = q^{-1}\sum_{k=1}^{d} z_n^{(k)}\beta^{-k}.$$

Since the latter equation has integral coefficients and is satisfied by β, it is also satisfied by each conjugate β_j, $2 \leq j \leq d$,

$$r_n^{(j)} = \beta_j^n\left(q^{-1}\sum_{i=0}^{d-1} p_i\beta_j^i - \sum_{k=1}^{n} x_k\beta_j^{-k}\right) = q^{-1}\sum_{k=1}^{d} z_n^{(k)}\beta_j^{-k}. \blacksquare$$

7.2. Beta-expansions

We resume the proof of Proposition 7.2.19. Let $V_n = qR_n$. The $(V_n)_{n\geq 1}$ have bounded norm, since $\max_{1\leq j\leq d}\{\sup_n\{|r_n^{(j)}|\}\} < +\infty$. As the matrix B is invertible, for every $n \geq 1$,

$$\|Z_n\| = \|(z_n^{(1)},\ldots,z_n^{(d)})\| = \max\{|z_n^{(j)}|\mid 1 \leq j \leq d\} < +\infty$$

so there exist p and $m \geq 1$ such that $Z_{m+p} = Z_m$, hence $r_{m+p} = r_m$ and the β-expansion of x is eventually periodic. ∎

On the other hand, there is a gap between Pisot and Perron numbers as shown by the following result.

PROPOSITION 7.2.21. *If S_β is sofic then β is a Perron number.*

Proof. With the automaton $\mathcal{A}_\beta$ defined in the proof of Theorem 7.2.13 one associates a matrix $M = M_\beta$ by taking for $M[i, j]$ the number of edges from state q_i to state q_j, that is, if $d_\beta(1) = t_1 \cdots t_N(t_{N+1} \cdots t_{N+p})^\omega$,

$$M[i, 1] = t_i,$$
$$M[i, i+1] = 1 \text{ for } i \neq N+p,$$
$$M[N+p, N+1] = 1,$$

and other entries are equal to 0.

Claim 1. The matrix M is primitive: $M^{N+p} > 0$, since $M^{N+p}[i, j]$ is equal to the number of paths of length $N+p$ from q_i to q_j in the strongly connected automaton $\mathcal{A}_\beta$.

Claim 2. The characteristic polynomial of M is equal to

$$K(X) = X^{N+p} - \sum_{i=1}^{N+p} t_i X^{N+p-i} - X^N + \sum_{i=1}^{N} t_i X^{N-i}$$

and β is one of its roots: this can be checked by a straightforward computation.

When $d_\beta(1) = t_1 \cdots t_m$ is finite, the matrix associated with the automaton is simpler, it is the companion matrix of the polynomial $K(X) = X^m - t_1 X^{m-1} - \cdots - t_m$, which is primitive, since $M^m > 0$.

Since $\beta > 1$ is an eigenvalue of a primitive matrix, by the Perron–Frobenius theorem, β is strictly greater in modulus than its algebraic conjugates. ∎

Thus when β is a nonintegral rational number (for instance 3/2), the β-shift S_β cannot be sofic.

EXAMPLE 7.2.22. There are Perron numbers which are neither Pisot nor Salem numbers and such that the β-shift is of finite type: for instance the root $\beta \sim 3.616$ of $X^4 - 3X^3 - 2X^2 - 3$ satisfies $d_\beta(1) = 3203$, and β has a conjugate $\gamma \sim -1.096$.

REMARK 7.2.23. If β is a Perron number with a real conjugate > 1, then $d_\beta(1)$ cannot be eventually periodic.

In fact, suppose that $d_\beta(1) = t_1 \cdots t_N (t_{N+1} \cdots t_{N+p})^\omega$, and that β has a conjugate $\gamma > 1$. Since β is a zero of the polynomial $K(X)$ of $\mathbb{Z}[X]$, γ is also a zero of this polynomial. Thus $d_\gamma(1) = d_\beta(1)$, and by Corollary 7.2.10, $\gamma = \beta$. For instance the quadratic Perron number $\beta = (5 + \sqrt{5})/2$ has a real conjugate > 1, and thus S_β is not sofic.

7.3. U-representations

We now consider another generalization of the notion of numeration system, which only allows us to represent the natural numbers. The base is replaced by an infinite sequence of integers. The basic example is the well-known Fibonacci numeration system.

7.3.1. Definitions

Let $U = (u_n)_{n \geq 0}$ be a strictly increasing sequence of integers with $u_0 = 1$. A *representation in the system U* – or a *U-representation* – of a nonnegative integer N is a finite sequence of integers $(d_i)_{k \geq i \geq 0}$ such that

$$N = \sum_{i=0}^{k} d_i u_i.$$

Such a representation will be written $d_k \cdots d_0$, most significant digit first.

Among all possible U-representations of a given nonnegative integer N one is distinguished and called the *normal U-representation of N*: it is sometimes called the *greedy* representation, since it can be obtained by the following greedy algorithm : given integers m and p let us denote by $q(m, p)$ and $r(m, p)$ the quotient and the remainder of the Euclidean division of m by p. Let $k \geq 0$ be such that $u_k \leq N < u_{k+1}$ and let $d_k = q(N, u_k)$ and $r_k = r(N, u_k)$, and, for $i = k-1, \ldots, 0$, $d_i = q(r_{i+1}, u_i)$ and $r_i = r(r_{i+1}, u_i)$. Then $N = d_k u_k + \cdots + d_0 u_0$. The normal U-representation of N is denoted by $\langle N \rangle_U$.

By convention the normal representation of 0 is the empty word ε. Under the hypothesis that the ratio u_{n+1}/u_n is bounded by a constant as

7.3. U-representations

n tends to infinity, the integers of the normal U-representation of any integer N are bounded and contained in a *canonical* finite alphabet A associated with U.

EXAMPLE 7.3.1. Let $U = \{2^n \mid n \geq 0\}$. The normal U-representation of an integer is nothing else than its 2-ary standard expansion.

EXAMPLE 7.3.2. Let $F = (F_n)_{n \geq 0}$ be the sequence of Fibonacci numbers (see Example 1.4.2). The canonical alphabet is equal to $A = \{0, 1\}$. The normal F-representation of the number 15 is 100010; another representation is 11010.

An equivalent definition of the notion of normal U-representation is the following one.

LEMMA 7.3.3. *The word $d_k \cdots d_0$, where each d_i, for $k \geq i \geq 0$, is a nonnegative integer and $d_k \neq 0$, is the normal U-representation of some integer if and only if for each i, $d_i u_i + \cdots + d_0 u_0 < u_{i+1}$.*

Proof. If $d_k \cdots d_0$ is obtained by the greedy algorithm, $r_{i+1} = d_i u_i + \cdots + d_0 u_0 < u_{i+1}$ by construction. ∎

As for beta-expansions, the U-representation obtained by the greedy algorithm is the greatest one for some order we define now. Let v and w be two words. We say that $v < w$ if $|v| < |w|$ or if $|v| = |w|$ and there exist letters $a < b$ such that $v = uav'$ and $w = ubw'$. This order is sometimes called "radix order" or "genealogic order", or even "lexicographic order" in the literature, although the definition is slightly different from the usual definition of lexicographic order on finite words (see Chapter 1).

PROPOSITION 7.3.4. *The normal U-representation of an integer is the greatest in the radix order of all the U-representations of that integer.*

Proof. Let $d = d_k \cdots d_0$ be the normal U-representation of N, and let $w = w_j \cdots w_0$ be another representation. Since $u_k \leq N < u_{k+1}$, $k \geq j$. If $k > j$, then $d > w$. If $k = j$, suppose $d < w$. Thus there exists i, $k \geq i \geq 0$, such that $d_i < w_i$ and $d_k \cdots d_{i+1} = w_k \cdots w_{i+1}$. Hence $d_i u_i + \cdots + d_0 u_0 = w_i u_i + \cdots + w_0 u_0$, but $d_i u_i + \cdots + d_0 u_0 \leq (w_i - 1)u_i + d_{i-1}u_{i-1} + \cdots + d_0 u_0$, so $u_i + w_{i-1}u_{i-1} + \cdots + w_0 u_0 \leq d_{i-1}u_{i-1} + \cdots + d_0 u_0 < u_i$ since d is normal, which is absurd. ∎

The order between natural numbers is given by their radix order between their normal U-representations.

PROPOSITION 7.3.5. *Let M and N be two nonnegative integers, then $M < N$ if and only if $\langle M \rangle_U < \langle N \rangle_U$.*

Proof. Let $v = v_k \cdots v_0 = \langle M \rangle_U$ with $u_k \le M < u_{k+1}$, and $w = w_j \cdots w_0 = \langle N \rangle_U$ with $u_j \le N < u_{j+1}$, and suppose that $v < w$. Then $k \le j$. If $k < j$, $u_{k+1} \le u_j$, and $M < N$. If $k = j$, there exists i such that $v_i < w_i$ and $v_k \cdots v_{i+1} = w_k \cdots w_{i+1}$. Hence

$$M = v_k u_k + \cdots + v_0 u_0$$
$$\le w_k u_k + \cdots + w_{i+1} u_{i+1} + (w_i - 1) u_i + v_{i-1} u_{i-1} + \cdots + v_0 u_0$$
$$< w_k u_k + \cdots + w_{i+1} u_{i+1} + w_i u_i \le N$$

since $v_{i-1} u_{i-1} + \cdots + v_0 u_0 < u_i$ by Lemma 7.3.3, thus $M < N$. ∎

7.3.2. The set of normal U-representations

The set of normal U-representations of all the nonnegative integers is denoted by $L(U)$.

EXAMPLE 7.3.2 (*continued*). Let F be the sequence of Fibonacci numbers. The set $L(F)$ is the set of words without the factor 11, and not beginning with a 0,

$$L(F) = 1\{0,1\}^* \setminus \{0,1\}^* 11 \{0,1\}^* \cup \varepsilon.$$

First, the analogue of Theorem 7.2.9 is the following result.

PROPOSITION 7.3.6. *The set $L(U)$ is the set of words over A such that each suffix of length n is less in the radix order than $\langle u_n - 1 \rangle_U$.*

Proof. Let $v = v_k \cdots v_0$ be in $L(U)$, and $0 \le n \le k+1$. By Lemma 7.3.3 $v_{n-1} u_{n-1} + \cdots + v_0 u_0 \le u_n - 1$, and by Proposition 7.3.5, $v_{n-1} \cdots v_0 \le \langle u_n - 1 \rangle_U$. The converse is immediate. ∎

An important case is when $L(U)$ is recognizable by a finite automaton, as is the case for the usual numeration systems. We first give a necessary condition.

Recall that a formal series with coefficients in $\mathbb{N}$ is said to be N-*rational* if it belongs to the smallest class containing polynomials with coefficients in $\mathbb{N}$, and closed under addition, multiplication and the star operation, where F^* is the series $1 + F + F^2 + F^n + \cdots = 1/(1-F)$, F being a series such that $F(0) = 0$. An N-rational series is necessarily $\mathbb{Z}$-rational, and thus can be written $P(X)/Q(X)$, with $P(X)$ and $Q(X)$ in $\mathbb{Z}[X]$, and $Q(0) = 1$. Therefore the sequence of coefficients of an N-rational series satisfies a linear recurrence relation with coefficients in $\mathbb{Z}$.

7.3. U-representations

It is classical that, if L is recognizable by a finite automaton, then the series $f_L(X) = \sum_{n \geq 0} \ell_n X^n$, where ℓ_n denotes the number of words of length n in L, is $\mathbb{N}$-rational (see Berstel and Reutenauer 1988).

PROPOSITION 7.3.7. *If the set $L(U)$ is recognizable by a finite automaton, then the series $U(X) = \sum_{n \geq 0} u_n X^n$ is $\mathbb{N}$-rational, and thus the sequence U satisfies a linear recurrence with integral coefficients.*

Proof. Let ℓ_n be the number of words of length n in $L(U)$. The series $f_{L(U)}(X) = \sum_{n \geq 0} \ell_n X^n$ is $\mathbb{N}$-rational. We have $u_n = \ell_n + \cdots + \ell_0$, because the number of words of length $\leq n$ in $L(U)$ is equal to the number of natural numbers smaller than u_n, whose normal representation has length $n + 1$. Thus $U(X) = f_{L(U)}(X)/(1 - X)$, and it is $\mathbb{N}$-rational. ∎

When the sequence U satisfies a linear recurrence with integral coefficients, we say that U defines a *linear numeration system*.

To determine sufficient conditions on the sequence U for the set $L(U)$ to be recognizable by a finite automaton is a difficult question (see Problem 7.3.1). It is strongly related to the theory of β-expansions where β is the dominant root of the characteristic polynomial of the linear recurrence of U. Nevertheless, there is a case where the set $L(U)$ and the factors of the beta-shift coincide. This means that the dynamical systems generated by the beta-expansions of real numbers and by normal U-representations of integers are the same.

It is obvious that if a word of the form $v0^n$ belongs to $L(U)$ then v itself is a word of $L(U)$, but the converse is not true in general. We will say that a set $L \subset A$ is *right-extendable* if the following property holds:

$$v \in L \Rightarrow v0 \in L.$$

THEOREM 7.3.8. *Let $U = (u_n)_{n \geq 0}$ be a strictly increasing sequence of integers, with $u_0 = 1$, and such that $\sup\{u_{n+1}/u_n\} < +\infty$, and let A be the canonical alphabet. There exists a real number $\beta > 1$ such that $L(U) = F(D_\beta)$ if and only if $L(U)$ is right-extendable. In that case, if $d_\beta^*(1) = (d_i)_{i \geq 1}$, the sequence U is determined by*

$$u_n = d_1 u_{n-1} + \cdots + d_n u_0 + 1.$$

Proof. Clearly, if $L(U) = F(D_\beta)$ for some $\beta > 1$, then $L(U)$ is right-extendable.

Conversely, suppose that $L(U)$ is right-extendable. For each n, denote

$$\langle u_n - 1 \rangle_U = d_1^{(n)} \cdots d_n^{(n)}.$$

Since $L(U)$ is right-extendable, for each $k < n$, $d_1^{(k)} \cdots d_k^{(k)} 0^{n-k} \in L(U)$, and thus $d_1^{(k)} \cdots d_k^{(k)} \leq d_1^{(n)} \cdots d_k^{(n)}$. Therefore $d_1^{(k)} \cdots d_k^{(k)} = d_1^{(n)} \cdots d_k^{(n)}$ because $d_1^{(k)} \cdots d_k^{(k)}$ is the greatest word of length k in the radix order.

Let $d_n = d_n^{(n)}$; then $d_n d_{n+1} \cdots \leq d_1 d_2 \cdots$. Let $d = (d_i)_{i \geq 1}$. If there exists m such that $d = \sigma^m(d)$ then d is periodic. Let m be the smallest such index. In that case, put $t_1 = d_1, \ldots, t_{m-1} = d_{m-1}$, $t_m = d_m + 1$, $t_i = 0$ for $i > m$. In case d is not periodic, put $t_i = d_i$ for every i. Then the sequence $(t_i)_{i \geq 1}$ satisfies $t_n t_{n+1} \cdots < t_1 t_2 \cdots$ for all $n \geq 2$, and thus by Corollary 7.2.10 there exists a unique $\beta > 1$ such that $d_\beta(1) = (t_i)_{i \geq 1}$.

Let us show that $L(U) = F(D_\beta)$. Recall that

$$D_\beta = \{s \mid \forall p \geq 0, \sigma^p(s) < d_\beta^*(1) = (d_i)_{i \geq 1}\};$$

hence

$$F(D_\beta) = \{v = v_k \cdots v_0 \mid \forall n, 0 \leq n \leq k, v_{n-1} \cdots v_0 \leq d_1 \cdots d_n = \langle u_n - 1 \rangle_U\}$$
$$= L(U)$$

by Proposition 7.3.6.

Now, since by definition $d_1 \cdots d_n = \langle u_n - 1 \rangle_U$, we get

$$u_n = d_1 u_{n-1} + \cdots + d_n u_0 + 1. \qquad \blacksquare$$

The numeration systems satisfying Theorem 7.3.8 will be called *canonical numeration systems associated with β*, and denoted by U_β. Note that if $d_\beta(1)$ is eventually periodic, then $L(U_\beta)$ is recognizable by a finite automaton and U_β satisfies a linear recurrence relation.

EXAMPLE 7.3.2 (*continued*). The Fibonacci numeration system is the canonical numeration system associated with the golden ratio.

7.3.3. Normalization in a canonical linear numeration system

We first give general definitions, valid for any linear numeration system defined by a sequence U. The *numerical value* in the system U of a representation $w = d_k \cdots d_0$ is equal to $\pi_U(w) = \sum_{i=0}^{k} d_i u_i$. Let C be a finite alphabet of integers. The *normalization* in the system U on C^* is the partial function

$$\nu_C : C^* \longrightarrow A^*$$

that maps a word w of C^* such that $\pi_U(w)$ is nonnegative onto the normal U-representation of $\pi_U(w)$.

7.3. U-representations

In what follows, we assume that $U = U_\beta$ is the canonical numeration system associated with a number β which is a Pisot number. Thus U satisfies an equation of the form

$$u_n = a_1 u_{n-1} + a_2 u_{n-2} + \cdots + a_m u_{n-m}, \quad a_i \in \mathbb{Z}, \quad a_m \neq 0, \quad n \geq m.$$

In that case, the canonical alphabet A associated with U is $A = \{0, \ldots, K\}$ where $K < \max\{u_{i+1}/u_i\}$. The polynomial $P(X) = X^m - a_1 X^{m-1} - \cdots - a_m$ will be called the *characteristic polynomial* of U.

We also make the hypothesis that P is exactly the minimal polynomial of β (in general, P is a multiple of the minimal polynomial).

Our aim is to prove the following result.

THEOREM 7.3.9. *Let $U = U_\beta$ be a canonical linear numeration system associated with a Pisot number β, and such that the characteristic polynomial of U is equal to the minimal polynomial of β. Then, for every alphabet C of nonnegative integers, the normalization on C^* is computable by a finite transducer.*

The proof is in several steps. Let $C = \{0, \ldots, c\}$, $\tilde{C} = \{-c, \ldots, c\}$, and let

$$Z(U, c) = \left\{ d_k \cdots d_0 \,\Big|\, d_i \in \tilde{C}, \; \sum_{i=0}^{k} d_i u_i = 0 \right\}$$

be the set of words on $\tilde{C}$ having numerical value 0 in the system U. We first prove a general result.

PROPOSITION 7.3.10. *If $Z(U, c)$ and $L(U)$ are recognizable by a finite automaton then ν_C is a function computable by a finite transducer.*

Proof. Let $f = f_n \cdots f_0$ and $g = g_k \cdots g_0$ be two words of C^*, with for instance $n \geq k$. We denote by $f \ominus g$ the word of $\tilde{C}^*$ equal to $f_n \cdots f_{k+1}(f_k - g_k) \cdots (f_0 - g_0)$. The graph of ν_C is equal to $\widehat{\nu_C} = \{(f, g) \in C^* \times A^* \mid g \in L(U), f \ominus g \in Z(U, c)\}$.

Let R be the graph of $\ominus$:

$$R = \left[\left(\bigcup_{a \in C} ((a, \varepsilon), a) \right)^* \cup \left(\bigcup_{a \in C} ((\varepsilon, a), -a) \right)^* \right] \left[\bigcup_{a, b \in C} ((a, b), a - b) \right]^*$$

R is a rational subset of $(C^* \times C^*) \times \tilde{C}^*$. Let us consider the set

$$R' = R \cap ((C^* \times L(U)) \times Z(U, c)) \subseteq (C^* \times A^*) \times \tilde{C}^*.$$

Then $\widehat{v_C}$ is the projection of R' on $C^* \times A^*$. As $L(U)$ and $Z(U,c)$ are rational by assumption, $(C^* \times L(U)) \times Z(U,c)$ is a recognizable subset of $(C^* \times A^*) \times \widetilde{C}^*$ as a Cartesian product of rational sets (see Berstel 1979b). Since R is rational, R' is a rational subset of $(C^* \times A^*) \times \widetilde{C}^*$. So, $\widehat{v_C}$ being the projection of R', $\widehat{v_C}$ is a rational subset of $C^* \times A^*$, that is, v_C is computable by a finite transducer. ∎

The core of the proof relies on the following result.

PROPOSITION 7.3.11. *Let U be a linear numeration system such that its characteristic polynomial is equal to the minimal polynomial of a Pisot number β. Then $Z(U,c)$ is recognizable by a finite automaton.*

Proof. Set $Z = Z(U,c)$ for short. We define on the set H of prefixes of Z the equivalence relation ζ as follows (m is the degree of P):

$$f \zeta g \Leftrightarrow [\forall n,\ 0 \le n \le m-1,\ \pi_U(f0^n) = \pi_U(g0^n)].$$

Let $f \zeta g$. It is clear that the sequences $(\pi_U(f0^n))_{n \ge 0}$ and $(\pi_U(g0^n))_{n \ge 0}$ satisfy the same recurrence relation as U. Since they coincide on the first m values, they are equal. Thus, for any $h \in \widetilde{C}$,

$$fh \in Z \Leftrightarrow \pi_U(f0^{|h|}) + \pi_U(h) = 0$$
$$\Leftrightarrow \pi_U(g0^{|h|}) + \pi_U(h) = 0$$
$$\Leftrightarrow gh \in Z$$

which means that f and g are right congruent modulo Z. If f and g are not in H, then $f \sim_Z g$ as well.

It remains to prove that ζ has finite index. This will be achieved by showing that there are only finitely many possible values of $\pi_U(f0^n)$ for $f \in H$ and for all $0 \le n \le m-1$. Recall that, if $\beta = \beta_1, \beta_2, \ldots, \beta_m$ are the roots of P, since P is minimal they are all distinct, and there exist complex constants $\lambda_1 > 0, \lambda_2, \ldots, \lambda_m$ such that for all $n \in \mathbb{N}$

$$u_n = \sum_{i=1}^{m} \lambda_i \beta_i^n.$$

If $f = f_k \cdots f_0$, let $\pi_\beta(f) = f_k \beta^k + \cdots + f_1 \beta + f_0$.
Claim 1. There exists η such that for all $f \in \widetilde{C}$

$$|\pi_U(f) - \lambda_1 \pi_\beta(f)| < \eta.$$

7.3. U-representations

We have

$$\pi_U(f) - \lambda_1 \pi_\beta(f) = \sum_{j=0}^{k} f_j u_j - \lambda_1 \sum_{j=0}^{k} f_j \beta^j$$

$$= \sum_{j=0}^{k} f_j \left(\sum_{i=1}^{m} \lambda_i \beta_i^j \right) - \lambda_1 \sum_{j=0}^{k} f_j \beta^j$$

$$= \sum_{j=0}^{k} f_j \left(\sum_{i=2}^{m} \lambda_i \beta_i^j \right).$$

Since β is a Pisot number, $|\beta_i| < 1$ for $2 \le i \le m$ and

$$|\pi_U(f) - \lambda_1 \pi_\beta(f)| < c \sum_{i=2}^{m} |\lambda_i| \frac{1}{1 - |\beta_i|} = \eta.$$

Claim 2. There exists γ such that for all $f \in H$, $|\pi_\beta(f)| < \gamma$.
Since $f \in H$ there exists $h \in \widetilde{C}$ such that $fh \in Z$. Thus

$$0 = \pi_U(f 0^{|h|}) + \pi_U(h) < \lambda_1 \pi_\beta(f 0^{|h|}) + \lambda_1 \pi_\beta(h) + 2\eta$$
$$< \lambda_1 \pi_\beta(f) \beta^{|h|} + \lambda_1 (c+1) \beta^{|h|} + 2\eta \beta^{|h|};$$

thus $\pi_\beta(f) > -c - 1 - 2\eta \lambda_1^{-1}$. Similarly $\pi_\beta(f) < c + 1 + 2\eta \lambda_1^{-1}$, hence $|\pi_\beta(f)| < c + 1 + 2\eta \lambda_1^{-1} = \gamma$.

Claim 3. There exists δ such that for all $f \in H$, for all $0 \le n \le m-1$

$$|\pi_U(f 0^n)| < \delta.$$

We have

$$|\pi_U(f 0^n)| \le |\pi_U(f 0^n) - \lambda_1 \pi_\beta(f 0^n)| + |\lambda_1 \pi_\beta(f 0^n)|$$
$$< \eta + |\lambda_1 \pi_\beta(f)| \beta^n$$
$$< \eta + \lambda_1 \gamma \beta^n;$$

hence $|\pi_U(f 0^n)| < \delta = \eta + \lambda_1 \gamma \beta^{m-1}$.

Thus there are only finitely many possible values of $\pi_U(f 0^n)$ for $f \in H$ and for all $0 \le n \le m-1$, therefore ζ has finite index, and $Z(U, c)$ is rational. ∎

Proof of the theorem. Since U is canonical for a Pisot number, $L(U)$ is recognizable by a finite automaton. The result follows from Proposition 7.3.10 and Proposition 7.3.11. ∎

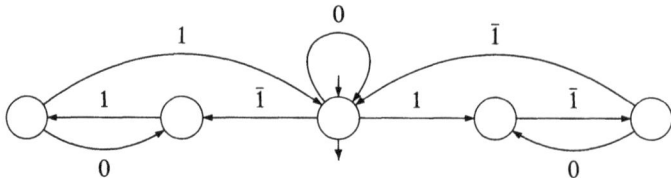

Figure 7.3. Automaton recognizing the set of words on $\{-1,0,1\}$ having value 0 in the Fibonacci numeration system.

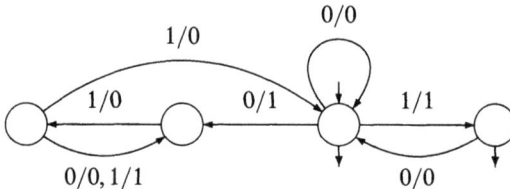

Figure 7.4. Normalization on $\{0,1\}$ in the Fibonacci numeration system.

COROLLARY 7.3.12. *Under the same hypothesis as in Theorem 7.3.9, addition of integers represented in the canonical linear numeration system U_β is computable by a finite transducer.*

Proof. The canonical alphabet being $A = \{0,\ldots,K\}$, take $C = \{0,\ldots,2K\}$ in Theorem 7.3.9. ∎

EXAMPLE 7.3.2 (*continued*). Let F be the sequence of Fibonacci numbers. The characteristic polynomial of F is $X^2 - X - 1$, and it is the minimal polynomial of the Pisot number $\beta = (1 + \sqrt{5})/2$. Figure 7.3 shows the automaton recognizing the set $Z(F, 1)$ of words on the alphabet $\{-1, 0, 1\}$ having numerical value 0 in the Fibonacci numeration system.

Figure 7.4 shows a finite transducer realizing the normalization on $\{0, 1\}$ in the Fibonacci numeration system. For simplicity, we assume that input and output words have the same length.

The result stated in Theorem 7.3.9 can be extended to the case where U is not the canonical numeration system associated with a Pisot number β, but where the characteristic polynomial of U is still equal to the minimal polynomial of β. There is a partial converse to this result; see Notes.

7.4. Representation of complex numbers

The usual method of representing real numbers by their decimal or binary expansions can be generalized to complex numbers. It is possible (see the Problems section) to represent complex numbers with an integral base and complex digits, but we present here results when the base is some complex number.

7.4.1. Gaussian integers

In this subsection we focus on representing complex numbers using integral digits. The set of *Gaussian integers*, denoted by $\mathbb{Z}[i]$, is the set $\{a+bi \mid a,b \in \mathbb{Z}\}$. The base β will be chosen as a Gaussian integer. It is quite natural to extend properties satisfied by integral bases for real numbers, namely the fact that integers coincide with numbers having a zero fractional part. More precisely, given a base β of modulus > 1 and an alphabet A of digits that are Gaussian integers, we will say that (β, A) is an *integral numeration system* for the field of complex numbers $\mathbb{C}$ if every Gaussian integer z has a *unique integer* representation of the form $d_k \cdots d_0$ such that $z = \sum_{j=0}^{k} d_j \beta^j$, with $d_j \in A$. We shall see later that, in that case, every complex number has a representation.

We first show preliminary results. A set $A \subset \mathbb{Z}[i]$ is a *complete residue system* for $\mathbb{Z}[i]$ modulo β if every element of $\mathbb{Z}[i]$ is congruent modulo β to a unique element of A. The *norm* of a Gaussian integer $z = x + yi$ is $N(z) = x^2 + y^2$. The following result is well known in elementary number theory.

THEOREM 7.4.1 (Gauss). *Let $\beta = a+bi$ be a nonzero Gaussian integer, and let N be the norm of β. If a and b are coprime, then a complete residue system for $\mathbb{Z}[i]$ modulo β is the set*

$$\{0,\ldots,N-1\}.$$

If $\gcd(a,b) = \lambda$, *a complete residue system for $\mathbb{Z}[i]$ modulo β is the set*

$$\{p+iq \mid p=0,1,\ldots,(N/\lambda)-1,\ q=0,1,\ldots,\lambda-1\}.$$

We use it in the following circumstances.

PROPOSITION 7.4.2. *Suppose that every Gaussian integer has an integer representation in (β, A). Then this representation is unique if and only if A is a complete residue system for $\mathbb{Z}[i]$ modulo β that contains 0.*

Proof. Let us suppose that A is a complete residue system containing 0, and let $d_k \cdots d_0$ and $c_p \cdots c_0$ be two representations of z in (β, A). One can suppose $d_0 \neq c_0$. Then $c_0 - d_0 = \beta(d_k\beta^{k-1} + \cdots + d_1 - c_p\beta^{p-1} - \cdots - c_1)$, thus d_0 and c_0 are congruent modulo β, and are elements of A, thus they are equal, which is absurd.

Conversely, suppose that every Gaussian integer z has a unique representation of the form $d_k \cdots d_0$, with digits d_j in A. Then z is congruent to d_0 modulo β, thus the digit set A must contain a complete residue system.

Now let c and d be two digits of A that are congruent modulo β. Then $c - d = \beta q$ with q in $\mathbb{Z}[i]$. Let $q_n \cdots q_0$ be the representation of q. Hence c has two representations, c itself and $q_n \cdots q_0 d$. ∎

If we require the digits to be natural numbers, the base must be a Gaussian integer $\beta = a + bi$ with a and b coprime, and the choice is drastically restricted.

THEOREM 7.4.3. *Let β be a Gaussian integer of norm N, and let $A = \{0, \ldots, N-1\}$. Then (β, A) is an integral numeration system for the complex numbers if and only if $\beta = -n \pm i$, for some $n \geq 1$.*

Proof. First let $\beta = a + bi$, a and b coprime, and let $A = \{0, \ldots, a^2 + b^2 - 1\}$. Suppose that $a > 0$. We shall show that the Gaussian integer $z = (1-a) + ib$ has no representation. Suppose on the contrary that z has a representation $d_k \cdots d_0$. Let $y = z(1-\beta) = a^2 + b^2 - 2a + 1$. Since $a > 0$, y belongs to A. But $y = d_0 + (d_1 - d_0)\beta + \cdots + (d_k - d_{k-1})\beta^k - d_k\beta^{k+1}$. Thus y is congruent to d_0 modulo β, and so $y = d_0$. It follows that $d_1 - d_0 = 0, \ldots, d_k - d_{k-1} = 0$, $d_k = 0$, so for $0 \leq j \leq k$, $d_j = 0$. Thus $y = 0$ and $a = 1$, $b = 0$. But $\beta = 1$ is not the base of a numeration system.

If $a = 0$ and $b = \pm 1$, then $\beta = \pm i$ is not a base either. If $a = 0$ and $|b| \geq 2$, the digit set is $\{0, \ldots, b^2 - 1\}$. If $b > 0$ then i has no integer representation, since $\langle i \rangle_\beta = 10 \cdot (b^2 - b)$. If $b < 0$, then $-i$ has no integer representation (see Exercise 7.4.2.)

Let now $a < 0$ and $b \neq \pm 1$. Suppose that a Gaussian integer z has a representation $d_k \cdots d_0$. Then $\operatorname{Im} z = d_k \operatorname{Im} \beta^k + \cdots + d_1 \operatorname{Im} \beta$. Since $\operatorname{Im} \beta = b$ is a divisor of $\operatorname{Im} \beta^k$ for all k, b divides $\operatorname{Im} z$. Take $z = i$. Since $b \neq \pm 1$, there is a contradiction.

Let now $\beta = -n + i$, $n \geq 1$, and thus $A = \{0, \ldots, n^2\}$. It remains to prove that any $z \in \mathbb{Z}[i]$ has an integer representation in (β, A). Let

7.4. Representation of complex numbers

$z = x + iy$, x and y in $\mathbb{Z}$. We have $z = c + d\beta$, with $d = y$ and $c = x + ny$. From the equality $\beta^2 + 2n\beta + n^2 + 1 = 0$, it is possible to write z as $z = d_3\beta^3 + d_2\beta^2 + d_1\beta + d_0$ with $d_i \in \mathbb{N}$.

Let $z = d_k\beta^k + \cdots + d_0$, with $d_i \in \mathbb{N}$, and $k \geq 3$, and let $d = d_k \cdots d_0 \in \mathbb{N}^*$. Denote by S the sum-of-digits function

$$S : \mathbb{C} \times \mathbb{N}^* \longrightarrow \mathbb{N}$$
$$(z, d) \longmapsto S(z, d) = d_k + \cdots + d_0.$$

In the following we will use the fact that $n^2 + 1 = \beta^3 + (2n-1)\beta^2 + (n-1)^2\beta$, that is, $\langle n^2 + 1 \rangle_\beta$ is equal to the word $1(2n-1)(n-1)^2 0$, and that the sum of digits of these two representations is the same and equal to $n^2 + 1$. By the Euclidean division by $n^2 + 1$, $d_0 = r_0 + q_0(n^2 + 1)$ with $0 \leq r_0 \leq n^2$, thus $z = r_0 + (d_1 + q_0(n-1)^2)\beta + (d_2 + q_0(2n-1))\beta^2 + (d_3 + q_0)\beta^3 + d_4\beta^4 + \cdots + d_k\beta^k = d_0^{(1)} + \cdots + d_k^{(1)}\beta^k$. Clearly $S(z, d) = S(z, d^{(1)})$, where $d^{(1)} = d_k^{(1)} \cdots d_0^{(1)}$.

Let $z_1 = d_1^{(1)} + \cdots + d_k^{(1)}\beta^{k-1}$; then $S(z_1, d^{(1)}) \leq S(z, d)$, and the inequality is strict if and only if $r_0 \neq 0$. Repeating this process, we get $z = \beta z_1 + r_0$, $z_1 = \beta z_2 + r_1, \ldots, z_{j-1} = \beta z_j + r_{j-1}$, with for $0 \leq i \leq j - 1$, $r_i \in A$, and $S(z, d) \geq S(z_1, d^{(1)}) \geq \cdots \geq S(z_{j-1}, d^{(j-1)})$.

Since the sequence $(S(z_j, d^{(j)}))_j$ of natural numbers is decreasing, there exists a p such that, for every $m \geq 0$, $S(z_p, d^{(p)}) = S(z_{p+m}, d^{(p+m)})$, thus β^m divides z_p for every m, therefore $z_p = 0$. So we get

$$\langle z \rangle_\beta = r_{p-1} \cdots r_0.$$

Let now $\beta = -n - i$. Using the result for the conjugate $\bar{\beta} = -n + i$, we have

$$\langle \bar{z} \rangle_{\bar{\beta}} = r_{p-1} \cdots r_0$$

for every Gaussian integer $\bar{z}$. Hence

$$\langle z \rangle_\beta = r_{p-1} \cdots r_0$$

for every Gaussian integer z. ∎

From this result, one can deduce that every complex number is representable in this system.

THEOREM 7.4.4. *If $\beta = -n \pm i$, $n \geq 1$, and $A = \{0, \ldots, n^2\}$, every complex number has a representation (not necessarily unique) in the numeration system (β, A).*

Proof. Let $z = x + iy$, x and y in $\mathbb{R}$, be a fixed arbitrary complex number. For $k \geq 0$, let $\beta^k = u_k + iv_k$. Then

$$z = \frac{(x+iy)(u_k + iv_k)}{\beta^k} = \frac{p_k + iq_k}{\beta^k} + \frac{r_k + is_k}{\beta^k}$$

where $xu_k - yv_k = p_k + r_k$, $xv_k + yu_k = q_k + s_k$, with p_k and q_k in $\mathbb{Z}$, and $|r_k| < 1$, $|s_k| < 1$. Let

$$z_k = \frac{p_k + iq_k}{\beta^k}, \quad y_k = \frac{r_k + is_k}{\beta^k}.$$

Since $y_k \to 0$ when $k \to \infty$, $\lim_{k \to \infty} z_k = z$. Since $p_k + iq_k$ is a Gaussian integer, by Theorem 7.4.3,

$$\langle p_k + iq_k \rangle_\beta = d_{t(k)}^{(k)} \cdots d_0^{(k)}.$$

Thus

$$z_k = d_{t(k)}^{(k)} \beta^{t(k)-k} + \cdots + d_0^{(k)} \beta^{-k}.$$

So

$$|d_{t(k)}^{(k)} \beta^{t(k)-k} + \cdots + d_k^{(k)}| \leq |z_k| + \frac{d_{k-1}^{(k)}}{|\beta|} + \cdots + \frac{d_0^{(k)}}{|\beta|^k}$$

$$\leq |z| + |y_k| + n^2 \left(\frac{1}{|\beta|} + \frac{1}{|\beta|^2} + \cdots \right)$$

$$\leq |z| + |y_k| + \frac{n^2}{|\beta| - 1} \leq c$$

where c is a positive constant not depending on k.

Since the representation of a Gaussian integer is unique, and since $\mathbb{Z}[i]$ is a discrete lattice, i.e. is an additive subgroup such that any bounded part contains only a finite number of elements, $t(k) - k$ has an upper bound. Let M be an integer such that $t(k) - k \leq M$. Then we can write z_k in the form

$$z_k = a_M^{(k)} \beta^M + \cdots + a_0^{(k)} + a_{-1}^{(k)} \beta^{-1} + a_{-2}^{(k)} \beta^{-2} + \cdots$$

where $a_j^{(k)} \in A$ for $M \geq j$. Let $b_M \in A$ be an integer such that $a_M^{(k)} = b_M$ for infinitely many k's. Let D_M be the subset of those k's such that $a_M^{(k)} = b_M$. Let $b_{M-1} \in A$ be an integer such that $a_{M-1}^{(k)} = b_{M-1}$ for infinitely many k's in D_M, and let D_{M-1} be the set of those k's. Repeating this process a set sequence $(D_\ell)_{\ell \geq M}$ such that $D_M \supseteq D_{M-1} \supseteq \cdots$ and such that for all $k \in D_\ell$, $a_j^{(k)} = b_j$ for each $\ell \leq j \leq M$ is constructed. Let

7.4. Representation of complex numbers

$k_1 < k_2 < \cdots$ be an infinite sequence such that $k_j \in D_{M-j+1}$ for $j \geq 1$. Since

$$z_{k_j} = b_M \beta^M + \cdots + b_{M-j+1}\beta^{M-j+1} + a^{(k_j)}_{M-j}\beta^{M-j} + a^{(k_j)}_{M-j-1}\beta^{M-j-1} + \cdots$$

we get $z_{k_j} \to \sum_{\ell \leq M} b_\ell \beta^\ell$ when $j \to \infty$. Since $\lim_{k\to\infty} z_k = z$, we have

$$\langle z \rangle_\beta = b_M \cdots b_0 . b_{-1} b_{-2} \cdots .$$ ∎

EXAMPLE 7.4.5. In Figure 7.5 is shown the set obtained by considering complex numbers having a zero integer part and a fractional part of length less than a fixed bound in their $(-1+i)$-expansion. This set actually tiles the plane.

Figure 7.5. Base $-1+i$ tile with fractal boundary.

Let C be a finite alphabet of Gaussian integers. The normalization on C^* is the function
$$\nu_C : C^* \longrightarrow A^*$$
$$c_k \cdots c_0 \longmapsto \langle \sum_{j=0}^k c_j \beta^j \rangle_\beta .$$

As for standard representations of integers (see Proposition 7.1.3), normalization is a right subsequential function, and in particular addition is right subsequential.

PROPOSITION 7.4.6. *For any finite alphabet C of Gaussian integers, the normalization in base $\beta = -n + i$ restricted to the set $C^* \setminus 0C^*$ is a right subsequential function.*

Proof. Let $m = \max\{|c-a| \mid c \in C, a \in A\}$, and let $\gamma = m/(|\beta|-1)$. First observe that, if $s \in \mathbb{Z}[i]$ and $c \in C$, there exist unique $a \in A$ and $s' \in \mathbb{Z}[i]$

such that $s+c = \beta s' + a$, because A is a complete residue system modulo β. Furthermore, if $|s| < \gamma$, then $|s'| \le (|s| + |c-a|)/|\beta| < (\gamma + m)/|\beta| = \gamma$.

Consider the subsequential finite transducer $(\mathscr{A}, \omega)$ over $C^* \times A^*$, where $\mathscr{A} = (Q, E, 0)$ is defined as follows. The set of states is $Q = \{s \in \mathbb{Z}[i] \mid |s| < \gamma\}$. Since $\mathbb{Z}[i]$ is a discrete lattice, Q is finite.

$$E = \{s \xrightarrow{c/a} s' \mid s + c = \beta s' + a\}.$$

Observe that the edges are "letter to letter". The terminal function is defined by $\omega(s) = \langle s \rangle_\beta$. The transducer is subsequential because A is a complete residue system.

Now let $c_k \cdots c_0 \in C^*$ and $z = \sum_{j=0}^{k} c_j \beta^j$. Setting $s_0 = 0$, there is a unique path

$$s_0 \xrightarrow{c_0/a_0} s_1 \xrightarrow{c_1/a_1} s_2 \xrightarrow{c_2/a_2} \cdots \xrightarrow{c_{k-1}/a_{k-1}} s_k \xrightarrow{c_k/a_k} s_{k+1}.$$

We get $z = a_0 + a_1 \beta + \cdots + a_k \beta^k + s_{k+1} \beta^{k+1}$, and thus $\langle z \rangle_\beta = \omega(s_{k+1}) a_k \cdots a_0$. ∎

7.4.2. Representability of the complex plane

In general, the question of deciding whether, given a base β and a set of digits A, every complex number is representable is difficult. A sufficient condition is given by the following result.

THEOREM 7.4.7. *Let β be a complex number of modulus greater than 1, and let A be a finite set of complex numbers containing zero. If there exists a bounded neighborhood V of zero such that $\beta V \subset V + A$, then every complex number z has a representation of the form*

$$z = \sum_{j \le m} d_j \beta^j$$

with m in $\mathbb{Z}$ and digits d_j in A.

Proof. Let z be in $\mathbb{C}$. There exists an integer $k \ge 0$ such that $\beta^{-k} z \in V$, thus it is enough to show that every element of V is representable. Let z be in V. A sequence $(z_j)_{j \ge 0}$ of elements of V is constructed as follows. Let $z_0 = z$. As $\beta V \subset V + A$, if z_j is in V, there exist d_{j+1} in A and z_{j+1} in V such that

$$z_{j+1} = \beta z_j - d_{j+1}.$$

Hence the sequence $(z_j)_{j \geq 0}$ is such that

$$z = d_1 \beta^{-1} + \cdots + d_j \beta^{-j} + z_j \beta^{-j}$$

and since V is bounded, by letting j tend to infinity,

$$z = \sum_{j \geq 0} d_j \beta^{-j}. \qquad \blacksquare$$

Problems

Section 7.1

7.1.1 Prove that addition in the standard beta-ary system is not left subsequential.

7.1.2 Give a right subsequential transducer realizing the multiplication by a fixed integer, and a left subsequential transducer realizing the division by a fixed integer, in the standard beta-ary system.

7.1.3 Prove the well-known fact that a number is rational if and only if its β-expansion in the standard β-ary system is eventually periodic.

7.1.4 Show that any real number can be represented without a sign using a negative base β, where β is an integer ≤ -2, and digit alphabet $\{0, \ldots, |\beta| - 1\}$. Integers have unique integer representations. Addition of integers is a right subsequential function.

7.1.5 Show that one can represent any real number without a sign using base 3, and digit alphabet $\{\bar{1}, 0, 1\}$. Integers have a unique integer representation. Addition of integers is a right subsequential function. Generalize this result to integral bases greater than 3.

Section 7.2

7.2.1 Show that the code Y defined in the proof of Proposition 7.2.11 is finite if and only if $d_\beta(1)$ is finite, and is recognizable by a finite automaton if and only if $d_\beta(1)$ is eventually periodic.

7.2.2 If every rational number of $[0, 1]$ has an eventually periodic β-expansion, then β must be a Pisot or a Salem number. (See Schmidt 1980.)

7.2.3 Normalization in base β. (See Frougny 1992, Berend and Frougny 1994.)

1. Let $s = (s_i)_{i \geq 1}$ and denote by $\pi_\beta(s)$ the real number $\sum_{i \geq 1} s_i \beta^{-i}$. Let C be a finite alphabet of integers. The canonical alphabet is $A = \{0, \ldots, \lfloor \beta \rfloor\}$. The *normalization function* on C

$$v_C : C^{\mathbb{N}} \longrightarrow A^{\mathbb{N}}$$

is the partial function which maps an infinite word s over C, such that $0 \leq \pi_\beta(s) \leq 1$, onto the β-expansion of $\pi_\beta(s)$.
A transducer is said to be *letter to letter* if the edges are labeled by pairs of letters.
Let $C = \{0, \ldots, c\}$, where c is an integer ≥ 1. Show that normalization v_C is a function computable by a finite letter-to-letter transducer if and only if the set

$$Z(\beta, c) = \left\{ s = (s_i)_{i \geq 0} \, \middle| \, s_i \in \mathbb{Z}, \, |s_i| \leq c, \, \sum_{i \geq 0} s_i \beta^{-i} = 0 \right\}$$

is recognizable by a finite automaton.

2. Prove that the following conditions are equivalent:
(i) normalization $v_C : C^{\mathbb{N}} \longrightarrow A^{\mathbb{N}}$ is a function computable by a finite letter-to-letter transducer on any alphabet C of nonnegative integers;
(ii) $v_{A'} : A'^{\mathbb{N}} \longrightarrow A^{\mathbb{N}}$, where $A' = \{0, \ldots, \lfloor \beta \rfloor + 1\}$, is a function computable by a finite letter-to-letter transducer;
(iii) β is a Pisot number.

Section 7.3

**7.3.1 (See Hollander 1998.) Let U be a linear recurrent sequence of integers such that $\lim_{n \to \infty} (u_{n+1}/u_n) = \beta$ for real $\beta > 1$.
1. Prove that if $d_\beta(1)$ is not finite or eventually periodic then $L(U)$ is not recognizable by a finite automaton.
2. If $d_\beta(1)$ is eventually periodic, $d_\beta(1) = t_1 \cdots t_N (t_{N+1} \cdots t_{N+p})^\omega$, set

$$B(X) = X^{N+p} - \sum_{i=1}^{N+p} t_i X^{N+p-i} - X^N + \sum_{i=1}^{N} t_i X^{N-i}.$$

Similarly, if $d_\beta(1)$ is finite, $d_\beta(1) = t_1 \cdots t_m$, set

$$B(X) = X^m - \sum_{i=1}^{m} t_i X^{m-i}.$$

Note that $B(X)$ is dependent on the choice of N and p (or m).
Any such polynomial is called an *extended beta polynomial* for β.
Prove that
(i) if $d_\beta(1)$ is eventually periodic, then $L(U)$ is recognizable by
a finite automaton if and only if U satisfies an extended beta
polynomial for β,
(ii) if $d_\beta(1)$ is finite, then

- if U satisfies an extended beta polynomial for β then $L(U)$ is recognizable by a finite automaton,

- if $L(U)$ is recognizable by a finite automaton then U satisfies a polynomial of the form $(X^m - 1)B(X)$ where $B(X)$ is an extended polynomial for β and m is the length of $d_\beta(1)$.

Section 7.4

7.4.1 1. Show that every Gaussian integer can be uniquely represented using base 3 and digit set $A = \{\bar{1}, 0, 1\} + i\{\bar{1}, 0, 1\} = \{0, 1, -1, i, -i, 1+i, 1-i, -1+i, -1-i\}$. If each digit is written in the form

$$0 = {}^0_0,\ 1 = {}^1_0,\ -1 = {}^{\bar{1}}_0,\ i = {}^0_1,\ -i = {}^0_{\bar{1}},$$

$$1+i = {}^1_1,\ 1-i = {}^1_{\bar{1}},\ -1+i = {}^{\bar{1}}_1,\ -1-i = {}^{\bar{1}}_{\bar{1}}$$

then for any representation the top row represents the real part and the bottom row is the imaginary part. Every complex number is representable.

2. Show that every complex number can be represented using base 2 and the same digit set A, but that the representation of a Gaussian integer is not unique.

7.4.2 Prove that every Gaussian integer has a unique representation of the form $d_k \cdots d_0.d_{-1}$ in base $\beta = \pm bi$, where b is an integer ≥ 2, and the digits d_j are elements of $A = \{0, \ldots, b^2 - 1\}$. Every complex number is representable. (See Knuth 1988.)

7.4.3 Show that every complex number can be represented using base 2 and digit set $A = \{0, 1, \zeta, \zeta^2, \zeta^3\}$, where $\zeta = \exp(2i\pi/4)$. These representations are called *polygonal* representations. (See Duprat, Herreros, and Kla 1993.)

7.4.4 Let β be a complex number of modulus > 1, and let A be a finite digit set containing 0. Let W be the set of fractional parts of complex numbers, $W = \{\sum_{j \geq 1} d_j \beta^{-j} \mid d_j \in A\}$.

1. Show that W is the only compact subset of $\mathbb{C}$ such that $\beta W = W + A$.

2. Show that if the set W is a neighborhood of zero, then every complex number has a representation with digits in A.

7.4.5 Let β be a complex number of modulus > 1, and let A be a finite digit set containing 0. An infinite sequence $(d_j)_{j \geq 1}$ of $A^\mathbb{N}$ is a *strictly proper* representation of a number $z = \sum_{j \geq 1} d_j \beta^{-j}$ if it is the greatest in the lexicographic order of all the representations of z with digits in A. It is *weakly proper* if each finite truncation is strictly proper. Let $W = \{\sum_{j \geq 1} d_j \beta^{-j} \mid d_j \in A\}$. Show that, if β is a complex Pisot number, the set of weakly proper representations of elements of W is recognizable by a finite automaton. (See Thurston 1989, Kenyon 1992, Petronio 1994.)

*7.4.6 Representation of algebraic number fields. (See Gilbert 1981, 1994, Kátai and Kovács 1981.)

Let β be an algebraic integer of modulus > 1, and let A be a finite set of elements of $\mathbb{Z}[\beta]$ containing zero. We say that (β, A) is an *integral numeration system* for the field $\mathbb{Q}(\beta)$ if every element of $\mathbb{Z}[\beta]$ has a unique integer representation of the form $d_k \cdots d_0$ with d_j in A.

1. Let $P(X) = X^m + p_{m-1}X^{m-1} + \cdots + p_0$ be the minimal polynomial of β. The *norm* of β is $N(\beta) = |p_0|$. Show that a complete residue system of elements of $\mathbb{Z}[\beta]$ modulo β is the set $\{0, \ldots, N(\beta) - 1\}$.

2. Suppose that every element of $\mathbb{Z}[\beta]$ has a representation in (β, A). Prove that this representation is unique if and only if A is a complete residue system for $\mathbb{Z}[\beta]$ modulo β that contains zero.

3. Suppose that (β, A) is an integral numeration system. Show that every element of the field $\mathbb{Q}(\beta)$ has a representation in (β, A).

4. Show that (β, A) is an integral numeration system if and only if β and all its conjugates have moduli greater than 1 and there is no positive integer q for which

$$d_{q-1}\beta^{q-1} + \cdots + d_0 \equiv 0 \pmod{\beta^q - 1}$$

with d_j in A for $0 \leq j \leq q$.

5. Now suppose that β is a quadratic algebraic integer, and let $A = \{0, \ldots, |p_0| - 1\}$. Prove that (β, A) is an integral numeration system for $\mathbb{Q}(\beta)$ if and only if $p_0 \geq 2$ and $-1 \leq p_1 \leq p_0$.

Notes

Concerning the representation of numbers in classical or less classical numeration systems, there is always something to learn in Knuth 1988. Representation in integral base with signed digits was popularized in computer arithmetic by Avizienis 1961 and can be found earlier in a work of Cauchy 1840.

We have not presented here p-adic numeration, or the representation of real numbers by their continued fraction expansions (see Chapter 2 for this last topic).

The notion of beta-expansion is due to Rényi 1957. Its properties were essentially set up by Parry 1960, in particular Theorem 7.2.9. Coded systems were introduced by Blanchard and Hansel 1986. The result on the entropy of the beta-shift is due to Ito and Takahashi 1974. The links between the beta-expansion of 1 and the nature of the beta-shift are presented in Ito and Takahashi 1974 and in Bertrand-Mathis 1986. Connections with Pisot numbers are to be found in Bertrand 1977 and Schmidt 1980. It is also known that normalization in base β is computable by a finite transducer on any alphabet if and only if β is a Pisot number; see Problem 7.2.3. If β is a Salem number of degree 4 then $d_\beta(1)$ is eventually periodic; see Boyd 1989. It is an open problem for degree ≥ 6. Perron numbers are introduced in Lind 1984. There is a survey on the relations between beta-expansions and symbolic dynamics by Blanchard (1989). In Solomyak 1994 and in Flatto, Lagarias, and Poonen 1994 is proved the following property: if $d_\beta(1)$ is eventually periodic, then the algebraic conjugates of β have modulus strictly less than the golden ratio. Beta-expansions also appear in the mathematical description of quasicrystals; see Gazeau 1995.

The representation of integers with respect to a sequence U is introduced in Fraenkel 1985. The fact that, if $L(U)$ is recognizable by a finite automaton, then the sequence U is linearly recurrent is due to Shallit 1994. We follow the proof of Loraud 1995. The converse problem is treated by Hollander 1998, see Problem 7.3.1. Canonical numeration systems associated with a number β come from Bertrand-Mathis 1989. Normalization in linear numeration systems linked with Pisot numbers is studied in Frougny 1992, Frougny and Solomyak 1996, and with the use of congruential techniques, in Bruyère and Hansel 1997. Moreover, if the sequence U has a characteristic polynomial which is the minimal polynomial of a Perron number which is not Pisot, then normalization cannot be computed by a finite transducer on every alphabet (Frougny and Solomyak 1996).

A famous result on sets of natural numbers recognized by finite automata is the theorem of Cobham 1969. Let k be an integer ≥ 2. A set X of positive integers is said to be k-*recognizable* if the set of k-representations of numbers of X is recognizable by a finite automaton. Two numbers k and l are said to be *multiplicatively independent* if there exist no positive integers p and q such that $k^p = l^q$. Cobham's theorem then states: if X is a set of integers which is both k-recognizable and l-recognizable in two multiplicatively independent bases k and l, then X is eventually periodic. There is a multidimensional version of Cobham's theorem due to Semenov 1977. The original proofs of these two results are difficult, and several other proofs have been given, some of them using logic (see Michaux and Villemaire 1996). There are many works on generalizations of the Cobham and Semenov theorems (see Fabre 1994, Bruyère and Hansel 1997, Point and Bruyère 1997, Fagnot 1997, Hansel 1998). In Durand 1998 there is a version of the Cobham theorem in terms of substitutions. We give now one result related to the concepts presented in Section 7.3. Let U be an increasing sequence of integers. A set X of positive integers is U-recognizable if the set of normal U-representations of numbers of X is recognizable by a finite automaton. Let β and β' be two multiplicatively independent Pisot numbers, and let U and U' be two linear numeration systems whose characteristic polynomials are the minimal polynomials of β and β' respectively. For every $n \geq 1$, if $X \subset \mathbb{N}^n$ is U- and U'-recognizable then X is definable in $\langle \mathbb{N}, + \rangle$ (Bès 2000). When $n = 1$, the result says that X is eventually periodic.

Theorem 7.4.3 on bases of the form $-n \pm i$, n integer ≥ 1, is due to Kátai and Szabó 1975. There is a more algorithmic proof, as well as results on the sum-of-digits function for base $\beta = -1 + i$, in Grabner, Kirschenhofer, and Prodinger 1998. Normalization in complex bases is studied in Safer 1998. Theorem 7.4.7 appeared in Thurston 1989, as well as the result on complex Pisot bases presented in Problem 7.4.5. Representation of complex numbers in imaginary quadratic fields is studied in Kátai 1994. We have not discussed here beta-automatic sequences. Results on these topics can be found in Allouche et al. 1997, particularly for the case $\beta = -1 + i$.

The numeration in complex bases is strongly related to fractals and tilings. Self-similar tilings of the plane in relation with complex Pisot bases are discussed in Thurston 1989, Kenyon 1992 and Petronio 1994. In Gilbert 1986, the fractal dimension of tiles obtained in some bases such as $-n + i$ is computed. A general survey has been written by Bandt 1991.

CHAPTER 8

Periodicity

8.0. Introduction

Periodicity is an important property of words that has applications in various domains. The first significant results on periodicity are the theorem of Fine and Wilf and the critical factorization theorem. These two results refer to two kinds of phenomena concerning periodicity: the theorem of Fine and Wilf considers the simultaneous occurrence of different periods in one finite word, whereas the critical factorization theorem relates local and global periodicity of words. Starting from these basic results the study of periodicity has grown along both directions. This chapter contains a systematic and self-contained exposition of this theory, including very recent results.

In Section 8.1 we analyze the structure of the set of periods of one finite word. This section includes a proof of the theorem of Fine and Wilf and also a generalization of this result to words having three periods. We next give the characterization of Guibas and Odlyzko concerning those sets of integers that yield the periods that can simultaneously occur in a single finite word. Another property is further investigated (similar to the one stated by the theorem of Fine and Wilf) in which the occurrence of two periods in a word of a certain length forces the word to have a shorter period only in a prefix (or suffix) of the word. The golden ratio appears in such a result as an extremal value of a parameter involved in the property. This section also contains some results concerning the squares that can appear as factors in a word. This is a prelude to the next section, since squares describe a special kind of local periodicity.

In Section 8.2 we investigate the relation between local and global periodicity. Local periodicity is described in terms of repetitions. A repetition occurring in a word is not in general a factor of the word, nor is it necessarily a square, but it may be of rational (not only integer) order. Moreover the repetition is referred to a "point" of the word and it is important to consider the relative positions of the repetition and that of the point at which the repetition is detected. Thus we distinguish between central repetitions and left (or right) repetitions. The study of central repetitions leads to the critical factorization theorem, of which we here report a new short proof. In such a result we need repetitions of order greater than or equal to 2 and the value 2 is proved to be tight. The study of left repetitions leads to a similar result, but in this case we need repetitions of order greater than φ^2, where φ denotes the golden ratio, and the value φ^2 is tight for such a result.

The last section is devoted to infinite words. By extending the ideas and the results of Section 8.2, we characterize recurrence, periodicity and eventual periodicity of infinite words in terms of local periods. In some of these results the golden ratio again plays a central role.

8.1. Periods in a finite word

8.1.1. Definitions and basic properties

Let $w = a_1 a_2 \cdots a_n$ be a word of length n over the alphabet A.

Recall from Subsection 1.2.1 that a positive integer $p \leq |w|$ is a *period* of w if $a_{i+p} = a_i$ for $i = 1, \ldots, n - p$. The smallest period p of w is called *the* period of w and it is denoted by $p(w)$. From the definition it follows that, if v is a factor of w, then $p(v) \leq p(w)$.

The positive rational number $|w|/p(w)$ is called the *order* of w and it is denoted by ord(w). If u is the prefix of length $p(w)$ of w, we can write $w = u^\rho$ where $\rho =$ ord(w), and we say that w is a *rational power* of u. Notice that a rational power u^ρ is defined only if $|u|\rho$ is an integer. For instance, $p(abaaba) = 3$, ord($abaaba$) $= 2$ and the word $abaaba$ can be uniquely written $abaaba = (aba)^2$. As another example, $p(ababaaba) = 5$, ord($ababaaba$) $= 1.6$ and the word $ababaaba$ can be written in a unique way as $ababaaba = (ababa)^{1.6}$.

A word v that is both a prefix and a suffix of another word w, with $v \neq w$, is called a *border* of w. It is easy to see that if v is a border of w, then $|w| - |v|$ is a period of w and, conversely, if p is a period of w, then the prefix v of w of length $|w| - p$ is a border of w. The empty string ε is a border of any string w. If there exists a nonempty border v of w then w is called *bordered*, otherwise it is called *unbordered*.

It is easy to verify that a word is unbordered if and only if ord(w) = 1, or, equivalently, if and only if $|w| = p(w)$.

The following three lemmas will often be used in this chapter. We invite the reader to spend some time in reading the proof of these lemmas

8.1. Periods in a finite word

together with Problem 8.1.1, to get acquainted with the basic tools and ideas used in this chapter.

LEMMA 8.1.1. *Let w be a word having two periods p and q, with $q < p \leq |w|$. Then the suffix and the prefix of w of length $|w| - q$ both have period $p - q$.*

Proof. We prove only that the prefix of w of length $|w| - q$ has period $p - q$, the proof for the suffix being analogous. Since $|w| - q \geq p - q$, we have to prove that

$$a_{i+p-q} = a_i, \quad i = 1, \ldots, n - p.$$

Let i be such that $1 \leq i \leq n - p$. Thus $1 \leq i + p - q \leq n - q$. Since w has period p, one has that $a_i = a_{i+p}$. Since w has period q and $1 \leq i + p - q \leq n - q$, one has that $a_{i+p-q} = a_{i+p}$. ∎

LEMMA 8.1.2. *Let u, v, w be words such that uv and vw have period p and $|v| \geq p$. Then the word uvw has period p.*

Proof. Let $uvw = a_1 \cdots a_n$, $u = a_1 \cdots a_l$, $v = a_{l+1} \cdots a_j$, $w = a_{j+1} \cdots a_n$. By the hypothesis $j - l \geq p$. Let i be an integer with $1 \leq i \leq n - p$. We have to prove that $a_i = a_{i+p}$.

If $i \leq j - p$, since uv has period p, then $a_i = a_{i+p}$. If $i > j - p$, since $j - l \geq p$, then $i \geq l + 1$. Since vw has period p, we have $a_i = a_{i+p}$. ∎

LEMMA 8.1.3. *Suppose that w has period q and that there exists a factor v of w with $|v| \geq q$ that has period r, where r divides q. Then w has period r.*

Proof. Let $w = a_1 \cdots a_n$ and let $v = a_h \cdots a_k$, with $1 \leq h < k \leq n$ and $k - h + 1 \geq q$. Let us suppose that $i \equiv j \pmod{r}$, $1 \leq i, j \leq n$. We have to prove that $a_i = a_j$. Since, by hypothesis, $k - h + 1 \geq q$, for any integers i, j, there exist i', j' with $h \leq i', j' \leq k$ such that $i \equiv i' \pmod{q}$ and $j \equiv j' \pmod{q}$. Since $i \equiv j \pmod{r}$ and since r divides q, one has $i' \equiv j' \pmod{r}$. But w has period q and thus $a_i = a_{i'}$ and $a_j = a_{j'}$. Finally, since v has period r, one gets $a_{i'} = a_{j'}$ and the lemma is proved. ∎

8.1.2. The theorem of Fine and Wilf

This subsection is devoted to the theorem of Fine and Wilf and some generalizations. It is a classical and basic result on periodicity and one of its proofs is reported in Lothaire, 1983. The proof we give here can be considered as a first step of the proof of a more general result that is stated and proved in the next subsection.

The proof reported here is closely related to Euclid's algorithm computing the greatest common divisor of two integers. In particular, we use the fact that, given two positive integers p, q, with $q < p$, $\gcd(p, q) = \gcd(p - q, q)$.

THEOREM 8.1.4 (Fine and Wilf). *Let w be a word having periods p and q, with $q \leq p$. If $|w| \geq p + q - \gcd(p, q)$, then w also has period $\gcd(p, q)$.*

Proof. Set $r = \gcd(p, q)$. The proof is by induction on the integer $n = (p + q)/r$. For $n = 2$, $q = p = r$ and the statement is trivially satisfied.

Let us consider the case $n > 2$. This, in particular, implies that $q < p$. Suppose that the statement holds for all integers smaller than n. Consider a word w having periods p and q and such that $|w| \geq p + q - r$. Denote by u the prefix of w of length q and set $w = uv$.

By Lemma 8.1.1, v has period $p - q$ and, since v is a factor of w, v also has period q. Moreover one has

$$|v| = |w| - q \geq (p - q) + q - r = (p - q) + q - \gcd(p - q, q).$$

By the inductive hypothesis, v also has period $\gcd(p-q, q) = \gcd(p, q) = r$. Since $p - q \geq r$, then $|v| = |w| - q \geq (p - q) + q - r \geq q$. By Lemma 8.1.3, the word w also has period r and this concludes the proof. ∎

REMARK 8.1.5. The bound given in Theorem 8.1.4 is tight, as shown by the word $w = abaababaaba$. It has period 5 and period 8, length $5 + 8 - 2 = 11$ and it does not have period $\gcd(5, 8) = 1$. An infinite family of words proving that the bound is tight is the family of central words considered in Chapter 2.

Theorem 8.1.4 can be generalized to words having three periods. As in the previous case, the statement and the proof are closely related to Euclid's algorithm.

Let $p = (p_1, p_2, p_3)$ be a triple of nonnegative integers. If $p_1 \leq p_2 \leq p_3$, we call p an *ordered triple*. We denote by O the operator which, given an

8.1. Periods in a finite word

arbitrary triple, returns the corresponding ordered triple. We define two additional operators R and S on ordered triples by

$$R(p) = \begin{cases} (p_1, p_2 - p_1, p_3 - p_1), & \text{if } p_1 \neq 0, \\ (0, p_2, p_3 - p_2), & \text{if } p_1 = 0, \end{cases}$$

and $S(p) = O(R(p))$.

Given an ordered triple p, let us consider the sequence $(p^{(k)})_{k \geq 0}$ of ordered triples defined recursively as follows:

$$p^{(0)} = p, \quad p^{(k+1)} = S(p^{(k)}), \quad k \geq 0.$$

The elements of the triple $p^{(k)}$ are denoted by

$$p^{(k)} = (p_1^{(k)}, p_2^{(k)}, p_3^{(k)}).$$

Let us denote by $|p| = p_1 + p_2 + p_3$ the sum of the elements of the triple p. Set

$$m(p) = \min\{k \mid p_1^{(k)} = 0\}, \quad M(p) = \min\{k \mid p_1^{(k)} = p_2^{(k)} = 0\}.$$

With these notations, given the triple $p = (p_1, p_2, p_3)$, one has

$$\gcd(p_1, p_2, p_3) = |p^{(M(p))}|.$$

We also define a function h which plays an important role in the next result by

$$h(p_1, p_2, p_3) = |p^{(m(p))}|.$$

By definition, the function h satisfies the condition

$$h(p_1, p_2, p_3) = h(p_1, p_2 - p_1, p_3 - p_1).$$

EXAMPLE 8.1.6. Consider the triple $p = (7, 11, 13)$. Euclid's algorithm gives the following sequence of triples: $p^{(0)} = (7, 11, 13)$, $p^{(1)} = (4, 6, 7)$, $p^{(2)} = (2, 3, 4)$, $p^{(3)} = (1, 2, 2)$, $p^{(4)} = (1, 1, 1)$, $p^{(5)} = (0, 0, 1)$. We get $\gcd(7, 11, 13) = h(7, 11, 13) = 1$.

Given an ordered triple $p = (p_1, p_2, p_3)$ of non negative integers, we introduce the function

$$f(p_1, p_2, p_3) = \frac{1}{2}[p_1 + p_2 + p_3 - 2\gcd(p_1, p_2, p_3) + h(p_1, p_2, p_3)].$$

Notice that $f(p_1, p_2, p_3)$ is greater than or equal to $\gcd(p_1, p_2, p_3)$.

THEOREM 8.1.7. *Let w be a word over the alphabet A having three periods p_1, p_2 and p_3, with $p_1 \leq p_2 \leq p_3$. If $|w| \geq f(p_1, p_2, p_3)$, then w also has period $\gcd(p_1, p_2, p_3)$.*

REMARK 8.1.8. The statement of this theorem includes, as a particular case, the statement of the theorem of Fine and Wilf. Indeed, the condition that a word w has periods p, q, with $p \leq q$, corresponds to the triple $(0, p, q)$. Since, by definition, $h(0, p, q) = p + q$, it follows that

$$f(0, p, q) = \frac{1}{2}[p + q - 2\gcd(p,q) + p + q] = p + q - \gcd(p, q).$$

Proof of theorem. We shall prove the theorem by induction on the integer

$$n = p_1(p_1 + p_2 + p_3).$$

The case $n = 0$ corresponds to the classical Fine and Wilf theorem (see also Remark 8.1.8).

Let us now suppose that the statement is true for all ordered triples $q = (q_1, q_2, q_3)$ such that $m = q_1(q_1 + q_2 + q_3) < n$ and consider an ordered triple $p = (p_1, p_2, p_3)$ such that $p_1(p_1 + p_2 + p_3) = n$.

Let w be a word having periods p_1, p_2 and p_3 and length $|w| \geq f(p_1, p_2, p_3)$. Let u be the prefix of w of length p_1:

$$w = uv, \qquad \text{with } |u| = p_1, \qquad |v| = |w| - p_1.$$

By Lemma 8.1.1, the word v has periods p_1, $p_2 - p_1$, $p_3 - p_1$ and length $|v| = |w| - p_1$. Since $|w| \geq f(p_1, p_2, p_3)$, one has

$$\begin{aligned}|v| = |w| - p_1 &\geq f(p_1, p_2, p_3) - p_1 \\ &= \tfrac{1}{2}[p_1 + p_2 + p_3 - 2\gcd(p_1, p_2, p_3) + h(p_1, p_2, p_3)] - p_1 \\ &= \tfrac{1}{2}[p_1 + (p_2 - p_1) + (p_3 - p_1) \\ &\qquad - 2\gcd(p_1, p_2 - p_1, p_3 - p_1) + h(p_1, p_2 - p_1, p_3 - p_1)] \\ &= f(p_1, p_2 - p_1, p_3 - p_1).\end{aligned}$$

By the inductive hypothesis v also has period

$$\gcd(p_1, p_2 - p_1, p_3 - p_1) = \gcd(p_1, p_2, p_3).$$

By Lemma 8.1.3, the word $w = uv$ has period $\gcd(p_1, p_2, p_3)$. This concludes the proof. ∎

REMARK 8.1.9. The bound given in Theorem 8.1.7 is tight, as shown by the following example. An infinite family of words proving the tightness is considered in Problem 8.1.8 (see also Remark 8.1.5).

EXAMPLE 8.1.10. Consider the word
$$w = abacabaabacaba.$$
The word w has length $|w| = 14$ and periods 7, 11, 13. Since $\gcd(7, 11, 13) = h(7, 11, 13) = 1$ (see Example 8.1.6), one has
$$f(7, 11, 13) = \frac{1}{2}(7 + 11 + 13 - 2 + 1) = 15.$$
Then w is a word having periods 7, 11, 13; its length is $|w| = f(7, 11, 13) - 1$ and w has no period $\gcd(7, 11, 13)$.

8.1.3. Structure of the periods of a word

In this subsection, we give the structure of the set of periods of a single finite word. As a consequence, we obtain, for any word w, a word w' over a binary alphabet that has exactly the same set of periods as w.

Let $\Pi(w)$ denote the set of all periods of w, with 0 included. For instance, if $\hat{w} = abcabcadefgabcabca$, then $\Pi(\hat{w}) = \{0, 11, 14, 17, 18\}$ and if $\underline{w} = aaaaaaaaaaaaaaaaaa$, then $\Pi(\underline{w}) = \{0, 1, \ldots, 18\}$.

Clearly, if p is a period of w then any multiple of p that is smaller than or equal to the length of w is also a period of w. Indeed, since 1 is a period of $\underline{w}$ in the example above, all positive integers smaller than or equal to $|\underline{w}|$ are also periods of $\underline{w}$.

If $\Pi(w) = \{0 = p_0 < p_1 < \cdots < p_s = |w|\}$, we define the sequence of differences
$$\delta_h = p_h - p_{h-1}, \quad 1 \le h \le s.$$

If p is a period of w and q is a period of the suffix of length $|w| - p$ of w then $p + q$ is also a period of w. Therefore (Problem 8.1.1), for any positive integer k such that $p + kq \le |w|$, the integer $p + kq$ is a period of w. This fact and Lemma 8.1.1 imply (Problem 8.1.9) that the sequence of differences δ_h is nonincreasing.

In the example above, the integers 11 and 14 are periods of $\hat{w}$. By Lemma 8.1.1, the integer $14 - 11 = 3$ is a period of the suffix of length $|\hat{w}| - 11 = 7$ of $\hat{w}$. Therefore $17 = 11 + 6$ must also be a period of $\hat{w}$.

We know that the sequence of the differences δ_h is nonincreasing, but some more conditions must be added in order to characterize the set of periods of a single word. They appear as conditions (iii) and (iv) in the next theorem.

THEOREM 8.1.11. *Let $\Pi = \{0 = p_0 < p_1 < \cdots < p_s = n\}$ be a set of integers and let $\delta_h = p_h - p_{h-1}, 1 \le h \le s$. Then the following conditions are equivalent.*

(i) *There exists a word w over a two-letter alphabet with $\Pi(w) = \Pi$.*
(ii) *There exists a word w with $\Pi(w) = \Pi$.*
(iii) *For each h, such that $\delta_h \leq n - p_h$, one has*
   (a) $p_h + k\delta_h \in \Pi$, *for* $k = 1, \ldots, \lfloor (n-p_h)/\delta_h \rfloor$, *and*
   (b) *if* $\delta_{h+1} < \delta_h$, *then* $\delta_h + \delta_{h+1} > n - p_h + \gcd(\delta_h, \delta_{h+1})$.
(iv) *For each h, such that $\delta_h \leq n - p_h$, one has*
   (a) $p_h + \delta_h \in \Pi$ *and*
   (b) *if* $\delta_h = k\delta_{h+1}$, *for some integer k then $k = 1$.*

Proof. Trivially condition (i) implies condition (ii). Let us prove that condition (ii) implies condition (iii).

Let us now suppose that there exists a word $w = a_1 \cdots a_n$ with $\Pi(w) = \Pi$ and that $\delta_h \leq n - p_h$. By Lemma 8.1.1, δ_h is a period of the suffix of w of length $n - p_{h-1}$. Hence for any $k > 0$ and $i > p_h$ such that $i + k\delta_h \leq n$, one has that $a_i = a_{i+k\delta_h}$. Since p_h is a period of w, for any $i > p_h$, $a_{i-p_h} = a_i$. Setting $j = i - p_h$ we have that for any $j > 0$ and for any k such that $j + p_h + k\delta_h \leq n$, one has that $a_j = a_{j+p_h+k\delta_h}$, i.e., $p_h + k\delta_h \in \Pi$, for $k = 1, \ldots, \lfloor (n-p_h)/\delta_h \rfloor$ and (iii)(a) is proved.

Let us suppose by contradiction that $\delta_h \leq n - p_h$, that $\delta_{h+1} < \delta_h$ and that $\delta_h + \delta_{h+1} \leq n - p_h + \gcd(\delta_h, \delta_{h+1})$. By Lemma 8.1.1, δ_h is a period of the suffix u of w of length $n - p_{h-1}$ and δ_{h+1} is a period of the suffix v of w of length $n - p_h$. Clearly v is a suffix of u. Hence, both δ_h and δ_{h+1} are periods of v. Since $|v| = n - p_h \geq \delta_{h+1} + \delta_h - \gcd(\delta_h, \delta_{h+1})$, by Theorem 8.1.4 v also has period $r = \gcd(\delta_h, \delta_{h+1})$. Since $|v| = n - p_h \geq \delta_h$ and since u has period δ_h, by Lemma 8.1.3, u also has period r. This fact implies that $p_{h-1} + r$ is a period of w. But $p_{h-1} + r < p_{h-1} + \delta_h = p_h$, contradicting the fact that p_h is the smallest period greater than p_{h-1}.

Let us prove that condition (iii) implies condition (iv). Trivially condition (iii)(a) implies condition (iv)(a).

Let us suppose by contradiction that $\delta_h \leq n - p_h$ and that $\delta_h = k\delta_{h+1}$, for some integer $k > 1$. Then $\delta_{h+1} < \delta_h$ and also $\delta_{h+1} = \gcd(\delta_h, \delta_{h+1})$. Therefore, by condition (iii)(b), $\delta_h + \delta_{h+1} > n - p_h + \gcd(\delta_h, \delta_{h+1}) = n - p_h + \delta_{h+1}$, i.e., $\delta_h > n - p_h$ which is a contradiction.

Let us finally prove that condition (iv) implies condition (i). We prove that if condition (iv) holds, then there exists a binary string w, such that $\Pi(w) = \Pi$.

Let $\Pi_h = \{p - p_h \mid p \in \Pi \text{ and } p \geq p_h\}$, $0 \leq h \leq s$. For instance, if $\Pi = \{0, 11, 14, 17, 18\}$, then $\Pi_4 = \{0\}$, $\Pi_3 = \{0, 1\}$, $\Pi_2 = \{0, 3, 4\}$, $\Pi_1 = \{0, 3, 6, 7\}$. Clearly, $\Pi_0 = \Pi$.

We prove by induction on $h = s, \ldots, 0$ that there exist binary strings w_h, such that $\Pi(w_h) = \Pi_h$. Then, w_0 is the required string with $\Pi(w_0) = \Pi$. Notice that $|w_h| = n - p_h$.

8.1. Periods in a finite word

For the basis of the induction we have that $w^s = \epsilon$, since $\Pi_s = \{0\}$. Assume now that there exists a string w_h, such that $\Pi(w_h) = \Pi_h$. There are two cases.

Case 1: $\delta_h > n - p_h$. We claim that there exists a sequence $a_1, \ldots, a_{\delta_h - |w_h|}$ of letters in the same binary alphabet as w_h such that the word $w_{h-1} = w_h a_1 \cdots a_{\delta_h - |w_h|} w_h$ has no periods of length smaller than δ_h.

The proof of this claim is by induction on $\delta_h - |w_h|$. Suppose $\delta_h - |w_h| = 1$. Consider the two words $w_h x w_h$ and $w_h y w_h$ with $x \neq y$ the two different letters in the binary alphabet of w_h. If, by contradiction, both words $w_h x w_h$ and $w_h y w_h$ have a period smaller than $\delta_h = |w_h x| = |w_h y|$ then, by Problem 8.1.4, they must be equal, which is impossible because they differ in the central position.

Inductive step of the claim: Suppose that the binary alphabet is $\{x, y\}$. Suppose that the claim is true for $\delta_h - |w_h| = n - 1$ and suppose also that, by contradiction, putting a letter x or a letter y between positions $a_{\lceil n/2 \rceil}$ and $a_{\lceil n/2 \rceil + 1}$ we get two different words that each have a period smaller than or equal to $|w_h| + \lceil n/2 \rceil + 1$. These two periods cannot be equal, because of the different letters in the same position, and so one must be smaller than $|w_h| + \lceil n/2 \rceil + 1$.

These words have length $2|w_h| + n + 1$ and the sum of the two periods is smaller than or equal to $2|w_h| + n + 1$. By Problem 8.1.4, they must be equal, which is impossible because they differ in the central position. This concludes the proof of the claim.

By the claim one has $\Pi(w_{h-1}) = \{\delta_h + p \mid p \in \Pi_h\} \cup \{0\} = \Pi_{h-1}$, and the inductive step is proved.

Case 2. If $\delta_h \leq n - p_h$, then let $w_{h-1} = a_1 \cdots a_{\delta_h} w_h$ where $a_1 \cdots a_{\delta_h}$ is the prefix of length δ_h of w_h. Since $p_h + \delta_h \in \Pi$, we get that $\delta_h \in \Pi_h$ and, by inductive hypothesis, δ_h is a period for w_h. Hence δ_h is also a period of w_{h-1}. Consequently, by Problem 8.1.1 and Lemma 8.1.1, $\delta_h + p$ is a period of w_{h-1}, for some integer $p \geq 0$, if and only if p is a period of w_h. This is equivalent to saying, by the inductive hypothesis, that $\Pi(w_{h-1}) \cap \{\delta_h, \ldots, |w_{h-1}|\} = \{\delta_h + p \mid p \in \Pi_h\}$.

Assume, by contradiction, that $\Pi(w_{h-1}) \neq \Pi_{h-1}$ and, consequently, there exists a period $t < \delta_h$ in $\Pi(w_{h-1}) \setminus \Pi_{h-1}$. We have that both t and $\delta_h - t$ must be periods of w_h. The first is a period because w_h is a suffix of w_{h-1}. The second is a period because both δ_h and t are periods of w_{h-1} and because, by Lemma 8.1.1, the suffix of w_{h-1} of length $|w_{h-1}| - t > |w_{h-1}| - \delta_h = |w_h|$ has period $\delta_h - t$ and contains w_h as factor.

Since by the induction hypothesis, δ_{h+1} is the shortest nonzero period of w_h, $\delta_{h+1} \leq t$ and $\delta_{h+1} \leq \delta_h - t$, and, consequently, $\delta_{h+1} + t \leq \delta_h$ and $\delta_{h+1} + \delta_h - t \leq \delta_h$.

By the theorem of Fine and Wilf and by the minimality of δ_{h+1},

$\delta_{h+1} = \gcd(\delta_{h+1}, t)$ and $\delta_{h+1} = \gcd(\delta_{h+1}, \delta_h - t)$, i.e., δ_{h+1} divides t and also divides $\delta_h - t$. Hence δ_{h+1} also divides $t + \delta_h - t = \delta_h$ and this contradicts condition (iv)(b) because $\delta_{h+1} \le t < \delta_h$. ∎

Let us now give an example of how to construct a word w' over a binary alphabet $\{a, b\}$ such that $\Pi(w') = \{0, 11, 14, 17, 18\}$.

Notice that the set $\Pi = \{0, 11, 14, 17, 18\}$ satisfies condition (iv) of the preceding theorem. It is indeed the set of periods of the word $\hat{w} = abcabcadefgabcabca$.

We have that $\Pi_4 = \{0\}$, $\Pi_3 = \{0, 1\}$, $\Pi_2 = \{0, 3, 4\}$, $\Pi_1 = \{0, 3, 6, 7\}$. Clearly, always $\Pi_0 = \Pi$.

Moreover we know that $\delta_4 = 18 - 17 = 1$, $\delta_3 = 17 - 14 = 3$, $\delta_2 = 14 - 11 = 3$, $\delta_1 = 11 - 0 = 11$.

We inductively construct words w_h for $h = 4, 3, 2, 1, 0$, where $w_0 = w'$ is the required word. Recall that $|w_h| = 18 - p_h$.

The word $w_4 = \epsilon$ and $\Pi(w_4) = \Pi_4 = \{0\}$. Since $\delta_4 = 1 > 18 - p_4 = 18 - 18 = 0$ we are in the first case of previous proof, and so there exists a sequence of $\delta_4 - |w_4| = 1 - 0 = 1$ letter(s) $a_1, \ldots, a_{\delta_4 - |w_4|}$ (in this case just one letter a_1) such that $w_3 = w_4 a_1 w_4$. In this case both letters a and b can be chosen to be a_1 in order to have $\Pi(w_3) = \Pi_3 = \{0, 1\}$. Let us choose $a_1 = a$, and so $w_3 = a$.

Since $\delta_3 = 3 > 18 - p_3 = 18 - 17 = 1$, we are in the first case of the preceding proof, and so there exists a sequence of $\delta_3 - |w_3| = 3 - 1 = 2$ letters $a_1, \ldots, a_{\delta_3 - |w_3|}$ (in this case two letters, a_1 and a_2) such that $w_2 = w_3 a_1 a_2 w_3 = aa_1 a_2 a$ has no period smaller than δ_3. The letter a_1 must be chosen such that $w_3 a_1 w_3$ has no period smaller than $|w_3| + 1 = 2$. There is only one possibility, which is that $a_1 = b$. The letter a_2 must be chosen such that $w_3 a_1 a_2 w_3 = aba_2 a$ has no period smaller than $|w_3| + 2 = 3$. In this case both letters a and b can be chosen in order to have $\Pi(w_2) = \Pi_2 = \{0, 3, 4\}$. Let us choose $a_1 = b$, and so $w_2 = abba$.

Notice that the order that we use to choose the letters in the general sequence $a_1, \ldots, a_{\delta_h - |w_h|}$ is not the usual order and follows the inductive proof of always choosing the "central" letter, i.e., $a_1, a_{\delta_h - |w_h|}, a_2, a_{\delta_h - |w_h| - 2}, \ldots, a_{\lceil (n - p_h)/2 \rceil}$ if $(n - p_h)$ is odd, while if it is even the last letter to be chosen is $a_{\lceil (n - p_h)/2 \rceil + 1}$. In the preceding situation, when there are only two letters, this order coincides with the usual one.

Since $\delta_2 = 3 \le 18 - p_2 = 18 - 14 = 4$, we are in the first case of the preceding proof and so $w_1 = abbw_2 = abbabba$, because abb is the prefix of length δ_2 of w_2 (whose length is $18 - p_2$). Indeed $\Pi(w_1) = \Pi_1 = \{0, 3, 6, 7\}$.

Since $\delta_1 = 11 > 18 - p_1 = 18 - 11 = 7$, we are in the first case of the preceding proof and so there exists a sequence of $\delta_1 - |w_1| = 11 - 7 = 4$

8.1. Periods in a finite word

letters a_1, a_2, a_3, a_4 such that $w_0 = w_1 a_1 a_2 a_3 a_4 w_1 = aa_1 a_2 a$ has no period smaller than δ_1. For brevity we do not perform the inductive steps for choosing this sequence of letters, which can be chosen to be a, a, a, a, i.e., $w' = w_0 = abbabbaaaaaabbabba$.

It is easy to verify that $\Pi(w') = \{0, 11, 14, 17, 18\}$.

8.1.4. Golden ratio and periodicity

We present here a result that has some analogies with the theorem of Fine and Wilf and relates periodicity and the golden ratio. This result plays an important role in Subsection 8.2.2.

The theorem of Fine and Wilf states, roughly speaking, that if a word w has two periods p and q and it is long enough ($|w| \geq p + q - \gcd(p, q)$), then it has a shorter period ($\gcd(p, q)$).

In the next result, we start from a weaker hypothesis on the length of the word, in which the golden ratio appears, and we derive a weaker conclusion.

Indeed, denoting by φ the golden ratio, we suppose that $|w| > \varphi \max\{p, q\}$ and we derive that there exists only a suffix (and a prefix) having "shorter" period.

Let us recall that the golden ratio, denoted by φ, is the real positive root of the equation $x^2 - x - 1 = 0$, i.e., $\varphi = (\sqrt{5}+1)2 = 1.618\ldots$. Since, by definition, $\varphi^2 = \varphi + 1$, in the rest of the chapter we shall sometimes interchange φ^2 with $\varphi + 1$ without mention.

THEOREM 8.1.12. *Let x and y be nonempty words and let ρ and σ be positive rational numbers such that $\varphi < \rho < \sigma$. If $x^\rho = y^\sigma$, then there exist a nonempty word z and a rational number $\tau \geq \rho + 1$ such that z^τ is a suffix (a prefix) of $x^\rho = y^\sigma$.*

Proof. Set $w = x^\rho = y^\sigma$. If $\sigma \geq \rho + 1$, then the statement is trivially satisfied with $z = y$ and $\tau = \sigma$. Let us now suppose that $\sigma < \rho + 1$. Let $|x| = p$ and $|y| = q$. Then

$$|w| = p\rho = q\sigma < q(\rho + 1).$$

If $\rho \geq \varphi$, then, by the definition of the golden ratio φ, $\rho + 1 < \rho^2$. From the inequality $p\rho < q(\rho + 1)$, we derive $p\rho < q\rho^2$ and then $p - \rho q < 0$. By adding $\rho p - q$ to both sides of the last inequality one has

$$p - \rho q + \rho p - q < \rho p - q$$

which can be rewritten

$$(p - q)(\rho + 1) < \rho p - q.$$

By definition, w has periods p and q, with $q < p$. By Lemma 8.1.1, the suffix (prefix) v of w of length $|w| - q$ has period $p - q$. Denoting by τ the order of v, we have

$$\tau = \mathrm{ord}(v) = \frac{|v|}{p-q} = \frac{|w|-q}{p-q} = \frac{\rho p - q}{p-q}.$$

By the preceding inequality, $\tau > \rho + 1$, i.e., $v = z^\tau$, with $\tau > \rho + 1$. ∎

REMARK 8.1.13. The number φ is tight in Theorem 8.1.12. Recall from Section 1.2 that $(F_n)_{n \geq 1}$ denotes the sequence of Fibonacci numbers and that the sequence of Fibonacci words $(f_n)_{n \geq 1}$ is defined by the inductive rules $f_1 = b$, $f_2 = a$, and $f_{n+1} = f_n f_{n-1}$.

For any $n \geq 1$ consider the word v_{n+2} defined as the prefix of length $|f_{n+2}| - 2$ of f_{n+2}, i.e.,

$$v_{n+2} xy = f_{n+2},$$

where x, y are letters, $x \neq y$. It is known that v_{n+2} has periods F_n and F_{n+1}. Then v_{n+2} can be written

$$v_{n+2} = f_n^{\rho_n} = f_{n+1}^{\rho'_n},$$

where

$$\rho_n = \frac{F_{n+2} - 2}{F_n}, \quad \rho'_n = \frac{F_{n+2} - 2}{F_{n+1}},$$

and $\rho'_n < \varphi$.

One can prove that there does not exist any prefix (or suffix) of v_{n+2} having order $\geq \varphi + 1$ (Problem 8.2.6). The tightness of Theorem 8.1.12 is a consequence of the well-known property of Fibonacci numbers stating that, for any $\epsilon > 0$, there exists an integer m such that for any $n \geq m$

$$\varphi - \epsilon < \rho'_n.$$

Also notice that the statement of Theorem 8.1.12 is sharp in the sense that the prefix (or suffix) of w of the form z^τ, with $\tau \geq \rho + 1$, cannot in general coincide with the whole word w. Indeed the word

$$w = a^6 b a^6$$

has periods 7 and 8 and it can be written

$$w = (a^6 b)^\rho = (a^6 b a)^\sigma,$$

with $\sigma = 13/8 = 1.625$ and $\rho = 13/7$, i.e., $\varphi < \sigma < \rho$. The word w satisfies the hypothesis of the lemma and it has as a prefix (and a suffix) the word a^6 according to the lemma. However, w cannot be written as $w = z^\gamma$ with $\gamma \geq \sigma + 1$.

8.1. Periods in a finite word

8.1.5. Squares in a word

In this subsection we study squares that appear as prefixes or as factors in words. Recall that a *square* is a word w of the form $w = v^2$.

The occurrence of a square as a factor in a word can be considered as a particular kind of *local period* occurring in the word. A more general notion of local period will be discussed in Section 8.2. Here we study the squares that can appear as prefixes or as factors of a word.

The first result we prove states a fundamental inequality on the lengths of different squares that can occur as prefixes of the same word. It is used in the second result and also in Chapter 12, Subsection 12.1.7.

The second result gives a bound on the number of squares occurring as factors of a word.

In what follows we write $w < w'$ to denote that the word w is a prefix of w'. Recall also that a word w is called primitive if it is not a power of another word, i.e., there exists no word z such that $w = z^k$ for some integer k greater than 1.

LEMMA 8.1.14. *Let u, v, w be words such that u is primitive, $v \notin u^*$, and $uu < vv < ww$. Then $|u| + |v| \leq |w|$.*

Proof. By contradiction we assume that $|v| < |w| < |vu|$. First assume that $2|u| \leq |v|$. Then we have $uu \leq v < w$. Since $vv < ww$, the (second) word w has a prefix ru, where $1 \leq |r| < |u|$. Hence u is a internal factor of uu, which is impossible since u is a primitive word (see Problem 8.1.6).

Therefore we may assume that $u < v < uu$. Let $v = uz$. Then we have $zu = uy$ for some nonempty word y. Let x be the primitive root of z. Then we can write $z = x^\beta$ and $u = x^\alpha r$ for some integers $\alpha \geq \beta \geq 1$ and some nonempty word $r < x$. Hence we have $v = x^\alpha r x^\beta < w$ and $sx^\beta < w$, where s is a proper suffix of $x^\alpha r$. If $|s| \geq |x|$, then $x \leq s$. Hence $|s| \geq |x|$ implies $s = x^\gamma r$ for some integer $\gamma < \alpha$. Therefore $x^\gamma r x < x^{\alpha+1}$, which is impossible since $1 \leq |r| < |x|$. Thus, s is a proper prefix of x. If $|s| \leq |x^{\alpha-1}r|$, then $sx \leq x^\alpha r < x^{\alpha+1}$, which again is a contradiction. Therefore we may assume that $\alpha = \beta = 1$ and $|r| < |s|$. We have $u = xr$, $v = xrx$, and $w = xrxt$ for some t such that $ts = xr$. In particular, $1 < t < x$. Hence, $txr < xrx < xrxr$ and this means $tu < uu$, which is impossible since u is primitive. ∎

REMARK 8.1.15. Notice that the statement of preceding lemma is sharp. Let

$$u = abbab, \quad v = abbababb, \quad \text{and} \quad w = abbababbabbab.$$

The lengths are 5, 8, and 13. The words u, v, w are primitive and $uu < vv < ww$. The general scheme for such an example is

$$u = pr,\ v = prp,\ \text{and}\ w = prppr$$

where p and r are words such that $r < p$.

We now study how many squares a word can contain.

Let $w = a_1 \cdots a_n$. For $i = 1, \ldots, n$, let $s_i(w)$ be the number of squares that are prefixes of $a_i \cdots a_n$ and never appear as prefixes of $a_j \cdots a_n$, with $i < j \leq n$. For example in the word $w = abaababaababa$ we have $s_3(w) = 1$ since $aababaabab$ is a square beginning in position 3 which does not appear later on. The square aa also begins in position 3 but it also appears in position 8 and so it does not affect the value of $s_3(w)$.

THEOREM 8.1.16. *For every nonempty word x of length n, $s_i(x) \leq 2$ for all $i \in \{1, \ldots, n\}$.*

Proof. Suppose, by contradiction, that there exists a word x of length n such that $s_i(x) \geq 3$, for some $i \in \{1, \ldots, n\}$.

Then $x_i = a_i a_{i+1} \cdots a_n$ has three prefixes u^2, v^2, w^2, $uu < vv < ww$, which do not occur elsewhere in the word x. Let $p = |w|, q = |v|$ and $r = |u|$. If $p \geq 2r$, then u^2 occurs again at position $i + p$, because w occurs there, which is impossible. Thus $p < 2r$, and we have that $p > q > r > p/2$ and also $q < p < 2r < q + r < 2q$. It follows from Lemma 8.1.14 that u is nonprimitive.

Therefore there exists a primitive word y such that $u = y^k$ for some integer $k \geq 2$. If we set $r_1 = |y|$, then $r = kr_1$. Now we have that $yy < uu < vv$ with y primitive, so by Lemma 8.1.14 that $r_1 + q \leq p$. Note that r_1 is also a period of u^2 and, since $p < 2r$, w is a prefix of u^2 and so r_1 is a period of w. Since $p < 2q$, w is a prefix of v^2 and so q is also a period of w. Since $r_1 + q \leq p$ we can apply Theorem 8.1.4 and obtain that $\gcd(p, q) = d$ is also a period of w. Since the word y is primitive and d divides r_1, we have $d = r_1$. Hence r_1 divides q. Now $r = kr_1$, and, consequently, $q = (k + h)r_1$ for some integer $h \geq 1$. Therefore v^2 has length $2(t + s)r_1$ and u^2 has length $2tr_1$. It follows that u^2 also appears at position $r_1 + 1$, which is a contradiction. ∎

Notice that, since there are no squares beginning at the last position, $s_n(x) = 0$ for any word x of length n.

For any word $x = a_1 \cdots a_n$ let us denote by $SQ(x)$ the cardinality of the set of squares that are factors of x. It is easy to verify by definition and by the preceding remark that $SQ(x) = \sum_{i=1}^{n-1} s_i(x)$.

8.2. Local versus global periodicity

Hence, we have the following corollary of Theorem 8.1.16.

COROLLARY 8.1.17. *For any word x of length n, $SQ(x) \leq 2n - 2$.*

8.2. Local versus global periodicity

In this section we investigate the relationships between local and global periodicity of words. In Subsection 8.1.5 we have taken into account a particular kind of local period, i.e., squares occurring as factors in a word. Here we introduce a very general notion of local period in terms of *repetitions*. A repetition occurring in a word w is not in general a factor of w, nor it is necessarily a square, but its order can be an arbitrary rational number ρ. Moreover the repetition is here referred to a "point" of the word w and it is important to consider the relative positions of the repetitions and that of the point of the word w at which the repetition is detected.

In order to give the formal definitions, we first introduce the notion of pointed word. This is the appropriate notion to define local properties of a word.

Let $w = a_1 a_2 \cdots a_n$ be a word over the alphabet A. A *pointed word* is a pair $(a_1 \cdots a_i, a_{i+1} \cdots a_n)$, $1 \leq i < n$. The pointed word is also denoted by (w, i) and we refer to (w, i) as the word w at the point (or the position) i.

Let (x, y) be a pair of words. The pair (x, y) *matches* the pointed word (w, i), or simply matches the word w at the point i, if

$$A^* x \cap A^* a_1 \cdots a_i \neq \emptyset \quad \text{and} \quad y A^* \cap a_{i+1} \cdots a_n A^* \neq \emptyset.$$

Notice that the word $z = xy$ is not in general a factor of the word w and that the pair (x, y) specifies the relative position of the word z and the point i. So we can distinguish between *central repetitions* and *left* (or right) *repetitions*.

8.2.1. Central repetitions

A word w contains a *repetition* of order ρ having as *center* the point (or position) i, or shortly a *central repetition* of order ρ at the point i, if there exist a nonempty word z of order $\text{ord}(z) = \rho$ and a factorization $z = xy$, with $|x| = |y|$, such that the pair (x, y) matches w at the point i. This means that the point i is *central* with respect to the repetition z. The word z is called a central repetition of (w, i) and must have even length. This central repetition is *proper* (or *internal*) if x is a suffix of $a_1 \cdots a_i$ and

y is a prefix of $a_{i+1}\cdots a_n$. It is *left external* if $a_1\cdots a_i$ is a proper suffix of x. It is *right external* if $a_{i+1}\cdots a_n$ is a proper prefix of y.

Central repetitions of order 2 play an important role in this theory. By definition, a central repetition of order 2 at the point i of w is a word z of the form $z = x^2$ such that the pair (x, x) matches w at the point i. We say that w has a *square* having its center in the position i.

EXAMPLE 8.2.1. Given the word

$$w = abaababaabaaba$$

the pointed word $(w, 8)$ is the pair

$$(abaababa, abaaba).$$

The pair (aba, aba) matches the pointed word $(w, 8)$, and so the word *abaaba* is a central repetition of w at the point 8. It has order 2 and period 3. At the point 8 of w there is another central repetition of order 2, or a square having its center at it. It is the word aa and it has period 1. Both these repetitions are proper. The pointed word $(w, 7)$ is the pair

$$(abaabab, aabaaba).$$

The word *aabaababaabaabab* is a central repetition of $(w, 7)$ of order 2 and period 8. It is both left and right external. Since the pair $(abab, aaba)$ matches w at the point 7, the word *ababaaba* is a proper central repetition of $(w, 7)$ of order 1.6 and period 5.

As shown in the preceding example, a word can have different central repetitions of the same order at a given point. We are interested, for a given order, in detecting the central repetition of minimal period. This leads to the notion of minimal central repetition and of central local period.

For any real $\alpha > 1$, $c_\alpha(w, i)$ denotes the *central local period* (of order α) of the pointed word (w, i),

$$c_\alpha(w, i) = \min\{p(z) \mid z \text{ is a central repetition of } (w, i) \text{ of order } \geq \alpha\}.$$

The central repetition z of (w, i) such that $p(z) = c_\alpha(w, i)$ is called the *minimal* central repetition (of order α) of w at the point i.

It is immediate to verify that, if $\alpha < \beta$, then $c_\alpha(w, i) \leq c_\beta(w, i)$ and that for any α and any $i \geq 1$, $c_\alpha(w, i) \leq p(w)$.

In the special case $\alpha = 2$ one has

$$c_2(w, i) = \min\{|x| \mid x \neq \varepsilon \text{ and } (x, x) \text{ matches } w \text{ in the position } i\}.$$

8.2. Local versus global periodicity

EXAMPLE 8.2.2. $w = abaababaabaaba$

w	a	b	a	a	b	a	b	a	a	b	a	a	b	a
i		1	2	3	4	5	6	7	8	9	10	11	12	13
$c_2(w,i)$		2	3	1	5	2	2	8	1	3	3	1	3	2
$c_{1.6}(w,i)$		2	3	1	5	2	2	5	1	3	3	1	3	2

We denote by $P_\alpha(w)$ the maximum of the central local periods of order α) of w:

$$P_\alpha(w) = \max\{c_\alpha(w,i) \mid 1 \leq i < |w|\}.$$

A point (or position) i is *critical* if $c_\alpha(w,i) = P_\alpha(w)$: We denote by $C_\alpha(w)$ the set of critical points of w:

$$C_\alpha(w) = \{i \mid 1 \leq i < |w| \text{ and } c_\alpha(w,i) = P_\alpha(w)\}.$$

We denote further by $Z_\alpha(w)$ and $S_\alpha(w)$ the minimum and the maximum respectively of the critical points:

$$Z_\alpha(w) = \min C_\alpha(w), \quad S_\alpha(w) = \max C_\alpha(w).$$

EXAMPLE 8.2.2 (*continued*). For $w = abaababaabaaba$, $P_2(w) = 8$, $C_2(w) = \{7\}$, $Z_2(w) = S_2(w) = 7$, $P_{1.6}(w) = 5$, $C_{1.6}(w) = \{4,7\}$, $Z_{1.6}(w) = 4$, $S_{1.6}(w) = 7$.

Notice that the notion of critical point introduced in this chapter slightly differs from the one used in the literature, where a critical point i usually denotes a position where the local period of order 2, $c_2(w,i)$, is equal to the global period $p(w)$. The difference is motivated by the fact that we here take into account also repetitions of an arbitrary order $\alpha > 1$.

It is easy to verify that $c_\alpha(w,i) \leq p(w)$ for $\alpha > 1$ and $i = 1,\ldots,|w|-1$, i.e., the central local periods are smaller than or equal to the period. On the other hand, if α is sufficiently large, i.e., $\alpha \geq 2|w|$, it is possible to prove that $c_\alpha(w,i) = p(w)$ for all i, as stated in particular in the next proposition.

PROPOSITION 8.2.3. *Let* $k = \lceil \alpha/2 \rceil$. *If the period of w is smaller than or equal to k then in every position of i, one has $c_\alpha(w,i) = p(w)$. Hence, if $\alpha \geq 2|w|$ then every position is critical of order α.*

Proof. Let i, $1 \leq i < |w|$, be a position in $|w|$. If the central repetition of order α at the point i is both left and right external then $c_\alpha(w,i)$ is also a period of w and, consequently, $p(w) \leq c_\alpha(w,i)$ and, by the preceding remark, the thesis follows.

Suppose now that the central repetition of order α at the point i is either left or right internal or both. Suppose that it is left internal. We claim that $c_\alpha(w,i)$ divides $p(w)$. Indeed if $c_\alpha(w,i) = 1$ there is nothing to prove. Suppose that $c_\alpha(w,i) > 1$. In this case the part v of the central repetition of order α at the point i that is at the left of the point i has, by hypothesis, length greater than or equal to $2p(w)$. Then v has periods $p(w)$ and $c_\alpha(w,i)$. We can apply the theorem of Fine and Wilf and obtain that it has period $d = \gcd(p(w), c_\alpha(w,i))$. But d cannot properly divide $c_\alpha(w,i)$ by the minimality of $c_\alpha(w,i)$. Hence $d = c_\alpha(w,i)$ and the claim is proved. We can now apply Lemma 8.1.3 and obtain that $c_\alpha(w,i)$ is also a period of w and, consequently, $p(w) \leq c_\alpha(w,i)$ and, by the preceding remark, the thesis follows. ∎

The critical factorization theorem in particular states that for $\alpha = 2$ there exists at least one point such that the central local period detected at this point coincides with the (global) period of the word, i.e., there exists an integer j, $1 \leq j < |w|$, such that $c_{2(w,j)} = p(w)$.

An important step in the proof of the critical factorization theorem is the following proposition.

PROPOSITION 8.2.4. *If $z = x^2$ is the square of minimal length having its center at position j of w, $1 \leq j < |w|$, then x is unbordered.*

Proof. If there exists a nonempty border t of x, i.e., t is both a prefix and a suffix of x, then t^2 is a square having its center at the position j of w that is shorter than x^2, contradicting the definition of x. ∎

THEOREM 8.2.5 (Critical factorization theorem). *Let w be a word having length $|w| \geq 2$. In every sequence of $l \geq \max\{1, p(w) - 1\}$ consecutive positions there is a critical one and, moreover, $P_2(w) = p(w)$.*

Proof. The proof is by induction on $P_2(w)$. Suppose that $P_2(w) = 1$. Since for all natural numbers i, $1 \leq i < |w|$, $c_{2(w,i)} = 1$, one has $a_i = a_{i+1}$. If $a = a_1$ and $n = |w|$, then w is of the form $w = a^n$, $p(w) = 1 = P_2(w)$ and all positions are critical.

Let us suppose that the statement of the proposition holds true for all words w' such that $P_2(w') \leq k - 1$, $k > 1$. Let w be a word having $P_2(w) = k$. We prove the following properties.
  (i) If j is a critical position and $j+1, \ldots, j+l$ are not critical then
      $P_2(w) > l + 1$.

8.2. Local versus global periodicity

(ii) If j is a critical position and $j-l,\ldots,j-1$ are not critical then $P_2(w) > l+1$.

As an immediate consequence of the preceding two properties one has the following.

(iii) Every sequence of at least $P_2(w) - 1$ consecutive positions contains a critical one.

Let us recall that for any position j of w one has $c_2(w,j) \le p(w)$, and so $P_2(w) \le p(w)$. Hence property (iii) implies the first part of the theorem.

In order to prove (i) let us consider the word $u = a_{j+1} \cdots a_{j+l}a_{j+l+1}$. Since any central repetition at point $j+i$ of w is a repetition having its center at the point i of u one has

$$c_2(u,i) \le c_2(w, j+i), \quad i = 1,\ldots, l.$$

Since no position $j+i$ of w, with $i = 1,\ldots, l$, is a critical position, one has that

$$c_2(u,i) < k, \quad i = 1,\ldots, l.$$

As a consequence, $c_2(u,i) < k$ for $i = 1,\ldots,l$, and then $P_2(u) < k$. By inductive hypothesis $p(u) = P_2(u) < k$.

Let $z = x^2$ be the square of minimal length having its center at position j of w. Since by hypothesis position j is critical, one has that $c_2(w,j) = P_2(w) = k$, and $|x| = k$.

By Proposition 8.2.4 the word x is unbordered. If x is a prefix of the word $u = a_{j+1} \cdots a_{j+l}a_{j+l+1}$ then $p(x) \le p(u) < k$, which is a contradiction. Hence u is a proper prefix of x and, consequently, $P_2(w) = k = |x| > |u| = l+1$.

The proof of (ii) is analogous by taking $u = a_{j-l} \cdots a_{j-1}$. In order to complete the proof of the theorem we must prove that

(iv) $p(w) = P_2(w)$.

As noticed above, we have always that $P_2(w) \le p(w)$. It remains to prove that $P_2(w) \ge p(w)$. Let i be a position such that $1 \le i < i+P_2(w) \le |w|$. By property (iii) there exists a critical position j in the set $\{i,\ldots, i+P_2(w) - 1\}$. There exists then a square x^2 having its center at position j with $|x| = P_2(w)$. Note that $a_i \cdots a_{i+P_2(w)}$ is a factor of x^2, and, consequently, $a_i = a_{i+P_2(w)}$. Therefore $P_2(w)$ is a period of w and then $P_2(w) \ge p(w)$, and this concludes the proof. ∎

COROLLARY 8.2.6. *Let w be a word of length $|w| \ge 2$ and $p(w) > 1$. We have that $Z_2(w) < P_2(w)$, i.e., the central repetition at point $Z_2(w)$ is left external. We have also that $|w| - P_2(w) < S_2(w)$, i.e., the central repetition at point $S_2(w)$ is right external.*

COROLLARY 8.2.7. *Let $w = a_1 \cdots a_n$, be a word of length n. Given i, j, $1 \le i < j \le n$, if $c_2(w, h) < c_2(w, j)$ for any h such that $i \le h < j$, then $c_2(w, j) > j - i + 1$.*

Proof. Let $v = a_i \cdots a_j$. One has that $c_2(v, h) \le c_2(w, h) < c_2(w, j)$, for $i \le h < j$. According to Theorem 8.2.5 we have that $p(v) < c_2(w, j)$. Let u^2 be the square of length $2c_2(w, j)$ having its center at position j of w. According to Proposition 8.2.4, u is an unbordered word. Hence u cannot be a suffix of v longer than $p(v)$. Therefore v is a proper suffix of u and $|u| = c_2(w, j) > j - i + 1$. ∎

In Example 8.2.2, $P_2(w)$ is, according to Theorem 8.2.5, exactly the period of w. Moreover the unique critical point of w is 7 and it satisfies the conditions (iii) and (iv) in the proof of the theorem. The same example shows that the theorem does not hold true for $\alpha = 1.6$. Indeed $P_{1.6}(w) = 5 \ne p(w) = 8$. The following example shows that the value $\alpha = 2$ is tight.

EXAMPLE 8.2.8. For any $\epsilon > 0$, consider the word $ba^{m-1}ba^m b$, with m such that $2m/(m+1) \ge 2 - \epsilon$. The unique critical point of order 2 is the point $m+1$, corresponding to the pair $(ba^{m-1}b, a^m b)$. The minimal central repetition of order 2 at such a point is the word $a^m ba^{m-1} ba^m ba^{m-1} b$, which has period $2m + 1$, according to the critical factorization theorem. However, the minimal central repetition of order $2 - \epsilon$ at the same point is the word $u = a^{m-1}ba^m$. Indeed u has period $m+1$ and order $2m/(m+1) > 2 - \epsilon$. It is easy to verify that such a point is also a critical point of order $2 - \epsilon$, and then

$$P_{2-\epsilon}(ba^{m-1}ba^m b) = m + 1 \ne p(ba^{m-1}ba^m b) = 2m + 1.$$

Statements (ii), (iii) and (iv) in the proof of Theorem 8.2.5 as given are sharp. Indeed the word $a^m ba^m ba^m$, $m \ge 1$, has period $m + 1$ and exactly four critical points, $m, m+1, 2m+1$ and $2m+2$, corresponding to the pairs $(a^m, ba^m ba^m)$, $(a^m b, a^m ba^m)$, $(a^m ba^m, ba^m)$ and $(a^m ba^m b, a^m)$ respectively.

8.2.2. Left repetitions

In the previous subsection we considered central repetitions, i.e., we required the repetition occurring at the point i of a word w to be such that this point is the center of the repetition

In this subsection we take into account a new notion of repetition, in which we require that the repetition occurs at a given point "immediately to the *left* of that point". The symmetric case of repetitions occurring

8.2. Local versus global periodicity

"immediately to the right of a given point" is similar and it is not explicitly considered here.

For technical reasons it is convenient, in the case of left repetitions, to change a bit the definition of a pointed word (w, i) and to allow that the positive integer i ranges from 1 to $|w|$ (in the case of central repetitions i ranges from 1 to $|w| - 1$).

A word $w = a_1 \cdots a_n$ contains a *left repetition* of order p at the point i ($1 \leq i \leq n$), if there exists a word z of order $\text{ord}(z) = p$ such that

$$A^* z \cap A^* a_1 \cdots a_i \neq \emptyset.$$

The word z is called a left repetition of (w, i). It is *external* if $a_1 \cdots a_i$ is a proper suffix of z. It is *proper* (or *internal*) if z is a suffix of $a_1 \cdots a_i$.

EXAMPLE 8.2.9. Given the word

$$w = abaababaabaaba,$$

aabaab is an external left repetition of $(w, 5)$. It has order 2 and period 3. The words *aa* and *baabaa* are both proper left repetitions of $(w, 12)$ of order 2 and periods 1 and 3 respectively. The word *abab* is a proper left repetition of $(w, 7)$ of order 2 and period 2, whereas the word *ababaababaabab* is an external left repetition of $(w, 7)$ of order 2.8 and period 5.

The preceding example shows that a word can have different left repetitions of the same order at a given point. As in the case of central repetitions, we are interested, for a given order, to detect the left repetition of minimal period.

For any real $\alpha > 1$, $l_\alpha(w, i)$ denotes the *left local period* (of order α) of the pointed word (w, i):

$$l_\alpha(w, i) = \min\{p(z) \mid z \text{ is a left repetition of } (w, i) \text{ of order } \geq \alpha\}.$$

The left repetition z of (w, i) such that $p(z) = l_\alpha(w, i)$ is called the *minimal left repetition* (of order α) of w at the point i.

It is immediate to verify that, if $\alpha < \beta$, then $l_\alpha(w, i) \leq l_\beta(w, i)$.

EXAMPLE 8.2.10. $w = abaababaabaaba$

w	a	b	a	a	b	a	b	a	a	b	a	a	b	a
i	1	2	3	4	5	6	7	8	9	10	11	12	13	14
$\ell_2(w, i)$	1	2	2	1	3	3	2	2	1	5	3	1	3	3
$\ell_{2.65}(w, i)$	1	2	2	3	3	3	5	5	5	5	5	8	3	3

We denote by $Q_\alpha(w)$ the maximum of the left local periods (of order α) of w:

$$Q_\alpha(w) = \max\{l_\alpha(w,i) \mid 1 \le i \le |w|\}.$$

A point (or position) i is (left) *critical* if $l_\alpha(w,i) = Q_\alpha(w)$. We denote by $K_\alpha(w)$ the set of (left) critical points of w:

$$K_\alpha(w) = \{i \mid 1 \le i \le |w| \text{ and } l_\alpha(w,i) = Q_\alpha(w)\}.$$

We denote further by $T_\alpha(w)$ and $R_\alpha(w)$ the minimum and the maximum of the (left) critical points:

$$T_\alpha(w) = \min K_\alpha(w), \quad R_\alpha(w) = \max K_\alpha(w).$$

EXAMPLE 8.2.10 (*continued*). For $w = abaababaabaaba$, $Q_2(w) = 5$, $K_2(w) = \{10\}$, $T_2(w) = R_2(w) = 10$, $Q_{2.65}(w) = 8$, $K_{2.65}(w) = \{12\}$, $T_{2.65}(w) = R_{2.65}(w) = 12$.

It is easy to verify that $l_\alpha(w,i) \le p(a_1 \cdots a_i) \le p(w)$ for $\alpha > 1$ and $i = 1,\ldots,|w|$, i.e., the left local periods are smaller than or equal to the period.

Contrary to the case of central repetitions, it is not possible, for left repetitions, to determine a fixed value of the parameter α (not depending on the length of the word w) such that the period of w coincides with the left local period of order α detected at some point i of w. The following example illustrates this fact and the differences between central local periods and left local periods.

EXAMPLE 8.2.11. Let $\alpha \ge 2$ be a real number and let $w = ba^t$, with $t > \alpha$.

The period of w is $p(w) = t+1$. Let $k = \lceil \alpha \rceil$ be the smallest integer greater than or equal to α. The following table gives the values of $c_\alpha(w,i)$ and $l_\alpha(w,i)$ respectively.

w	b	a	a	a	$\cdots$	a	a	a	a	$\cdots$	a	a	a
i	1	2	3		$\cdots$	$k-1$	k	$k+1$		$\cdots$	$t-1$	t	$t+1$
$c_\alpha(w,i)$	$t+1$		$\cdots$			$t+1$	1	1	1	$\cdots$		1	1
$l_\alpha(w,i)$		1	2	3	$\cdots$	$k-1$	1	1	1	$\cdots$	1	1	1

In the preceding example $P_\alpha(w) = t+1 = p(w)$, according to the critical factorization theorem, but $Q_\alpha(w) = k-1 \neq p(w)$.

In spite of this, the main theorem of this subsection states that, for a suitable value of the parameter α, $Q_\alpha(w)$ is equal to the period of

8.2. Local versus global periodicity

the prefix of w of length $R_\alpha(w)$. For instance, in the preceding example, $R_\alpha(w) = k-1$ and the prefix of w of length $R_\alpha(w)$ is ba^{k-2}. Its period is $k-1$ and it coincides with $Q_\alpha(w)$. Notice that $\alpha = 2$ does not suffice to establish this relationship. Indeed consider the word $w' = aw$, where w is the word in Example 8.2.2. The values of the function $l_2(w', i)$ are given in the following table.

w'	a	a	b	a	a	b	a	b	a	a	b	a	a	b	a
i	1	2	3	4	5	6	7	8	9	10	11	12	13	14	15
$l_2(w', i)$	1	2	3	3	1	3	3	2	2	1	5	3	1	3	3

$Q_2(w') = 5$ and the prefix of w' of length $R_2(w') = 11$ has period 8, as one can easily verify.

The parameter required for the main theorem of this subsection is related to the golden ratio (see Subsection 8.1.4).

The following theorem states in particular that for $\alpha = \varphi^2 = \varphi + 1 = 2.618\ldots$ the maximal values of local periods $Q_\alpha(w)$ coincide with the global period of the prefix of w of length $R_\alpha(w)$. For convenience of notation the subscript φ^2 will often be omitted. So by $l(w, i), K(w), T(w), R(w)$ we will refer to $l_{\varphi^2}(w, i), K_{\varphi^2}(w), T_{\varphi^2}(w), R_{\varphi^2}(w)$ respectively. In the same way, in what follows, by left repetitions we will refer to left repetitions of order φ^2.

A fundamental step in the proof of the next theorem is given by the following lemma, which uses Theorem 8.1.12 and explains the role of the golden ratio φ in the result.

LEMMA 8.2.12. *If the minimal left repetition at the point i is proper, then* $l(w, i - l(w, i)) \geq l(w, i)$.

Proof. Let z be the minimal repetition at the point i of order $\geq \varphi^2$. Then $z = x^\gamma$, with $|x| = p(z)$ and γ, the order of z, a rational number greater than φ^2. By definition, $l(w, i) = |x|$.

Since this repetition is proper, z is a suffix of $a_1 \cdots a_i$. Consider the word $a_1 \cdots a_j$, with $j = i - |x| = i - l(w, i)$. Since x^γ is a suffix of $a_1 \cdots a_i$, $x^{\gamma - 1}$ is a suffix of $a_1 \cdots a_j$. From the inequality $\gamma > \varphi^2 = \varphi + 1$, it follows that $\gamma - 1 = \rho > \varphi$.

Let t be the minimal left repetition at the point j of order $\geq \varphi^2$. By definition
$$l(w, i - l(w, i)) = l(w, j) = p(t).$$
Since x^ρ is a suffix of $a_1 \cdots a_j$, either t is a suffix of x^ρ or x^ρ is a suffix of t.

If t is a suffix of x^ρ, then t is a *proper* suffix of $x^{\rho+1} = x^\gamma = z$. Hence t is also a suffix of $a_1 \cdots a_i$, i.e., t is a left repetition at the point i of order $\geq \varphi^2$, with $p(t) < p(z)$. This contradicts the minimality of z. Then x^ρ is a suffix of t.

Assume now that $l(w, i - l(w, i)) < l(w, i)$. Then $p(t) < p(z) = |x|$ and the word x^ρ can also be written $x^\rho = y^\sigma$, for some word y and some rational σ, with $|y| = p(t) < |x|$, and then $\sigma > \rho > \varphi$. The word $x^\rho = y^\sigma$ satisfies the conditions of Theorem 8.1.12. It follows that x^ρ has a suffix of the form u^τ, with $0 < |u| < |x|$ and $\tau > \varphi^2$, contradicting the hypothesis that $z = x^\gamma = x^{\rho+1}$ is the minimal left repetition at the point i of order $\geq \varphi^2$. ∎

THEOREM 8.2.13. *Let* $w = a_1 a_2 \cdots a_n$ *be a nonempty word. One has:*
  (i) $p(a_1 \cdots a_{R(w)}) = Q(w)$.
  (ii) *If* r, s ($r < s$) *are consecutive elements of* $K(w)$, *then* $s - r \leq Q(w)$.
  (iii) $T(w) < \varphi^2 Q(w)$.

Proof. Let us first prove (iii), which states that $T(w) < \varphi^2 Q(w)$, i.e., that the minimal repetition at the point $T(w)$ is external. Indeed, if we assume that this repetition is proper, then, by Lemma 8.2.12,

$$l(w, T(w) - Q(w)) \geq l(w, T(w)) = Q(w),$$

contradicting the fact that $T(w)$ is the least critical point.

Let us now prove (ii). Let us consider two consecutive elements r, s of $K(w)$, with $r < s$. If the minimal left repetition at the point s is proper, then, by Lemma 8.2.12, $s - Q(w)$ is also an element of $K(w)$. Since r and s are consecutive elements of $K(w)$, one has $r \geq s - Q(w)$, i.e., $s - r \leq Q(w)$.

In the opposite case, i.e., if the minimal left repetition at the point s is left external, the minimal left repetition at the point r is also left external. One has that $p(a_1 \cdots a_s) = l(w, s) = Q(w)$. Indeed, by definition, $l(w, s) \leq p(a_1 \cdots a_s)$. Moreover, if the minimal left repetition at the position s is external, then $l(w, s)$ is a period of $a_1 \cdots a_s$, i.e., $p(a_1 \cdots a_s) \leq l(w, s)$. Analogously, one has the same result for position r, i.e., $p(a_1 \cdots a_r) = l(w, r) = Q(w)$.

Let us suppose, by contradiction, that $s - r > Q(w)$. Then the word $a_1 \cdots a_{r+Q(w)}$ is a prefix of $a_1 \cdots a_s$ and it has as prefix the word $a_1 \cdots a_r$. Hence $Q(w) = p(a_1 \cdots a_r) \leq p(a_1 \cdots a_{r+Q(w)}) \leq p(a_1 \cdots a_s) = Q(w)$, i.e., the word $a_1 \cdots a_{r+Q(w)}$ has period $Q(w)$. It follows that $a_1 \cdots a_r$ is a suffix of $a_1 \cdots a_{r+Q(w)}$ and then $l(w, r) \leq l(w, r + Q(w))$. Since $l(w, r) = Q(w)$, one has $r + Q(w) \in K(w)$. This contradicts the hypothesis that r, s are consecutive elements of $K(w)$ and $s - r > Q(w)$.

8.2. Local versus global periodicity

Let us finally prove statement (i). The proof is by induction on the integer $m = \text{Card}(K(w))$. We first prove the statement for $m = 1$. In this case $T(w) = R(w)$ and, as a consequence of (iii), the minimal left repetition at the point $T(w)$ is left external, i.e., $p(a_1 \cdots a_{R(w)}) = Q(w)$.

We now have to prove the inductive step. Let us suppose that the statement (i) is true for all words v such that $\text{Card}(K(v)) = m$, and consider a word $w = a_1 \cdots a_n$ such that $\text{Card}(K(w)) = m + 1$. Let

$$\hat{R}(w) = \max\{i \mid i \in K(w) \text{ and } i < R(w)\}$$

denote the greatest among the critical points of w smaller than $R(w)$. By the inductive hypothesis

$$p(a_1 \cdots a_{\hat{R}(w)}) = Q(w).$$

By statement (ii), $R(w) - \hat{R}(w) \le Q(w)$. On the other hand, since the minimal left repetition of order $\ge \varphi^2$ at the point $R(w)$ has period $Q(w)$, we have that $Q(w)$ is a period of $a_{\hat{R}(w)-Q(w)+1} \cdots a_{R(w)}$. Set

$$u_1 = a_1 \cdots a_{\hat{R}(w)-Q(w)},$$
$$u_2 = a_{\hat{R}(w)-Q(w)+1} \cdots a_{\hat{R}(w)},$$
$$u_3 = a_{\hat{R}(w)+1} \cdots a_{R(w)}.$$

We know that $u_1 u_2$ and $u_2 u_3$ have period $Q(w)$ and that $|u_2| = Q(w)$. By Lemma 8.1.2, $u_1 u_2 u_3 = a_1 \cdots a_{R(w)}$ has period $Q(w)$. This concludes the proof of the theorem. ∎

REMARK 8.2.14. Notice that, in the case of central repetitions, the smallest critical point $Z_2(w)$ is bounded above by $P_2(w)$, whereas, for left repetitions, the smallest critical point $T(w)$ is bounded above by $Q(w)$ times φ^2. The following example shows that $Q(w)$ is not an upper bound for $T(w)$. Consider the word $w = a^{m-1}ba^mba^mb$ ($m > 3$). The least element of $K(w)$ is $T(w) = 2m + 1$ whereas the maximal local left period is $Q(w) = m + 1$. The last two critical positions of w are $2m + 3$ and $3m + 2$. Their distance is $m - 1 = Q(w) - 2$ and this is the best possible. Indeed it is possible to improve statement (ii) of Theorem 8.2.13 by proving the following tight one. In every sequence of $d \ge \max\{1, Q(w) - 2\}$ consecutive positions between $T(w)$ and $R(w)$ there is a left critical one (of order φ^2) (Problem 8.2.10).

REMARK 8.2.15. The constant φ^2 is tight in Theorem 8.2.13. Indeed, for any $\epsilon > 0$ it is possible to prove that there exist $\bar{n}$ and a constant $D(\epsilon)$

such that for any $n > \bar{n}$ the maximum $Q_{\varphi^2-\epsilon}(f_n)$ of the left local periods of order $\varphi^2 - \epsilon$ of the n-th Fibonacci word (defined in Remark 8.1.13) is smaller than $D(\epsilon)$. Moreover, for any ϵ, the sequence of the maximum of left critical points in f_n $(R_{\varphi^2-\epsilon}(f_n))_{n\in\mathbb{N}}$ is not bounded. Let w_n be the prefix of f_n of length $R_{\varphi^2-\epsilon}(f_n)$. The sequence of periods of w_n, $(p(w_n))_{n\in\mathbb{N}}$ is not bounded (Problem 8.2.11).

8.3. Infinite words

In this section we will consider applications of the results of the previous section to the case of one-sided and two-sided infinite words. In particular, characterizations of recurrent, periodic and eventually periodic infinite words are given.

8.3.1. Recurrence and periodicity

Recall from Section 1.2 that a one-sided infinite word $x = x_0 x_1 \cdots$ is *periodic* if there exists a positive integer p such that $x_i = x_{i+p}$, for all $i \in \mathbb{N}$. The smallest p satisfying the preceding condition is called the period of x.

A one-sided infinite word $x = x_0 x_1 \cdots$ is *eventually periodic* if there exist two positive integers k, p such that $x_i = x_{i+p}$, for all $i \geq k$. An infinite word is *aperiodic* if it is not eventually periodic.

A one-sided infinite word x is *recurrent* if any factor occurring in x has an infinite number of occurrences.

Notice that a one-sided infinite word is periodic if and only if it is recurrent and eventually periodic.

A two-sided infinite word $x = \cdots x_{-1} x_0 x_1 \cdots$ is *periodic* if there exists a positive integer p such that $x_i = x_{i+p}$, for all $i \in \mathbb{Z}$. The smallest p satisfying the preceding condition is called the period of x.

Concerning the notion of local period, the definitions of the previous sections extend to one-sided and to two-sided infinite words but there are some natural differences.

In the case of a one-sided infinite word x, for any order α, there could exist integers j such that there are no central repetitions of order α at position j. In this case the value of $c_\alpha(x, j)$ is $+\infty$.

Notice further that any central repetition cannot be right external. As an example consider the one-sided word $x = x_0 x_1 x_2 x_3 \cdots$ with $x_i \in \{a, b\}$ defined by $x_0 = a$ and for any $i \geq 1$, $x_i = b$ (i.e., $x = abbbbbbb\cdots$). At position 0 of x, for any $\alpha > 1$ there exists no central repetition of order α, and, consequently, $c_\alpha(x, 0) = +\infty$.

A more sophisticated example is the following one.

8.3. Infinite words

EXAMPLE 8.3.1. Let f be the infinite word of Fibonacci (see Section 1.2). For any position j, $c_2(f, j)$ is finite and the square of minimal length having its center at position j is external if and only if $j = F_n - 2$ for some Fibonacci number F_n as proved in the next proposition.

PROPOSITION 8.3.2. *In the infinite word of Fibonacci f, there exists a square having its center at any position and the square of minimal length having its center at position j is external if and only if $j = F_n - 2$ for some Fibonacci number F_n. Moreover, when $j = F_n - 2$, the minimal length $c_2(f, j)$ of a square having its center at position j is F_n.*

Proof. Recall that if F_n, $n \in \mathbb{N}$, is the sequence of Fibonacci numbers defined by $F_0 = 1$, $F_1 = 1$, $F_{n+1} = F_n + F_{n-1}$ for $n \geq 1$, one has that $|f_n| = F_n$ for any $n > 0$.

We will prove that for any $n \geq 2$ and for any position $j \leq F_n - 2$ one has that there exists a square having its center at j and the square of minimal length having its center at position j is external if and only if $j = F_k - 2$ for some Fibonacci number F_k, $k \leq n$. Moreover, when $j = F_k - 2$, the value of $c_2(f, j)$ is F_k.

The proof of this is by induction on n. The base of the induction is easily verified for $n = 2, 3$. Let us suppose the preceding statement is true for $n > 3$ and let us prove it for $n + 1$. By inductive hypothesis the statement is true for any j up to $F_n - 2$.

We have that $f_{n+1}f_{n+1}$ is a prefix of f and, moreover, we have that $f_{n+1} = f_n f_{n-1} = f_{n-1}f_{n-2}f_{n-2}f_{n-3}$. Hence (f_{n-2}, f_{n-2}) matches position $F_n - 1$. Moreover, since f_{n-2} is a prefix of $f_{n-3}f_{n+1}$, it follows that for any position j, $F_n - 1 \leq j \leq F_n + F_{n-2}$, one has that there is a repetition of order 2 of length $2F_{n-2}$ having its center at position j, i.e., $c_2(f, j) \leq F_{n-2}$.

Let us consider now j such that $F_n + F_{n-2} \leq j \leq F_n + F_{n-1} - 2 = F_{n+1} - 2$. These positions belong to the occurrence of f_{n-1} in the prefix $f_{n+1}f_{n+1} = f_n f_{n-1}f_{n+1}$ of f. Since $f_{n-1}f_{n-1}$ is also a prefix of f and $f_n f_{n-1}f_{n-1}$ a prefix of $f_n f_{n-1}f_{n+1}$, the inductive hypothesis gives us the information that for any position j, with $F_n + F_{n-2} \leq j \leq F_{n+1} - 2$, there exists an internal square having its center at j, with the exception of the position $j = F_n + F_{n-1} - 2 = F_{n+1} - 2$. Moreover, again by inductive hypothesis, we know that for both positions there is "almost" a square centered at it. More precisely there would be a square if the $(j + 1)$th letter of f were equal to the $(j - F_{n-1} + 1)$th letter of f. And it is not difficult to prove that this is false. By inductive hypothesis there is no central square at j of length $\leq 2F_{n-1}$. If there were an internal central square, by a classical result, it would have length a Fibonacci number, i.e. F_n, because $F_{n+1} > j + 1$. But again it is not difficult to prove that the

$(j+1)$th letter of f is different from the $(j-F_n+1)$th letter of f, and so for $j = F_{n+1} - 2$ there are no internal central squares. But, since $f_{n+1}f_{n+1}$ is a prefix of f, it is easy to see that $c_2(f, j) = F_{n+1}$, and this concludes the proof. ∎

In the case of a two-sided infinite word x also, a fortiori there could exist integers j such that there are no central repetitions of order α at position j, i.e., $c_\alpha(x, j) = +\infty$. However, in the case of a two-sided infinite word x all the central repetitions at every position j such that $c_\alpha(x, j)$ is finite are internal. As an example consider the two-sided infinite word $y = \cdots y_{-2}y_{-1}y_0 y_1 y_2 y_3 \cdots$ with $y_i \in \{a, b\}$ defined by $y_0 = a$ and for any $i \neq 0$, $y_i = b$ (i.e., $y = \cdots bbbbabbbb \cdots$). At position 0 and position -1 of y, for any $\alpha > 1$ there exist no central repetitions of order α and for all other positions there exists a square having it for a center.

The following proposition is an easy consequence of Lemma 8.1.3 and its proof is left to the reader.

PROPOSITION 8.3.3. *Suppose that x is an infinite word that has period q and that there exists a factor v of x with $|v| \geq q$ that has period d, where d divides q. Then x has period d.*

The periodicity of an infinite word x strongly depends on whether the sequence $c_\alpha(x, j)$ of local periods is bounded or not. Let

$$M_\alpha(x) = \sup\{c_\alpha(x, j) \mid j \in \mathbb{N}\}.$$

The following theorem is a consequence of the critical factorization theorem.

THEOREM 8.3.4. *An infinite word x is periodic if and only if $M_2(x)$ is finite. Moreover the period of x is equal to $M_2(x)$.*

Proof. If x is periodic then trivially in any position there exists a square having it for a center and the sequence of local periods $(c_2(x, i))_{i \in \mathbb{N}}$ is bounded by the period P of x, i.e., $M_2(x) \leq P$. If $M_2(x) < P$ then take a factor v of length $2P$ of x. Clearly P is a period of v and $P_2(v) \leq M_2(x) < P$. By the critical factorization theorem $P_2(v)$ is a period of v and $P_2(v) < P$. By the theorem of Fine and Wilf v also has period $d = \gcd(P, P_2(v)) < P$. Since d divides P, by Proposition 8.3.3, d is also a period of x, contradicting the minimality of P.

Let us prove the "if" part of the proposition. Let Z be a position where the sequence $(c_2(x, i))_{i \in \mathbb{N}}$ reaches its maximum $M_2(x)$. We have to

8.3. Infinite words

prove that for any $i \in \mathbb{N}$, $x_i = x_{i+M_2(x)}$. Take the factor $v = x_r \cdots x_s$ of x where

$$r = \min\{i, Z - M_2(x)\} \text{ and } s = \max\{i + M_2(x), Z + M_2(x)\}.$$

In the preceding definition if $r < 0$ we consider v defined as $v = x_0 \cdots x_s$.

It is easy to see that position Z is also a critical position for v and its central local period is again $M_2(x)$. This implies that $M_2(x) = P_2(v)$. Hence, by the critical factorization theorem, $p(v) = P_2(v) = M_2(x)$ and, so, $x_i = x_{i+M_2(x)}$. ∎

The proof of the following theorem is analogous to the proof of Theorem 8.3.4 and it is left to the reader.

THEOREM 8.3.5. *A two-sided infinite word x is periodic if and only if $M_2(x)$ is finite. Moreover the period of x is equal to $M_2(x)$.*

In both Theorems 8.3.4 and 8.3.5 the constant 2 is tight. Indeed, for any $\epsilon > 0$, we can construct one-sided and two-sided infinite words that are nonperiodic and in any position have a central repetition of order $2 - \epsilon$, as shown in the next example.

EXAMPLE 8.3.6. For any $\epsilon > 0$ let m be a positive integer such that for any $n \geq m$, $\epsilon > 2/n$. Let v_n be the finite word defined by $v_n = a^{n^2} b^n$, and let y_m the infinite word obtained by concatenating $v_m, v_{m+1}, v_{m+2}, \ldots$.

In any position of y_m there is a central repetition of order $2 - \epsilon$. Indeed, if the square aa and the square bb are not central in position j then either j is a position between the concatenation of v_n and v_{n+1} for some n or j is the position between the sequence of a's and b's inside a word v_n for some n.

In the first case the pair $(a^{n^2-n}b^n, a^{n^2})$ matches position j and in the second case the pair $(b^{n-1}a^{n^2}, b^n a^{n^2-1})$ matches position j. Both $a^{n^2-n}b^n a^{n^2}$ and $b^{n-1}a^{n^2} b^n a^{n^2-1}$ have period $n^2 + n$. The first has length $2n^2$ and the second $2n^2 + 2n - 2 \geq 2n^2$. In both cases the order α of this repetition is greater than or equal to

$$\frac{2n^2}{n^2+n} = \frac{2n}{n+1} = 2 - \frac{2}{n+1} > 2 - \epsilon.$$

One can define a two-sided infinite word $x_m = \cdots x_{-1}x_0x_1 \cdots$ starting from the previously defined one-sided infinite word $y_m = y_0y_1 \cdots$ by the rule $x_i = y_{|i|}$. It is easy to check that also at any position of x there is a central repetition of order $2 - \epsilon$.

THEOREM 8.3.7. *Let x be a one-sided infinite word. If x is recurrent then at any position there is a central repetition of order α, for any α such that $1 < \alpha \leq 2$. Conversely, for any $\alpha \geq 2$, if at any position there is a central repetition of order α, then x is recurrent. In particular x is recurrent if and only if at any position there is a central repetition of order 2.*

Proof. Let us suppose that x is recurrent. If we prove that at every position j there exists a square having it for a center then, a fortiori, there is a central repetition of order α for any $\alpha < 2$. Let $k > 0$ be the position where the prefix $x_0 \cdots x_j$ occurs for the second time, i.e., $x_0 \cdots x_j = x_k \cdots x_{k+j}$. If we set $v = x_{j+1} \cdots x_{k+j}$ then it is not difficult to see that (v, v) matches position j and so $z = v^2$ is a square having its center at position j.

Suppose now that at every position of x there exists a central repetition of order $\alpha \geq 2$. In particular, at every position of x there exists a square having it for a center. If the sequence of central local periods is bounded then, by Theorem 8.3.4, x is periodic and so recurrent. If the sequence of central local periods is not bounded then there exists a sequence $(j_i)_{i \in \mathbb{N}}$ of positions such that for any i, $c_2(x, j_i) > c_2(x, h)$ for any $h < j_i$. For any i consider the finite word $v = x_0 \cdots x_{j_i + c_2(x, j_i)}$. It is not difficult to prove that position j_i is the least critical position for v and its central local period is again $c_2(x, j_i)$, i.e., $j_i = Z_2(v)$ and $c_2(x, j_i) = c_2(v, j_i)$. By Corollary 8.2.6 the minimal square $z = u^2$ having its center at j_i is left external in v. This means that the prefix $x_0 \cdots x_{j_i}$ is a suffix of u and then it is also suffix of $x_0 \cdots x_{j_i + c_2(x, j_i)}$, i.e., the prefix $x_0 \cdots x_{j_i}$ occurs a second time in x. Since the sequence $(j_i)_{i \in \mathbb{N}}$ of positions is an increasing sequence, we find a sequence of prefixes of x of increasing length that have second occurrences in x. This fact easily implies that any prefix of x has a second occurrence, i.e., x is recurrent. ∎

The value 2 is tight in both directions of the preceding theorem. For any $\epsilon > 0$ it is known that there exists a one-sided recurrent infinite word x that is $(1 + \epsilon)$-power-free. For $\alpha \geq 2 + 2\epsilon$, the word x has no central repetition of order α in any position.

Conversely, for fixed m, the word $y = a^m baaaaa \cdots$, i.e., the word $y = y_0 y_1 \cdots$ with $y_i = a$ if $i \neq m$ and $y_m = b$, has in every position a central repetition of order $2 - (1/m)$ and it is not recurrent.

For two-way infinite words there are no similar characterizations. Any square-free recurrent two-sided infinite word obviously fails to have a square having its center at any position. Notice that in a square-free recurrent one-sided infinite word, at any position the minimal central repetition of order 2 exists (according to Theorem 8.3.7) and it must be left external.

8.3. Infinite words

In the following characterizations of periodic infinite words, the local property here considered is related to unbordered words.

Let x be an infinite word (one- or two-sided). Denote by $U(x)$ the maximal length of unbordered factors of x if such length exists, infinite otherwise:

$$U(x) = \sup\{|v| \mid v \text{ is an unbordered factor of } x\}.$$

PROPOSITION 8.3.8. *Let x be a one-sided infinite word. If $U(x)$ is finite then x is recurrent.*

Proof. Let f be a finite prefix of x. We prove, by induction on $|f|$, that if f has no second occurrence in x then $U(x) = \infty$.

If $|f| = 1$, i.e., if $f = a$ with $a \in A$, then all the prefixes of x are unbordered and, consequently, $U(x) = \infty$.

Let us suppose that the statement holds for $|f| = n \geq 1$ and consider the case $|f| = n + 1$. The prefix f can be written as $f = av$, with $a \in A$ and $|v| = n$. Let W be the set of prefixes of x having v as suffix. We distinguish two cases.

Case 1: W is finite. Let w be an element of W of maximal length and let w' be such that $w = w'v$. Let y be the infinite word defined by the relation $x = w'y$. The prefix of y of length n is v and v has no second occurrence in y. By the inductive hypothesis $U(y) = \infty$. Since $U(y) \leq U(x)$, the statement follows.

Case 2: W is infinite. Let w be an arbitrary element of W having length greater than $2n + 2$. By definition we can write

$$w = avw'v.$$

Let au be the longest unbordered prefix of av. The word au is not a suffix of the word $g = avw'u$, since the last letter of w' is different from a (otherwise $f = av$ would have a second occurrence in w and hence in x). Let $s(g)$ denote the shortest border of g. It is easy to verify that $s(g)$ is unbordered.

Now $s(g)$ cannot be a proper prefix of au. Indeed, in this case, $s(g)$ is also a suffix of u, contradicting the hypothesis that au is unbordered.

Also, $s(g)$ is not equal to au, since au is not a suffix of g. Moreover it is not possible that $s(g)$ is a prefix of av that has as prefix au, since au is the longest unbordered prefix of av.

It follows that av is a prefix of $s(g)$.

Since av has no second occurrence in g, we have that $s(g) = g$, i.e., g is unbordered. We conclude that there exist unbordered factors of x of increasing length, i.e., $U(x) = \infty$. ∎

THEOREM 8.3.9. *Let x be a one-sided infinite word. Then x is periodic if and only if $U(x)$ is finite. Moreover the period of x is $U(x)$.*

Proof. Suppose that $U(x)$ is finite. By Proposition 8.3.8 x is recurrent. By Theorem 8.3.7 at every position of x there exists a square having its center at it. Let $z_i = u_i^2$ be the square of minimal length having its center at position i. By Proposition 8.2.4 applied to the finite prefix w of x having length $i + |u_i|$, we know that u_i is unbordered. By hypothesis, it follows that $|u_i| \le U(x) < \infty$, i.e., the sequence of central local periods is bounded by $U(x)$. Therefore, by Theorem 8.3.4, x is periodic and the period is smaller than or equal to $U(x)$. The period P of x cannot be smaller than $U(x)$, otherwise, if v is an unbordered factor of x of length $|v| = U(x)$, v has period $P < |v|$ and hence it is bordered, a contradiction.

If x is periodic with period P then, by Theorem 8.3.4, $P = M_2(x)$, i.e., there exists a position j such that the minimal square $z = uu$ having its center at position j is such that $|u| = M_2(x) = P$. By Proposition 8.2.4, the word u is unbordered. Since z is right internal, u is a factor of x, i.e., $U(x) \ge P$. But, as proved above, the period P of x cannot be smaller than $U(x)$ and, consequently, $P = U(x)$. ∎

COROLLARY 8.3.10. *Let $x = \cdots a_{-1}a_0 a_1 \cdots$ be a two-sided infinite word. One has that x is periodic if and only if $U(x)$ is finite. Moreover the period of x is $U(x)$.*

Proof. Suppose that $U(x)$ is finite. Let i be an integer. We have to prove that $a_i = a_{i+U(x)}$. Pick an unbordered factor $v = a_j \cdots a_{j+U(x)-1}$ of x having length $U(x)$ and define $m = \min\{i, j\}$. The one-sided infinite word $x' = a_m a_{m+1} a_{m+2} \cdots$ is such that $U(x') = U(x)$ because it has v as a factor, and, by Theorem 8.3.9, it has period $U(x)$. Since $i \ge m$, one has $a_i = a_{i+U(x)}$.

Suppose that x is periodic with period P and let $x' = a_0 a_1 a_2 \cdots$ be the one-sided infinite word that is the right suffix of x starting from position 0. The word x' also has period P (see also Problem 8.3.1) and, by Theorem 8.3.9, $P = U(x')$. Since x is periodic any unbordered word v that is a factor of x is also a factor of x'. It follows that $U(x) = U(x') = P$. ∎

The following theorem summarizes some of the results presented in this subsection.

THEOREM 8.3.11. *Let x be a (one-sided or two-sided) infinite word. The following conditions are equivalent:*
  (i) *x is periodic and P is its minimal period;*
  (ii) *$M_2(x) = P$;*
  (iii) *$U(x) = P$.*

8.3. Infinite words

8.3.2. Characterizations of eventually periodic words

We now give a characterization of one-sided infinite eventually periodic words. This characterization property is similar to the property characterizing one-sided recurrent infinite words as described in Theorem 8.3.7. Now we require something less (for all "large enough" positions j there is a square having j for a center) and something more (the minimal central repetition at position j is internal if j is "large enough").

Notice further that here we do not explicitly require, as in Theorem 8.3.4, that the sequence $(c_2(x, j))_{j \in \mathbb{N}}$, is bounded. This condition is actually obtained (see Lemma 8.3.13 below) as a consequence of the existence of an internal repetition for any large enough position j.

THEOREM 8.3.12. *A one-sided infinite word $x = x_0 x_1 x_2 \cdots$ is eventually periodic if and only if there exists a number k such that for any $j \geq k$ there exists a suffix of $x_0 \cdots x_j$ that is also a prefix $x_{j+1} x_{j+2} \cdots$, i.e., at any position $j \geq k$ there exists a proper central repetition of order 2.*

The proof of this theorem is based on the following lemma.

LEMMA 8.3.13. *If there exists a number k such that at any position $j \geq k$ there exists a proper central repetition of order 2, then the sequence of local periods $(c_2(x, j))_{j \geq k}$ is bounded.*

Proof. The proof is by contradiction. Let us suppose that the sequence of central local periods at positions $j \geq k$ is not bounded. By hypothesis $(j - c_2(x, j)) > 0$ for any $j \geq k$.

Let j_1 be such that $(j_1 - c_2(x, j_1))$ assumes the minimal value between all $j \geq k$ and let j_2 be the least position greater than j_1 such that $c_2(x, j_2) > c_2(x, j_1)$.

Consider the word $v = x_{j_1 - c_2(x, j_1) + 1} x_{j_1 - c_2(x - j_1) + 2} \cdots x_{j_2 + c_2(x - j_2)}$.

In the rest of this proof, for simplicity, with abuse of notation, we will denote by t the position $t - (j_1 - c_2(x, j_1))$ of v. It is easy to verify that $c_2(x, j_1) = c_2(v, j_1)$.

Also notice that the minimal square having its center at position j_2 of v is not right external. This square cannot be left external by the minimality of j_1 and the fact that, for any position t of v, $c_2(x, t) \geq c_2(v, t)$. Since this square is not left external one has $c_2(v, j_2) = c_2(x, j_2) > c_2(x, j_1) \geq c_2(v, j_1)$.

Since for any position t of v, $c_2(x, t) \geq c_2(v, t)$, and since j_2 is the least position greater than j_1 such that $c_2(x, j_2) > c_2(x, j_1) = c_2(v, j_1)$, one has that for any position t of v with $j_1 \leq t < j_2$, $c_2(v, t) \leq c_2(v, j_1)$. If $j_1 - c_2(x, j_1) + 1 \leq t < j_1$, we also have, by construction of v, that $c_2(v, t) \leq c_2(v, j_1)$.

By Corollary 8.2.7, $c_2(v, j_2) > j_2 - (j_1 - c_2(x, j_1) + 1) + 1$, i.e., $j_2 - c_2(v, j_2) < j_1 - c_2(x, j_1)$, contradicting the minimality of j_1. ∎

Proof of Theorem 8.3.12. If x is eventually periodic then we can write $x = wy$, where y is a one-sided infinite word that is periodic with period P. Hence, if we set $k = |w| + P$, it is easy to check that at any position $j \geq k$ of x there exists a central repetition of order 2 that is internal.

Let us prove the "if" part. Let us write $x = uy$, where $|u| = k$. Let us consider a position i of y and the corresponding position $i + k$ of x. Since y is a suffix of x one has that at any position $i \geq 0$ there exists a central repetition of order 2 and $c_2(y, i) \leq c_2(x, i + k)$. Hence, by Lemma 8.3.13 the sequence of local periods $(c_2(y, i))_{i \in \mathbb{N}}$ is bounded and, by Theorem 8.3.4, y is periodic. ∎

By the fact that an infinite word is periodic if and only if it is recurrent and eventually periodic, and by Theorem 8.3.7, one has

COROLLARY 8.3.14. *A one-sided infinite word x is periodic if and only if at any position there is a central repetition of order 2 and this repetition is external only for finitely many positions.*

REMARK 8.3.15. Notice that, in the proof of the preceding theorem, one cannot bound the period of y as function of k, as shown by the one-sided infinite word $y_n = ba^n ba^n ba^n \cdots$ where n is any positive natural number. In this word the number k is 2 and this word has period n.

EXAMPLE 8.3.6 (*continued*). The word y_1 previously defined with $m = 1$ shows that the number 2 is tight in Lemma 8.3.13. Indeed for any ϵ it is easy to see that there exists a constant $k(\epsilon)$ such that at any position $j \geq k(\epsilon)$ there exists a central repetition of order $2 - \epsilon$, but the sequence of local periods at positions $j \geq k(\epsilon)$ is not bounded.

The same word y_1 also shows that the constant 2 is tight in Theorem 8.3.12, because y_1 is not eventually periodic.

A more sophisticated example, showing that the constant 2 is tight in Theorem 8.3.12, is given by the infinite word of Fibonacci f defined in Example 8.3.1. The word f is not eventually periodic but it is possible to prove that for any ϵ there exists a constant $k(\epsilon)$ such that at any position $j \geq k(\epsilon)$ there exists a central repetition of order $2 - \epsilon$, and that the sequence of local periods at positions $j \geq k(\epsilon)$ is bounded (Problem 8.3.3).

An analogue of Theorem 8.3.12 does not hold for two-sided infinite words, as shown by the next example.

8.3. Infinite words

EXAMPLE 8.3.16. For any $\alpha > 1$ one can construct a nonperiodic two-sided infinite word $x_\alpha = \cdots x_{-1}x_0x_1\cdots$ such that at any position there exists a central repetition of order α.

Consider the sequence of all integers $0, -1, 1, -2, 2, -3, 3, \ldots, -i, i, \ldots$. Our construction inductively fixes letters in the word x_α in order to have a central repetition at position n_j, where $n_j = (-1)^j \lceil (j/2) \rceil$ is the jth element of the preceding sequence.

First, let $k = \lceil (\alpha/2) \rceil$ be the smallest integer greater than or equal to $\alpha/2$ and set $x_{-k} = x_{-k+1} = \cdots = x_0 = x_1 = \cdots = x_{k-1} = a$ and $x_k = b$.

By construction at position 0 there is a central repetition of order α. Suppose that we have fixed letters from position s_j up to position t_j such that at all positions $n_0, \ldots, n_j$ there exists a central repetition of order α that is internal to the word $x_{s_j} \cdots x_{t_j}$. Since position n_{j+1} is adjacent to position n_{j-1} one has $s_j \le n_{j+1} \le t_j$.

Let us denote $u = x_{s_j}x_{s_j+1}\cdots x_{n_{j+1}}$ and $v = x_{n_{j+1}+1}\cdots x_{t_j}$. Suppose that $w = uv$ has period P and that a is the letter such that wa has period greater than P. Set $x_{t_j+1} = a$. Now assign letters from the position $s_{j+1} = n_{j+1} - k|vau|$ to the position $t_{j+1} = n_{j+1} + 1 + k|vau|$ so that

$$(vau)^k = x_{s_{j+1}}x_{s_{j+1}+1}\cdots x_{n_{j+1}} \text{ and } (vau)^k = x_{n_{j+1}+1}\cdots x_{t_{j+1}}.$$

Notice that this assignment is compatible with the previous assignment and that at position n_{j+1} there exists a central repetition of order α. Notice further that, since $x_{s_{j+1}}x_{s_{j+1}+1}\cdots x_{t_{j+1}}$ has wa as factor, its period is strictly greater than the period P of w. Using this property it is not difficult to prove that the infinite word x_α is nonperiodic.

As regards left repetitions, in the case of one-sided infinite words x, for any order α, the minimal left repetition of order α is defined at any position j, i.e., $l_\alpha(x, j)$ is defined for any $\alpha > 1$ and for any $j \in \mathbb{N}$. In the case of two-sided infinite words x, there could exist integers j such that there are no left repetitions of order α at position j and, consequently, l_α is not defined at position j. We can define, analogously to the central case,

$$L_\alpha(x) = \sup\{l_\alpha(x, j) \mid j \in \mathbb{N}\}.$$

THEOREM 8.3.17. *Let $x = a_1 a_2 \cdots$ be a one-sided infinite word. $L_{\varphi^2}(x)$ is finite if and only if the word x is eventually periodic, i.e., $x = wy$, with y periodic. Moreover, if P is the period of y then $P \le L_{\varphi^2}(x)$.*

Proof. If x is eventually periodic then we can write $x = wy$ where y is a one-sided infinite word that is periodic with period P. Hence, if we set $k = |w| + 3P$, it is easy to check that at any position $j \ge k$ of x there exists a left repetition of order $3 > \varphi^2$ that is internal and that $l_{\varphi^2}(x, j) \leq P$. Hence $L_{\varphi^2}(x) \leq \max\{P, Q_{\varphi^2}(w)\}$, where $Q_{\varphi^2}(w)$ is the maximum of the left local periods of order φ^2 of the prefix w of x of length k.

Let us prove the "only if" part. The proof is by induction on $Q = L_{\varphi^2}(x)$.

If $Q = 1$ then for any $j \geq 3$, $a_1 \cdots a_j$ ends with a cube of period 1, i.e., it ends with a^3 where a is a letter. Trivially the infinite word x is periodic with period 1.

Let us suppose the statement is true for any Q', $1 \leq Q' \leq Q$. Two cases are possible.

Case 1: there are infinitely many positions j such that $l_{\varphi^2}(x, j) = Q$. We want to prove that x has period Q. Let us consider position i, $1 \leq i$. We have to prove that $a_i = a_{i+Q}$. Take j such that $j \geq i + Q$ and $l_{\varphi^2}(x, j) = Q$. By Theorem 8.2.13(i), the word $a_1 \cdots a_j$ has period Q and, consequently $a_i = a_{i+Q}$.

Case 2: there are finitely many numbers j such that $l_{\varphi^2}(x, j) = Q$. Let s be the greatest of these numbers and let $x' = a_{s+1} a_{s+2} \cdots$ be the suffix of x that starts at position $s + 1$. Since the maximum of the sequence $(l_{\varphi^2}(x, i))_{i \in \mathbb{N}}$ is a number $Q' < Q$ one has that x' satisfies the inductive hypothesis and consequently it is eventually periodic with period $P \leq Q' < Q$. Since x' is a suffix of x, x is also eventually periodic with period $P < Q$. ∎

We now give another characterization of one-sided infinite eventually periodic words that is very similar to the characterization given in Theorem 8.3.12.

THEOREM 8.3.18. *A one-sided infinite word $x = a_0 a_2 \cdots$ is eventually periodic if and only if there exists a number k such that for any $j \geq k$ there exists a suffix of $a_0 \cdots a_j$ of order greater than φ^2.*

Proof. If the one-sided infinite word $x = a_0 a_2 \cdots$ is eventually periodic then there exist natural numbers $M > 0$, $Q > 0$, such that the infinite word $x' = a_M a_{M+1} \cdots$ has period Q. Since $\varphi^2 < 3$, at any position j, $j \geq M + 3Q$, there exists a left repetition of order φ^2 that is internal.

Let us suppose now that there exists a number k such that at any position $j \geq k$ there exists a left repetition of order φ^2 that is internal. If the sequence $(l_{\varphi^2}(x, i))_{i \in \mathbb{N}}$ is not bounded then there exists an increasing sub-sequence of positions $(j_i)_{i \in \mathbb{N}}$ such that for any $i \in \mathbb{N}$ and any position $s < j_i$ one has that $l_{\varphi^2}(x, j_i) > l_{\varphi^2}(x, s)$. By Theorem 8.2.13(iii) applied to the words $a_1 \cdots a_{j_i}$, the minimal left repetition of order φ^2 at position j_i is external, a contradiction whenever $j_i > k$. Therefore the sequence

8.3. Infinite words

$(l_{\varphi^2}(x,i))_{i\in\mathbb{N}}$ is bounded and by applying Theorem 8.3.17 x is eventually periodic. ∎

REMARK 8.3.19. By the statement of Theorem 8.3.17 one has that any suffix y of x that is periodic has period P where P is bounded by $L_{\varphi^2}(x)$. But one cannot bound the length of the shortest prefix w of x such that $x = wy$ with y periodic, as shown by the one-sided infinite word $x_n = (aaab)^n aaaaaaaaaaaa\cdots$ where n is any positive natural number. In this word $L_{\varphi^2}(x_n) = 4$ and the word $w = (aaab)^n$ has length $4n$.

By the proof of Theorem 8.3.18 one can see that any suffix y of x that is periodic has period P where P is bounded by k where k is the number in the statement of the theorem but one cannot bound the length of the shortest prefix w such that $x = wy$ with y periodic. Indeed in the same example $x_n = (aaab)^n aaaaaaaaaaaa\cdots$ where n is any positive natural number, one can take $k = 14$ and the word $w = (aaab)^n$ has length $4n$.

The following example shows that in Theorem 8.3.18 the number φ^2 is tight.

EXAMPLE 8.3.1 (*continued*). Let f be the infinite word of Fibonacci. For any $\epsilon > 0$ there exists a constant $\bar{n}$ such that for any $n \geq \bar{n}$ there exists a left repetition of order $\varphi^2 - \epsilon$ at position n and f is not eventually periodic (Problem 8.3.4).

An analogue of Theorem 8.3.18 does not hold for two-sided infinite words, as shown by the next example.

EXAMPLE 8.3.16 (*continued*). For any $\beta > 1$, take $\alpha = 2\beta$. The word x_α has, by construction, at any position a left repetition of order β, which is obviously internal.

The following theorem summarizes some of the results presented in this subsection.

THEOREM 8.3.20. *Let $x = a_0 a_1 \cdots$ be a one-sided infinite word. The following conditions are equivalent:*
  (i) *x is eventually periodic;*
  (ii) *there exists an integer k such that, for any $j \geq k$, there exists a suffix of $a_0 \cdots a_j$ that is equal to a prefix of $a_{j+1} a_{j+2} \cdots$;*
  (iii) *there exists an integer h such that for any $j \geq h$ there exists a suffix of $a_0 \cdots a_j$ of order greater than φ^2.*

Problems

Section 8.1

8.1.1 A word $w = a_1 \cdots a_n$ has period $p \leq n$ if and only if for any integers i, j, $1 \leq i, j \leq n$
$$i \equiv j \pmod{p} \Rightarrow a_i = a_j.$$
Prove the following. If q is a period of w then for any positive integer k such that $kq \leq n$, kq is a period of w. If p is a period of w and q is a period of the suffix of length $n - p$ of w then $p + q$ is also a period of w. Therefore for any positive integer k such that $p + kq \leq n$, $p + kq$ is a period of w.

8.1.2 Consider the nondirected graph $G = (I_{p+q}, E)$, where the set of vertices $I_{p+q} = \{1, 2, 3, \ldots, p+q-1, p+q\}$ is the set of positive integers smaller than or equal to $p+q$ where p, q are integers such that $\gcd(p, q) = 1$. The arc $\{i, j\} \in E$ if and only if $|i - j| \in \{p, q\}$. Prove that the graph G is a cycle. Deduce that, if G_t is the graph obtained by G by eliminating vertex t and all arcs containing t, then G_t is a connected graph.

8.1.3 Let $w = a_1 a_2 \cdots a_{p+q-1}$ be a word that has period p and period q with $\gcd(p, q) = 1$.
Use the graph G_i with $i = p + q$ defined in the previous problem to give a new proof of Theorem 8.1.4 that works for the case where $\gcd(p, q) = 1$. (*Hint*: If there is an arc (i, j) in G_{p+q} then $a_i = a_j$.)

8.1.4 Let $w = a_1 \cdots a_n$ and $v = b_1 \cdots b_n$ be two words having the same length n, such that v has period p and w has period q with $p \neq q$ and $p + q \leq n$.
Suppose that there exists a position t, $1 \leq t \leq n$, such that for any position $i \neq t$, $1 \leq i \leq n$, one has that $a_i = b_i$ (i.e., the two words w and v coincide except, maybe, in position t). Show that w and v both have period $r = \gcd(p, q)$, and since $\gcd(p, q) \leq \min\{p, q\} \leq \lfloor n/2 \rfloor$, that $w = v$.
(*Hint*: Use the technique developed in the previous problems, and in particular, if $\gcd(p, q) = 1$ use the graph G_t.)

8.1.5 Prove the following statement.
A necessary and sufficient condition for two nonempty words u, v to be powers of the same word is that uv and vu contain a common left factor of length $|u| + |v| - \gcd(|u|, |v|)$.
(*Hint*: If $v < u$ then $uv < u^2$ and $vu < v^k$ for some $k \geq 2$. Use the theorem of Fine and Wilf.)

Problems

8.1.6   A factor of a word v is internal if it is not a suffix or a prefix of v. Prove that a word u is primitive if and only if u is not an internal factor of uu.

*8.1.7   Prove the equivalence of (i), (ii) and (iv) in Theorem 1.3.13 by using the theorem of Fine and Wilf.

**8.1.8   A triple (p_1, p_2, p_3) is called a *good triple* if

$$h(p_1, p_2, p_3) = \gcd(p_1, p_2, p_3) = 1.$$

Given a good triple (p_1, p_2, p_3) prove that there exists a word w of length $|w| = \frac{1}{2}(p_1 + p_2 + p_3 - 3) = f(p_1, p_2, p_3) - 1$ over an alphabet of three letters that has periods p_1, p_2 and p_3.

8.1.9   If $\Pi(w) = \{0 = p_0 < p_1 < \cdots < p_s = |w|\}$ and if $\delta_h = p_h - p_{h-1}$, $1 < h \leq s$, show that the sequence of the differences δ_h is a nonincreasing sequence. (*Hint*: Use Lemma 8.1.1 and Problem 8.1.1.)

Section 8.2

8.2.1   Fix $\alpha > 1$. If j is a critical point for w and i, $j < i < |w|$ (respectively $0 < i < j$) is not a critical position, show that the minimal central repetition of order α at position i is not left external (respectively right external).

8.2.2   Let $w = a_1 \cdots a_n$ and let $u = a_i \cdots a_j$ be a factor of w with $1 \leq i < j < n$. Show that either $c_2(w, j) < p(u)$ or $c_2(w, j) \geq |u| + 1$.

8.2.3   Let $w = a_1 \cdots a_n$ and let $v = a_h \cdots a_k$, $1 < h$, $k \leq n$, be a factor of w such that $p(v) \geq 2$ and the word $v' = a_{h-1}v$ has period $p(v') > p(v)$ (one cannot extend the word v to the left maintaining the period).
Show that if $Z_2(v)$ is the least critical position of v then $c_2(w, Z_2(v)) > p(v)$.

8.2.4   Find examples different from $a^m b a^m b a^m$ that show that statement (iii) in the proof of the critical factorization theorem 8.2.5 is tight. (*Hint*: Look for words in the set of central words defined in Chapter 2.)

8.2.5   Let $k = \lceil \alpha \rceil$. If the period of w is smaller than or equal to k show that every position of w is left critical (of order α).

*8.2.6   For any natural number n show that there does not exist any prefix (or suffix) of the nth Fibonacci word f_n having order $\geq \varphi + 1$.

8.2.7 Prove that if a square uu is a factor of a Fibonacci word then its length is a Fibonacci number and it is a conjugate of some other Fibonacci word.

8.2.8 Fix $\alpha > 1$. If j is a left critical point for w and i, $j < i < |w|$, is not a left critical position, show that the minimal left repetition of order α at position i is not left external.

8.2.9 Let $w = a_1 \cdots a_n$ and let $v = a_h \cdots a_k$, $1 < h, k \leq n$, be a factor of w such that $p(v) \geq 2$ and the word $v' = a_{h-1}v$ has period $p(v') > p(v)$ (one cannot extend the word v to the left maintaining the period).
Show that if $T_{\varphi^2}(v)$ is the least critical position of the word v then $l_{\varphi^2}(w, T_{\varphi^2}(v)) > p(v)$.

**8.2.10 Show that in every sequence of $d \geq \max\{1, Q(w)-2\}$ consecutive positions between $R(w)$ and $T(w)$, there is a left critical one (of order φ^2). (*Hint:* Use induction on the period $p(w)$ of w. The base of induction is $p(w) = 1, 2, 3$. In the inductive step suppose that there exist two consecutive critical points $r < s$ such that $s - r > Q(w) - 2$. Find the positions j, $r < j < s$, where $l_{\varphi^2}(w, j)$ reaches the maximum value. Use the inductive hypothesis, Theorem 8.2.13 and the previous two problems in order to find a contradiction.)

**8.2.11 Prove that the constant φ^2 is tight in Theorem 8.2.13. (*Hint:* Follow the suggestions in Remark 8.2.15.)

**8.2.12 Prove that any word w admits a factorization $w = vu$ such that there exists at most one internal left repetition of $w, |v|$ of order φ^2 and $|u| \leq Cp(v)$, with the real constant C smaller than or equal to 2.

**8.2.13 Prove or disprove the following open conjecture. Let v be a word of length $2n$ such that its prefix u of length n is an unbordered word and such that any of its unbordered factors have length $\leq n$. Then v is a square, i.e., $v = u^2$.

Section 8.3

8.3.1 Let x be a two-sided infinite word and suppose that P is the period of x, and v is a factor of x of length $|v| \geq 2P - 1$ or v is a one-sided infinite word that is a suffix of x. Then P is also the period of v. Find a periodic word x with minimal period P and a factor v of x with $|v| = 2P - 1$ such that the period of v is smaller than P.

8.3.2 Let x be a one-sided eventually periodic infinite word and let p be its (shortest) period. Let $m(1), m(2), \ldots$ be an increasing sequence of points such that $m(i+1) < 2m(i)$, and denote $p(i) = m(i+1) - m(i)$.
Prove that, if $c_2(x, m(i)) = p(i)$, then from a given rank, the sequence $(p(i))_{i \geq 1}$ is constant and equal to the period p of the word x.

*8.3.3 Let f be the infinite word of Fibonacci. Show that for any ϵ there exists a constant $k(\epsilon)$ such that at any position $j \geq k(\epsilon)$ there exists a central repetition of order $2 - \epsilon$, and the sequence of local periods at positions $j \geq k(\epsilon)$ is bounded.

*8.3.4 Let f be the infinite Fibonacci word. Show that for any $\epsilon > 0$ there exists a constant $\bar{n}$ such that for any $n \geq \bar{n}$ there exists a left repetition of order $\varphi^2 - \epsilon$ at position n and f is not eventually periodic.

8.3.5 A two-sided infinite word $y = \cdots a_{-1} a_0 a_1 \cdots$ is *eventually periodic to the right* with period P if there exists an integer j such that the one-sided infinite word $a_j a_{j+1} a_{j+2} \cdots$ is periodic with period P. It is *eventually periodic to the left* with period P if there exists an integer j such that the one-sided infinite word $a_j a_{j-1} a_{j-2} \cdots$ is periodic with period P.
Prove that if $L_{\varphi^2}(x)$ is finite then x is eventually periodic to the left with period P_1 and it is eventually periodic to the right with period P_2. Moreover $P_1 = L_{\varphi^2}(x) \leq P_2$ and if $P_1 = P_2$ then x is periodic with period P_1.
(*Hint*: Cf. the proof of Theorem 8.3.17.)

8.3.6 Show that, for each not eventually periodic one-sided infinite word w, there exists a one-sided infinite word w' such that

(i) each factor of w' is a factor of w,
(ii) w' does not begin with any $1 + \varphi$ powers.

**8.3.7 Let w be a one-sided infinite word such that all except finitely many prefixes have a square uu as suffix with $|u| \leq 4$. Prove that w is eventually periodic.
Give examples that show that if the condition on the length of u is relaxed to $|u| \leq 5$ the above conclusion does not hold any more.

**8.3.8 Let w be a one-sided infinite word such that all except finitely many prefixes v have a suffix u such that

(i) u appears a second time as a factor in v,
(ii) $|u|/|v| \geq c$ for some fixed constant c.

Find the smallest value (or the inf of those values) for c such that the preceding conditions imply that w is eventually periodic.

**8.3.9   Let w be a recurrent word and let $R(n)$ be the smallest integer such that any factor of w of length $R(n)$ contains all factors of w of length n. $R(n)$ is called the recurrence function of w. Prove that, if for all except finitely many n $R(n)/n < 2 + \varphi$, then w is periodic.

Notes

The original reference for Theorem 8.1.4 is Fine and Wilf 1965. Another proof can be found in Lothaire 1983. A good reference to Euclid's algorithm is Knuth 1988. The ideas used in Problem 8.1.3 can be found in Choffrut and Karhumäki 1997 and also independently in Giancarlo and Mignosi 1994. A solution to Problem 8.1.5 can be found in Lentin and Schützenberger 1967. A solution to Problem 8.1.4 can be found in Berstel and Boasson 1999. Theorem 8.1.7 is from Castelli et al. 1999, where one can also find a solution to Problem 8.1.8. The words described in this problem are words of Arnoux and Rauzy and can be considered as a generalization of Sturmian words to a three-letter alphabet (Chapter 2). A generalization of these results to the case of more than three periods is in Justin 2000. A notion of quasi-periodicity for words was introduced in Apostolico and Ehrenfeucht 1993. For recent contributions on this subject see Régnier and Mouchard 2000 and Brodal and Pedersen 2000 and references therein. For other extensions of the notion of periodicity see Carpi and Luca 2000 and references therein. A solution of Problem 8.1.7 can be found in Epifanio, Koskas, and Mignosi 1999. Many results and applications concerning generalizations to the multidimensional case of Theorem 8.1.4 have been developed, starting from the seminal works Amir and Benson 1992, Amir and Benson 1998.

The equivalence of points (i), (ii) and (iii) of Theorem 8.1.11 is proved in Guibas and Odlyzko 1981. Point (iv) and also some of the proofs are from Breslauer 1995. A simple proof of the equivalence of only (i) and (ii) can be found in Halava, Harju, and Ilie 2000. They also describe a linear time algorithm which, given a word, computes a binary one with the same set of periods. Some related results can be found in Rivals and Rahmann 2001 and in Régnier and Mouchard 2000.

Theorem 8.1.16 and its corollary are from Fraenkel and Simpson 1998. See also Crochemore, Hancart, and Lecroq 2001. Lemma 8.1.14 is from Crochemore and Rytter 1995. The simple and elegant proof given here is due to V. Diekert (Chapter 12 of the present book). Short surveys on related problems can be found in the introductions of the beautiful papers Kolpakov and Kucherov 1999a, Kolpakov and Kucherov 1999b. See also Crochemore and Rytter 1994, Czumaj and Gasieniec 2000 and references therein. A solution to Problem 8.1.6 can be found in Crochemore and Rytter 1995.

A weak form of the critical factorization theorem was first conjectured in Schützenberger 1976 and settled in Césari and Vincent 1978. Subsequent improvements in Duval 1979 led to the actual formulation (see also Duval 1982, Duval 1998). The proof reported here is from Duval, Mignosi, and Restivo 2001. Among the applications of the critical factorization theorem we cite Crochemore and Perrin 1991 and Breslauer, Jiang, and Jiang 1997. A solution to Problem 8.2.6 can be found in Mignosi and Pirillo 1992. A solution to Problem 8.2.7 can be found in Seébold 1985, and see Pirillo 1997 for further improvements. A solution to Problem 8.2.12 can be found in Mignosi, Restivo, and Salemi 1995.

Theorem 8.2.13 is proved in Mignosi et al. 1995 and in Mignosi, Restivo, and Salemi 1998.

Problem 8.2.13 states a long standing open conjecture in Duval 1982. This is the latest and strongest of a sequence of three conjectures stated by different authors. To our knowledge, no weaker versions of it have even been proved.

Theorem 8.3.9 and Corollary 8.3.10 are in Ehrenfeucht and Silberger 1979 (see also Assous and Pouzet 1979). The proof here reported, as well as Proposition 8.3.8, is inspired by Duval 1982. Almost all the remaining results in the last section are from Mignosi et al. 1995 and Mignosi et al. 1998, if they concern left repetitions, and from Duval et al. 2001 otherwise.

Solutions to Problems 8.3.6 and 8.3.7 can be found respectively in Holton and Zamboni 2000 and in Karhumäki, Lepistö, and Plandowski 1998a. See also Lepistö 1999 for related results. The article Holton and Zamboni 2000 is the last, for the moment, of a long sequence of articles and results concerning periodicities and Sturmian words (Chapter 2). It is worth noticing that in Sturmian words, periodicities are strongly related to the recurrence function defined in Problem 8.3.9, as shown by the two beautiful and independent papers Cassaigne 1999 and Vandeth 2000.

Problem 8.3.8 was posed, as a conjecture for a fixed constant c, in Shallit and Breibart 1996 and settled in Cassaigne 1997. Problem 8.3.9 was stated as conjecture in Rauzy 1983. It is linked in some way to the previous problem, as proved in Allouche and Bousquet-Mélou 1995.

CHAPTER 9

Centralizers of Noncommutative Series and Polynomials

9.0. Introduction

It is a well-known and not too difficult result of combinatorics on words that if two words commute under the concatenation product, then they are both powers of the same word: they have a common *root*. This fact is essentially equivalent to the following one: the centralizer of a nonempty word, that is, the set of words commuting with it, is the set of powers of the shortest root of the given word.

The main results of this chapter are an extension of this latter result to noncommutative series and polynomials: Cohn's and Bergman's centralizer theorems. The first asserts that the centralizer of an element of the algebra of noncommutative formal series is isomorphic to an algebra of formal series in one variable. The second is the similar result for noncommutative polynomials. Note that these theorems admit the following consequences: if two noncommutative series (resp. polynomials) commute, then they may both be expressed as a series (resp. a polynomial) in a third one. This formulation stresses the similarity with the result on words given above.

We begin with Cohn's theorem, since it is needed for Bergman's theorem. Its proof requires mainly a divisibility property of noncommutative series. The proof of Bergman's theorem is rather indirect: it uses the noncommutative Euclidean division of Cohn, the difficult result that the centralizer of a noncommutative polynomial is integrally closed in its field of fractions, its embeddability in a one-variable polynomial ring, which uses a pretty argument of combinatorics on words, and finally another result of Cohn characterizing free subalgebras of a one-variable polynomial algebra. The latter result is proved in the Appendix, since it is a result of commutative algebra, and for the sake of completeness, we have proved all the results on valuation rings which are needed for its proof. We have added a section on the defect theorem and free subalgebras: a result of Cohn, together with Bergman's centralizer theorem, shows that the defect theorem holds for two noncommutative polynomials. However, a counterexample of Bergman shows that it does not hold for more than two polynomials: this was proved by Kolotov, and for this, we give his theorem asserting that each free subalgebra of a free associative algebra is an anti-ideal.

9.1. Cohn's centralizer theorem

In all that follows, k will be a field (all fields are assumed to be commutative) and all algebras will be over k. Let X be an *alphabet*, that is, a set of noncommuting variables. We denote by $k\langle\!\langle X\rangle\!\rangle$ the algebra of noncommutative formal series in these variables with coefficients in k and by $k\langle X\rangle$ its subalgebra of noncommutative polynomials. We call an element of $k\langle X\rangle$ simply a *polynomial*.

THEOREM 9.1.1 (Cohn). *The centralizer of a nonscalar element in $k\langle\!\langle X\rangle\!\rangle$ is isomorphic to $k[[t]]$, for a single variable t.*

The isomorphism will be shown to be continuous, for the X- and t-adic topologies. In other words we shall prove that for some series b in the centralizer, with zero constant term, each element in the centralizer has a unique representation of the form $\sum_{n\geq 0}\alpha_n b^n$, with the α_n in k.

Recall Levi's lemma for words over X: if u, v, u', v' are words over X such that $uv' = vu'$ and that u is not shorter than v, then $u = vm, mv' = u'$ for some word m. We need the following lemma, which is the analogue of Levi's lemma for series. Recall the well-known fact that a series is invertible if and only if its constant term is nonzero. As usual, X^* is the free monoid generated by X, and the *length* $|w|$ of a word w is the number of letters appearing in it. In other words, the length of a word is its X-degree. If a is a series, we denote it by $a = \sum_{w \in X^*} a_w w$, where a_w is the coefficient of the word w in the series a. Denote by $v(a)$ the X-adic valuation of a, that is, the length of the shortest word w such that a_w is nonzero (with $v(a) = \infty$ if $a = 0$). Note that $v(ab) = v(a) + v(b)$.

LEMMA 9.1.2. *If a, b, a', b' are nonzero series such that $ab' = ba'$ and $v(a) \geq v(b)$, then $a = bq$ for some series q.*

Proof. If b' is an invertible series, then the conclusion follows with $q = a'b'^{-1}$.

In the general case, let m be a fixed word of shortest length appearing (with nonzero coefficient) in the series b'. This length is equal to $v(b')$. Since $v(a) + v(b') = v(b) + v(a')$ and $v(a) \geq v(b)$, we have $v(b') \leq v(a')$, hence each word appearing in a' has length at least the length of m.

Now, since $ab' = ba'$, for any word w the two sums $\sum_{uv'=wm} a_u b'_{v'}$ and $\sum_{vu'=wm} b_v a'_{u'}$, over all words u, v', v, u', are equal. We have seen that in order that $b'_{v'} \neq 0$ and $a'_{u'} \neq 0$, the length of v' and u' cannot be smaller than that of m. In this case, the equalities $uv' = wm$ and $vu' = wm$ imply, by Levi's lemma for words, $v' = v_1 m, u' = u_1 m$ and $uv_1 = w, vu_1 = w$. Since in a sum one may disregard vanishing terms, we obtain $\sum_{uv_1=w} a_u b'_{v_1 m} = \sum_{vu_1=w} b_v a'_{u_1 m}$, where the sum is now over all words u, v_1, v, u_1. Define the series B and A by $B_{v_1} = b'_{v_1 m}$ and $A_{u_1} = a'_{u_1 m}$. Then the previous equality shows that $aB = bA$.

Finally, note that the constant term of B is b'_m, hence is nonzero, by the choice of m, and we are done by the beginning of the proof. ∎

Proof of Theorem 9.1.1. Let Z be the centralizer in $k\langle\langle X \rangle\rangle$ of a nonscalar element a. The constant term of a is in k and by subtracting it, we may suppose a to have constant term 0. We claim that if c_1, c_2 are nonzero elements of Z with $v(c_2) \geq v(c_1)$ then $c_2 = c_1 d$ for some $d \in Z$.

The claim being assumed, let b be an element of Z such that $v(b)$ is positive and minimal. We show that each element c in Z has a unique decomposition $c = \sum_{n \geq 0} \alpha_n b^n$, with the α_n in k. This will imply that Z is isomorphic to $k[[t]]$.

Note that uniqueness is clear, since the valuation of $\alpha_m b^m + \alpha_{m+1} b^{m+1} + \cdots$ is, for nonzero α_m, equal to $mv(b)$. In order to prove existence, let $c \in Z$. Then for $\alpha_0 =$ constant term of c, we have $v(c - \alpha_0) > 0$, hence $v(c - \alpha_0) \geq v(b)$, by the minimality of $v(b)$. Suppose that we have found scalars $\alpha_0, \alpha_1, \ldots, \alpha_n$ such that

$$v(c - \alpha_0 - \alpha_1 b - \cdots - \alpha_n b^n) \geq (n+1)v(b). \qquad (*)$$

Since $(n+1)v(b) = v(b^{n+1})$, the series on the left-hand side of $(*)$ is, by the claim, equal to $b^{n+1} d$ for some series $d \in Z$. Now as before we choose a scalar α_{n+1} such that $v(d - \alpha_{n+1}) \geq 0$, hence $v(d - \alpha_{n+1}) \geq v(b)$, and we obtain $c - \alpha_0 - \alpha_1 b - \cdots - \alpha_n b^n - \alpha_{n+1} b^{n+1} = b^{n+1} d - \alpha_{n+1} b^{n+1} = b^{n+1}(d - \alpha_{n+1})$.

Therefore, this series has valuation $\geq (n+2)v(b)$, which concludes the induction step, and Formula $(*)$ holds for each n.

In Formula $(*)$, let n tend to ∞. Then we obtain that $c = \sum_{n \geq 0} \alpha_n b^n$, as desired.

9.2. Euclidean division and principal right ideals

It remains to prove the claim. Since a has zero constant term, we have $v(a^n) = nv(a) \geq v(c_2)$ for n large enough. Since c_1, c_2 are in Z, they commute with a, hence with a^n. Thus $a^n c_1 = c_1 a^n$ and $a^n c_2 = c_2 a^n$. From the latter equation and the lemma, we conclude that $a^n = c_2 q$, for some series q. Hence, we have $c_2 q c_1 = c_1 a^n$. Since $v(c_2) \geq v(c_1)$, the lemma implies that $c_2 = c_1 d$. Now $c_1 a d = a c_1 d = a c_2 = c_2 a = c_1 d a$, and canceling c_1, we obtain $ad = da$, hence d is in Z. ∎

COROLLARY 9.1.3. *The centralizer of any nonscalar element in $k\langle\langle X \rangle\rangle$ or $k\langle X \rangle$ is commutative.*

This means that if two polynomials (or series) commute with a third nonscalar one, then they commute with each other.

9.2. Euclidean division and principal right ideals

The next result is Cohn's Euclidean division in $k\langle X\rangle$ (a particular case of his *weak algorithm*).

THEOREM 9.2.1. *If a, b, a', b' are polynomials such that $ab' = ba'$ and b, b' are nonzero, then $a = bq + r$, $\deg(r) < \deg(b)$ for some unique polynomials q, r.*

Proof. We totally order the free monoid on X in the following way: let $<$ be a fixed order on X and define $u < v$ if either $|u| < |v|$, or $|u| = |v| = n$ and $u < v$ in the lexicographic order of X^n (from left to right). In any finite nonempty set of words the greatest element will be called its *leader*. Note that if u (resp. v) is the leader of A (resp. B), then $w = uv$ is the leader of the set AB consisting of all the products $u_1 v_1$, $u_1 \in A$, $v_1 \in B$, and that w has a unique such decomposition. Likewise we call the leader of a nonzero polynomial the leader of the words appearing in it. Then in a product of two nonzero polynomials, the leader is the product of the leaders.

Now let $ab' = ba'$. If $\deg(a) < \deg(b)$, we take $q = 0$, $r = a$ and we are done. Otherwise, let u, v' be the leaders of a, b' and v, u' be the leaders of b, a'. Then we must have $uv' = vu'$. Since $\deg(a) \geq \deg(b)$, we have $|u| \geq |v|$ and therefore by Levi's lemma for words, $u = vu_1$ for some word u_1.

Hence, for some scalar α, the polynomial $a - \alpha b u_1$ has a smaller leader than a. Now, we have $ab' = ba'$, hence $(a - \alpha b u_1)b' = b(a' - \alpha u_1 b')$, and we conclude by induction that $a - \alpha b u_1 = bq' + r$ for some polynomials

q', r with $\deg(r) < \deg(b)$. Thus $a = b(q' + \alpha u_1) + r$. Uniqueness is proved as in the commutative case. ∎

COROLLARY 9.2.2. *If I is a family of nonzero polynomials such that any two of them always have a nonzero right multiple, then the right ideal $Ik\langle X \rangle$ of $k\langle X \rangle$ is principal.*

Proof. We first prove that if two polynomials a, b have a nonzero right multiple, then the right ideal $ak\langle X \rangle + bk\langle X \rangle$ is principal. Indeed, we may suppose that $\deg(a) \geq \deg(b)$ and $ab' = ba'$, b, b' nonzero. Then the theorem shows that $a = bq + r$, with $\deg(r) < \deg(b)$. If $r = 0$, then the previous ideal is $bk\langle X \rangle$, and hence is principal. Otherwise, $rb' = b(a' - qb')$, r and b have a nonzero common right multiple (since r, b' are nonzero), and we conclude by induction that $rk\langle X \rangle + bk\langle X \rangle$ is principal. But since $a = bq + r$, the latter ideal is the same as the previous one, which concludes this part of the proof.

Now suppose that I is finite. We may suppose that it has at least two elements. Let $a \in I$ and $I' = I - \{a\}$. Then by induction, the right ideal $I'k\langle X \rangle$ is principal, equal to $bk\langle X \rangle$ say. We may choose some element c in I'. By hypothesis, a and c have a nonzero common right multiple, hence so have a and b, since $c \in bk\langle X \rangle$. By the first part, the right ideal $ak\langle X \rangle + bk\langle X \rangle$ is principal. But this ideal is equal to $ak\langle X \rangle + I'k\langle X \rangle = Ik\langle X \rangle$, which concludes the second part of the proof.

In the general case, for each nonempty finite subset I' of I, we have $I'k\langle X \rangle = ak\langle X \rangle$, for some nonzero polynomial a. Choose I' and a such that the latter has least possible degree. Then for any b in I, we have, by the previous part of the proof, $ak\langle X \rangle + bk\langle X \rangle = I'k\langle X \rangle + bk\langle X \rangle = (I' \cup b)k\langle X \rangle = ck\langle X \rangle$, for some polynomial c. By the minimality of the degree of a, we must have $\deg(a) \leq \deg(c)$. Since $a \in ck\langle X \rangle$, we conclude that $a = \alpha c$ for some nonzero scalar α. Hence $ak\langle X \rangle$ contains c, hence b, hence any element of I. Thus $ak\langle X \rangle = Ik\langle X \rangle$. ∎

9.3. Integral closure of the centralizer

The next result, due to Bergman, asserts that the centralizer is integrally closed, and is one of the key ingredients in his proof of the centralizer theorem.

THEOREM 9.3.1. *Let Z be the centralizer in $k\langle X \rangle$ of some nonscalar polynomial. Let $\bar{Z}$ be the integral closure of Z in its field of fractions. Then $\bar{Z} = Z$.*

9.3. Integral closure of the centralizer

Note that we take for granted that Z is commutative (Corollary 9.1.3). We shall use the following lemma.

LEMMA 9.3.2. *Let r be a polynomial and $R = \{a \in k\langle X\rangle \mid ra \in k\langle X\rangle r\}$. If $a, ab \in R$, with a nonzero, then $b \in R$.*

Note that $k\langle X\rangle r$ is a left ideal of $k\langle X\rangle$, but not a two-sided one in general. We leave to the reader the verification of the following fact (not needed in the proof): R is the largest subring of $k\langle X\rangle$ containing $k\langle X\rangle r$ as a two-sided ideal. This is called the *idealizer* of $k\langle X\rangle r$ in $k\langle X\rangle$.

Proof of lemma. We may suppose that r and b are nonzero. Since $a, ab \in R$, we have $ra = a'r$ and $rab = b'r$ for some polynomials a', b'. Hence $a'rb = rab = b'r$. Thus rb and r have a nonzero common left multiple. If we choose such a nonzero multiple of least degree, then it is of the form $crb = dr$ and c, d have no common left factor (otherwise, we may cancel it and lower the degree).

Now let t be a new variable, and $A = k[t], K = k(t)$, respectively, the algebras of polynomials and rational functions in t over k. We consider the ring $K\langle X\rangle$ of noncommutative polynomials in X over K and its subring $A\langle X\rangle$. Both have their degree function with respect to X and $A\langle X\rangle$ also has a degree with respect to t. Note that $A\langle X\rangle$ may be thought of as the algebra of polynomials in the variable t over the ring of coefficients $k\langle X\rangle$.

We have in $A\langle X\rangle$ the equality $cr(t - b) = (ct - d)r$. Viewing this equality in $K\langle X\rangle$, we obtain by Corollary 9.2.2 (with K replacing k) that $crK\langle X\rangle + (ct - d)K\langle X\rangle$ is a principal right ideal of $K\langle X\rangle$. Let $E \in K\langle X\rangle$ be a generator of this ideal. Then we have $cr = EF$, $ct - d = EG$, for some elements F, G in $K\langle X\rangle$. We claim that E, F, G may be chosen in $A\langle X\rangle$.

Let us assume this for the moment. We view each element of $A\langle X\rangle$ as a polynomial in the variable t over $k\langle X\rangle$. Then the equality $cr = EF$, together with $cr \in k\langle X\rangle$, implies that E is in $k\langle X\rangle$. Moreover, the equality $ct - d = EG$, together with $c, d, E \in k\langle X\rangle$, implies that E divides c and d on the left in $k\langle X\rangle$. Hence $E \in k$, since c, d have no common left factor.

This shows that the ideal $crK\langle X\rangle + (ct - d)K\langle X\rangle$ is equal to $K\langle X\rangle$. Hence we may express 1 as a right linear combination over $K\langle X\rangle$ of cr and $ct - d$. By multiplying by a suitable element ϕ in $k[t]$, we obtain a relation

$$crP + (ct - d)Q = \phi,$$

with P, Q in $A\langle X\rangle$. Viewing again the elements of $A\langle X\rangle$ as polynomials in t over $k\langle X\rangle$, with their t-degree, let p, q be the leading coefficients of

318   9. Centralizers of Noncommutative Series and Polynomials

P, Q. We may also suppose that the leading coefficient of ϕ is 1, and that ϕ has t-degree n. If the t-degree of P is $m > n$, then Q must be of degree $m - 1$ and we have, upon canceling c, $rp + q = 0$. Hence we have $cr(P - pt^m + bpt^{m-1}) + (ct - d)(Q - qt^{m-1}) = crP + (ct - d)Q - crpt^m + crbpt^{m-1} - cqt^m + dqt^{m-1} = \phi$, since $crbp = drp = -dq$. Hence in the relation above we may suppose that P has degree $\leq n$, and consequently Q has degree $\leq n - 1$: let q' be the coefficient of t^{n-1} in Q. Looking at the coefficient of t^n in the preceding relation, we obtain $crp + cq' = 1$, which implies $c(rp + q') = 1$, and so c is in k, since r, p, q' are in $k\langle X\rangle$.

Finally, we obtain $rb = c^{-1}dr$, which shows that $b \in R$.

It remains to prove the claim. This is very similar to Gauss' lemma for commutative polynomials. Call a nonzero polynomial $P \in A\langle X\rangle$ *primitive* if its coefficients have no nontrivial common divisor in A (A is a unique factorization domain as is $k[t]$). The product of two primitive polynomials is primitive: otherwise, let a be an irreducible common divisor of the coefficients of the product; then by taking the images of these three polynomials in (A/a), we obtain that this ring has zero-divisors, which is a contradiction. Now, if $P \in K\langle X\rangle$ is nonzero (K is the field of fractions of A), we may write $P = aQ$, where $a \in K$ and $Q \in A\langle X\rangle$ is primitive. This representation is unique up to a unit in A. Choose such a representation for each nonzero P: $P = c(P)P'$, $c(P) \in K$, $P' \in A\langle X\rangle$ primitive. $c(P)$ is called the *content* of P. Then $c(PQ) = c(P)c(Q)$ and $(PQ)' = P'Q'$, up to a unit in A: indeed, $c(P)c(Q)P'Q' = PQ = c(PQ)(PQ)'$, $P'Q'$, $(PQ)'$ are primitive, and we are done by uniqueness up to a unit of the representation.

Coming back to the claim, we had $PK\langle X\rangle + QK\langle X\rangle = EK\langle X\rangle$ for some nonzero polynomials $P, Q \in A\langle X\rangle$, $E \in K\langle X\rangle$. Hence, $P = EF$, $Q = EG$ with $F, G \in K\langle X\rangle$. Then, with equalities holding up to a unit of A (that is, a nonzero element of k), $P' = E'F'$, $Q' = E'G'$, $P = E'(c(P)F')$, $Q = E'(c(Q)G')$, $EK\langle X\rangle = E'K\langle X\rangle$, and we are done since $E', F', G' \in A\langle X\rangle$, and $c(P), c(Q) \in A$. ∎

Proof of Theorem 9.3.1. We show that if C is a subring of F (the field of fractions of Z), and a finitely generated Z-module, then C is contained in Z. This will imply the theorem, since $\bar{Z}$ is the union of such C.

Since C is a finitely generated sub-Z-module of F, there exists a common denominator for the elements of C, that is a nonzero element z_0 of Z such that $z_0 C \subset Z$. Let $I = z_0 C$. This is a C-module, since C is a subring of F. It is also an ideal of Z, since C is a Z-module.

The right ideal $Ik\langle X\rangle$ of $k\langle X\rangle$ is principal: indeed, two nonzero elements of I always have a nonzero common right multiple, I being a subset of Z, which is commutative by Corollary 9.1.3, and to this we apply Corollary 9.2.2. Hence $Ik\langle X\rangle = rk\langle X\rangle$, for some $r \in k\langle X\rangle$.

We have $Zr \subset rk\langle X\rangle$. Indeed, $Zr \subset Zrk\langle X\rangle = ZIk\langle X\rangle \subset Ik\langle X\rangle$ (since I is an ideal of $Z) = rk\langle X\rangle$.

Since r is nonzero (indeed $1 \in C$, hence $C \neq 0$), there exists a well-defined function $f_0: Z \to k\langle X\rangle$ such that $zr = rf_0(z)$ for any z in Z. Note that f_0 is an injective ring homomorphism.

Suppose that $c \in C$, with $c = z_1/z_2$, $z_i \in Z$. Then $rf_0(z_1) = z_1 r \in z_1 rk\langle X\rangle = z_1 Ik\langle X\rangle \subset z_1 CIk\langle X\rangle = z_2 cCIk\langle X\rangle \subset z_2 CIk\langle X\rangle \subset z_2 Ik\langle X\rangle$ (since I is a C-module) $= z_2 rk\langle X\rangle = rf_0(z_2)k\langle X\rangle$.

This shows that $f_0(z_1) \in f_0(z_2)k\langle X\rangle$, and there is a function $f: C \to k\langle X\rangle$ such that

$$f_0(z_1) = f_0(z_2)f(c), \text{ where } c = z_1/z_2. \tag{*}$$

This function is well defined, since f_0 is a homomorphism. Furthermore, f extends f_0 (take $z_2 = 1$ in (*)) and f is an injective homomorphism $C \to k\langle X\rangle$: indeed, by (*), $f(c)$ commutes with every element of $f_0(Z)$, since Z is commutative and f_0 is a homomorphism, hence we have

$$f_0(z_1 z_2' + z_1' z_2) = f_0(z_1)f_0(z_2') + f_0(z_1')f_0(z_2)$$
$$= f_0(z_2)f(c)f_0(z_2') + f_0(z_2')f(c')f_0(z_2)$$
$$= f_0(z_2)f_0(z_2')(f(c) + f(c')) = f_0(z_2 z_2')(f(c) + f(c')).$$

Since $c + c' = (z_1 z_2' + z_1' z_2)/(z_2 z_2')$, we deduce that $f(c+c') = f(c) + f(c')$. Similarly, f preserves multiplication.

Let $R = \{a \in k\langle X\rangle \mid ra \in k\langle X\rangle r\}$. Then we know by the above lemma that $a, ab \in R$, with a nonzero, implies $b \in R$. Note that $f_0(z) \in R$. For c as above, and $a = f_0(z_2), b = f(c)$, we have $a \in R$, $ab = f_0(z_1) \in R$, hence $b = f(c) \in R$.

This implies that for each c in C, we may define $g(c) \in k\langle X\rangle$ by the condition $rf(c) = g(c)r$. Then g is an injective homomorphism $C \to k\langle X\rangle$, and for $z \in Z$, we have $zr = rf_0(z) = rf(z) = g(z)r$, which implies that g is the identity on Z. Hence $g: C \to k\langle X\rangle$ extends the identity mapping $Z \to k\langle X\rangle$, and we obtain that $g(C)$ is a commutative subring of $k\langle X\rangle$ containing Z. Since Z is a centralizer, it is necessarily a maximal commutative subring of $k\langle X\rangle$, thus $g(C) = Z$, and thus $C = Z$, because g is injective. ∎

9.4. Homomorphisms into $k[t]$

The next result will allow us, still following Bergman, to embed the centralizer into a one-variable polynomial algebra.

THEOREM 9.4.1. *Let Z be a finitely generated subalgebra of $k\langle X\rangle$. Then there exists a nontrivial homomorphism $Z \to k[t]$.*

Proof. Let X^+ denote the set $X^* \setminus 1$, the set of nonempty words on X. Denote by X^ω the set of right-infinite words on X, order it lexicographically from left to right (where X is totally ordered) and for $u \in X^+$, denote by u^ω the infinite word $uuu\cdots$. For any set M of nonempty words, one has $u^\omega = v^\omega$ for each $u, v \in M$ if and only if the words in M are powers of the same word, which is unique if of minimum length.

Let Y be a finite set of polynomials generating the subalgebra Z and let $m \in X^+$ be such that m^ω is the maximum of all u^ω, for all nonempty words u appearing with nonzero coefficient in all elements of Y. Take m of minimum length, and let $M = \{m^n \mid n \geq 0\}$. For each $a = \sum_{u \in X^*} a_u u \in Z$, let $f(a) = \sum a_u u$, where the sum is over $u \in M$. Then f is a linear mapping from Z into the subalgebra of $k\langle X\rangle$ generated by m, which is isomorphic to a one-variable polynomial algebra. We show below that f is an algebra homomorphism, necessarily nontrivial since some element of Y involves a word of the form m^n, $n \geq 1$.

We claim that for any nonempty words u, v such that $u^\omega \leq m^\omega$ and $v^\omega \leq m^\omega$, one has $(uv)^\omega \leq m^\omega$. If moreover, one of the two former inequalities is strict, then the last is, too. Indeed, either $(uv)^\omega < (vu)^\omega$, and then $(vu)^\omega = v(uv)^\omega < v(vu)^\omega = v^2(uv)^\omega < v^2(vu)^\omega < \cdots < v^\omega$ (by taking the limit), hence $(uv)^\omega < m^\omega$, or $(vu)^\omega < (uv)^\omega$, and then similarly $(uv)^\omega = u(vu)^\omega < u(uv)^\omega < \cdots < u^\omega$, hence $(uv)^\omega < m^\omega$, or $(uv)^\omega = (vu)^\omega$, hence uv, vu are powers of the same word and thus equal, so that u, v are also powers of the same word, which implies $(uv)^\omega = u^\omega = v^\omega \leq m^\omega$. This ends the proof of the claim.

In order to finish the proof, it is enough to show that f preserves products. This will follow from the fact that $u, v \in M$ implies that $uv \in M$, and from the following fact: if u, v are words appearing in elements of Z, then $uv \in M$ implies that u and v are in M. We may suppose $u, v \neq 1$. Note first that each nonempty word w appearing in an element of Z is a product of elements appearing in Y. Hence, by the claim, $w^\omega \leq m^\omega$. Hence, again by the claim, if $u^\omega < m^\omega$ or if $v^\omega < m^\omega$, then $(uv)^\omega < m^\omega$, hence $uv \notin M$, which contradicts the assumption.

This proves the preceding fact. ∎

9.5. Bergman's centralizer theorem

THEOREM 9.5.1. *The centralizer of a nonscalar polynomial is isomorphic to $k[t]$.*

We first prove a lemma.

9.5. Bergman's centralizer theorem

LEMMA 9.5.2. *The centralizer of a nonscalar polynomial p is a finitely generated subalgebra of $k\langle X\rangle$, and also a finitely generated $k[p]$-module.*

Proof. We show first that if $\bar{p}$ is a nonscalar homogeneous polynomial, then there exists a homogeneous polynomial q such that each homogeneous polynomial commuting with $\bar{p}$ is a scalar multiple of some power of q. Indeed, the centralizer of $\bar{p}$ in $k\langle\langle X\rangle\rangle$ is of the form $k[[s]]$, for some series s with zero constant term (Theorem 9.1.1). Let q be the lowest homogeneous part of s. Now let r be a homogeneous polynomial commuting with $\bar{p}$. Then $r = \sum_{n\geq 0} \alpha_n s^n$, for some scalars α_n. In this sum, the lowest homogeneous part is $\alpha_n q^n$, where n with $\alpha_n \neq 0$ is chosen as small as possible. Thus by homogeneity $r = \alpha_n q^n$, which concludes the first part of the proof.

Now, let p be any nonscalar polynomial, of degree n. Let Z be its centralizer in $k\langle X\rangle$ and denote by $\bar{p}$ the highest homogeneous part of p. By what we have just seen, there exists a homogeneous polynomial q such that each homogeneous polynomial commuting with $\bar{p}$ is a scalar multiple of some power of q.

For $i = 0,\ldots,n-1$ such that some element of degree $\equiv i \pmod{n}$ exists in Z, let p_i denote such an element of least degree. If r is in Z, we may find l and i such that r and $p_i p^l$ have the same degree. Both polynomials are in Z, so that their highest homogeneous part commutes with $\bar{p}$, and they are scalar multiples of some power of q, necessarily of the same power. Hence for some scalar α, $r - \alpha p_i p^l$ is of degree less than that of r, and we conclude by induction that Z is spanned over k by the polynomials $p_i p^l$.

This shows that Z is finitely generated as an algebra, and also a finitely generated $k[p]$-module. ∎

Proof of Theorem 9.5.1. Let Z be the centralizer in $k\langle X\rangle$ of a nonscalar polynomial p. By Corollary 9.1.3, Z is commutative. We know by the lemma that Z is a finitely generated subalgebra of $k\langle X\rangle$. Hence by Theorem 9.4.1, there exists a nontrivial homomorphism $f: Z \to k[t]$, that is, we have $f(Z) \neq k$. Since by the lemma, Z is a finitely generated $k[p]$-module, it is of transcendence degree 1 over k. Hence f must be injective, otherwise $f(Z)$ would be of transcendence degree 0 over k (since noninjective homomorphisms decrease the transcendence degree) and thus would be equal to k (since no nonscalar polynomial is algebraic over k). We conclude that Z is isomorphic to $f(Z)$.

Now, by Theorem 9.3.1, Z, and hence $f(Z)$, is integrally closed. This implies by Theorem 9.7.3 in the Appendix that $f(Z)$, and hence Z, is isomorphic to $k[t]$. ∎

9.6. Free subalgebras and the defect theorem

The next result is due to Cohn.

THEOREM 9.6.1. *If a, b are elements of $k\langle\langle X\rangle\rangle$, without constant term, satisfying a nontrivial relation $S(a, b) = 0$ for some noncommutative series S in two variables, then a, b commute.*

Using Bergman's theorem (Theorem 9.5.1), we obtain the following corollary.

COROLLARY 9.6.2. *If P, Q are two polynomials in $k\langle X\rangle$ which do not freely generate a subalgebra of $k\langle X\rangle$, then they lie in a subalgebra of $k\langle X\rangle$ generated by a single polynomial.*

This is the defect theorem for two polynomials. We show below that it does not hold for more than two polynomials. Note that a similar result holds for two series, instead of polynomials (one has to use Cohn's theorem, Theorem 9.1.1).

Proof of corollary. We may suppose that P, Q have zero constant term, and that P is nonscalar. If P, Q do not generate a free subalgebra, then we have $S(P, Q) = 0$ for some nonzero noncommutative polynomial S in two variables. Hence, by Theorem 9.6.1, they commute. Hence Q lies in the centralizer of P and we are done by Theorem 9.5.1, since this centralizer is generated by a single polynomial. ∎

Proof of Theorem 9.6.1. We show the result by contradiction, and induction on $v(ab - ba)$. Suppose that a, b do not commute. Then $a, b \ne 0$. Let $S(u, v)$ be a noncommutative series in two variables u, v, of smallest possible valuation, such that $S(a, b) = 0$. Then we may write $S(u, v) = \alpha + uS_u(u, v) + vS_v(u, v)$ and we have the relation $\alpha + ab' + ba' = 0$, where $a' = S_u(a, b), b' = S_v(a, b)$ are elements of $k\langle\langle X\rangle\rangle$. Then α must be equal to 0, since a, b have no constant term, and a', b' must be nonzero, by the minimal choice of S. We may apply Lemma 9.1.2: assuming without loss of generality that $v(a) \ge v(b)$, we have $a = bq$ for some series q in $k\langle\langle X\rangle\rangle$. Let $q = \beta + q_1$, where β is the constant term of q. Then $0 \ne ab - ba = bqb - bbq = b(qb - bq) = b(q_1b - bq_1)$, so that $v(q_1b - bq_1) < v(ab - ba)$. Moreover, $S(a, b) = 0$ implies a similar nontrivial relation for q_1 and b, since $a = \beta b + bq_1$, and gives the desired contradiction. ∎

Consider the following example, due to Bergman: $f = xyxz + xy, g = xyx, h = zxyx + yx$. Then one has $fg = xyxzxyx + xyxyx = gh$. Hence

9.6. Free subalgebras and the defect theorem

f, g, h do not freely generate a subalgebra of $k\langle x, y, z\rangle$. We show that the three polynomials f, g, h do not belong to a free subalgebra generated by two polynomials in $k\langle x, y, z\rangle$. This will imply that the defect theorem does not hold in general.

First, we need to prove a necessary condition satisfied by free subalgebras. Following Kolotov, we say that a subalgebra A of $k\langle X\rangle$ is an *anti-ideal* if for $a \in k\langle X\rangle$ and any nonzero $b, c \in A$, $ab, ca \in A$ implies $a \in A$. The next result is due to Kolotov.

THEOREM 9.6.3. *If A is a free subalgebra of $k\langle X\rangle$, then it is an anti-ideal.*

Proof. We know by Theorem 9.2.1 that one can perform the Euclidean division of u by v in $k\langle X\rangle$ whenever u, v have a nonzero right multiple in $k\langle X\rangle$. We claim that if u, v are in the free subalgebra A and if they have a nonzero right multiple in A, then the quotient and the remainder are also in A. This being assumed, let $a \in k\langle X\rangle$ and $b, c \in A$ with $b, c \neq 0$ and $ab, ca \in A$. We may suppose that a is nonzero. Since ca and c have the nonzero right multiple cab, and since ca, c, cab, b, ab are all in A, the claim implies that the quotient of the division of ca by c, which is a, is in A.

It remains to prove the claim. Suppose that A is freely generated by a set Y of polynomials in $k\langle X\rangle$, and denote by Deg the degree function in A with respect to this set Y. Let u, v in A have a nonzero right multiple in A. Then we have by Theorem 9.2.1 applied to $k\langle X\rangle$ and $k\langle Y\rangle$ that $u = vq + r$, $u, v \in k\langle X\rangle$, $\deg(r) < \deg(v)$ and $u = vQ + R$, $Q, R \in A$, $\mathrm{Deg}(R) < \mathrm{Deg}(v)$. We prove by induction on $n = \mathrm{Deg}(v)$ that $Q = q, R = r$, which will imply the claim. If $n = 0$, this is clear. If $R = 0$, the result is also clear, since the quotient and remainder of Euclidean division in $k\langle X\rangle$ are unique. Hence we may suppose that $R \neq 0$. We know that $uv' = vu'$ for some nonzero u', v' in A. Thus $(vQ + R)v' = vu'$, which implies $v(u' - Qv') = Rv'$. All the elements involved in this equation are in A, and since $\mathrm{Deg}(R) < \mathrm{Deg}(v)$, the induction hypothesis implies that $v = Rq' + r'$, with $q', r' \in A$ and $\deg(r') < \deg(R)$, $\mathrm{Deg}(r') < \mathrm{Deg}(R)$. Since $\mathrm{Deg}(R) < \mathrm{Deg}(v)$, q' is not a constant, thus $\deg(R) < \deg(R) + \deg(q') = \deg(v)$. This implies by uniqueness of quotient and remainder that $Q = q, R = r$. ∎

We come back to the preceding example: suppose that f, g, h lie in a free subalgebra A. Then, since $fg = gh$, the claim in the preceding proof shows that $f = gq + r, q, r \in A, \deg(r) < \deg(g)$. Since evidently $f = gz + xy, \deg(xy) = 2 < 3 = \deg(g)$, we have by uniqueness that $z, xy \in A$. Furthermore, $g = (xy)x = x(yx), xy, yx = h - zg$ all lie in A, so that $x \in A$, since A is an anti-ideal. For the same reason, $xy, yx \in A$ implies that $y \in A$. Thus A contains x, y, z and is thus equal to $k\langle x, y, z\rangle$,

9. Centralizers of Noncommutative Series and Polynomials

which cannot be generated by two elements (since its commutative image $k[x, y, z]$ has transcendence degree 3).

9.7. Appendix: some commutative algebra

In this section, all rings and fields are commutative, without zero divisors, and all algebras are over the field k. We begin with Lüroth's theorem. Here t is a variable.

THEOREM 9.7.1 (Lüroth). *If F is subfield of $k(t)$ properly containing k, then it is isomorphic to $k(t)$.*

We need a lemma.

LEMMA 9.7.2. *If $u = f(t)/g(t) \in k(t)$ is nonscalar, with f, g relatively prime in $k[t]$, then t is algebraic over $k(u)$, of degree $\deg(u) = \max\{\deg(f), \deg(g)\}$.*

Proof. Note that t is a root of the polynomial $f(x) - ug(x)$ in $k(u)[x]$, where x is a new variable. Since this is a nonzero polynomial in x (otherwise u is scalar), we deduce that t is algebraic over $k(u)$. Note that $k(t, x)$ is of transcendence degree 2 over k, so that $k(u, x)$ is too. Hence u, x are algebraically independent. If the preceding polynomial is not irreducible in $k(u)[x]$, then it may be factorized in $k(u)[x]$, and by Gauss' lemma, also in $k[u, x]$. Since it is linear in u, one factor must be independent of u, which is impossible because f, g are relatively prime. Hence, the polynomial is irreducible, and since its x-degree is $\max\{\deg(f), \deg(g)\}$, the lemma follows. ∎

Proof of Theorem 9.7.1. There exists $u \in F \setminus k(t)$. By the lemma, t is algebraic over $k(u)$. Hence t is also algebraic over F. Let the minimal polynomial of t over F be

$$\phi(x) = x^n + u_1(t)x^{n-1} + \cdots + u_n(t),$$

where $u_i(t) \in F$. By multiplying the rational fractions $u_i(t)$ by their lowest common denominator, we obtain a polynomial

$$\Phi(x, t) = v_0(t)x^n + v_{n-1}(t)x^{n-1} + \cdots + v_n(t),$$

where the $v_i(t)$ are polynomials without common divisor, and $v_0 \neq 0$. This implies that Φ, considered as element of $k[t][x]$, is primitive. The u_i are not all constant, and we choose j such that $u_j \notin k$. By the lemma, t has degree $m = \deg(u_j)$ over $k(u_j)$, while its degree over F is n.

9.7. Appendix: some commutative algebra

Thus $m = [k(t):k(u_j)] = [k(t):F][F:k(u_j)] = n[F:k(u_j)]$, and to complete the proof, we need only show that $m = n$, for then $F = k(u_j)$. Write $u_j = a(t)/b(t)$ with relatively prime polynomials a, b. Note that the t-degree of Φ is $\geq \deg(a), \deg(b)$. We may assume b is monic and we have by the lemma $m = \max\{\deg(a), \deg(b)\}$. The polynomial $a(x) - u_j b(x)$ in $F[x]$ has t as a root, so that it is divisible by $\phi(x)$ in $F[x]$. Hence we obtain $a(x) - u_j(t)b(x) = q(x)\phi(x)$, $q(x) \in F[x]$. Replace $u_j(t)$ in terms of $a(t), b(t)$ and multiply by $b(t)$. Then we obtain

$$a(x)b(t) - a(t)b(x) = Q(x,t)\Phi(x,t), \qquad (*)$$

where Q is a polynomial in x. Since Φ and $a(x)b(t) - a(t)b(x)$ are polynomials in $k[t][x]$, and the first is primitive, we deduce by Gauss' lemma that Q is also such a polynomial. The polynomial $a(x)b(t) - a(t)b(x)$ has degree m in t, and Φ has degree at least m in t. This implies that Q is independent of t. Suppose that Q depends on x. Then it has a zero, α, in some extension of k. Thus $a(\alpha)b(t) - a(t)b(\alpha) = 0$. If $b(\alpha) = 0$, then $a(\alpha) = 0$, which implies that a, b are both divisible by the minimal polynomial of α over k, and they are not relatively prime. Hence $b(\alpha) \neq 0$. We thus have $u_j(t) = a(t)/b(t) = a(\alpha)/b(\alpha)$, and $u_j(t)$ is algebraic over k, hence t is too and we have a contradiction. This shows that Q is also independent of x. Hence $m = n$ by comparing the x-degrees of the two sides in $(*)$. ∎

Finally, we shall prove Cohn's result characterizing the free subalgebras of $k[t]$.

THEOREM 9.7.3. *A subalgebra of $k[t]$ is free (and then isomorphic to k or $k[t]$) if and only if it is integrally closed.*

In order to prove the theorem, we need some results of valuation theory. Before that, recall that a *local ring* is a ring R which has a unique maximal ideal M, which is (necessarily) the set of noninvertible elements of R. If S is a commutative integral domain, and P a prime ideal of S, a classical construction in ring theory is the *local ring* at P, which is $S_P = \{a/b \mid a, b \in S, b \notin P\}$ (we view S as a subring of its field of fractions). Then S_P is a local ring, with maximal ideal $PS_P = \{a/b \mid a, b \in S, a \in P, b \notin P\}$. Observe that $PS_P \cap S = P$, since P is prime.

Recall also that a subring R of a field F is a *valuation ring* if for any x in F, either x or x^{-1} is in R.

THEOREM 9.7.4. (i) *If R is a valuation subring of the field F, then it is a local ring with maximal ideal $\{x \in R \mid x^{-1} \notin R\}$.*

(ii) *Each proper valuation subalgebra of $k(t)$ is either of the form $R_p = \{f/g \mid f \in k[t], g \in k[t] \setminus pk[t]\}$ for some irreducible polynomial p in $k[t]$, or of the form $D = \{a/b \mid a, b \in k[t], \deg(b) \geq \deg(a)\}$.*

(iii) (Chevalley) *If R is a subring of a field F and P a prime ideal of R, then there exists a valuation subring S of F such that S contains R and $M \cap R = P$, where M is the maximal ideal of S.*

(iv) (Krull) *If $F \subset G$ is a field extension, and R a valuation ring of F, then there exists a valuation ring S of G such that $R = S \cap F$.*

(v) (Krull) *If a ring is integrally closed, then it is an intersection of valuation subrings of its field of fractions.*

Note that the converse of part (v) is also true, but we shall not use it.

Proof of theorem. (i) It is enough to show that $\{x \in R \mid x = 0 \text{ or } x^{-1} \notin R\}$ is an ideal of R. If x, y are in this set and r is in R, then rx is also, otherwise, $r^{-1}x^{-1}$ is in R, which implies that x^{-1} is in R, which is a contradiction. Furthermore, we have either $x^{-1}y \in R$, or $y^{-1}x \in R$, since R is a valuation ring. In the first case, we have $y = xr$ for some $r \in R$, hence $x + y = x(1 + r)$ is in the preceding set. The other case is symmetric.

(ii) Let R be a proper valuation subalgebra of $k(t)$. Suppose first that R contains $k[t]$. If p, q are relatively prime polynomials such that p/q is in R, then $ap + bq = 1$ for some polynomials a, b, and thus $1/q = ap/q + b$ is in R. This implies that if p, q are distinct irreducible polynomials, then at least one of $1/p, 1/q$ is in R: indeed, since R is a valuation subring, either p/q is in R, hence $1/q$ is too, or q/p is, and $1/p$ too. Thus for at most one irreducible polynomial, $1/p$ is not in R. Since $R \neq k(t)$, there is exactly one such p, and $R = R_p$.

Suppose now that R does not contain $k[t]$. Hence R contains $x = 1/t$, thus R contains $k[x]$. By what we have just seen, R is of the form $S = \{a/b \mid a \in k[x], b \in k[x] \setminus qk[x]\}$, for some irreducible polynomial $q(x) \in k[x]$. Now, if $q = x$, it is easy to see that $S = D$, and if $q \neq x$, then $S = R_p$, where $p(t)$ is the reciprocal polynomial of $q(x)$.

(iii) We may suppose that $P \neq 0$. Let $\mathscr{F}$ denote the family of local subrings S of F, with maximal ideal M, such that S contains R and that $M \cap R = P$. This family is nonempty, since it contains R_P. Indeed R_P contains R, has the unique maximal ideal $M = PR_P$, and $M \cap R = P$, since P is prime.

This family $\mathscr{F}$ is inductive, as the reader may verify. Hence, by Zorn's lemma, there is a maximal element S in this family. We show that S is the required valuation ring. Suppose by contradiction that S is not a valuation subring of F. Then for some x in F, we have $x \notin S, x^{-1} \notin S$. For $S' = S[x]$ or $S' = S[x^{-1}]$, we have $S' \neq S$. Suppose that $S'M \neq S'$, where M denotes the unique maximal ideal of S. Then $S'M$ is contained

9.7. Appendix: some commutative algebra

in some maximal ideal M' of S', and $S'_{M'}$ would be in the family $\mathscr{F}$. Indeed $M'S'_{M'} \cap S = M'S'_{M'} \cap S' \cap S = M' \cap S$ is an ideal of S containing M, hence is equal to M by maximality of M (since $1 \notin M'S'_{M'}$). Thus $M'S'_{M'} \cap R = M'S'_{M'} \cap S \cap R = M \cap R = P$. This contradicts the maximality of S in the family $\mathscr{F}$, and we conclude that $S'M = S'$. This implies that one can write

$$1 = \sum_{0 \le i \le m} a_i x^i, \quad 1 = \sum_{0 \le i \le n} b_i x^{-i},$$

with a_i, b_i in M. We choose n, m minimal. Then if for example $n \le m$, we have $x^n = \sum_{0 \le i \le n} b_i x^{n-i}$, hence $(1 - b_0) x^n = \sum_{1 \le i \le n} b_i x^{n-i}$. Since $b_0 \in M$, $1 - b_0$ is not in M (otherwise $M = S$), it hence is invertible in S (since S is a local ring), and this shows that x^n is a linear combination with coefficients in M of $1, x, \ldots, x^{n-1}$. Hence $x^{n+1}, \ldots, x^m$ may also be written as such a linear combination, and using the first equality above, we would have $1 = \sum_{1 \le i \le n-1} c_i x^i$, for c_i in M, which contradicts the minimality of m. This contradiction shows that we cannot have $x, x^{-1} \notin S$, and S must be a valuation ring.

(iv) Let P be the unique maximal ideal of R. It is a prime ideal, hence we may apply part (iii), and obtain that there is a valuation subring S of G, with maximal ideal M, such that S contains R and $P = M \cap R$. Then $S \cap F$ contains R. Suppose that there exists some element x in $S \cap F$ and not in R. Then x^{-1} is in R, and by part (i) even in P, hence in M, which contradicts by part (i) the fact that x is in S.

(v) Let R be integrally closed in its field of fractions F, and let $a \in F \setminus R$. We show that there is a valuation subring S of F such that S contains R but not a (this will imply part (v)). Note that $a \notin R[a^{-1}]$, otherwise $a = \sum_{0 \le i \le n} r_i a^{-i}$, for some r_i in R, which implies $a^{n+1} = \sum_{0 \le i \le n} r_i a^{n-i}$, hence a is integral over R, and thus belongs to R, which is a contradiction. Consider the family of subrings S of F such that $R[a^{-1}] \subset S$ and $a \notin S$. This family is nonempty, and is inductive. Hence, by Zorn's lemma, it has a maximal element, S say. We now show that S is a valuation subring of F.

Suppose that $x \in F, x \notin S$. Then S is strictly included in $S[x]$, which implies by maximality of S that $a \in S[x]$. Hence $a = s_0 x^n + \cdots + s_{n-1} x + s_n$, which implies after multiplying by $a^{-1} x^{-n}$ that $x^{-n}(1 - s_n a^{-1}) = s_{n-1} a^{-1} x^{1-n} + \cdots + s_0 a^{-1}$. We claim that S is a local ring. This being assumed for the moment, a^{-1} must be in its unique maximal ideal, hence $1 - s_n a^{-1}$ is invertible in S, which implies that x^{-1} is integral over S. Let S' denote the integral closure of S in F. If S were strictly included in S', then we would have $a \in S'$, by maximality of S. Hence $a^m = t_1 a^{m-1} + \cdots + t_m$, for some t_i in S, and $a = t_1 + \cdots + t_m a^{1-m} \in S$ (since $a^{-1} \in S$), which is a contradiction. Thus $S = S'$, $x^{-1} \in S$ and S is a valuation ring.

It remains to prove the claim. Since $a \notin S$, Sa^{-1} is strictly included in S. Hence for some maximal ideal P of S, one has $Sa^{-1} \subset P$. Then $S \subset S_P$, and $a \notin S_P$ (otherwise $a = b/c$ with $b, c \in S$, $c \notin P$, hence $c = a^{-1}b \in P$, which is a contradiction). Thus, by the maximality of S, we have $S = S_P$ and S is a local ring. ∎

Proof of Theorem 9.7.3. Let R be a free subalgebra of $k[t]$. Since it is free and commutative, it can be only k or isomorphic to $k[t]$. In both cases, it is integrally closed, as is well known.

Conversely, suppose that R is an integrally closed subalgebra of $k[t]$. We may suppose that R is not equal to k. By Theorem 9.7.4(v), R is an intersection of valuation subalgebras of its field of fractions F. Now, by Theorem 9.7.4(iv), each valuation subalgebra of F is of the form $S \cap F$, where S is a valuation subalgebra of $k(t)$. Hence R is the intersection of F with the family of valuation subalgebras of $k(t)$ containing R. These valuation subalgebras are given in Theorem 9.7.4(ii): note that, with the notations of this theorem, $k[t] \subset R_p$, which implies that R_p contains R. Hence the only valuation subalgebra which possibly does not contain R is D, the second case of Theorem 9.7.4(ii). But certainly D cannot contain R, since the intersection of all valuation subalgebras of $k(t)$ is k, and we would have $R = k$, which was excluded. Hence R is the intersection of all valuation subalgebras of F, except $D \cap F$.

Now by Lüroth's theorem (Theorem 9.7.1), $F = k(x)$, for some x in $k(t) \setminus k$. Note that $k[x]$ is the intersection of all valuation subalgebras of $k(x)$, except $E = \{a/b \mid a, b \in k[x], \deg_x(a) \leq \deg_x(b)\}$ (a consequence of Theorem 9.7.4(ii). Also note that x is in all valuation subalgebras of $k(x)$, except E (ibid.). Suppose that x is not in $D \cap F$: since x is in all valuation subalgebras of F except E, we must have $E = D \cap F$, and this implies that $R = k[x]$. If x is in D, then we claim that $y = 1/(x - \alpha) \notin D$, for suitable $\alpha \in k$. Then we have $F = k(y)$, and we conclude as before that $R = k[y]$.

For the claim, note that $x = f/g$, with $\deg(f) \leq \deg(g)$. For some $\alpha \in k$, $x - \alpha = f'/g$, with $\deg(f') < \deg(g)$. Then $g/f' \notin D$, which proves the claim and completes the proof. ∎

Notes

Bergman's centralizer theorem was conjectured by Cohn (Cohn 1963, p. 348). For its proof, we have followed the original proof (Bergman 1969), with the help of Cohn 1985. We did not make any real improvement, but hope to give a larger audience to this result and its proof. That is why we have also included in the Appendix all the results of commutative algebra which are needed in order to prove Cohn's characterization of free subalgebras of $k[t]$. For Lüroth's theorem, we have followed Cohn 1991, pp. 172–174. For valuation ring theory, we have followed Ribenboim 1965. For the proof of Theorem 9.3.1, we have followed Melançon 1993, which also gives a computational version of Cohn's weak algorithm. The proof of Lemma 9.3.2 is due to Cohn. It would be a real challenge to give to the centralizer theorem a proof that would be simpler that the one found by Bergman thirty years ago.

A result of Cohn (Theorem 9.1.1) allows one to deduce the defect theorem for two polynomials from the centralizer theorem. Then a result of Kolotov implies that the defect theorem does not hold in general (he works out an example of three polynomials due to Bergman). His result gives a necessary condition for a subalgebra to be free, which is similar to the condition of *stability*, necessary and sufficient for a submonoid of a free monoid to be free (see Chapter 6). Kolotov's condition is not sufficient in general, although it is in the one-variable case (see Kolotov 1978 and Cohn 1985).

No simple condition characterizing free subalgebras of a free associative algebra is known. It seems likely that there is no such condition, in view of Bergman 1979 (whose title is clear enough, so requires no further comments from the author of these lines).

Note that the centralizer theorem and the defect theorem have analogues in free monoids, free groups and free Lie algebras. See Theorem 6.2.1, Lyndon and Schupp 1977 (in the free group, it is a consequence of the theory of Nielsen transformations, see especially Prop. 1.2.2 and Prop. 1.2.5) and Reutenauer 1993 Th. 2.2.10. A defect theorem (in the terminology of Lentin) in an algebraic structure is a theorem that asserts that if n elements generate a substructure which is not free, then they lie in some substructure generated by $n-1$ elements. For the free field the centralizer theorem is not known in general, but partial results exist; see Cohn 1985, Section 7.7.8, and Cohn 1978.

CHAPTER 10

Transformations on Words and q-Calculus

10.0. Introduction

When sorting was systematically studied in the 1960s and 1970s, in particular for comparing the different methods used in practice, it was essential to go back to the classics, to the works by MacMahon and especially to his treatise on combinatory analysis. He had made an extensive study of the distributions of several *statistics* on permutations, or more generally, on "permutations" with repeated elements, simply called *words* in what follows. The most celebrated of those statistics is probably the classical *number of inversions* which stands for a very natural measurement of how far a permutation is from the identity. There are several other statistics relevant to sorting or to statistical theory, such as the *number of descents*, the *number of excedences*, the *major index*, and more recently the *Denert statistic*.

MacMahon had already calculated the distributions of the early statistics and proved that some of them were equally distributed on each class of rearrangements of a given word. Let us state one of his basic results. To this end suppose that X is a finite nonempty set, referred to as an *alphabet*. For convenience, take X to be the subset $\{1, 2, \ldots, r\}$ ($r \geq 1$) of the positive integers, equipped with its standard ordering. Let $\mathbf{c} = (c_1, c_2, \ldots, c_r)$ be a sequence of r nonnegative integers and v be the nondecreasing word $v = 1^{c_1} 2^{c_2} \ldots r^{c_r}$, i.e., $v = y_1 y_2 \ldots y_m$ with $m = c_1 + c_2 + \cdots + c_r$ and $y_1 = \cdots = y_{c_1} = 1$, $y_{c_1+1} = \cdots = y_{c_1+c_2} = 2, \ldots,$ $y_{c_1+\cdots+c_{r-1}+1} = \cdots = y_m = r$. The class of all rearrangements of v, i.e., the class of all the words w that can be obtained from v by permuting its letters in some order, will be denoted by $R(\mathbf{c})$.

If $w = x_1 x_2 \ldots x_m$ is such a word, the *number of excedences*, exc w,

10.0. Introduction

and the *number of descents*, des w, and also the *major index* of w are classically defined as

$$\begin{aligned}
\text{exc } w &= \text{Card}\{i \mid 1 \leq i \leq m, x_i > y_i\}, \\
\text{des } w &= \text{Card}\{i \mid 1 \leq i \leq m-1, x_i > x_{i+1}\}, \\
\text{maj } w &= \sum \{i \mid 1 \leq i \leq m-1, x_i > x_{i+1}\}.
\end{aligned} \qquad (10.0.1)$$

Let $A_{\mathbf{c}}^{\text{exc}}(t)$ (resp. $A_{\mathbf{c}}^{\text{des}}(t)$) be the generating polynomial for the class $R(\mathbf{c})$ by the statistic "exc" (resp. "des"), i.e.,

$$A_{\mathbf{c}}^{\text{exc}}(t) = \sum_w t^{\text{exc } w}, \qquad A_{\mathbf{c}}^{\text{des}}(t) = \sum_w t^{\text{des } w} \quad (w \in R(\mathbf{c})).$$

MacMahon showed that those two polynomials were equal for every $\mathbf{c}$. More explicitly he showed that the generating functions for those two families of polynomials had the same analytic expression. This raises the question of providing methods for deriving those analytic expressions. This will be done in the first part of this chapter in the more general set-up of q-calculus, as not only single statistics will be considered, but pairs of statistics.

Now saying that the preceding two polynomials are equal for every $\mathbf{c}$ implies that the two statistics exc and des are *equidistributed* on each rearrangement class $R(\mathbf{c})$. Proving this equidistribution property in a *bijective* manner means that a bijection ϕ on each rearrangement class $R(\mathbf{c})$ is to be constructed with the property that

$$\text{exc } w = \text{des } \phi(w) \qquad (10.0.2)$$

holds for every w.

This brings up the matter of the second part of this chapter: does there exist a systematic way for constructing those bijections? We shall see that a large class of those bijections can be constructed by means of a *straightening* algorithm on biwords which is based on a *commutation rule* itself defined on the biwords. Although any commutation rule can be integrated in the algorithm, our attention will be focused on the *contextual* commutation that serves to construct a bijection Φ mapping a pair of statistics onto another pair. Instead of property (10.0.2) we shall have

$$(\text{exc}, \text{den}) \, w = (\text{des}, \text{maj}) \, \Phi(w), \qquad (10.0.3)$$

where maj and den are the *major index* and the *Denert statistic* (further defined in Section 10.11), respectively.

For every class $R(\mathbf{c})$ introduce the two generating polynomials

$$A_{\mathbf{c}}^{\text{exc,den}}(t, q) = \sum_{w \in R(\mathbf{c})} t^{\text{exc } w} q^{\text{den } w}, \qquad A_{\mathbf{c}}^{\text{des,maj}}(t, q) = \sum_{w \in R(\mathbf{c})} t^{\text{des } w} q^{\text{maj } w}.$$

An analytical expression for $A_c^{\text{des,maj}}(t,q)$ was already derived by MacMahon (see Section 10.2). But there is no direct way for proving that the polynomial $A_c^{\text{exc,den}}(t,q)$ is equal to that analytical expression. Thus the construction of the bijection Φ is crucial.

After recalling the fundamental material on q-calculus in Section 10.1 we present the *MacMahon Verfahren* which is a rearrangement method that has been generalized in various contexts. In Section 10.3 we discuss an insertion technique that makes possible the derivation of a recurrence relation for generating polynomials for words and in Section 10.4 we show how to go from a recurrence relation to an identity between q-series. All the calculations made in Sections 10.2–4 involve the generating polynomials $A_c^{\text{des,maj}}(t,q)$.

The second part of the chapter (Sections 10.5–11) is devoted to the construction of the main algorithm. It involves the introduction of commutation rules on biwords that serve to construct both bijections ϕ and Φ. We conclude with the proofs of the equidistribution properties (10.0.2), (10.0.3).

10.1. The q-binomial coefficients

We use the following notation on q-calculus. First, $(a;q)_n$ denotes the *q-ascending factorial*

$$(a;q)_n = \begin{cases} 1, & \text{if } n = 0, \\ (1-a)(1-aq)\ldots(1-aq^{n-1}), & \text{if } n \geq 1. \end{cases}$$

Here a and q are any symbols, variables, or real or complex numbers. The *q-binomial coefficient* (or the Gaussian polynomial) is defined by

$$\begin{bmatrix} n \\ k \end{bmatrix} = \begin{cases} \dfrac{(q;q)_n}{(q;q)_k (q;q)_{n-k}}, & \text{if } 0 \leq k \leq n, \\ 0, & \text{otherwise.} \end{cases} \quad (10.1.1)$$

The following properties of the q-binomial coefficients are straightforward and given without proof:

$$\begin{bmatrix} n \\ 0 \end{bmatrix} = \begin{bmatrix} n \\ n \end{bmatrix} = 1; \quad \begin{bmatrix} n \\ k \end{bmatrix} = \begin{bmatrix} n \\ n-k \end{bmatrix}; \quad (10.1.2)$$

$$\begin{bmatrix} n \\ k \end{bmatrix} = \begin{bmatrix} n-1 \\ k \end{bmatrix} + q^{n-k} \begin{bmatrix} n-1 \\ k-1 \end{bmatrix}; \quad (10.1.3)$$

$$\begin{bmatrix} n \\ k \end{bmatrix} = \begin{bmatrix} n-1 \\ k-1 \end{bmatrix} + q^k \begin{bmatrix} n-1 \\ k \end{bmatrix}; \quad (10.1.4)$$

$$\lim_{q \to 1} \begin{bmatrix} n \\ k \end{bmatrix} = \binom{n}{k}. \quad (10.1.5)$$

10.1. The q-binomial coefficients

The q-binomial coefficient has a combinatorial interpretation in terms of nondecreasing sequences of integers, as stated in the next proposition, where $\mathbf{a} = (a_1, \ldots, a_n)$ denotes a nonincreasing sequence of nonnegative integers and where $\|\mathbf{a}\| = a_1 + \cdots + a_n$.

PROPOSITION 10.1.1. *For each pair of nonnegative integers (k, n) we have*

$$\begin{bmatrix} k+n \\ k \end{bmatrix} = \sum_{k \geq a_1 \geq \cdots \geq a_n \geq 0} q^{\|\mathbf{a}\|} = \sum_{n \geq b_1 \geq \cdots \geq b_k \geq 0} q^{\|\mathbf{b}\|}. \quad (10.1.6)$$

Proof. The fact that the above two summations are equal follows from the symmetry of the q-binomial coefficient $\begin{bmatrix} k+n \\ k \end{bmatrix}$ in k and n. Denote the first summation by $D(k, n)$ and let $D(0, 0) = 1$. Then $D(n, 0) = D(0, k)$ for every $n \geq 1$ and $k \geq 1$. Next, for k and $n \geq 1$

$$D(k, n) = \sum_{\mathbf{a},\, a_n = 0} q^{\|\mathbf{a}\|} + \sum_{\mathbf{a},\, a_n \geq 1} q^{\|\mathbf{a}\|}.$$

Let $b_i = a_i - 1$ $(i = 1, \ldots, n)$ in the second summation. Then

$$D(k, n) = \sum_{k \geq a_1 \geq \cdots \geq a_{n-1} \geq 0} q^{\|\mathbf{a}\|} + \sum_{k-1 \geq b_1 \geq \cdots \geq b_n \geq 0} q^{n+\|\mathbf{b}\|}$$
$$= D(k, n-1) + q^n D(k-1, n).$$

This shows that $D(k, n)$ satisfies the recurrence relation (10.1.2), (10.1.3) for the q-binomial coefficient $\begin{bmatrix} k+n \\ k \end{bmatrix}$. ∎

Proposition 10.1.1 provides the generating function for the nonincreasing sequences of integers bounded from above. There is also a formula for sequences without upper bound, as explained next. For each integer $n \geq 0$ consider the expansion

$$\frac{1}{(t; q)_{1+n}} = \sum_{s \geq 0} \sum_{m \geq 0} t^s q^m p(s, m). \quad (10.1.7)$$

The coefficient $p(s, m)$ is equal to the number of sequences of nonnegative integers $(i_0, i_1, \ldots, i_n)$ such that $i_0 + i_1 + \cdots + i_n = s$ and $1 \cdot i_1 + 2 \cdot i_2 + \cdots + n \cdot i_n = m$. Consequently, $p(s, m)$ is equal to the number of nonincreasing sequences $\mathbf{a} = (a_1, a_2, \ldots, a_s)$ such that $n \geq a_1 \geq \cdots \geq a_s \geq 0$ and $\|\mathbf{a}\| = m$. It follows from Proposition 10.1.1 that for each $s \geq 0$

$$\sum_{m \geq 0} q^m p(s, m) = \sum_{n \geq a_1 \geq \cdots \geq a_s \geq 0} q^{\|\mathbf{a}\|} = \sum_{s \geq a_1 \geq \cdots \geq a_n \geq 0} q^{\|\mathbf{a}\|}, \quad (10.1.8)$$

so that

$$\frac{1}{(t; q)_{1+n}} = \sum_{s \geq 0} t^s \sum_{s \geq a_1 \geq \cdots \geq a_n \geq 0} q^{\|\mathbf{a}\|} = \sum_{s \geq 0} t^s \begin{bmatrix} n+s \\ n \end{bmatrix}. \quad (10.1.9)$$

10.2. The MacMahon Verfahren

Let $A_{\mathbf{c}}(t,q) = A_{\mathbf{c}}^{\text{des,maj}}(t,q)$ be the generating polynomial for the class $R(\mathbf{c})$ by the pair (des, maj). Those two statistics have been defined in (10.0.1). By convention, $A_{\mathbf{c}}(t,q) = 1$, if $\mathbf{c}$ is the null sequence. In this section we shall derive the identity

$$\frac{1}{(t;q)_{1+\|\mathbf{c}\|}} A_{\mathbf{c}}(t,q) = \sum_{s \geq 0} t^s \begin{bmatrix} c_1 + s \\ s \end{bmatrix} \cdots \begin{bmatrix} c_r + s \\ s \end{bmatrix}, \qquad (10.2.1)$$

by means of the so-called *MacMahon Verfahren*.

First let us derive a symmetry property for the polynomials $A_{\mathbf{c}}(t,q)$. For each permutation σ of the set of letters $\{1,2,\ldots,r\}$ denote by $\sigma\mathbf{c}$ the sequence $(c_{\sigma(1)}, c_{\sigma(2)}, \ldots, c_{\sigma(r)})$, so that $R(\sigma\mathbf{c})$ is the class of all the rearrangements of the word $1^{c_{\sigma(1)}} 2^{c_{\sigma(2)}} \ldots r^{c_{\sigma(r)}}$.

THEOREM 10.2.1. *For each permutation σ of the set $\{1,2,\ldots,r\}$ the distributions of the pair* (des, maj) *over $R(\mathbf{c})$ and over $R(\sigma\mathbf{c})$ are identical. In other words, $A_{\mathbf{c}}(t,q) = A_{\sigma\mathbf{c}}(t,q)$.*

Proof. It suffices to prove the property when σ is a transposition $(i, i+1)$ of two adjacent integers ($1 \leq i \leq r-1$). Consider a word w in $R(\mathbf{c})$ and write all its factors of the form $(i+1)i$ in bold-face; then replace all the maximal factors of the form $i^a(i+1)^b$, with $a \geq 0$, $b \geq 0$, that do not involve any bold-face letters by $i^b(i+1)^a$. Finally, rewrite all the bold-face letters in ordinary type. Clearly, the transformation is a bijection that maps each word w in $R(\mathbf{c})$ onto a word w' in $R((i, i+1)\mathbf{c})$ with the property that (des, maj) w = (des, maj) w'. ∎

To derive identity (10.2.1) we proceed as follows. By (10.1.9) the left-hand side of (10.2.1) is equal to the sum of the series

$$\sum t^{s' + \text{des } w} q^{\|\mathbf{a}\| + \text{maj } w},$$

extended over the triples $(s', \mathbf{a}, w)$, where s' is a nonnegative integer, where $\mathbf{a}$ is a nonincreasing sequence of length $\|\mathbf{c}\|$ such that $s' \geq a_1 \geq \cdots \geq a_{\|\mathbf{c}\|} \geq 0$ and where $w \in R(\mathbf{c})$.

By (10.1.6) the right-hand side of (10.2.1) is the sum of the series

$$\sum t^s q^{\|\mathbf{a}^{(1)}\| + \cdots + \|\mathbf{a}^{(r)}\|}$$

extended over all sequences $(s, \mathbf{a}^{(1)}, \ldots, \mathbf{a}^{(r)})$, where s is a nonnegative integer and where $\mathbf{a}^{(1)} = (a_{1,1}, \ldots, a_{1,c_1})$, $\ldots$, $\mathbf{a}^{(r)} = (a_{r,1}, \ldots, a_{r,c_r})$ are nonincreasing sequences of integers all comprised between s and 0.

10.2. The MacMahon Verfahren

To prove that the sums of those two series are equal it suffices to build a bijection $(s, \mathbf{a}^{(1)}, \ldots, \mathbf{a}^{(r)}) \mapsto (s', \mathbf{a}, w)$ having the properties

$$s = s' + \operatorname{des} w \quad \text{and} \quad \|\mathbf{a}^{(1)}\| + \cdots + \|\mathbf{a}^{(r)}\| = \|\mathbf{a}\| + \operatorname{maj} w. \qquad (10.2.2)$$

The construction of the bijection is an updated version of a bijection already derived by MacMahon that has been generalized in several contexts. The rearrangement method described below is usually referred to as the *MacMahon Verfahren*.

Form the two-row matrix

$$\begin{pmatrix} a_{1,1} & \cdots & a_{1,c_1} & a_{2,1} & \cdots & a_{2,c_2} & \cdots & a_{r,1} & \cdots & a_{r,c_r} \\ 1 & \cdots & 1 & 2 & \cdots & 2 & \cdots & r & \cdots & r \end{pmatrix}$$

and rearrange its columns in such a way that the relative orders of the columns with the same bottom entries are preserved and the entire top row is *nonincreasing*. Let

$$\begin{pmatrix} v \\ w \end{pmatrix} = \begin{pmatrix} y_1 & y_2 & \cdots & y_{\|\mathbf{c}\|} \\ x_1 & x_2 & \cdots & x_{\|\mathbf{c}\|} \end{pmatrix} \qquad (10.2.3)$$

be the resulting matrix (remember that $c_1 + \cdots + c_r = \|\mathbf{c}\|$.) From the preceding method of rearrangement we have $y_k = y_{k+1} \Rightarrow x_k \leq x_{k+1}$, or equivalently

$$x_k > x_{k+1} \Rightarrow y_k > y_{k+1}. \qquad (10.2.4)$$

The top row of the matrix (10.2.3) is a word $v = y_1 y_2 \ldots y_{\|\mathbf{c}\|}$ of length $\|\mathbf{c}\|$ which is the unique nonincreasing rearrangement of the juxtaposition product $\mathbf{a}^{(1)} \ldots \mathbf{a}^{(r)}$. The bottom row of the matrix (10.2.3) is a word $w = x_1 x_2 \ldots x_{\|\mathbf{c}\|}$ that belongs to $R(\mathbf{c})$.

For $i = 1, 2, \ldots, \|\mathbf{c}\|$ let z_i be the number of descents in the right factor $x_i x_{i+1} \ldots x_{\|\mathbf{c}\|}$ of w, that is to say, the number of indices j such that $i \leq j \leq \|\mathbf{c}\| - 1$ and $x_j > x_{j+1}$. In particular,

$$z_1 = \operatorname{des} w. \qquad (10.2.5)$$

Also, by the very definition of the major index,

$$\operatorname{maj} w = z_1 + z_2 + \cdots + z_{\|\mathbf{c}\|}. \qquad (10.2.6)$$

Now condition (10.2.5) implies that the word $\mathbf{a} = a_1 a_2 \ldots a_{\|\mathbf{c}\|}$ defined by

$$a_i = y_i - z_i \quad (i = 1, 2, \ldots, \|\mathbf{c}\|) \qquad (10.2.7)$$

is *nonincreasing*; moreover, its letters are *nonnegative*. Then define

$$z' = s - \operatorname{des} w.$$

As $s \geq y_1 = \max\{a_{i,j}\}$ and $z_1 = \operatorname{des} w$, we deduce that

$$s' = s - \operatorname{des} w \geq y_1 - z_1 \geq 0$$

and also

$$\|\mathbf{a}^{(1)}\| + \cdots + \|\mathbf{a}^{(r)}\| = \sum_i y_i = \sum_i a_i + \sum_i z_i = \|\mathbf{a}\| + \operatorname{maj} w.$$

The two conditions (10.2.2) are fulfilled. The bijection

$$(s, \mathbf{a}^{(1)}, \ldots, \mathbf{a}^{(r)}) \mapsto (s', \mathbf{a}, w)$$

is fully described and is completely reversible. Identity (10.2.1) is then established. ∎

EXAMPLE 10.2.2. We illustrate the preceding construction with an example. Start with the sequence $(s, \mathbf{a}^{(1)}, \ldots, \mathbf{a}^{(r)})$ defined by $r = 3$; $\mathbf{a}^{(1)} = 6, 5, 1, 1, 0, 0$; $\mathbf{a}^{(2)} = 5, 4, 1, 1$; $\mathbf{a}^{(3)} = 3, 1$ and $s = 7$. The rearrangement of the matrix

$$\begin{pmatrix} 6 & 5 & 1 & 1 & 0 & 0 & 5 & 4 & 1 & 1 & 3 & 1 \\ 1 & 1 & 1 & 1 & 1 & 1 & 2 & 2 & 2 & 2 & 3 & 3 \end{pmatrix}$$

as in (10.2.3) yields

$$\begin{pmatrix} 6 & 5 & 5 & 4 & 3 & 1 & 1 & 1 & 1 & 1 & 0 & 0 \\ 1 & 1 & 2 & 2 & 3 & 1 & 1 & 2 & 2 & 3 & 1 & 1 \end{pmatrix}.$$

Hence

$$v = 6, 5, 5, 4, 3, 1, 1, 1, 1, 1, 0, 0;$$
$$w = 1, 1, 2, 2, 3, 1, 1, 2, 2, 3, 1, 1;$$
$$z = 2, 2, 2, 2, 2, 1, 1, 1, 1, 1, 0, 0;$$
$$\mathbf{a} = 4, 3, 3, 2, 1, 0, 0, 0, 0, 0, 0, 0;$$
$$\operatorname{des} w = 2; \quad s' = s - \operatorname{des} w = 5.$$

Therefore

$$\mathbf{a}^{(1)} + \mathbf{a}^{(2)} + \mathbf{a}^{(3)} = 6 + 5 + 1 + 1 + 5 + 4 + 1 + 1 + 3 + 1 = 28$$
$$= \|\mathbf{a}\| + \operatorname{maj} w = (4 + 3 + 3 + 2 + 1) + (5 + 10) = 28.$$

If $u_1, u_2, \ldots, u_r$ are r commuting variables, it is convenient to use the notations $\mathbf{u}^{\mathbf{c}} = u_1^{c_1} u_2^{c_2} \cdots u_r^{c_r}$ and $(\mathbf{u}; q)_{s+1} = (u_1; q)_{s+1} \cdots (u_r; q)_{s+1}$. Below,

10.3. The insertion technique

the summations of the form $\sum_\mathbf{c}$ are extended to all sequences $\mathbf{c} = (c_1, \ldots, c_r)$ of r nonnegative integers, including the null sequence.

Form the *factorial* generating function

$$A(t, q; \mathbf{u}) = \sum_\mathbf{c} A_\mathbf{c}(t, q) \frac{\mathbf{u}^\mathbf{c}}{(t; q)_{1+\|\mathbf{c}\|}} \qquad (10.2.8)$$

for the polynomials $A_\mathbf{c}(t, q)$. It follows from (10.2.1) that

$$A(t, q; \mathbf{u}) = \sum_{s \geq 0} t^s \sum_\mathbf{c} \mathbf{u}^\mathbf{c} \begin{bmatrix} c_1 + s \\ s \end{bmatrix} \cdots \begin{bmatrix} c_r + s \\ s \end{bmatrix}$$

$$= \sum_{s \geq 0} t^s \left(\sum_{c_1} u_1^{c_1} \begin{bmatrix} c_1 + s \\ s \end{bmatrix} \right) \cdots \left(\sum_{c_r} u_r^{c_r} \begin{bmatrix} c_r + s \\ s \end{bmatrix} \right),$$

so that by (10.1.9)

$$A(t, q; \mathbf{u}) = \sum_{s \geq 0} \frac{t^s}{(\mathbf{u}; q)_{s+1}}. \qquad (10.2.9)$$

Conversely, it is clear that (10.2.9) implies (10.2.1). We then have two ways for expressing the polynomials $A_\mathbf{c}(t, q)$. In the next section we will see another expression for those polynomials by means of a recurrence relation.

10.3. The insertion technique

When deriving a recurrence relation for generating polynomials over permutation groups of order $n = 1, 2, \ldots$, the insertion technique is of frequent use: starting with a permutation of order n we study the modification brought to the underlying statistic when the letter $(n + 1)$ is *inserted* into the $(n + 1)$ slots of the permutation. With words with repetitions some transformations called *word marking* in what follows must be made on the initial word.

Write

$$A_\mathbf{c}(t, q) = \sum_{s \geq 0} A_{\mathbf{c},s}(q) t^s, \qquad (10.3.1)$$

so that $A_{\mathbf{c},s}(q)$ is the generating polynomial for the words $w \in R(\mathbf{c})$ such that $\operatorname{des} w = s$ by the major index. It will be convenient to use the notations $[s]_q = 1 + q + q^2 + \cdots + q^{s-1}$ and $\mathbf{c} + 1_j = (c_1, \ldots, c_j + 1, \ldots, c_r)$ for each $j = 1, 2, \ldots, r$ and each sequence $\mathbf{c} = (c_1, c_2, \ldots, c_r)$.

PROPOSITION 10.3.1. *With* $\|\mathbf{c}\| = c_1 + \cdots + c_r$ *and* $1 \le j \le r$ *the following relations hold:*

$$(1 - q^{c_j+1})A_{\mathbf{c}+1_j}(t, q)$$
$$= (1 - tq^{1+\|\mathbf{c}\|})A_{\mathbf{c}}(t, q) - q^{c_j+1}(1 - t)A_{\mathbf{c}}(tq, q); \quad (10.3.2)$$

$$[c_j + 1]_q A_{\mathbf{c}+1_j,s}(q)$$
$$= [c_j + 1 + s]_q A_{\mathbf{c},s}(q) + q^{s+c_j}[1 + \|\mathbf{c}\| - s - c_j]_q A_{\mathbf{c},s-1}(q). \quad (10.3.3)$$

Proof. The latter identity is equivalent to the former one, so that only (10.3.3) is to be proved. By Theorem 10.2.1 this relation is equivalent to the relation formed when j is replaced by any integer in $\{1, \ldots, r\}$. It is convenient to prove the relation for $j = 1$ which reads

$$(1 + q + \cdots + q^{c_1})A_{\mathbf{c}+1_1,s}(q)$$
$$= (1 + q + \cdots + q^{c_1+s})A_{\mathbf{c},s}(q) + (q^{c_1+s} + \cdots + q^{\|\mathbf{c}\|})A_{\mathbf{c},s-1}(q). \quad (10.3.4)$$

Consider the set $R^*(\mathbf{c} + 1_1, s)$ of 1-*marked* words, i.e., rearrangements w^* of $1^{c_1+1} \ldots r^{c_r}$ with s descents such that exactly one letter equal to 1 has been marked. Each word $w \in R(\mathbf{c} + 1_1)$ that has s descents gives rise to $c_1 + 1$ marked words $w^{(0)}, \ldots, w^{(c_1)}$. Define

$$\text{maj}^* w^{(i)} = \text{maj } w + n_1,$$

where n_1 is the number of letters equal to 1 to the *right* of the marked 1. Then clearly

$$\sum_{i=0}^{c_1} \text{maj}^* w^{(i)} = (1 + q + \cdots + q^{c_1}) \text{maj } w.$$

Hence

$$(1 + q + \cdots + q^{c_1})A_{\mathbf{c}+1_1,s}(q) = \sum_{w \in R^*(\mathbf{c}+1_1,s)} q^{\text{maj}^* w}.$$

Let $m = \|\mathbf{c}\|$ and let the word $w = x_1 x_2 \ldots x_m \in R(\mathbf{c})$ have s descents. Say that w has $m + 1$ slots $x_i x_{i+1}$, $i = 0, \ldots, m$ (where $x_0 = 0$ and $x_{m+1} = \infty$ by convention). Call the slot $x_i x_{i+1}$ *green* if either $x_i x_{i+1}$ is a descent, $x_i = 1$, or $i = 0$. Call the other slots *red*. Then there are $1 + s + c_1$ green slots and $m - s - c_1$ red slots. Label the green slots $0, 1, \ldots, c_1 + s$ from right to left, and label the red slots $c_1 + s + 1, \ldots, m$ from left to right.

For example, with $r = 3$, the word $w = 2, 2, 1, 3, 2, 1, 2, 3, 3$ has three descents and ten slots. As $c_1 = 2$, there are eight green slots and two red slots, labeled as follows:

```
slot     0 | 2 | 2 | 1 | 3 | 2 | 1 | 2 | 3 | 3 | ∞
label      5   6   4   3   2   1   0   7   8   9
```

10.4. The (t,q)-factorial generating functions

Denote by $w^{(i)}$ the word obtained from w by inserting a marked 1 into the ith slot. Then it may be verified that

$$\text{des } w^{(i)} = \begin{cases} \text{des } w, & \text{if } i \leq c_1 + s, \\ \text{des } w + 1, & \text{otherwise,} \end{cases} \quad (10.3.5)$$

$$\text{maj}^* w^{(i)} = \text{maj } w + i. \quad (10.3.6)$$

EXAMPLE 10.3.2. Consider the above word w. The following table shows the values of des and maj* on $w^{(i)}$. Descents are indicated by $\frown$ and the marked **1** is written in bold-face.

i	$w^{(i)}$	des $w^{(i)}$	maj* $w^{(i)}$
0	2 2⌢1 3⌢2⌢1 **1** 2 3 3	3	11
1	2 2⌢1 3⌢2⌢1 **1** 2 3 3	3	12
2	2 2⌢1 3⌢1 2⌢1 2 3 3	3	13
3	2 2⌢1 **1** 3⌢2⌢1 2 3 3	3	14
4	2 2⌢**1** 1 3⌢2⌢1 2 3 3	3	15
5	**1** 2 2⌢1 3⌢2⌢1 2 3 3	3	16
6	2⌢1 2⌢1 3⌢2⌢1 2 3 3	4	17
7	2 2⌢1 3⌢2⌢1 2⌢**1** 3 3	4	18
8	2 2⌢1 3⌢2⌢1 2 3⌢**1** 3	4	19
9	2 2⌢1 3⌢2⌢1 2 3 3⌢**1**	4	20

So each word $w \in R(\mathbf{c})$ with s descents and maj $w = n$ gives rise to c_1+s+1 marked words in $R^*(\mathbf{c} + 1_1, s)$ with maj* equal to $n, n+1, \ldots, n+c_1+s$; and to $m - s - c_1$ marked words in $R^*(\mathbf{c} + 1_1, s+1)$ with maj* equal to $n+c_1+s+1, \ldots, n+m$. Hence a word w in $R(\mathbf{c})$ with $s-1$ descents gives rise to $m-s+1-c_1$ marked words in $R^*(\mathbf{c}+1_1, s)$ with maj* equal to maj $w + c_1 + s, \ldots,$ maj $w + m$. This now proves relation (10.3.4). ∎

10.4. The (t,q)-factorial generating functions

In the previous section we have seen that formulas (10.2.1) and (10.2.9) implied each other. The purpose of this section is to show that the recurrence formula (10.3.2) is also equivalent to (10.2.1) and (10.2.9). This is achieved by a manipulation of q-series we shall describe in full details.

As defined in (10.2.8) consider the factorial generating function

$$A(t,q;\mathbf{u}) = \sum_{\mathbf{c}} \frac{\mathbf{u}^{\mathbf{c}}}{(t;q)_{1+\|\mathbf{c}\|}} A_{\mathbf{c}}(t,q) \tag{10.4.1}$$

and consider the partial q-difference

$$D_{u_r} = A(t,q;u_1,\ldots,u_r) - A(t,q;u_1,\ldots,u_{r-1},u_rq).$$

Directly from (10.4.1) we obtain

$$D_{u_r} = \sum_{\mathbf{c}} (1 - q^{c_r+1}) \frac{\mathbf{u}^{\mathbf{c}+1_r}}{(t;q)_{2+\|\mathbf{c}\|}} A_{\mathbf{c}+1_r}(t,q)$$

$$= \sum_{\mathbf{c}} (1 - tq^{\|\mathbf{c}\|+1}) \frac{\mathbf{u}^{\mathbf{c}+1_r}}{(t;q)_{2+\|\mathbf{c}\|}} A_{\mathbf{c}}(t,q)$$

$$- \sum_{\mathbf{c}} q^{c_r+1}(1-t) \frac{\mathbf{u}^{\mathbf{c}+1_r}}{(t;q)_{2+\|\mathbf{c}\|}} A_{\mathbf{c}}(tq,q).$$

Now use the recurrence relation (10.3.2). We get

$$\sum_{\mathbf{c}} (1 - tq^{\|\mathbf{c}\|+1}) \frac{\mathbf{u}^{\mathbf{c}+1_r}}{(t;q)_{2+\|\mathbf{c}\|}} A_{\mathbf{c}}(t,q) = \sum_{\mathbf{c}} \frac{\mathbf{u}^{\mathbf{c}+1_r}}{(t;q)_{1+\|\mathbf{c}\|}} A_{\mathbf{c}}(t,q)$$
$$= u_r A(t,q;\mathbf{u},)$$

and

$$\sum_{\mathbf{c}} q^{c_r+1}(1-t) \frac{\mathbf{u}^{\mathbf{c}+1_r}}{(t;q)_{2+\|\mathbf{c}\|}} A_{\mathbf{c}}(tq,q) = \sum_{\mathbf{c}} \frac{\mathbf{u}^{\mathbf{c}+1_r} q^{c_r+1}}{(tq;q)_{1+\|\mathbf{c}\|}} A_{\mathbf{c}}(tq,q)$$
$$= u_r q A(tq,q;u_1,\ldots,u_{r-1},u_rq,).$$

Hence

$$A(t,q;\mathbf{u}) - A(t,q;u_1,\ldots,u_{r-1},u_rq,)$$
$$= u_r A(t,q;\mathbf{u}) - u_r q A(tq,q;u_1,\ldots,u_{r-1},u_rq,). \tag{10.4.2}$$

The (partial) q-difference equation with respect to each u_i ($i = 1,\ldots,r$) has the form

$$A(t,q;\mathbf{u}) - A(t,q;u_1,\ldots,u_iq,\ldots,u_r)$$
$$= u_i A(t,q;\mathbf{u}) - u_i q A(tq,q;u_1,\ldots,u_iq,\ldots,u_r). \tag{10.4.3}$$

10.5. Words and biwords

Now let
$$A(t,q;\mathbf{u}) = \sum_{s\geq 0} t^s G_s(\mathbf{u},q).$$

From (10.4.3) we get

$$\sum_{s\geq 0} t^s(1-u_i) G_s(\mathbf{u},q) = \sum_{s\geq 0} t^s(1-u_i q^{s+1}) G_s(u_1,\ldots,u_i q,\ldots,u_r,q).$$

Taking the coefficients of t^s in both sides yields the relation

$$G_s(\mathbf{u},q) = \frac{1-u_i q^{s+1}}{1-u_i} G_s(u_1,\ldots,u_i q,\ldots,u_r,,q), \tag{10.4.4}$$

for $i=1,\ldots,r$. Now put

$$F_s(\mathbf{u},q) = G_s(\mathbf{u},q)(\mathbf{u};q)_{s+1}. \tag{10.4.5}$$

From Equation (10.4.4) we deduce that for $i=1,\ldots,r$

$$F_s(\mathbf{u},q) = F_s(u_1,\ldots,u_i q,\ldots,u_r,q). \tag{10.4.6}$$

But $F_s(\mathbf{u},q)$ can be expressed as $F_s(\mathbf{u},q) = \sum_{\mathbf{c}} \mathbf{u}^{\mathbf{c}} F_{s,\mathbf{c}}(q)$, where $F_{s,\mathbf{c}}(q)$ is a power series in nonnegative powers of q. Fix $\mathbf{c}$ and let a be a nonzero component of $\mathbf{c}$. Then Equation (10.4.6) implies that $F_{s,\mathbf{c}}(q) = q^a F_{s,\mathbf{c}}(q)$. Therefore, $F_{s,\mathbf{c}}(q) = 0$. Hence $F_s(\mathbf{u},q) = F_{s,0}(q)$. It remains to evaluate $F_{s,0}(q)$. But from (10.4.5)

$$F_{s,0}(q) = F_s(\mathbf{u},q)\Big|_{\mathbf{u}=0} = G_s(\mathbf{u},q)(\mathbf{u};q)_{s+1}\Big|_{\mathbf{u}=0} = G_s(0,q) = 1,$$

as $\sum_{s\geq 0} t^s G_s(0,q) = A(t,q;0) = \frac{1}{(t;q)_1} = \sum_{s\geq 0} t^s$. Thus $G_s(\mathbf{u},q) = \frac{1}{(\mathbf{u};q)_{s+1}}$
by (10.4.5). This proves the identity (10.2.9). Conversely showing that (10.2.9) $\Rightarrow$ (10.3.2) is much simpler, for (10.2.9) implies (10.4.3) in an easy manner and from (10.4.3) the recurrence relation (10.3.2) can be reached without any difficulty.

10.5. Words and biwords

The rest of this chapter is devoted to the construction of a class of bijections on each class $R(\mathbf{c})$ based on specific commutation rules. We will see that by means of the so-called Cartier–Foata rule and the contextual rule two bijections ϕ and Φ can be constructed having properties (10.0.2) and (10.0.3), respectively.

Keep the same alphabet $X = \{1, 2, \ldots, r\}$. A *biword* is an ordered pair of words of the same length, written as $\alpha = (h, b)$ ("h" stands for "high" and "b" for "bottom") or as

$$\alpha = \binom{h}{b} = \binom{h_1 h_2 \ldots h_m}{b_1 b_2 \ldots b_m}.$$

For easy reference we shall sometimes indicate the *places* $1, 2, \ldots, m$ of the letters on the top of the biword:

$$\begin{bmatrix} \text{id} \\ h \\ b \end{bmatrix} = \begin{bmatrix} 1 & 2 & \ldots & m \\ h_1 & h_2 & \ldots & h_m \\ b_1 & b_2 & \ldots & b_m \end{bmatrix}.$$

The word h (resp. b) is the *top* (resp. *bottom*) word of the biword (h, b). Each biword $\binom{h}{b}$ can also be seen as a word whose letters are the *biletters* $\binom{h_1}{b_1}, \ldots, \binom{h_m}{b_m}$. The integer m is the *length* of the biword w. A triple $(h, b; i)$ where i is an integer satisfying $1 \le i \le m - 1$ is called a *pointed biword*. When h and b are rearrangements of each other, the biword (h, b) is said to be a *circuit*.

Two classes of *circuits* will play a special role. First, we introduce the *standard circuits* $\Gamma(b)$ which are circuits of the form $\binom{\bar{b}}{b}$, where $\bar{b}$ is the nondecreasing rearrangement of the word b with respect to the standard ordering. Clearly Γ maps each word onto a standard circuit in a bijective manner.

The second class of circuits is defined as follows. A nonempty word $b = b_m b_1 \ldots b_{m-2} b_{m-1}$ is said to be *dominated*, if $b_m > b_1, b_m > b_2, \ldots, b_m > b_{m-1}$. The *right-to-left cyclic shift* of b is defined to be the word $\delta b = b_1 b_2 \ldots b_{m-1} b_m$. A biword of the form $\binom{\delta b}{b}$ with b dominated is called a *dominated cycle*.

As is known or easily verified, each word b is the juxtaposition product $u^1 u^2 \ldots$ of dominated words whose first letters $\text{pre}(u^1), \text{pre}(u^2), \ldots$ are in nondecreasing order:

$$\text{pre}(u^1) \le \text{pre}(u^2) \le \cdots. \tag{10.5.1}$$

That factorization, called the *increasing factorization* of b, is unique.

Given the increasing factorization $u^1 u^2 \ldots$ of a word b, we can form the juxtaposition product

$$\Delta(b) = \binom{\delta u^1 \; \delta u^2 \; \ldots}{u^1 \; u^2 \; \ldots} \tag{10.5.2}$$

of the dominated cycles. Clearly Δ maps each word onto a product of dominated cycles satisfying Inequalities (10.5.1), in a bijective manner.

Such a product, written as a biword (10.5.2), will be called a *well-factorized circuit*.

EXAMPLE 10.5.1. Consider the word $b = 2,2,1,3,5,3,4,5,1$. The standard circuit associated with b reads

$$\Gamma(b) = \begin{pmatrix} 1 & 1 & 2 & 2 & 3 & 3 & 4 & 5 & 5 \\ 2 & 2 & 1 & 3 & 5 & 3 & 4 & 5 & 1 \end{pmatrix}.$$

It has an increasing factorization given by $2|21|3|534|51$, so that the corresponding well-factorized circuit reads

$$\Delta(b) = \begin{pmatrix} 2 & 1\,2 & 3 & 3\,4\,5 & 1\,5 \\ 2 & 2\,1 & 3 & 5\,3\,4 & 5\,1 \end{pmatrix}.$$

As will be seen the next bijections on words can be viewed as composition products

$$b \mapsto \Gamma(b) \mapsto \Delta(c) \mapsto c,$$

where the mapping $\Gamma(b) \mapsto \Delta(c)$ will be described as a sequence of *commutations* on circuits.

10.6. Commutations

Suppose given a four-variable Boolean function $Q(x,y;z,t)$ (also written as $Q\begin{pmatrix} x,y \\ z,t \end{pmatrix}$) defined on quadruples of letters in X. The *commutation* Com induced by the Boolean function $Q(x,y;z,t)$ is defined to be a mapping that maps each pointed biword $(h,b;i)$ onto a biword $(h',b') = \text{Com}(h,b;i)$ with the following properties: if

$$\begin{pmatrix} h \\ b \end{pmatrix} = \begin{pmatrix} h_1 h_2 \cdots h_m \\ b_1 b_2 \cdots b_m \end{pmatrix} \quad \text{and} \quad \begin{pmatrix} h' \\ b' \end{pmatrix} = \begin{pmatrix} h'_1 h'_2 \cdots h'_{m'} \\ b'_1 b'_2 \cdots b'_{m'} \end{pmatrix},$$

then

(C0) $m' = m$;

(C1) $h'_j = h_j$, $b'_j = b_j$ for every $j \neq i, i+1$;

(C2) $h'_{i+1} = h_i$, $h'_i = h_{i+1}$ (the ith and $(i+1)$th letters of the top word are transposed);

(C3) $b'_i = b_i$ and $b'_{i+1} = b_{i+1}$ if $Q(h_i, h_{i+1}; b_i, b_{i+1})$ true; $b'_i = b_{i+1}$ and $b'_{i+1} = b_i$ if $Q(h_i, h_{i+1}; b_i, b_{i+1})$ false.

We can also describe the commutation by the following pair of mappings:

$$h = h_1 \cdots h_{i-1} h_i h_{i+1} h_{i+2} \cdots h_m \mapsto h' = h_1 \cdots h_{i-1} h_{i+1} h_i h_{i+2} \cdots h_m;$$
$$b = b_1 \cdots b_{i-1} b_i b_{i+1} b_{i+2} \cdots b_m \mapsto b' = b_1 \cdots b_{i-1} z\, t\, b_{i+2} \cdots b_m;$$

where either $z = b_i$, $t = b_{i+1}$ if $Q(h_i, h_{i+1}; b_i, b_{i+1})$ true, or $z = b_{i+1}$, $t = b_i$ if $Q(h_i, h_{i+1}; b_i, b_{i+1})$ false.

DEFINITION 10.6.1. A Boolean function $Q(x, y; z, t)$ is said to be *bisymmetric* if it is symmetric in the two sets of parameters $\{x, y\}$, $\{z, t\}$.

LEMMA 10.6.2. *The commutation* Com *induced by a bisymmetric Boolean function* $Q(x, y; z, t)$ *is involutive, i.e., if* $(h', b') = \text{Com}(h, b; i)$, *then* $(h, b) = \text{Com}(h', b'; i)$.

The proof of the lemma is a simple verification and will be omitted. In the rest of the chapter we will assume that all the four-variable Boolean functions $Q(x, y; z, t)$ are bisymmetric.

Two extreme cases are worth being mentioned, when Q is the Boolean function Q_{true} "always true" (resp. Q_{false} "always false"). The commutation Com_{true}, associated with Q_{true}, permutes only the *i*th and $(i+1)$th letters of the *top word* b, while $\text{Com}_{\text{false}}$, associated with Q_{false}, permutes the *i*th and $(i+1)$th *biletters* of the *biword* (h, b).

Sorting a biword is defined as follows. Again consider a biword (h, b) of length m and let $(h', b') = \text{Com}(h, b; i)$ with $1 \leq i \leq m-1$. If $1 \leq j \leq m-1$, we can form the pointed biword $(h', b'; j)$ and further apply the commutation Com to $(h', b'; j)$. We obtain the biword $\text{Com}(h', b'; j) = \text{Com}(\text{Com}(h, b; i); j)$, which we shall denote by $\text{Com}(h, b; i, j)$. By induction $\text{Com}(h, b; i_1, \ldots, i_n)$ can be defined, where $(i_1, \ldots, i_n)$ is a given sequence of integers less than m.

As each commutation always permutes two adjacent letters within the *top* word (condition (C2)), we can transform each biword (h, b) into a biword (h', b') whose top word h' is *nondecreasing* by applying a sequence of commutations. We can also say that for each biword (h, b) there exists a sequence $(i_1, \ldots, i_n)$ of integers such that the top word in the resulting biword $\text{Com}(h, b; i_1, \ldots, i_n)$ is nondecreasing. Such a biword is called a *minimal* biword and the sequence $(i_1, \ldots, i_n)$ a *commutation sequence*.

When using the commutations Com_{true} and $\text{Com}_{\text{false}}$ we always reach the same minimal biword, but the commutation sequence is not unique. With an arbitrary commutation Com neither the minimal biword, nor the commutation sequence, is necessarily unique. We then define a particular commutation sequence $(i_1, \ldots, i_n)$ called the *minimal sequence* by the following two conditions:

(i) it is of minimum length;
(ii) it is minimal with respect to the lexicographic order.

Clearly the minimal sequence is uniquely defined by those two conditions and depends only on the top word h in (h,b). The minimal biword derived from (h,b) by using the minimal sequence is called the *straightening* of the biword (h,b). The derivation is described in the following algorithm SORTB.

Algorithm SORTB: **sorting a biword**: Given a biword (h,b) and a commutation Com the following algorithm transforms (h,b) into its straightening (h',b').

Prototype $(h',b') :=$ SORTB$(h,b,$ Com$)$.

1. Let $(h',b') := (h,b)$.

2. If h' is nondecreasing, RETURN (h',b').

3. Else, let j be the smallest integer such that $h'(j) > h'(j+1)$. Then let $(h',b') :=$ Com$(h',b';j)$. Go to (2).

EXAMPLE 10.6.3. Consider the biword

$$\begin{bmatrix} \mathrm{id} \\ h \\ b \end{bmatrix} = \begin{bmatrix} 1\ 2\ 3\ 4\ 5\ 6\ 7\ 8\ 9 \\ 2\ 1\ 2\ 3\ 3\ 4\ 5\ 1\ 5 \\ 2\ 2\ 1\ 3\ 5\ 3\ 4\ 5\ 1 \end{bmatrix}.$$

The sequence of the indices j that occur in algorithm SORTB applied to the biword is

1, which transforms h' into 1,2,2,3,3,4,5,1,5,
7,   "   1,2,2,3,3,4,1,5,5,
6,   "   1,2,2,3,3,1,4,5,5,
5,   "   1,2,2,3,1,3,4,5,5,
4,   "   1,2,2,1,3,3,4,5,5,
3,   "   1,2,1,2,3,3,4,5,5,
2,   "   1,1,2,2,3,3,4,5,5,

so that the minimal sequence is 1,7,6,5,4,3,2, and accordingly the final word h' is 1,1,2,2,3,3,4,5,5. Notice that the final word b' depends on the commutation rule Com.

10.7. The two commutations

We shall introduce two commutations associated with two specific Boolean functions Q.

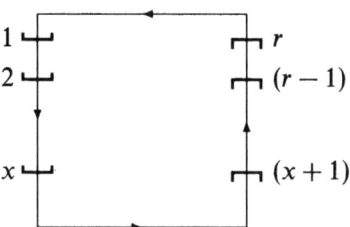

Figure 10.1. Cyclic interval.

10.7.1. *The Cartier–Foata commutation.* We denote by Com_{CF} the commutation induced by the following Boolean function Q_{CF}:

$$Q_{CF}\begin{pmatrix} x,y \\ z,t \end{pmatrix} \text{ true if and only if } x = y. \qquad (10.7.1)$$

10.7.2. *The contextual commutation.* For each letter x let $x^+ = x + \frac{1}{2}$ and denote by Com_H the commutation induced by the following Boolean function Q_H:

$$Q_H\begin{pmatrix} x,y \\ z,t \end{pmatrix} \text{ true iff } (z - x^+)(z - y^+)(t - x^+)(t - y^+) > 0. \qquad (10.7.2)$$

Notice that both Q_{CF} and Q_H are bisymmetric, so that $Q_{CF}^2 = Q_H^2 =$ the identity map.

The second commutation can also be defined by means of the following "cyclic intervals". Place the r elements $1, 2, \ldots, x, (x + 1), \ldots, (r - 1), r$ on a circle or on a square (!) counterclockwise and place a bracket on each of those elements as shown in Figure 10.1. For $x, y \in X$ ($x \neq y$) the cyclic interval $]\!]x, y]\!]$ is the subset of all the elements that lie between x and y when the circle is read counterclockwise. The brackets (in the French notation) indicate if the extremities of the interval are to be included or not.

For instance, suppose $1 < x < r$. Then $]\!]1, x]\!] = \{2, \ldots, x\}$ (the origin 1 excluded, but the end x included), while $]\!]x, 1]\!] = \{x + 1, \ldots, r, 1\}$ (x excluded but 1 included); finally, let $]\!]x, x]\!] = \emptyset$.

PROPOSITION 10.7.1. *The Boolean function* $Q_H\begin{pmatrix} x,y \\ z,t \end{pmatrix}$ *is true if and only if both z, t are in* $]\!]x, y]\!]$ *or neither in* $]\!]x, y]\!]$.

10.8. The main algorithm

The proof is a lengthy but easy verification and is omitted. Notice that the condition (10.7.2) is efficient in programming while the other condition involving cyclic intervals is more adapted for human beings!

10.8. The main algorithm

This is denoted by **T** and is defined for any Boolean function Q. Let Com be the commutation induced by Q. Then **T** transforms each word b into a rearrangement c of b.

Prototype $c := \mathbf{T}(b, \text{Com})$.

1. Let h be the nondecreasing rearrangement of b. Form the standard circuit $\Gamma(b) = (h, b)$,

$$\begin{bmatrix} \text{id} \\ h \\ b \end{bmatrix} = \begin{bmatrix} 1 & 2 & \cdots & m \\ h_1 & h_2 & \cdots & h_m \\ b_1 & b_2 & \cdots & b_m \end{bmatrix},$$

let $c := b$ and α be the empty cycle $\alpha = \begin{bmatrix} \cdot \\ \cdot \\ \cdot \end{bmatrix}$.

2a. If all the places $1, 2, \ldots, m$ occur in α, RETURN c (the juxtaposition product of the bottom words in α).

2b. Else, let D be the greatest *place* not occurring in α.

2c. Let M be the greatest *letter* in h not in α, so that $h_D = M$ and the initial biword has been changed into

$$\begin{bmatrix} \text{id} \\ h \\ c \end{bmatrix} = \begin{bmatrix} * & \cdots & * & D \\ * & \cdots & * & M \\ * & \cdots & * & * \end{bmatrix} \overbrace{\begin{bmatrix} * & \cdots & * \\ * & \cdots & * \\ * & \cdots & * \end{bmatrix} \cdots \begin{bmatrix} * & \cdots & * \\ * & \cdots & * \\ * & \cdots & * \end{bmatrix}}^{\alpha}.$$

3a. Let $B := c_D$.

3b. If $B = M$, terminate the dominated cycle $\begin{bmatrix} * & \cdots & * \\ * & \cdots & M \\ M & \cdots & * \end{bmatrix}$ and add it to the left of α, so that the new α reads $\begin{bmatrix} * & \cdots & * \\ * & \cdots & M \\ M & \cdots & * \end{bmatrix} \alpha$. Go to 2a.

3c. Else, look for the greatest *place* $j \leq D - 1$ such that $B = h_j$; in short

$$\begin{bmatrix} \text{id} \\ h \\ c \end{bmatrix} = \begin{bmatrix} \cdots & j & \cdots & D & \cdots & * \\ \cdots & B & \cdots & * & \cdots & M \\ \cdots & * & \cdots & B & \cdots & * \end{bmatrix} \begin{bmatrix} * & \cdots & * \\ * & \cdots & * \\ * & \cdots & * \end{bmatrix} \cdots \begin{bmatrix} * & \cdots & * \\ * & \cdots & * \\ * & \cdots & * \end{bmatrix}.$$

If $j \leq D - 2$, apply the commutation

$$(h, c) = \text{Com}(h, c; j, j + 1, \cdots, D - 2),$$

so that $h_{D-1} = B$ after running the commutation; in short,

$$\begin{bmatrix} \text{id} \\ h \\ c \end{bmatrix} = \begin{bmatrix} \cdots & j & \cdots & D-1 & D & \cdots & * \\ \cdots & * & \cdots & B & * & \cdots & M \\ \cdots & * & \cdots & * & B & \cdots & * \end{bmatrix} \begin{bmatrix} * & \cdots & * \\ * & \cdots & * \\ * & \cdots & * \end{bmatrix} \cdots \begin{bmatrix} * & \cdots & * \\ * & \cdots & * \\ * & \cdots & * \end{bmatrix}.$$

3d. Let $D := D - 1$ and go to 3a.

We can verify that each step in the preceding algorithm is feasible. For example, the place j in step 3c is well defined: at this stage $\binom{h}{c}$ is the product of the left factor (in square brackets) $\begin{bmatrix} h' \\ c' \end{bmatrix}$ by α and h' is necessarily a rearrangement of c'.

Define the two transformations

$$\phi(b) = \text{T}(b, \text{Com}_{CF}) \quad \text{and} \quad \Phi(b) = \text{T}(b, \text{Com}_H). \tag{10.8.1}$$

EXAMPLE 10.8.1. Consider the word $b = 2, 1, 2, 3, 3, 5, 4, 5, 1$ and the circuit

$$\Gamma(b) = \begin{bmatrix} \text{id} \\ h \\ b \end{bmatrix} = \begin{bmatrix} 1 & 2 & 3 & 4 & 5 & 6 & 7 & 8 & 9 \\ 1 & 1 & 2 & 2 & 3 & 3 & 4 & 5 & 5 \\ 2 & 1 & 2 & 3 & 3 & 5 & 4 & 5 & 1 \end{bmatrix}$$

and calculate the image of b under ϕ and Φ.

For the first transformation we easily obtain

$$\begin{bmatrix} \text{id} \\ h \\ c \end{bmatrix} \mapsto \begin{bmatrix} 1 & 2 & 3 & 4 & 5 & 6 & 7 & 8 & 9 \\ 2 & 3 & 4 & 5 & 3 & 2 & 1 & 1 & 5 \\ 2 & 3 & 4 & 5 & 5 & 3 & 2 & 1 & 1 \end{bmatrix},$$

so that $c = \phi(b) = 2\,3\,4\,5\,5\,3\,2\,1\,1$.

10.9. The inverse of the algorithm

For the second we indicate all the commutations needed in bold-face:

$$\begin{bmatrix}h\\b\end{bmatrix} = \begin{bmatrix}1\,1\,2\,2\,3\,3\,4\,5\,5\\2\,1\,2\,3\,3\,5\,4\,5\,1\end{bmatrix} \mapsto \begin{bmatrix}1\,\mathbf{2}\,\mathbf{1}\,2\,3\,3\,4\,5\,5\\2\,\mathbf{2}\,\mathbf{1}\,3\,3\,5\,4\,5\,1\end{bmatrix} \mapsto \begin{bmatrix}1\,2\,2\,\mathbf{1}\,3\,3\,4\,5\,5\\2\,2\,1\,3\,3\,5\,4\,5\,1\end{bmatrix}$$
$$\mapsto \begin{bmatrix}1\,2\,2\,3\,\mathbf{1}\,3\,4\,5\,5\\2\,2\,1\,3\,3\,5\,4\,5\,1\end{bmatrix} \mapsto \begin{bmatrix}1\,2\,2\,3\,3\,\mathbf{1}\,4\,5\,5\\2\,2\,1\,3\,\mathbf{5}\,\mathbf{3}\,4\,5\,1\end{bmatrix} \mapsto \begin{bmatrix}1\,2\,2\,3\,3\,4\,\mathbf{1}\,5\,5\\2\,2\,1\,3\,5\,3\,4\,5\,1\end{bmatrix}$$
$$\mapsto \begin{bmatrix}1\,2\,2\,3\,3\,4\,5\,1\,5\\2\,2\,1\,3\,5\,3\,4\,5\,1\end{bmatrix} \mapsto \begin{bmatrix}1\,2\,2\,3\,3\,4\,5\,1\,5\\2\,2\,1\,3\,5\,3\,4\,5\,1\end{bmatrix}$$
$$\mapsto \begin{bmatrix}1\,2\,2\,3\,3\,4\,5\,1\,5\\2\,2\,1\,3\,5\,3\,4\,5\,1\end{bmatrix} \mapsto \begin{bmatrix}\mathbf{2}\,\mathbf{1}\,2\,3\,3\,4\,5\,1\,5\\2\,2\,1\,3\,5\,3\,4\,5\,1\end{bmatrix}$$

so that $c = \Phi(b) = 2\,2\,1\,3\,5\,3\,4\,5\,1$.

10.9. The inverse of the algorithm

Given a commutation Com, the following algorithm denoted by $\mathbf{T}^{-1}$ transforms a word c into a word b such that $b = \mathbf{T}^{-1}(c, \text{Com})$.

Prototype $b := \mathbf{T}^{-1}(c, \text{Com})$.

1. Let $i := 1$, $S := c_1$.

2a. If $i = \text{length}(c)$, let $h_i := S$, $(h, b) := \text{SORTB}(h, c, \text{Com})$. RETURN b.

2b. Else, let $B := c_{i+1}$.

2c. If $B \geq S$, let $h_i := S$, $S := B$. Else, let $h_i := B$.

3. Let $i := i + 1$. Go to 2a.

Now examine the algorithm **T**. Before returning c in step 2a the algorithm provides the juxtaposition product $\alpha = \gamma^1 \gamma^2 \ldots$ of cycles. Let $u^1, u^2, \ldots$ be the bottom words of those cycles and let $\text{pre}(u^1)$, $\text{pre}(u^2)$, ... be the first letters of those bottom words. Steps 2c and 3b say that each cycle γ^i was terminated as soon as $\text{pre}(u^i)$ was greater than all the other letters in the cycle. Accordingly, all the cycles γ^i are *dominated*. Furthermore, $\text{pre}(u^1) \leq \text{pre}(u^2) \leq \cdots$.

Thus $u^1 u^2 \ldots$ is the increasing factorization of c (in the terminology of Section 10.6), while $\alpha = \gamma^1 \gamma^2 \ldots = \begin{pmatrix} \delta u^1 & \delta u^2 & \ldots \\ u^1 & u^2 & \ldots \end{pmatrix}$ is the *increasing product of dominated cycles*, i.e., α is equal to the well-factorized circuit $\Delta(c)$. We can say that the algorithm **T** maps b onto $\Gamma(b)$, then transforms each *standard circuit* $\Gamma(b)$ into a *well-factorized circuit* $\Delta(c)$, the word c

being a rearrangement of b. Let U be the mapping $U: \Gamma(b) \mapsto \Delta(c)$, so that **T** is the composition product

$$b \mapsto \Gamma(b) \stackrel{U}{\mapsto} \Delta(c) \mapsto c. \tag{10.9.1}$$

As each commutation applied to a pointed biword is involutive, U and therefore **T** are bijective.

Further examine the algorithm $\mathbf{T}^{-1}$ and let $u^1 u^2 \cdots$ be the increasing factorization of c as a product of dominated words. Once we have reached step 2a, verified that the test $i = \text{length}(c)$ was positive and executed $h_i := S$, the biword (h, c) is exactly the well-factorized circuit

$$\Delta(c) = \binom{h}{c} = \binom{\delta u^1 \; \delta u^2 \; \cdots}{u^1 \quad u^2 \quad \cdots}.$$

Thus the algorithm $\mathbf{T}^{-1}$ first builds up the well-factorized circuit $\Delta(c)$ and applies the algorithm SORTB to $\Delta(c)$ to produce a standard circuit $\Gamma(b)$, so that $\mathbf{T}^{-1}$ may be represented as the sequence

$$c \mapsto \Delta(c) \xmapsto{\text{SORTB}} \Gamma(b) \mapsto b. \tag{10.9.2}$$

Again as each local commutation applied to a pointed biword is involutive, $\mathbf{T}^{-1}$ is a bijection.

Finally, to prove that **T** and $\mathbf{T}^{-1}$ are inverse to each other, we simply examine the algorithm **T**. The commutations are made only in steps 2c and 3c. In both steps the reverse operation can be written as

$$(h, c) := \text{SORTB}(h, c, \text{Com}).$$

We have then proved the following property.

PROPERTY 10.9.1. *The algorithms* **T** *and* $\mathbf{T}^{-1}$ *are inverse to each other, i.e., for each word b we have*

$$\mathbf{T}^{-1}(\mathbf{T}(b, \text{Com}), \text{Com}) = \mathbf{T}(\mathbf{T}^{-1}(b, \text{Com}), \text{Com}) = b.$$

REMARK 10.9.2. The algorithms **T** and $\mathbf{T}^{-1}$ are valid for each bisymmetric Boolean function Q. However, only the Cartier–Foata and the contextual commutations will be used to derive the next results on statistics on words.

10.10. Statistics on circuits

Let $C(X)$ denote the set of all circuits. Remember that a circuit is a pair of words $\alpha = \binom{h}{b}$, where $h = y_1 y_2 \ldots y_m$ and $b = x_1 x_2 \ldots x_m$ are rearrangements of each other and h is *not* necessarily nondecreasing. Two circuits α and β are said to be *H-equivalent*, written $\alpha \sim \beta$, if one can be obtained from the other by a sequence of commutations Com_H (see 10.7.2).

The two statistics des and maj for each circuit $\alpha = \binom{h}{b}$ are defined as follows. They depend only on the bottom word b. First let $\text{des}\,\alpha = \text{des}\,b$. Then the statistic maj is based on the notion of *cyclic interval*, as introduced in Section 10.7. Put $x_{m+1} = \infty$ (an auxiliary letter greater than every letter of X). Then for each $i = 1, 2, \ldots, m$ define q_i to be the number of j such that $1 \leq j \leq i - 1$ and $x_j \in \,]x_i, x_{i+1}]$. The sequence $(q_1, q_2, \ldots, q_m)$ is said to be the maj-*coding* of α. Define

$$\text{maj}\,\alpha = q_1 + q_2 + \cdots + q_m. \tag{10.10.1}$$

Now given the commutation Com_H we can apply the algorithm SORTB of Section 10.6 to each circuit α. It produces a standard circuit β to which the rearrangement U defined in (10.9.1) can be further applied to derive a well-factorized circuit γ:

$$\alpha \xmapsto{\text{SORTB}} \beta \xmapsto{U} \gamma. \tag{10.10.2}$$

Let Ψ denote the mapping $\alpha \mapsto \gamma$. Because of (10.9.1) and (10.9.2) we have $\Psi(\alpha) = \alpha$ if α is well-factorized. In particular, Ψ is surjective.

THEOREM 10.10.1. *There exists at most one bivariate statistic (f, g) defined on $C(X)$ having the following two properties:*
  (i) $\alpha \sim \alpha' \Rightarrow (f, g)\,\alpha = (f, g)\,\alpha'$;
  (ii) *if α is well factorized, then*

$$(f, g)\,\alpha = (\text{des}, \text{maj})\,\alpha. \tag{10.10.3}$$

Proof. Both algorithms SORTB and U involve sequences of commutations Com_H, so that if $\gamma = \Psi(\alpha)$, we have $(f, g)\,\alpha = (f, g)\,\gamma = (\text{des}, \text{maj})\,\gamma$. ∎

Our next task is to give an explicit definition of the pair (f, g). For each circuit $\alpha = \binom{h}{b}$ with $h = y_1 y_2 \ldots y_m$ and $b = x_1 x_2 \ldots x_m$ define $\text{exc}\,\alpha$ to be the number of integers i such that $1 \leq i \leq m$ and $x_i > y_i$. For each place i $(1 \leq i \leq m)$ define p_i to be the number of j such that $1 \leq j \leq i - 1$ and $x_j \in \,]x_i, y_i]$. The sequence $(p_1, p_2, \ldots, p_m)$ is said to be the den-*coding* of α. Furthermore, define

$$\text{den}\,\alpha = p_1 + p_2 + \cdots + p_m. \tag{10.10.4}$$

EXAMPLE 10.10.2. The circuit

$$\alpha = \begin{bmatrix} \text{id} \\ h' \\ b' \end{bmatrix} = \begin{bmatrix} 1\ 2\ 3\ 4\ 5\ 6\ 7\ 8\ 9 \\ 2\ 1\ 2\ 3\ 3\ 4\ 5\ 1\ 5 \\ 2\ 2\ 1\ 3\ 5\ 3\ 4\ 5\ 1 \end{bmatrix}$$

was already considered in Example 10.6.3. It has an excedence at places 2, 5, 8, so that $\operatorname{exc} \alpha = 4$. For its den-coding we first have $p_1 = p_2 = 0$. As $2 \in \,]1,2]$, we have $p_3 = 2$. Then $p_4 = 0$. As $\,]5,3] = \{1,2,3\}$, $p_5 = 4$. Next $p_6 = 0$. Also $\,]4,5] = \{5\}$, so that $p_7 = 1$ and $\,]5,1] = \{1\}$, so that $p_8 = 1$. As $\,]1,5] = \{2,3,4,5\}$, we get $p_9 = 7$. Thus $\operatorname{den} \alpha = 0+0+2+0+4+0+1+1+7 = 15$.

THEOREM 10.10.3. *The pair* (exc, den) *is the unique bivariate statistic defined on* $C(X)$ *having properties* (i) *and* (ii) *of Theorem* 10.10.1.

Proof. Proving that $\alpha \sim \alpha' \Rightarrow (\operatorname{exc}, \operatorname{den}) \alpha = (\operatorname{exc}, \operatorname{den}) \alpha'$ is lengthy but easy, as the property is to be proved only when α and α' differ by a commutation Com_H. The proof is omitted.

To show that $(\operatorname{exc}, \operatorname{den}) \alpha = (\operatorname{des}, \operatorname{maj}) \alpha$ when α is well factorized proceed as follows.

Let $\begin{pmatrix} a_2 & \ldots & a_{i+1} & a_{i+2} & \ldots & a_k & a_1 \\ a_1 & \ldots & a_i & a_{i+1} & \ldots & a_{k-1} & a_k \end{pmatrix}$ and $\begin{pmatrix} b_2 & \ldots \\ b_1 & \ldots \end{pmatrix}$ be two successive dominated cycles in the increasing factorization of α, so that

$$\alpha = \begin{pmatrix} \ldots & a_2 & \ldots & \boxed{a_{i+1}} & a_{i+2} & \ldots & a_k & \boxed{a_1} & b_2 & \ldots \\ \ldots & a_1 & \ldots & \boxed{a_i} & \boxed{a_{i+1}} & \ldots & a_{k-1} & \boxed{a_k} & \boxed{b_1} & \ldots \end{pmatrix}.$$

Inside each dominated cycle a pair like (a_i, a_{i+1}) occurs horizontally and vertically, so that there is a descent $a_i a_{i+1}$ if and only if there is an excedence $\binom{a_{i+1}}{a_i}$. Furthermore, the letters in w to the left of a_i that fall into the cyclic interval $\,]a_i, a_{i+1}]$ bring the same contribution to both $\operatorname{maj} \alpha$ and $\operatorname{den} \alpha$. If $\binom{a_{i+1}}{a_i}$ is the jth biletter of α (when read from left to right), we have $p_j = q_j$ in the notations used in (10.10.4)) and (10.10.1).

At the end of a dominated cycle we have to compare the contributions of the horizontal pair (a_k, b_1) with the contribution of the vertical pair $\binom{a_1}{a_k}$. But $a_k < a_1 \leq b_1$ by definition of the increasing factorization, so that (a_k, a_1) is never an excedence and (a_k, b_1) never a descent.

Now if $a_1 = b_1$, the two cyclic intervals $\,]a_k, b_1]$ and $\,]a_k, a_1]$ that serve in the calculation of $\operatorname{maj} \alpha$ and $\operatorname{den} \alpha$ are identical. If $a_1 < b_1$, there is no letter x in w to the left of b_1 such that $a_1 < x \leq b_1$. For any two sets A, B let $A + B$ denote the union of A and B when the intersection $A \cap B$ is empty. As

$$\,]a_k, b_1] = \,]a_k, a_1] + \{x \mid a_1 < x \leq b_1\},$$

there are as many letters to the left of a_k falling into the interval $]\!]a_k, b_1]\!]$ as letters falling into $]\!]a_k, a_1]\!]$.

Suppose that α is of length m and take up again the notations of (10.10.1) and (10.10.2). It remains to compare q_m and p_m. Let $\binom{y_m}{x_m}$ be the rightmost biletter of α. The letter y_m is necessarily equal to the greatest letter occurring in w. Hence the cyclic intervals $]\!]x_m, \infty]\!]$ used for evaluating q_m and $]\!]x_m, y_m]\!]$ for evaluating p_m are equal. ∎

As the transformation U is a sequence of commutations Com_H we have

$$(\text{exc}, \text{den})\,\alpha = (\text{exc}, \text{den})\,U(\alpha) = (\text{des}, \text{maj})\,U(\alpha). \tag{10.10.5}$$

The above development can be reproduced for the commutation Com_{CF}. However, the proofs are far simpler. In the same manner, we can prove that the statistic "exc" is the unique statistic having the following properties:
 (i) $\alpha \sim \alpha'$ (for Com_{CF}) $\Rightarrow$ $\text{exc}\,\alpha = \text{exc}\,\alpha'$;
 (ii) if α is well factorized, then $\text{exc}\,\alpha = \text{des}\,\alpha$.
Hence, if Com_{CF} is used, we have

$$\text{exc}\,\alpha = \text{exc}\,U(\alpha) = \text{des}\,U(\alpha). \tag{10.10.6}$$

10.11. Statistics on words and equidistribution properties

To get the definitions of $\text{des}\,w$, $\text{maj}\,w$, $\text{exc}\,w$ and $\text{den}\,w$ for a word w we simply form the *standard circuit* $\Gamma(w)$ and put

$$\begin{array}{ll} \text{des}\,w = \text{des}\,\Gamma(w), & \text{maj}\,w = \text{maj}\,\Gamma(w), \\ \text{exc}\,w = \text{exc}\,\Gamma(w), & \text{den}\,w = \text{den}\,\Gamma(w). \end{array} \tag{10.11.1}$$

The definitions given for $\text{des}\,w$ and $\text{exc}\,w$ are identical with the definitions given in the introduction to this chapter. The definition of $\text{den}\,w$ is new, while that of $\text{maj}\,w$ differs from the definition given in the introduction. However, we have the following result.

THEOREM 10.11.1. *The statistic* $\text{maj}\,w$ *given in (10.11.1) and the statistic* $\text{maj}\,w$ *given in the introduction are identical.*

This theorem is easy to prove by induction on the length of the word.

The *excedence index* of w is defined as the sum, $\text{excindex}\,w$, of all i such that i is an excedence in w. When a certain correcting term is added to $\text{excindex}\,w$, we get the second definition of $\text{den}\,w$. To fully describe that correcting term we need further definitions. For each word $w = x_1 x_2 \ldots x_m$ let

$$\begin{aligned} \text{inv } w &= \text{Card}\{1 \leq i < j \leq m \mid x_i > x_j\}, \\ \text{imv } w &= \text{Card}\{1 \leq i < j \leq m \mid x_i \geq x_j\}. \end{aligned} \qquad (10.11.2)$$

Now if $\text{exc } w = e$, let $i_1 < i_2 < \cdots < i_e$ be the increasing sequence of the excedences of w and let $j_1 < j_2 < \cdots < j_{m-e}$ be the complementary sequence. Form the two subwords

$$\text{Exc } w = x_{i_1} x_{i_2} \ldots x_{i_e}, \qquad \text{Nexc } w = x_{j_1} x_{j_2} \ldots x_{j_{m-e}}.$$

Then the *Denert statistic* of w is also defined to be

$$\text{den } w = \text{excindex } w + \text{imv Exc } w + \text{inv Nexc } w. \qquad (10.11.3)$$

THEOREM 10.11.2. *For every word w the two definitions of* den w *occurring in (10.11.1) and (10.11.3) are identical.*

Surprisingly this theorem is not easy to prove; see the Notes below.

THEOREM 10.11.3. *The transformations ϕ and Φ defined in (10.8.1) have the equidistribution properties*

$$\text{exc}(w) = \text{des } \phi(w) \quad \text{and} \quad (\text{exc, den}) w = (\text{des, maj}) \Phi(w).$$

Proof. As shown in (10.8.1) both transformations ϕ and Φ are defined by means of the main algorithm **T** which itself is defined by the chain $b \mapsto \Gamma(b) \overset{U}{\mapsto} \Delta(c) \mapsto c$ (see (10.9.1)).

If Com_{CF} is used, then $\text{exc } \Gamma(b) = \text{des } \Delta(c)$ by (10.10.6). On the other hand, $\text{exc } b = \text{exc } \Gamma(b)$ by (10.11.1) and $\text{des } \Delta(c) = \text{des } c$, as the definition of des depends only on the bottom word c of the circuit. Thus $\text{exc } b = \text{des } c$.

If Com_H is used, then $(\text{exc, den}) \Gamma(b) = (\text{des, maj}) \Delta(c)$ by (10.10.5). Also $(\text{exc, den}) b = (\text{exc, den}) \Gamma(b)$ by (10.11.1) and $(\text{des, maj}) \Delta(c) = (\text{des, maj})(c)$, as the definition of (des, maj) depends only on the bottom word c. Hence $(\text{exc, den}) b = (\text{des, maj})(c)$. ∎

EXAMPLE 10.11.4. Again consider the word $b = 2, 1, 2, 3, 3, 5, 4, 5, 1$ and its standard circuit $\Gamma(b) = \begin{bmatrix} \text{id} \\ h \\ b \end{bmatrix} = \begin{bmatrix} 1\ 2\ 3\ 4\ 5\ 6\ 7\ 8\ 9 \\ 1\ 1\ 2\ 2\ 3\ 3\ 4\ 5\ 5 \\ 2\ 1\ 2\ 3\ 3\ 5\ 4\ 5\ 1 \end{bmatrix}$. Then $\text{exc } b = 3$.
Using the definition (10.11.3) for the Denert statistic we find $\text{den } b = (1 + 4 + 6) + \text{imv}(2, 3, 5) + \text{inv}(1, 2, 3, 4, 5, 1) = 11 + 0 + 4 = 15$.

The images $\phi(b) = 2,3,4,5,5,3,2,1,1$ and $\Phi(b) = 2,2,1,3,5,3,4,5,1$ have been determined in Example 10.8.1. Observe that des $\phi(b) = 3 =$ exc b. The word $\Phi(b)$ also has three descents. Furthermore, its major index is equal to $2+5+8 = 15$, so that (exc, den) $b = $ (des, maj) $\Phi(b) = (3, 15)$.

Problems

Section 10.1

10.1.1 (The q-binomial theorem) Prove the following. Using the notation $(a; q)_\infty = \prod_{j=0}^{\infty}(1 - aq^j)$ the q-binomial theorem reads

$$\sum_{n=0}^{\infty} \frac{(a;q)_n}{(q;q)_n} u^n = \frac{(au;q)_\infty}{(u;q)_\infty}.$$

The symbols q and u can be taken as complex numbers such that $|q| < 1$, $|u| < 1$ or as variables. In the latter case the preceding identity holds in the algebra of formal power series in two variables with coefficients in a given ring. See Andrews 1976, Theorem 2.1, or Gasper and Rahman 1990, 1.3. Consider the following special cases. For $a = 0$,

$$\sum_{n=0}^{\infty} \frac{u^n}{(q;q)_n} = \frac{1}{(u;q)_\infty} = e_q(u) \text{ (the first } q\text{-exponential)}.$$

For $u \to -u/a$, $a \to \infty$,

$$\sum_{n=0}^{\infty} \frac{q^{\binom{n}{2}} u^n}{(q;q)_n} = (-u;q)_\infty = E_q(u) \text{ (the second } q\text{-exponential)}.$$

With $a = q^{k+1}$,

$$\sum_{n=0}^{\infty} \begin{bmatrix} k+n \\ n \end{bmatrix} u^n = \frac{1}{(u;q)_{k+1}},$$

and with $a = q^{-k}$, $u \to -uq^k$,

$$\sum_{n=0}^{\infty} q^{\binom{n}{2}} \begin{bmatrix} k \\ n \end{bmatrix} u^n = (-u;q)_k.$$

Extraction of coefficients of u^n in the next to the last identity gives

$$\begin{bmatrix} k+n \\ k \end{bmatrix} = \sum_{k \geq a_1 \geq \cdots \geq a_n \geq 0} q^{a_1 + \cdots + a_n},$$

and in the last one

$$q^{\binom{n}{2}}\begin{bmatrix}k\\n\end{bmatrix} = \sum_{k-1\geq a_1>\cdots>a_n\geq 0} q^{a_1+\cdots+a_n}.$$

This provides another proof of Proposition 10.1.1.

Section 10.2

10.2.1 For each Ferrers diagram λ with m boxes (see Section 5.1) and each vector $\mathbf{c} = (c_1, c_2, \ldots, c_r)$ of positive integers such that $c_1 + c_2 + \cdots + c_r = m$ let $\mathcal{K}(\lambda, \mathbf{c})$ denote the set of Young tableaux containing c_1 1's, c_2 2's, $\ldots$, c_r r's. Let σ be a permutation of the set $\{1, 2, \ldots, r\}$. The symmetry argument that is carried over the proof of Theorem 10.2.1 can be used to construct a one-to-one correspondence between $\mathcal{K}(\lambda, \mathbf{c})$ and $\mathcal{K}(\lambda, \sigma \mathbf{c})$. Proceed as follows. Let $\mathbf{c} = (c_1, \ldots, c_i, c_{i+1}, \ldots, c_r)$ and $\mathbf{c}' = (c_1, \ldots, c_{i+1}, c_i, \ldots, c_r)$ differ only by a transposition of two adjacent terms and consider a tableau T in $\mathcal{K}(\lambda, \mathbf{c})$ in its planar representation (as in Section 5.1). Write all the pairs $i, i+1$ in bold-face whenever those two integers occur in the same column with $(i+1)$ just above i. The remaining i's and $(i+1)$'s in T occur as horizontal blocks $i^a j^b$ ($a \geq 0$, $b \geq 0$). We define a bijection $T \mapsto T'$ of $\mathcal{K}(\lambda, \mathbf{c})$ onto $\mathcal{K}(\lambda, \mathbf{c}')$ by replacing each block $i^a j^b$ in T by $i^b j^a$ and rewriting the vertical pairs $i, i+1$ in ordinary type. This argument provides another proof of the symmetry of the Schur function (see Section 5.4).

10.2.2 (The MacMahon Verfahren revisited) Let $U = (S_<, S_\leq, L_<, L_\leq)$ be a partition of the alphabet $X = \{1, \ldots, r\}$ such that $S_< \cup S_\leq = \{1, \ldots, h\}$ (the *small* letters) and $L_< \cup L_\leq = \{h+1, \ldots, r\}$ (the *large* letters) for a certain h ($0 \leq h \leq r$). Let $w = x_1 x_2 \ldots x_m$ be a word in the alphabet and let $x_{m+1} = h + \frac{1}{2}$. An integer i such that $1 \leq i \leq m$ is said to be a *U-descent* in w, if either $x_i > x_{i+1}$, or $x_i = x_{i+1}$ and $x_i \in S_\leq \cup L_\leq$. Let $\text{des}_U w$ (resp. $\text{maj}_U w$) denote the *number* (resp. the *sum*) of the U-descents in w. For each sequence $\mathbf{c} = (c_1, \ldots, c_r)$ consider the generating polynomial for the class $R(\mathbf{c})$ by the pair $(\text{des}_U, \text{maj}_U)$, i.e., $A_\mathbf{c}^U(t, q) = \sum_w t^{\text{des}_U w} q^{\text{maj}_U w}$ ($w \in R(\mathbf{c})$). The identity to be proved reads

$$\frac{1}{(t;q)_{1+\|\mathbf{c}\|}} A_\mathbf{c}^U(t,q) = \sum_{s\geq 0} t^s \prod_{i\in S_<}\begin{bmatrix}c_i+s\\c_i\end{bmatrix} \prod_{i\in S_\leq} q^{\binom{c_i}{2}}\begin{bmatrix}s+1\\c_i\end{bmatrix}$$

$$\times \prod_{i\in L_<} q^{c_i}\begin{bmatrix}c_i+s-1\\c_i\end{bmatrix} \prod_{i\in L_\leq} q^{\binom{c_i+1}{2}}\begin{bmatrix}s\\c_i\end{bmatrix},$$

Problems 357

and can be derived as follows. As in Section 10.2 the left-hand side is equal to the sum of the series $\sum t^{s'+\mathrm{des}_U w} q^{\|\mathbf{a}\|+\mathrm{maj}_U w}$ over all triples $(s', \mathbf{a}, w)$. By using the last two identities of Problem 10.1.1 the right side is equal to the sum of the series $\sum t^s q^{\|\mathbf{a}^{(1)}\|+\cdots+\|\mathbf{a}^{(r)}\|}$, where each $\mathbf{a}^{(i)} = (a_{i,1}, \ldots, a_{i,c_i})$ is a sequence of integers satisfying $s \geq a_{i,1} \geq \cdots \geq a_{i,c_i} \geq 0$, if $i \in S_<$; $s \geq a_{i,1} > \cdots > a_{i,c_i} \geq 0$, if $i \in S_\leq$; $s \geq a_{i,1} \geq \cdots \geq a_{i,c_i} \geq 1$, if $i \in L_<$; $s \geq a_{i,1} > \cdots > a_{i,c_i} \geq 1$, if $i \in L_\leq$. The bijection $(s', \mathbf{a}, w) \mapsto (s, \mathbf{a}^{(1)}, \ldots, \mathbf{a}^{(r)})$ such that $s = s' + \mathrm{des}_U w$ and $\|\mathbf{a}\| + \mathrm{maj}_U w = \|\mathbf{a}^{(1)}\| + \cdots + \|\mathbf{a}^{(r)}\|$ can be constructed by rewriting the MacMahon Verfahren developed in Section 10.2 almost verbatim. (See Foata and Krattenthaler 1995.)

10.2.3 Show that the identity derived in Problem 10.2.2 is equivalent to the following identity between q-series:

$$\sum_{\mathbf{c}} A_{\mathbf{c}}^U(t, q) \frac{\mathbf{u}^{\mathbf{c}}}{(t; q)_{1+\|\mathbf{c}\|}} = \sum_{s \geq 0} t^s \frac{\prod_{i \in S_\leq}(-u_i; q)_{s+1} \prod_{i \in L_\leq}(-q u_i; q)_s}{\prod_{i \in S_<}(u_i; q)_{s+1} \prod_{i \in L_<}(q u_i; q)_s}.$$

(See Foata and Krattenthaler 1995.)

10.2.4 Write the preceding identity as

$$\sum_{\mathbf{c}} A_{\mathbf{c}}^U(t, q) \frac{\mathbf{u}^{\mathbf{c}}}{(t; q)_{1+\|\mathbf{c}\|}} = \sum_{s \geq 0} t^s a_s(\mathbf{u}; q)$$

and let

$$a_\infty(\mathbf{u}; q) = \frac{\prod_{i \in S_\leq}(-u_i; q)_\infty \prod_{i \in L_\leq}(-q u_i; q)_\infty}{\prod_{i \in S_<}(u_i; q)_\infty \prod_{i \in L_<}(q u_i; q)_\infty}.$$

Prove the following. The sequence $(a_s(\mathbf{u}; q))$ $(s \geq 0)$ converges to $a_\infty(\mathbf{u}; q)$ in the topology of the formal power series in the variables $u_1, \ldots, u_r$. Let $a_{-1}(\mathbf{u}; q) = 0$; then the sequence $(a_s(\mathbf{u}; q) - a_{s-1}(\mathbf{u}; q))$ $(s \geq 0)$ is summable of sum $a_\infty(\mathbf{u}; q)$. As we have $(1 - t) \sum_s t^s a_s(\mathbf{u}; q) = \sum_s t^s (a_s(\mathbf{u}; q) - a_{s-1}(\mathbf{u}; q))$, it makes sense to multiply the identity in Problem 10.2.3 by $(1 - t)$ and make $t = 1$. This yields

$$\sum_{\mathbf{c}} A_{\mathbf{c}}^U(1, q) \frac{\mathbf{u}^{\mathbf{c}}}{(q; q)_{\|\mathbf{c}\|}} = a_\infty(\mathbf{u}; q).$$

(See Foata and Krattenthaler 1995.)

Section 10.3

10.3.1 Take up again the notations of Problem 10.2.2 with the further assumption that the subalphabets $S_\le$ and $L_<$ are empty, so that $\{1,\ldots,h\}$ (resp. $\{(h+1),\ldots,r\}$) is the set of small (resp. large) letters. With $1 \le h < r$ derive the following recurrence relations for the polynomials $A_\mathbf{c}^U(t,q)$ by using the insertion technique:

$$(1 - q^{c_h+1}) A_{\mathbf{c}+1_h}^U(t,q) = (1 - tq^{1+\|\mathbf{c}\|}) A_\mathbf{c}^U(t,q)$$
$$- q^{c_h+1}(1-t) A_\mathbf{c}^U(tq,q);$$
$$(1 - q^{-c_r-1}) A_{\mathbf{c}+1_r}^U(t,q) = -(1 - tq^{1+\|\mathbf{c}\|}) A_\mathbf{c}^U(t,q)$$
$$- q^{-c_r}(1-t) A_\mathbf{c}^U(tq,q).$$

(See Clarke and Foata 1995a.)

Section 10.4

10.4.1 With the specializations of Problem 10.3.1 for U the identity written in Problem 10.2.3 becomes

$$\sum_\mathbf{c} A_\mathbf{c}^U(t,q) \frac{\mathbf{u}^\mathbf{c}}{(t;q)_{1+\|\mathbf{c}\|}} = \sum_{s \ge 0} t^s \frac{\prod_{h+1 \le i \le r} (-q u_i; q)_s}{\prod_{1 \le i \le h} (u_i; q)_{s+1}}.$$

Derive the latter identity directly from the recurrence relations in Problem 10.3.1. (See Clarke and Foata 1995a.)

10.4.2 (q-Eulerian polynomials). With the specializations of Problem 10.3.1 for U and for $\mathbf{c} = 1^r$ the identity in Problem 10.2.2 becomes

$$\frac{1}{(t;q)_{1+r}} A_{1^r}^U(t,q) = \sum_{s \ge 0} t^s ([s+1]_q)^h q^{r-h} ([s]_q)^{r-h}.$$

Let $A_r^h(t,q) = A_{1^r}^U(t,q)$ $(0 \le h \le r)$ and form

$$A_r(x,y;t,q) = \sum_{h=0}^r \binom{r}{h} x^{r-h} y^h A_r^h(t,q).$$

Show that

$$\sum_{r \ge 0} \frac{u^r}{r!} \frac{A_r(x,y;t,q)}{(t;q)_{r+1}} = \sum_{s \ge 0} t^s \exp(u(xq[s]_q + y[s+1]_q)).$$

Problems 359

For $h = r$ the polynomial $A_r^r(t,q)$ is the traditional q-Eulerian polynomial for the symmetric group $\mathscr{S}_r$ by the pair (des, maj). (See Carlitz 1975.)

10.4.3 (The t-extension of the q-evaluation of a tableau). With the notations of Problem 5.5.1 let des T be the number of the recoils of a tableau T of shape λ with m boxes. Prove the following identity, the t-extension of the identity in Problem 5.5.1.4:

$$\sum_{T \in \mathrm{STab}(\lambda)} t^{\mathrm{des}\, T} q^{\mathrm{maj}\, T} = (t;q)_{m+1} \sum_{k \geq 0} t^k s_\lambda(1, q, q^2, \ldots, q^k).$$

(See Désarménien and Foata 1985, theorem 4.1.)

10.4.4 (The Schur function method). Prove the following. Again let $A_{\mathbf{c}}(t,q) = A_{\mathbf{c}}^{\mathrm{des,maj}}(t,q)$. To each word $w \in R(\mathbf{c})$ there corresponds a unique pair of tableaux (P, Q) such that $\mathrm{ev}(P) = \mathbf{c}$ and Q is a standard tableau such that $\mathrm{des}\, w = \mathrm{des}\, Q$ and $\mathrm{maj}\, w = \mathrm{maj}\, Q$ (see Problem 5.5.1). Hence

$$\sum_{\mathbf{c}} A_{\mathbf{c}}(t,q) \frac{\mathbf{u}^{\mathbf{c}}}{(t;q)_{1+\|\mathbf{c}\|}} = \sum_{\mathbf{c}} \frac{\mathbf{u}^{\mathbf{c}}}{(t;q)_{1+\|\mathbf{c}\|}} \sum_{|\lambda|=\|\mathbf{c}\|} \sum_{(P,Q)} t^{\mathrm{des}\, Q} q^{\mathrm{maj}\, Q}$$

$$= \sum_{\mathbf{c}} \sum_{|\lambda|=\|\mathbf{c}\|} \sum_{P} \mathbf{u}^{\mathrm{ev}(P)} \times \frac{1}{(t;q)_{1+\|\mathbf{c}\|}} \sum_{Q} t^{\mathrm{des}\, Q} q^{\mathrm{maj}\, Q}$$

$$= \sum_{\lambda} s_\lambda(u_1, \ldots, u_r) \times \sum_{k \geq 0} t^k s_\lambda(1, q, q^2, \ldots, q^k),$$

by the definition of a Schur function (see Definition 5.4.1) and Problem 10.4.3. The last product is equal to $\sum_{s \geq 0}(t^s/(\mathbf{u};q)_{s+1})$ by using the Cauchy identity (see Theorem 5.4.2) with the alphabets $\xi \leftarrow \{u_1, \ldots, u_r\}$ and $\eta \leftarrow \{1, q, \ldots, q^k\}$. (See Foata 1995.)

Section 10.5

The remaining problems refer to the last seven sections of this chapter and will be numbered 10.5.n.

10.5.1 (Euler–Mahonian statistics). Prove the following. As seen in Problem 10.4.2, the polynomial $A_r^r(t,q) = A_{1^r}(t,q)$ is the q-Eulerian polynomial that can be interpreted as the generating function for the symmetric group $\mathscr{S}_r$ by the pair (des, maj). Let $A_r^r(t,q) = \sum_{s \geq 0} A_{r,s}^r(q) t^s$. With $\mathbf{c} = 1^{r-1}$, $j = r$, $c_r = 0$ the recurrence relation (10.3.3) specializes into

$$A_{r,s}^r(q) = [1+s]_q A_{r-1,s}^{r-1}(q) + q^s [r-s]_q A_{r-1,s-1}^{r-1}(q), \qquad (*)$$

for $1 \leq s \leq r-1$ with the initial conditions $A_{r,0}^r(q) = 1$ for all $r \geq 0$ and $A_{r,s}^r(q) = 0$ for $s \geq r$. The first values of the polynomials $A_{r,s}^r(q)$ read

$s=$	0	1	2	3
$r=0$	1			
1	1			
2	1	q		
3	1	$2q + 2q^2$	q^3	
4	1	$3q + 5q^2 + 3q^3$	$3q^3 + 5q^4 + 3q^5$	q^6

Let $E = (E_r)$ ($r \geq 0$) be a family of finite sets such that Card $E_r = r!$ for every $r \geq 0$. A family $(f, g) = (f_r, g_r)$ ($r \geq 0$) is said to be *Euler–Mahonian* on E, if $f_0 = g_0 = 0$, $f_1 = g_1 = 1$ and if for every $r \geq 2$ both f_r and g_r are integral-valued functions defined on E_r and there exists a bijection $\psi_r : (w', j) \mapsto w$ of $E_{r-1} \times [0, r-1]$ onto E_r having the following properties:

$$g_r(w) = g_{r-1}(w') + j;$$
$$f_r(w) = \begin{cases} f_{r-1}(w'), & \text{if } 0 \leq j \leq f_{r-1}(w'), \\ f_{r-1}(w') + 1, & \text{if } f_{r-1}(w) + 1 \leq j \leq r - 1. \end{cases} \quad (**)$$

Each pair (f_r, g_r) is called an *Euler–Mahonian statistic* on E_r.

1. Let (f, g) be Euler–Mahonian on E and for every triple (r, s, l) let $A_{r,s,l}^r$ be the number of elements $w \in E_r$ such that $f_r(w) = l$ and $g_r(w) = s$ and form $A_{r,s}^r(q) = \sum_l A_{r,s,l}^r q^l$. Then the family $(A_{r,s}^r(q))$ satisfies the above recurrence relation ($*$).

2. A word $w = x_1 \cdots x_r$ of length r is said to be *subexcedent* if its letters are integral numbers and satisfy $0 \leq x_i \leq i - 1$ for all $i = 1, \ldots, r$. Denote by SE_r the set of those words. Let the *sum* of w be defined by sum $w = x_1 + \cdots + x_r$ and its *Eulerian value*, eul w, by eul $= 0$ if w is of length 1, and for $r \geq 2$

$$\text{eul } x_1 \cdots x_r = \begin{cases} \text{eul } x_1 \cdots x_{r-1}, & \text{if } x_r \leq \text{eul } x_1 \cdots x_{r-1}, \\ \text{eul } x_1 \cdots x_{r-1} + 1, & \text{if } x_r \geq \text{eul } x_1 \cdots x_{r-1} + 1. \end{cases}$$

Then the pair (sum, eul) is an Euler–Mahonian statistic on SE_r for every $r \geq 0$. The bijection ψ_r is easy to imagine.

3. Let $r \geq 2$ and let $w' = x_1 x_2 \ldots x_{r-1}$ be a permutation of $12 \ldots (r-1)$ having s descents. Let $x_0 = 0$, $x_r = \infty$ and for

each $i = 0, 1, \ldots, (r-1)$ label the r slots $x_i x_{i+1}$ as follows: $x_{r-1} x_r$ gets label 0, then reading the permutation *from right to left* label $1, 2, \ldots, s$ the s descents $x_i > x_{i+1}$; then reading *from left to right* label $(s+1), \ldots, r-1$ the $(r-1-s)$ nondescents $x_i < x_{i+1}$ $(0 \leq i \leq r-2)$. If the slot $x_i x_{i+1}$ gets label j define $\psi_r(w', j) = x_1 \ldots x_i r x_{i+1} \ldots x_{r-1}$. Then with $(f, g) = (\text{des}, \text{maj})$ the mapping ψ_r has the properties (**), so that (des, maj) is an Euler–Mahonian statistic on each permutation group $\mathcal{S}_r$. (See Carlitz 1954, Rawlings 1981.)

4. Let $r \geq 2$ and let $w' = x_1 x_2 \ldots x_{r-1}$ be a permutation of $12 \ldots (r-1)$ having s excedences. Let $(x_{i_1} > \cdots > x_{i_s})$ be the *decreasing* sequence of the excedence values $x_k > k$ and let $(x_{i_{s+1}} < \cdots < x_{i_{r-1}})$ be the *increasing* sequence of the nonexcedence values $x_k \leq k$. By convention, let $x_{i_0} = r$.
Define $\psi_r(w, 0) = x_1 x_2 \ldots x_{r-1} r$. If $1 \leq j \leq s$ (resp. $s+1 \leq j \leq r-1$) replace each letter x_{i_m} in w' such that $1 \leq m \leq j$ (resp. such that $1 \leq m \leq s$) by $x_{i_{m-1}}$, leave the other letters invariant and insert x_{i_j} (resp. x_{i_s}) into the i_jth place in w'. Let $w = \psi_r(w', j)$ be the permutation thereby obtained.
For example, $w' = 32541$ has the $s = 2$ excedences $x_3 = 5 > 3$, $x_1 = 3 > 1$ (in decreasing order) and three nonexcedences $x_5 = 1 \leq 5$, $x_2 = 2 \leq 2$, $x_4 = 4 \leq 4$ (in increasing order), so that $(i_1, i_2, i_3, i_4, i_5) = (3, 1, 5, 2, 4)$. With $j = 1$ we have $i_j = 3$ and $x_3 = 5$. To obtain $\psi_6(w', 1)$ replace $x_{i_1} = 5$ by $x_{i_0} = 6$, leave the other letters invariant and insert $x_{i_1} = 5$ into the i_1th=3rd place. Thus $\psi_6(w', 1) = 325641$. For $j = 3$ we have $i_j = 5$ and $x_5 = 1$. As $j = 3 > s = 2$, replace $x_{i_1} = x_3$ by $x_{i_0} = 6$, then $x_{i_2} = x_1 = 3$ by $x_{i_1} = 5$, leave the other letters invariant and insert $x_{i_s} = x_{i_2} = x_1 = 3$ into the i_3th = 5th place to yield $\psi_6(w', 3) = 526431$.
With $(f, g) = (\text{exc}, \text{den})$ the mapping ψ_r has the properties (**), so that (exc, den) is an Euler–Mahonian statistic on each permutation group $\mathcal{S}_r$. (See Han 1990b.)

5. Let (f, g) be an Euler–Mahonian family on $E = (E_r)$. For each $w \in E_r$ $(r \geq 2)$ let $\psi_r^{-1}(w) = (w', j_r)$, $\psi_{r-1}^{-1}(w') = (w'', j_{r-1}), \ldots, \psi_2^{-1}(w^{(r-2)}) = (w^{(r-1)}, j_2)$ and $j_1 = 0$; the word $\Psi(w) = j_1 j_2 \cdots j_{r-1} j_r$ is a subexcedent word and Ψ is a bijection of E_r onto SE_r such that $f(w) = \text{sum } \Psi(w)$ and $g(w) = \text{eul } \Psi(w)$. The bijection Ψ is called the (f, g)-*coding* of E_r.
Let $\Psi_{(\text{des},\text{maj})}$ (resp. $\Psi_{(\text{exc},\text{den})}$) be the (des, maj)-coding (see part 3.) (resp. the (exc, den)-coding (see part 4.) of $\mathcal{S}_r$. Then

$\Theta = \Psi^{-1}_{(\text{des,maj})} \circ \Psi_{(\text{exc,den})}$ is a bijection of $\mathscr{S}_r$ onto itself that satisfies $(\text{exc}, \text{den}) w = (\text{des}, \text{maj}) \Theta(w)$. (See Han 1990b.)

10.5.2 Prove the following. With the assumptions of Problem 10.3.1 the alphabet $X = \{1, \ldots, r\}$ is split into two disjoint parts, the set $S = \{1, \ldots, h\}$ of the small letters and $L = \{h+1, \ldots, r\}$, the set of the large letters. A U-descent of the word $w = x_1 \ldots x_m$ is an integer i such that $1 \le i \le m$ and either $x_i > x_{i+1}$, or $x_i = x_{i+1} \in L$ (by convention, $x_{m+1} = h + \frac{1}{2}$.) Denote by $\text{des}_U w$ (resp. $\text{maj}_U w$) the number (resp. the sum) of the U-descents in w. Now if $y_1 y_2 \ldots y_m$ is the nondecreasing rearrangement of the word w let $\text{exc}_U w$ be the number of i such that $x_i > y_i$, or $x_i = y_i \in L$. The definition of den_U requires the introduction of three further statistics. The U-excedence index of w is defined as the sum, $\text{excindex}_U w$, of all i such that i is an U-excedence in w. Also let

$$\text{inv}_U w = \text{Card}\{1 \le i < j \le m \mid x_i > x_j \text{ or } x_i = x_j \ge h+1\}$$
$$+ \text{Card}\{1 \le i \le m \mid x_i \ge h+1\},$$
$$\text{imv}_U w = \text{Card}\{1 \le i < j \le m \mid x_i > x_j \text{ or } x_i = x_j \le h\}.$$

If $\text{exc}_U w = e$, let $i_1 < i_2 < \cdots < i_e$ be the increasing sequence of the U-excedences of w and let $j_1 < j_2 < \cdots < j_{m-e}$ be the complementary sequence. Form the two subwords $\text{Exc}_U w = x_{i_1} x_{i_2} \ldots x_{i_e}$, $\text{Nexc}_U w = x_{j_1} x_{j_2} \ldots x_{j_{m-e}}$. Then the U-Denert statistic, $\text{den}_U w$, of w is defined to be

$$\text{den}_U w = \text{excindex}_U w + \text{imv}_U \text{Exc}_U w + \text{inv}_U \text{Nexc}_U w.$$

When the set L of large letters is empty, all the statistics without any subscript U that were defined in the chapter are recovered. The algorithm **T** described in Section 10.8 can be adequately modified to make up a bijection Φ of each rearrangement class $R(\mathbf{c})$ onto itself having the property

$$(\text{exc}_U, \text{den}_U) w = (\text{des}_U, \text{maj}_U) \Phi(w)$$

for every word w in $R(\mathbf{c})$. Thus the generating polynomial for $R(\mathbf{c})$ by the pair $(\text{exc}_U, \text{den}_U)$ is the polynomial $A_\mathbf{c}^U(t, q)$ whose factorial generating polynomial is shown in Problem 10.4.1. (See Foata and Han 1998 for the details of the construction of Φ; see Clarke and Foata 1994 for an earlier construction and Han 1995 for another equivalent definition for den_U.)

10.5.3 In Section 10.10 it is proved that if $\alpha = (h, b)$ and $\alpha' = (h', b')$ are H-equivalent, then $(\text{exc}, \text{den})\, \alpha = (\text{exc}, \text{den})\, \alpha'$. Show that the converse is true whenever the words h, h', b, b' are words without repetitions. (See Clarke 1997.)

Notes

With his treatise on combinatory analysis and his numerous papers MacMahon 1915, 1978 may be regarded as the initiator of the study of permutation statistics that includes methods for deriving analytical expressions for generating functions. In particular, he made a clever use of his master theorem (see MacMahon 1915, p. 97) that allowed him to show that the generating polynomials for each rearrangement class by the number of descents des and by the number of excedences exc were equal, so that des and exc are equidistributed on every rearrangement class. Back in the 1960s, as initiated by the late Marcel Paul Schützenberger, it was natural to prove such equidistribution properties in a *bijective* manner. The transformation ϕ that satisfies (10.0.2) was constructed in Foata 1965. A further presentation was made in Knuth 1973, pp. 24–29, a more algebraic version appeared in Cartier and Foata 1969 and also in Lothaire 1983, Chapter 10.

In studying the genus zeta function of local minimal hereditary orders Denert 1990 introduced a new permutation statistic, which was later christened "den". She observed and conjectured that the generating polynomials for each rearrangement class by the pairs (des, maj) and (exc, den) were equal. Foata and Zeilberger 1990 proved the conjecture for permutations by making use of the linear partial recurrence operator algebra. Then Han 1990a, 1990b proved the result combinatorially.

The definition of den for arbitrary words (with repetitions) is due to Han 1994 who also constructed a bijection Φ having the property (10.0.3) for an arbitrary rearrangement class. In the case of permutations the equivalence between the two definitions (10.11.1) and (10.11.3) of the Denert statistic was given in Foata and Zeilberger 1990. Another proof appeared in Clarke 1995. The general case for arbitrary words was derived by Han 1994, who also introduced the definition (10.10.1) of maj, which was basic for constructing the bijection Φ.

When the underlying alphabet X is partitioned into two subalphabets S of *small* letters and L of *large* letters, the classical permutation statistics can be further refined to take large inequalities into account (see Clarke and Foata 1994, 1995a, 1995b). Those statistics are denoted by des_k, exc_k, ... (or by des_U, exc_U, ... in Problem 10.5.2). It is also possible to derive explicit formulas for the generating polynomials by using the techniques developed in Sections 10.2–4. Furthermore, a bijection Φ_k of each rearrangement class can be constructed (see Clarke and Foata 1995a) such that $(\text{exc}_k, \text{den}_k) w = (\text{des}_k, \text{maj}_k) \Phi_k(w)$ holds identically. As shown in Foata and Han 1998 there is a common frame for constructing all the bijections ϕ, Φ, Φ_k based on the concept of biword commutation as presented in Sections 10.5–8.

When $\mathbf{c} = 1^r$ the generating polynomial for (des, maj) is the celebrated q-Eulerian polynomial $A_r(t, q)$ whose first study goes back to Carlitz 1954, 1959, 1975. Also see Problem 10.5.1. There is another class of q-Eulerian polynomials that can be introduced as generating functions for the permutation group by the pair (des, inv), where inv is the number of inversions (Stanley 1976).

The basic material on q-calculus can be found in Andrews 1976, Gasper and Rahman 1990. The MacMahon Verfahren takes its rise in MacMahon 1913. Formula (10.2.1) already appeared in MacMahon 1915, vol. 2, p. 211. Stanley 1972 and his disciple Reiner 1993 have extended the MacMahon Verfahren from the linear model used in this chapter and in Problem 10.2.2 to the poset environment and developed an adequate P-partition theory. There have been several papers that propose various techniques to derive analytical expressions for the permutation or word distributions, for example, Gessel 1977, Garsia 1979, Garsia and Gessel 1979, Gessel 1982, Zeilberger 1980 for the "comma'd" permutation technique; Fedou and Rawlings 1995 for an adjacency study; Foata and Han 1997 for an iterative method. A systematic permutation statistic study has been undertaken by Clarke, Steingrimsson, and Zeng 1997a, 1997b.

CHAPTER 11

Statistics on Permutations and Words

11.0. Introduction

This chapter is devoted, as was the previous one, to combinatorial properties of permutations, considered as words. The starting point in this subject is the bivalent status of permutations, which can be considered as products of cycles as well as a sequence of the first n integers written in disorder.

The fundamental results concerning this area are presented in Chapter 10 of Lothaire 1983. They consist essentially in two transformations on words. The first one (first fundamental transformation) is an encoding of the cycle decomposition of a permutation. The second one (second fundamental transformation) accounts for a statistical property of permutations, namely the equidistribution of the number of inversions and the inverse major index on permutations with a given shape.

In the previous chapter (Chapter 10) some other properties are presented, including basic facts on q-calculus and additional statistics on permutations.

In this chapter, we carry on with complements focused on two main aspects. The first one is a shortcut avoiding the second fundamental transformation by a simple evaluation of determinants. The second one is the analogy between the first fundamental transformation and the Lyndon factorization (the Gessel normalization).

The organization of the chapter is the following. After some preliminaries, Section 11.2 provides a determinantal expression for the commutative image of some sets of words. This expression is used in Sections 11.3 and 11.4, for evaluating respectively the inverse major index and the number of inversions of permutations with a given shape. In Section 11.5 the

Gessel normalization is introduced. In Section 11.6, it is then applied to evaluating the major index of permutations with a given cycle structure.

11.1. Preliminaries

Let us recall some notations and preliminary results.

In what follows, $[n]$ will denote the set of integers $\{1,\ldots,n\}$. We shall also use alphabets whose letters are the nonnegative integers. By N_k we mean the alphabet $\{0,1,\ldots,k\}$ equipped with the natural ordering. By letting k tend to infinity, we obtain the infinite alphabet $\mathbb{N}$. Quite obviously, if $k < r$, then $N_k \subset N_r$. In that way, a word on N_k is a finite sequence of nonnegative integers which do not exceed k.

To each N_k one can associate the set X_k of commuting indeterminates $\{x_0, x_1, \ldots, x_k\}$. Given a word $w = w_1 w_2 \ldots w_n$ on N_k, its *commutative image* is the commutative monomial $\mathrm{GS}(w) = x_{w_1} x_{w_2} \ldots x_{w_n}$. This definition can be extended to a set of words. If A is such a set, its *generating series* $\mathrm{GS}(A)$ is the sum of the commutative images of its elements. When A is infinite, so is its generating series. This does not yield any problem of convergence. When the number of letters (finite or infinite) is irrelevant, we shall simply write N for the alphabet and X for the associated indeterminates.

A particular – and fundamental – example is the following. Let a and k be integers. By $W_a(N_k)$ we denote the set of all nonincreasing words of length a over the alphabet N_k. Its commutative image, denoted by $S_a(X_k)$, is known as the complete symmetric function.

A favorite operation for deriving generating functions is the substitution of powers of a new indeterminate q in the generating series.

First, recall the definition of the *q-factorial* (see Chapter 10),

$$(q)_n = (1-q)(1-q^2)\cdots(1-q^n),$$

and define the *Gaussian polynomial* (or *q-binomial*) by

$$\begin{bmatrix} n \\ k \end{bmatrix} = (q)_n/(q)_k(q)_{n-k}.$$

Replacing x_i by q^i in generating series will provide generating functions for various statistics on words or permutations. In that respect, we shall need two preliminary results. The first one is a reformulation of Proposition 10.1.1.

11.2. Words with a given shape

PROPOSITION 11.1.1. *Let $S_a(\{1, q, q^2, \ldots, q^k\})$ be the result of the replacement of x_i by q^i in $S_a(X_k)$, $0 \leq i \leq k$. Then*

$$S_a(\{1, q, q^2, \ldots, q^k\}) = \begin{bmatrix} a+k \\ k \end{bmatrix}.$$

The next one is the limit of the former when k tends to infinity.

PROPOSITION 11.1.2. *Let $S_a(\{1, q, q^2, \ldots\})$ be the result of the replacement of x_i by q^i in $S_a(X_\mathbb{N})$, $i \geq 0$. Then*

$$S_a(\{1, q, q^2, \ldots\}) = 1/(q)_a.$$

The series $S_a(\{1, q, q^2, \ldots\})$ has a classical combinatorial interpretation. It is the generating function of the number of partitions of integers into parts less than a. A *partition* of an integer n is a nonincreasing sequence of positive integers (the *parts*) whose sum is n. Removing any restriction on the size of the parts yields the celebrated Euler generating function $1/(q)_\infty$ for the number of partitions.

Without the nonincreasing condition, the corresponding combinatorial object is called a *composition* of the integer n. So it is a sequence of integers whose sum is n.

We shall enumerate various statistics on permutations. Let us recall that a *permutation* of $[k]$ can be thought of either as a bijection from $[k]$ onto itself or as a word of length k with distinct letters in $[n]$. If $\sigma = \sigma(1)\sigma(2)\ldots\sigma(k)$ is such a permutation, an *inversion* of σ is a pair (i, j) such that $1 \leq i < j \leq k$ and $\sigma(i) > \sigma(j)$. The number of inversions of σ will be denoted by inv σ.

A *descent* of σ is an index i, $1 \leq i < k$, such that $\sigma(i) > \sigma(i+1)$. The set of the descents of σ, its *descent set*, is denoted by des σ. The sum of the elements of this set is the so-called *major index*, maj σ.

We say that i, $0 \leq i < k$, is a *backstep* of σ when the letter $i+1$ appears to the left of i in the word σ. This condition is equivalent to saying that i is a descent of the inverse of σ. The set of the descents of σ is designated by BS σ. As for descents, we can sum all the backsteps of σ to obtain imaj σ, the *inverse major index* of σ.

11.2. Words with a given shape

Each word $w = w_1 w_2 \cdots w_n$ can be uniquely factored as a product of maximal nonincreasing words, $w = u_1 u_2 \cdots u_k$, where each u_j is nonincreasing, its last letter being strictly smaller than the first letter of u_{j+1}. The respective lengths $a_1, a_2, \ldots$ of $u_1, u_2, \ldots$ constitute a composition **a** of n. This composition is called the *shape* of w.

EXAMPLE 11.2.1. Let $w = 1\,0\,2\,2\,0\,4\,7\,5\,1$. Its shape is $\mathbf{a} = (2,3,1,3)$.

Let $(b_1 \leq b_2 \leq \cdots \leq b_k)$ be a nondecreasing sequence of integers, $\mathbf{N} = (N_{b_1} \subseteq N_{b_2} \subseteq \cdots \subseteq N_{b_k})$ the sequence of corresponding alphabets and $\mathbf{a}$ a composition of n. Let us consider the set $W(\mathbf{a}, \mathbf{N})$ consisting of the words $w = u_1 u_2 \cdots u_k$ of shape $\mathbf{a}$ such that the factor u_i contains only letters in N_{b_i}.

Let us denote by $S(\mathbf{a}, \mathbf{X})$ the generating series of $W(\mathbf{a}, \mathbf{N})$. If all alphabets are equal to the same alphabet X, finite or infinite, we shall simplify the notation to $S(\mathbf{a}, X)$. If furthermore $\mathbf{a}$ is reduced to a single integer a, we have $S(\mathbf{a}, X) = S_a(X)$.

PROPOSITION 11.2.2. *Let $\mathbf{a} = (a_1, a_2, \ldots, a_k)$ be a shape, $b_1 \leq b_2 \leq \cdots \leq b_k$ and $\mathbf{X} = (X_{b_1}, X_{b_2}, \ldots, X_{b_k})$. Then $S(\mathbf{a}, \mathbf{X}) = \det(M)$, where M is the matrix $M = (M_{i,j})_{\{1 \leq i,j \leq k\}}$ with*

$$M_{i,j} = \begin{cases} S_{a_i + a_{i+1} + \cdots + a_j}(X_{b_i}) & \text{if } i \leq j, \\ 1 & \text{if } i = j+1, \\ 0 & \text{if } i > j+1. \end{cases} \quad (11.2.1)$$

Proof. Fully written, the matrix M is

$$M = \begin{pmatrix} S_{a_1}(X_{b_1}) & S_{a_1+a_2}(X_{b_1}) & \cdots & S_{a_1+\cdots+a_{k-1}}(X_{b_1}) & S_{a_1+\cdots+a_k}(X_{b_1}) \\ 1 & S_{a_2}(X_{b_2}) & \cdots & S_{a_2+\cdots+a_{k-1}}(X_{b_2}) & S_{a_2+\cdots+a_k}(X_{b_2}) \\ 0 & 1 & \cdots & S_{a_3+\cdots+a_{k-1}}(X_{b_3}) & S_{a_3+\cdots+a_k}(X_{b_3}) \\ \vdots & \vdots & \ddots & \vdots & \vdots \\ 0 & 0 & \cdots & 1 & S_{a_k}(X_{b_k}) \end{pmatrix}.$$

The proof is by induction on the length k of the shape $\mathbf{a}$ of w.

If $k = 1$, Equation (11.2.1) is trivially the definition of $S_a(X)$. For a general k, let us expand the determinant along its last row:

$$S(\mathbf{a}, \mathbf{X}) = S_{a_k}(X_{b_k}) \det(M_1) - \det(M_2), \quad (11.2.2)$$

where M_1 consists of the first $k-1$ rows and columns of M and M_2 is identical to M_1, except for its last column, which consists of the first $k-1$ rows of the last column of M, and so is

$$S_{a_i+a_{i+1}+\cdots+a_k}(X_{b_i}) \quad \text{for } 1 \leq i \leq k.$$

By the induction hypothesis,

$$\det(M_1) = S(\mathbf{a}_1, \mathbf{X}_1) = S((a_1, a_2, \ldots, a_{k-1}), (X_{b_1}, X_{b_2}, \ldots, X_{b_{k-1}})),$$

11.3. Backsteps of permutations with a given shape

and

$$\det(M_2) = S(\mathbf{a}_2, \mathbf{X}_1) = S((a_1, a_2, \ldots, a_{k-2}, a_{k-1} + a_k), (X_{b_1}, X_{b_2}, \ldots, X_{b_{k-1}})).$$

To the term $S_{a_k}(X_{b_k}) \det(M_1)$ of Equation (11.2.2) contribute exactly pairs of words made of a word w' in $W(\mathbf{a}_1, \mathbf{X}_1)$ and a nonincreasing word w'' of length a_k over the alphabet X_{b_k}. Let $w = w'w''$. Then either w is an element of $W(\mathbf{a}, \mathbf{X})$ if the last letter of w' is strictly smaller than the first letter of w'', or it is an element of $W(\mathbf{a}_2, \mathbf{X}_1)$ otherwise.

In the first case it contributes to $S(\mathbf{a}, \mathbf{X})$. In the latter case, it contributes to $S(\mathbf{a}_2, \mathbf{X}_1) = \det(M_2)$. This combinatorial interpretation yields precisely Equation (11.2.2). ∎

11.3. Backsteps of permutations with a given shape

Let $\sigma = \sigma(1)\sigma(2)\ldots\sigma(n)$ be a permutation of $[n]$. Let a_1 be the first index such that $\sigma(a_1) > \sigma(a_1 + 1)$. Let $a_1 + a_2$ be the second such index and, more generally, let $a_1 + a_2 + \cdots + a_i$ be the ith such index. (We suppose, by convention, that $\sigma(n+1) = 0$.) The sequence $\mathbf{a} = (a_1, a_2, \ldots, a_k)$ is a composition of n called the *shape* of σ. If the permutation is considered as a word with distinct letters, its shape as a word coincides with its shape as a permutation, provided that the order is reversed. This apparent complication actually keeps the results of the next sections reasonably simple to state.

EXAMPLE 11.3.1. Let $\sigma = 6\,8\,4\,5\,9\,3\,1\,2\,7$. Its shape is $\mathbf{a} = (2, 3, 1, 3)$.

Let $w = w_1 w_2 \cdots w_n$ be a word of shape $\mathbf{a}$. On $[n]$ define the total order $\preceq$ by $i \prec j$ iff $w_i > w_j$, or $w_i = w_j$ and $i < j$. Let $\sigma(i)$ be the rank of index i according to this total order. This implies that $i \prec j \Leftrightarrow \sigma(i) < \sigma(j)$. Then σ is a permutation called the *standard normalization* of w.

Let $\varphi(w)$ be the pair (σ, s) consisting of the standard normalization σ of w and of the nonincreasing reordering s of the letters of w. In the next theorem, as well as in Theorem 11.6.1, we shall say that the nonincreasing sequence $s = s_1 s_2 \ldots$ is *compatible* with a set E of positive integers $s_i > s_{i+1}$ whenever i is an element of E.

THEOREM 11.3.2. *Let* $\mathbf{a}$ *be a shape. Then* φ *is a bijection between the set of words* w *of shape* $\mathbf{a}$ *and the set of pairs* (σ, s), *where* σ *is a permutation of shape* $\mathbf{a}$ *and* s *a nonincreasing sequence compatible with* BS σ.

Proof. It is clear that the shape of σ and the shape of w are identical.

Suppose $i = \sigma(u)$ and $i + 1 = \sigma(v)$. Then $\sigma(u) < \sigma(v)$, hence $u < v$. Then, from the very definition of $\preceq$, we have $w_u \geq w_v$, which can also be written $w_{\sigma^{-1}(i)} \geq w_{\sigma^{-1}(i)}$. Since the application which, to i, associates $w_{\sigma^{-1}(i)}$ is nonincreasing, it is the nonincreasing rearrangement of w. This means that $s_i = w_{\sigma^{-1}(i)}$.

Suppose moreover that i is a backstep of σ, that is $u > v$. We already know that $u \prec v$. Those last two conditions are compatible only if $w_u > w_v$, which is the same as $w_{\sigma^{-1}(i)} > w_{\sigma^{-1}(i)}$ or $s_i > s_{i+1}$.

Conversely, suppose we are given a pair (σ, s) where s is compatible with BS σ. Let w be the word of length n given by $w_u = s_{\sigma(u)}$. Consider the order $\prec$ defined by $u \prec v$ iff $\sigma(u) < \sigma(v)$. We shall prove that this order is precisely the same as the order $\prec$ which has been defined earlier on w. This will prove that $\varphi(w) = (\sigma, s)$. To do so, suppose $\sigma(u) < \sigma(v)$. Since s is nonincreasing, we have $s_{\sigma(u)} \geq s_{\sigma(v)}$ and so $w_u \geq w_v$. Either $w_u > w_v$, which proves the point, or $w_u = w_v$.

In this latter case we have $s_{\sigma(u)} = s_{\sigma(v)}$. This implies (because s is compatible with BS σ) that none of $\sigma(u), \sigma(u) + 1, \ldots, \sigma(v) - 1$ is a backstep. Hence

$$u = \sigma^{-1}(\sigma(u)) < \sigma^{-1}(\sigma(u) + 1) < \cdots < \sigma^{-1}(\sigma(v) - 1) < \sigma^{-1}(\sigma(v)) = v.$$

We have then proved that $u \prec v$ and $w_u = w_v$ imply $u < v$, which completes the proof of the theorem. ∎

EXAMPLE 11.3.3. Let $w = 1\,0\,2\,2\,0\,4\,7\,5\,1$. Then $\varphi(w) = (\sigma, s)$ where

$$\sigma = 6\,8\,4\,5\,9\,3\,1\,2\,7,$$
$$s = 7\,5\,4\,2\,2\,1\,1\,0\,0.$$

The backsteps of σ and the corresponding elements of s appear in boldface. It can be noted that, although $s_1 > s_2$, the integer 1 is not a backstep of σ. The shape of both w and σ is $(2, 3, 1, 3)$.

COROLLARY 11.3.4. *Let* **a** *be a composition of n and A be a subset of $[n - 1]$. Denote by $D_A(\mathbf{a})$ the number of permutations of n of shape* **a** *whose backsteps are precisely the elements of A. Then*

$$S(\mathbf{a}, X_{\mathbb{N}}) = \sum_{A \subset [n-1]} D_A(\mathbf{a}) \sum_{s} GS(s), \tag{11.3.1}$$

where the second sum is over the set of all nonincreasing sequences s compatible with A.

11.3. Backsteps of permutations with a given shape

Proof. A word of shape **a** corresponds to a commutative monomial of $S(\mathbf{a}, X_N)$. Applying the bijection φ of Proposition 11.3.2 and summing over subsets of $[n-1]$ yields the corollary. ∎

It can be noted that the second sum in Corollary 11.3.4 is independent of the shape **a**. As those sums are linearly independent, the numbers $D_A(\mathbf{a})$ are uniquely determined by $S(\mathbf{a}, X_N)$. See Problem 11.6.2 for a use of this property.

As a corollary, we shall now derive the enumeration of permutations with shape **a** by imaj, by substituting successive powers q^i for the letters i.

COROLLARY 11.3.5. *Let **a** be a composition of n. Then*

$$\sum_\sigma q^{\mathrm{imaj}\,\sigma} = (q)_n S(\mathbf{a}, \{1, q, q^2, \ldots\}), \tag{11.3.2}$$

*where the sum is over the set of permutations of $[n]$ of shape **a**.*

Proof. Let A be a subset of $[n-1]$ and s be a nonincreasing sequence compatible with A. From s define a sequence d of nonnegative integers by

$$d_k = \begin{cases} s_k - s_{k+1} & \text{if } k \notin A \text{ and } 0 \le k < n, \\ s_k - s_{k+1} - 1 & \text{if } k \in A, \\ s_n & \text{if } k = n. \end{cases} \tag{11.3.3}$$

The set of sequences d of length n consisting of nonnegative integers is in that way in bijection with the set of nonincreasing sequences s of integers compatible with A. This bijection depends on A.

Let $\sum A$ denote the sum of the elements of A. Substituting q^i for i transforms $\mathrm{GS}(s)$ into q to the power $\sum_{1 \le k \le n} s_k$. Now, from Equation (11.3.3), one obtains

$$\sum_{1 \le k \le n} s_k = \sum_{1 \le k \le n} k d_k + \sum A.$$

Therefore,

$$\sum_s q^{s_1 + s_2 + \cdots + s_n} = q^{\sum A} \sum_d q^{d_1 + 2d_2 + \cdots + n d_n}.$$

The latter sum is the generating series for partitions into positive parts not greater than n, which is equal to $1/(q)_n$. This, combined with Corollary 11.3.4, yields the conclusion. ∎

EXAMPLE 11.3.6. In Example 11.3.3, we had

$$\sigma = 684593127,$$
$$s = 754221100,$$

and BS $\sigma = \{2,3,5,7\}$. The sequence d defined by Equation (11.3.3) is then

$$d = 201000000,$$

corresponding to the partition $1^2 3$.

PROPOSITION 11.3.7. *Let* $\mathbf{a} = (a_1, a_2, \ldots, a_k)$ *be a composition of* n. *Then* $\sum_\sigma q^{\text{imaj}\,\sigma} = (q)_n \det(N)$. *The sum is over the set of permutations of* $[n]$ *of shape* $\mathbf{a}$ *and* $N = (N_{i,j})_{\{1 \le i, j \le k\}}$ *with*

$$N_{i,j} = \begin{cases} 1/(q)_{(a_i + a_2 + \cdots + a_j)} & \text{if } i \le j, \\ 1 & \text{if } i = j+1, \\ 0 & \text{if } i > j+1. \end{cases} \quad (11.3.4)$$

Proof. Apply Proposition 11.2.2 in the case when all alphabets X_{b_i} are equal to X_N, then substitute q^i for i. From Proposition 11.1.2, one obtains precisely Formula (11.3.4). ∎

Using Proposition 11.3.7 one can obtain explicit expressions for the enumeration of permutations of special shapes by imaj. This has been done for alternating permutations and permutations with a given number of descents, thus leading to q-analogues of tangent and secant numbers as well as q-analogues of Eulerian polynomials. See Problems 11.3.2 and 11.3.3 for more specific details.

11.4. Inversions of permutations with a given shape

The inversions of a permutation can be conveniently taken care of by a simple encoding with particular words.

Let σ be a permutation of $[n]$. For $1 \le i \le n$, let ℓ_i be the number of indices $j < i$ such that $\sigma(j) > \sigma(i)$. The word $\ell = \ell_1 \ell_2 \cdots \ell_n$ will be called the *Lehmer encoding* of σ.

EXAMPLE 11.4.1. To the permutation

$$\sigma = 684593127$$

corresponds the Lehmer encoding

$$\ell = 002205662.$$

11.4. Inversions of permutations with a given shape

It can be noted that σ and ℓ have the same shape $\mathbf{a} = (2, 3, 1, 3)$. This point will be established in the next proposition.

PROPOSITION 11.4.2. *The Lehmer encoding is a bijection between the set of permutations of $[n]$ and the set of words ℓ of length n such that $0 \le \ell_i \le i-1$ for $1 \le i \le n$. This bijection has the following properties:*
  (i) *σ and ℓ have the same shape,*
  (ii) *the sum of the letters of ℓ is the number of inversions of σ.*

Proof. From the definition, it is clear that $0 \le \ell_i \le i$ for $1 \le i \le n$. Proving that the Lehmer encoding is a bijection is done by describing its inverse. This is left as an elementary exercise (To start, observe that the position of 1 in σ is the greatest index i such that $\ell_i = i - 1$.)

If $\sigma(i+1) < \sigma(i)$ then for any $j < i$ one has $\sigma(j) > \sigma(i) \Rightarrow \sigma(j) > \sigma(i+1)$. Therefore $\ell_{i+1} \ge \ell_i$. Since i is smaller than $i+1$ and $\sigma(i) > \sigma(i+1)$, one has in fact $\ell_{i+1} > \ell_i$. Conversely, if $\sigma(i+1) > \sigma(i)$ then for any $j < i$ one has $\sigma(j) > \sigma(i+1) \Rightarrow \sigma(j) > \sigma(i)$. Therefore $\ell(i+1) \le \ell(i)$. This proves that σ and ℓ have the same shape.

The second assertion is obvious from the definition of the Lehmer encoding. ∎

PROPOSITION 11.4.3. *Let $\mathbf{a} = (a_1, a_2, \ldots, a_k)$ be a shape, σ be a permutation of shape $\mathbf{a}$ and ℓ be the Lehmer encoding of σ. Denote by $\mathbf{N}$ the sequence $(N_0, N_{a_1}, N_{a_1+a_2}, \ldots, N_{a_1+a_2+\cdots+a_{k-1}})$. Then ℓ is an element of $W(\mathbf{a}, \mathbf{N})$. Furthermore, the Lehmer encoding induces a bijection between the set of permutations of shape $\mathbf{a}$ and $W(\mathbf{a}, \mathbf{N})$.*

Proof. Let $0 \le i < k$ and let r be 0 if $i = 0$ and $a_1 + a_2 + \cdots + a_i$ otherwise. As $\ell_{r+1} \le r$ (general property of Lehmer codes) and $\ell_{r+1} \ge \ell_{r+2} \ge \cdots \ge \ell_{r+a_{i+1}}$ (because ℓ has shape $\mathbf{a}$), it follows that, for $r+1 \le j \le r+a_{i+1}$, one has $0 \le \ell_j \le r$. This implies that ℓ is in $W(\mathbf{a}, \mathbf{N})$.

The conditions on the alphabets ensure that every element of $W(\mathbf{a}, \mathbf{N})$ is an admissible Lehmer encoding. The corresponding permutation has the same shape $\mathbf{a}$, which proves the bijection part of the proposition. ∎

Using the same arguments as at the end of Section 11.3, it is possible to derive the enumeration of permutations of special shapes by number of inversions. This leads to the following theorem, which can also be proved using Foata's "second fundamental transformation". See Lothaire 1983.

THEOREM 11.4.4 (Foata–Schützenberger). *Let $\mathbf{a}$ be a composition of n. Then, for any integer m, the number of permutations of shape $\mathbf{a}$ having m*

inversions is the same as the number of permutations of shape **a** having m backsteps.

Proof. Proposition 11.3.7 gives a determinantal expression for the generating polynomial of permutations of given shape by imaj. We shall establish a similar expression for the generating polynomial of the same permutations by number of inversions. Finally, we shall verify that both determinants are equal.

If σ is a permutation of shape **a**, its number of inversions $\text{inv}\,\sigma$ is the sum of the letters of ℓ (Proposition 11.4.2). This implies that the generating polynomial $\sum_\sigma q^{\text{inv}\,\sigma}$ is obtained by substituting q^i for i in $S(\mathbf{a}, X)$, where X is defined as in Proposition 11.4.3.

We use the determinantal formula of Proposition 11.2.2 $S(\mathbf{a}, X) = \det(M)$ where, in this particular case,

$$M_{i,j} = \begin{cases} S_{a_i + a_{i+1} + \cdots + a_j}(X_{a_1 + \cdots + a_{i-1}}) & \text{if } i \leq j, \\ 1 & \text{if } i = j+1, \\ 0 & \text{if } i > j+1. \end{cases} \quad (11.4.1)$$

Proposition 11.1.1 implies that, when substituting q^i for i in the entries of this determinant, its generic element becomes

$$\begin{cases} (q)_{a_1 + \cdots + a_j} / (q)_{a_1 + \cdots + a_{i-1}} (q)_{a_i + \cdots + a_j} & \text{if } i \leq j, \\ 1 & \text{if } i = j+1, \\ 0 & \text{if } i > j+1. \end{cases}$$

We can then factor $(q)_{a_1 + \cdots + a_j}$ in column j and $(q)^{-1}_{a_1 + \cdots + a_{i-1}}$ in row i.

Thus, the generating polynomial of permutations of shape **a** by number of inversions $\sum_\sigma q^{\text{inv}\,\sigma}$ is equal to

$$\prod_{1 \leq i, j \leq k} \frac{(q)_{a_1 + \cdots + a_j}}{(q)_{a_1 + \cdots + a_{i-1}}} \times \det(((q)_{a_i + \cdots + a_j})_{\{1 \leq i, j \leq k\}}),$$

with the convention that the generic term of the determinant is 1 if $i = j+1$ and 0 if $i > j+1$. After canceling we obtain

$$(q)_{a_1 + \cdots + a_k} \det((1/(q)_{a_i + \cdots + a_j})_{\{1 \leq i, j \leq k\}}),$$

which, since $a_1 + \cdots + a_k = n$, is Equation (11.3.4) of Proposition 11.3.7. ∎

11.5. Lyndon factorization and cycles of permutations

Recall from Subsection 1.2.1 that a *Lyndon word* of length n is a word $w = w_1 w_2 \cdots w_n$ whose letters are nonnegative integers, and which is strictly smaller lexicographically than its conjugates. As a consequence, a Lyndon word is primitive.

The following result, due to Lyndon, will play the role of the factorization of Section 11.2.

THEOREM 11.5.1 (Lyndon). *Any word w can be written uniquely as a nonincreasing product $w = u_1 u_2 \cdots u_k$ of Lyndon words.*

The lengths of the Lyndon factors constitute a partition λ of n, which will be called the *type* of w.

EXAMPLE 11.5.2. The word $w = 210120101$ of length 9 has the following Lyndon factorization:

$$w = 2 \ \ 1 \ \ 012 \ \ 01 \ \ 01.$$

The type of w is the partition $\lambda = 32211$ of 9.

The same Lyndon factorization can be performed on permutations (for the same technical reasons as in Section 11.3, the order has to be reversed). Start with a permutation τ considered as the concatenation of its values. Its Lyndon factorization (for the reverse order) is $\tau = \theta_1 \theta_2 \ldots \theta_k$. Consider each Lyndon word θ_j as the sequence of the values of cyclic permutation. Finally, let σ be the product (in the symmetric group) of those cycles.

This transformation from τ to σ is a bijection of the symmetric group onto itself. Its inverse is known as Foata's "first fundamental transformation". In the case of permutations, the first fundamental transformation can be easily described: Start with a permutation σ. Consider its decomposition as a product of cycles. Write each cycle with its greatest value first. Finally, concatenate those words in increasing order of their first elements. The resulting word is τ.

If σ is a permutation of $[n]$, the multiset of the lengths of its cycles is a partition of n, which will also be called the *type* of σ. The type of a permutation is characteristic of its conjugacy class in the symmetric group, but we shall not use this property in what follows.

EXAMPLE 11.5.3. Let

$$\tau = \tau(1)\tau(2)\ldots\tau(9) = 147328596.$$

As a word, its Lyndon factorization (for the reverse order) is

$$\tau = 1 \quad 4 \quad 7\,3\,2 \quad 8\,5 \quad 9\,6.$$

Hence the resulting permutation σ is the product of cycles

$$\sigma = (1)(4)(7\,3\,2)(8\,5)(9\,6).$$

As a word,
$$\sigma = \sigma(1)\sigma(2)\cdots\sigma(9) = 1\,7\,2\,4\,8\,9\,3\,5\,6.$$

We are now ready to describe the *Gessel normalization* of a word w, which will play the same role for the cycle decomposition as the standard normalization of Section 11.3 did for the shape.

Let $w = w_1 w_2 \cdots w_n$ be a word of length n and $w = u_1 u_2 \cdots u_k$ be its Lyndon decomposition. Suppose that the letter w_i of w is in the Lyndon factor $u_r = w_{r_1} w_{r_2} \cdots w_{r_s}$. Let $p(i) = w_i w_{i+1} \cdots w_{r_s} u_r^\omega$ be the infinite word obtained by writing the suffix of u_r starting with w_i and concatenating to it an infinite number of copies of u_r. Then, we can define the total order $\leq$ on $[n]$ by

$$i < j \text{ iff } \begin{cases} p(i) > p(j), \text{ or} \\ p(i) = p(j) \text{ and } i < j. \end{cases} \tag{11.5.1}$$

Let $\tau(i)$ be the rank of index i according to this total order. Now apply to τ the inverse of Foata's first fundamental transformation to obtain σ. This last permutation is the Gessel normalization of w.

We shall need the following technical lemma to prove Proposition 11.5.5.

LEMMA 11.5.4. *Let u and v be two words such that lexicographically $u > v$. Suppose that u is a Lyndon word. Then, as infinite words and for the lexicographic order, $u^\omega > v^\omega$.*

Proof. Let us say that u is strongly greater that v – which we write $u \gg v$ – when $u > v$ and v is not a proper prefix of u. The following statement is easy to prove: if $u \gg v$, then $ux > vy$ for any words x and y.

Thus, if $u \gg v$, the lemma is true even if u is not a Lyndon word.

Suppose now that v is a proper prefix of u. We can factor $u = v^k u'$ for some $k \geq 1$, where v is not a proper prefix of u'. (The case when u' is empty is to be excluded, as $u \neq v$ and u is Lyndon, hence primitive.) Then, being a Lyndon word, u is strictly smaller than any of its proper suffixes (Lothaire 1983, Proposition 5.1.2), that is $u < u'$. As $u > v$ by hypothesis, we have $u' > v$. As v is not a proper prefix of u', we have in fact $u' \gg v$. Then $u' u^\omega > v v^\omega$. Multiplying on the left by v^k yields $u^\omega > v^\omega$. ∎

11.5. Lyndon factorization and cycles of permutations

PROPOSITION 11.5.5. *A word w and its Gessel normalization σ have the same type.*

Proof. We shall prove that the Lyndon factorization of w and that of the permutation τ have corresponding Lyndon factors of the same length. That will imply the proposition since the factors of τ correspond to the cycles of σ.

Consider $w = u_1 u_2 \cdots u_k$. The factor u_r, written as a concatenation of letters, is equal to $w_{r_1} w_{r_2} \cdots w_{r_s}$. Let i be an index such that $r_1 < i \le r_s$. Then
$$p(i) = w_i \cdots w_{r_s} u_r^\omega.$$
Since u_r is a Lyndon word, $p(r_1) < p(i)$ so that $i < r_1$ and, consequently, $\tau(r_1) > \tau(i)$. This proves that $\tau(r_1)$ is strictly greater than any of the $\tau(i)$, $i = r_1 + 1, \ldots, r_s$. So $\tau(r_1)\tau(r_1 + 1) \cdots \tau(r_s)$ is eligible as a Lyndon factor of τ.

To finish the proof, we must show that the first letters of those potential factors of τ are increasing. Let $u_r = w_{r_1} w_{r_2} \cdots w_{r_s}$ and $u_f = w_{f_1} w_{f_2} \cdots w_{f_g}$ be two Lyndon factors of w such that $r_1 < f_1$. Then, either $u_r = u_f$. In that case $p(r_1) = p(f_1)$ and $r_1 < f_1$ so $r_1 < f_1$, hence $\tau(r_1) < \tau(f_1)$, which is what is expected. In the other case $u_r \ne u_f$. Then, for the lexicographic order $u_r > u_f$. We can then apply Lemma 11.5.4 so that $p(r_1) = u_r^\omega > p(f_1) = u_f^\omega$. This implies that $r_1 < f_1$ which in turn implies that $\tau(r_1) < \tau(f_1)$. This is exactly what was needed to complete the proof. ∎

EXAMPLE 11.5.6. Let $w = 210120101$ as in Example 11.5.2. Its Lyndon factorization is
$$w = 2 \quad 1 \quad 012 \quad 01 \quad 01.$$
For that w, we find

$$p(1) = 222222\cdots, \qquad p(2) = 111111\cdots,$$
$$p(3) = 012012\cdots, \qquad p(4) = 120120\cdots,$$
$$p(5) = 201201\cdots, \qquad p(6) = 010101\cdots,$$
$$p(7) = 101010\cdots, \qquad p(8) = 010101\cdots,$$
$$p(9) = 101010\cdots.$$

The order $\le$ on the indices is then
$$1 < 5 < 4 < 2 < 7 < 9 < 3 < 6 < 8.$$
Consequently,
$$\tau = 147328596,$$

which has the following Lyndon decomposition:

$$\tau = 1 \quad 4 \quad 7\,3\,2 \quad 8\,5 \quad 9\,6.$$

As a product of cycles,

$$\sigma = (1)(4)(7\,3\,2)(8\,5)(9\,6),$$

which, written as the sequence of its values, is

$$\sigma = 1\,7\,2\,4\,8\,9\,3\,5\,6.$$

11.6. Major index of permutations with a given cyclic type

In Section 11.5, a shape-preserving normalization was used to enumerate permutations with a given shape by backsteps. In this section, the type-preserving Gessel normalization will allow the same enumeration for permutations with a given type by descents.

Let w be a word and $\psi(w)$ be the pair (σ, s) consisting of the Gessel normalization of σ and of the nonincreasing reordering s of the letters of w.

THEOREM 11.6.1. *Let λ be a partition of n. Then ψ is a bijection between the set of words w of type λ and the set of pairs (σ, s), where σ is a permutation of type λ and s a nonincreasing sequence compatible with des σ.*

Proof. We already know (Proposition 11.5.5) that w and σ have the same type.

Suppose that $i < j$ and put $\tau^{-1}(i) = a$ and $\tau^{-1}(j) = b$. As $\tau(a) < \tau(b)$, we deduce from the definition of τ that $a \prec b$ and so $p(a) \geq p(b)$. This latter relation implies that the first letters of each infinite word also satisfy $w_a \geq w_b$. We have then proved

$$i < j \Rightarrow w_{\tau^{-1}(i)} \geq w_{\tau^{-1}(j)}.$$

As for Theorem 11.3.2, this proves that $s(i) = w_{\tau^{-1}(i)}$.

To prove that s is compatible with des σ, we shall prove the (apparently) more general result that if $i < j$ and $\sigma(i) = a > \sigma(j) = b$, then $s_i > s_j$. We already know that $s(i) \geq s(j)$.

Let us suppose then that $i < j$ (that is $a \prec b$) and $s_i = s_j$. This is the same as $w_a = w_b$. We also know that $p(a) \geq p(b)$. But $p(a) = w_a p(\bar{a})$ and $p(b) = w_b p(\bar{b})$. As $w_a = w_b$, it follows that $p(\bar{a}) \geq p(\bar{b})$. The index $\bar{a}$ is

11.6. Major index of permutations with a given cyclic type

the index of the element immediately following i cyclically in the Lyndon factor to which it belongs, so that $\tau(\bar{a}) = \sigma(i)$.

We face two possibilities:

If $p(\bar{a}) > p(\bar{b})$, then, by definition, $\bar{a} \prec \bar{b}$, which implies $\tau(\bar{a}) < \tau(\bar{b})$, which is also $\sigma(i) < \sigma(j)$.

If $p(\bar{a}) = p(\bar{b})$ then $p(a) = p(b)$. As $a \prec b$, we must have $a < b$. Then w_a and w_b are in two equal Lyndon factors of w and the one containing w_a precedes the one containing w_b. But $w_{\bar{a}}$ and $w_{\bar{b}}$ belong to the same factors, which implies $\bar{a} \prec \bar{b}$. Then, as in the previous case, $\sigma(i) < \sigma(j)$.

We have then proved that $i < j$ and $s_i = s_j$ imply $\sigma(i) < \sigma(j)$, which implies the compatibility of s with the set $\mathrm{des}\,\sigma$.

To prove that ψ is a bijection, start with σ, then construct τ by rearranging the cycles of σ using Foata's first fundamental transformation. Then define the word w by $w_a = s_{\tau(a)}$. To verify that $\psi(w) = (\sigma, s)$ we only have to verify that the Lyndon decompositions of w and of τ have corresponding factors of the same length.

To do so, given an index a, define a $p(a)$ from τ: if $\tau(a)$ belongs to the Lyndon factor $\theta_r = \tau(r_1)\cdots\tau(r_s)$, let $p(a) = w_a \cdots w_{r_s} w_{r_1} \cdots w_a \cdots$. The nonincreasing property of s ensures that, if $\tau(a) < \tau(b)$, then $w_a = s_{\tau(a)} \geq s_{\tau(b)} = w_b$.

Suppose then that $\tau(a)$ and $\tau(b)$ belong to the same Lyndon factor of τ and that $\tau(b)$ is the first (and greatest) letter of that factor. If $w_a > w_b$ we have $p(a) > p(b)$ which implies that the corresponding factor of w satisfies the minimality property of Lyndon words. If $w(a) = w(b)$ consider the cyclic successors $\bar{a}$ and $\bar{b}$. Then, as before, $\tau(\bar{a}) = \sigma(\tau(a))$. As s is compatible, we have $\sigma(\tau(a)) < \sigma(\tau(b))$, which is $\tau(\bar{a}) < \tau(\bar{b})$. If $w_{\bar{a}} > w_{\bar{b}}$, we can again conclude that $p(a) > p(b)$, otherwise we iterate until the iterated cyclic successor of a is b, which leads to a contradiction with the fact that θ_r is a Lyndon word.

The factors of w corresponding to the Lyndon factors of τ are consequently Lyndon words. We still have to prove that they constitute the Lyndon decomposition of w, i.e. that they are in nonincreasing order.

Suppose then that $\tau(a)$ and $\tau(b)$ do not belong to the same Lyndon factor of τ. Suppose that each of them is the first letter of its respective Lyndon factor, and $a < b$. This implies that $\tau(a) < \tau(b)$. If $w_a > w_b$ we have as above $p(a) > p(b)$ which implies from Lemma 11.5.4 that the corresponding factors of w are in nonincreasing order. Otherwise, if $w_a = w_b$, and for the same reason as above, either we find successors a' of a and b' of b such that, for the first time, $p(a') > p(b')$, and so $p(a) > p(b)$, or, in the other case, for all corresponding letters of both factors we have $p(a') = p(b')$ which implies (as Lyndon words are primitive) that the corresponding factors of w are equal, and so in nonincreasing order. ∎

EXAMPLE 11.6.2. Proceed with Example 11.5.6. We have found that

$$w = 210120101,$$
$$\tau = 147328596,$$
$$\sigma = 172489356 \text{ and}$$
$$s = 221111000.$$

The descent set of σ is $\{2, 6\}$. Notice that $s_2 > s_3$ and $s_6 > s_7$.

Since the descent set of a permutation is the backstep set of its inverse, and since both permutations have the same type, Theorem 11.6.1 can also be used for counting permutations with a given type by its backsteps.

Let $L(\lambda, X)$ the commutative image of the sum of all words on the alphabet N with type λ.

COROLLARY 11.6.3. *Let λ be a partition of n and A be a subset of $[n-1]$. Denote by $E_A(\lambda)$ the number of permutations with type λ whose descents are precisely the elements of A. Then*

$$L(\lambda, X) = \sum_{A \subset [n-1]} E_A(\lambda) \sum_s GS(s), \qquad (11.6.1)$$

where the second sum is over the set of all nonincreasing sequences s compatible with λ.

Proof. This is identical to that of Corollary 11.3.4 ∎

There is also an analogue of Corollary 11.3.5.

COROLLARY 11.6.4. *Let λ be a partition of n. Then*

$$\sum_\sigma q^{\text{maj}\,\sigma} = \sum_\sigma q^{\text{imaj}\,\sigma} = (q)_n L(\lambda, \{1, q, q^2, \ldots\}), \qquad (11.6.2)$$

where the sum is over the set of permutations of $[n]$ of type λ.

As a consequence, any information about $L(\lambda, X)$ can be used to derive information about the distribution of descents on permutations with type λ. No "general" formula, similar to the determinant of Proposition 11.2.2, is known. Some particular cases of types, such as cyclic permutations, involutions, *dérangements*, will be given as exercises.

Contrary to what happens for permutations with a given shape, the distribution of inversions is in general different from that of descents. There is no analogue of Theorem 11.4.4 for permutations with a given type.

Problems

Section 11.1

11.1.1 (The q-binomial theorem) The generating series for all words on the alphabet N is

$$E(X) = \sum_{a \geq 0} S_a(X) = \prod_{i \in N}(1 - x_i)^{-1}.$$

From this, derive the q-binomial theorem

$$\sum_{a \geq 0} \begin{bmatrix} a+k \\ k \end{bmatrix} t^a = \prod_{0 \leq i \leq k}(1 - tq_i)^{-1}$$

and Euler's q-exponential series

$$e(t;q) = \sum_{a \geq 0} \frac{t^a}{(q)_a} = \prod_{0 \leq i}(1 - tq_i)^{-1}.$$

See for example Andrews 1976.

Section 11.2

11.2.1 The simplest nontrivial case for Proposition 11.2.2 is when $\mathbf{a} = (k, n-k)$. In that case, show that

$$S(\mathbf{a}, X) = S_k(X)S_{n-k}(X) - S_n(X).$$

11.2.2 Consider alternating words of odd length n, i.e. words with shape $\mathbf{a} = (2, 2, \ldots, 2, 1)$. Denote by X' the set of indeterminates $\{x_1, x_2, \ldots\}$. Prove that the generating series $S(\mathbf{a}, X) = \tan(X)$ for alternating words satisfy the recurrence relation

$$\tan(X) - \tan(X') = x_0(1 + \tan(X)\tan(X')).$$

To do so, consider the last occurrence of the letter 0 in an alternating word. Note the similarity with the classical differential equation satisfied by the tangent function.
From this recurrence, by induction on the size of X, and by letting this size tend to infinity, derive the formula

$$\tan(X) = \frac{1}{i} \frac{E(iX) - E(-iX)}{E(ix) + E(-iX)}.$$

(By iX, we mean the set obtained by multiplying each indeterminate by the square root of -1 in the series $E(X)$ as defined in Problem 11.1.1.)

Similarly, for alternating words of even length, one obtains as generating series

$$\sec(X) = \frac{2}{E(iX) + E(-iX)}.$$

See Désarménien 1983.

11.2.3 Let $A_{n,k}(X)$ be the generating series for words of length n whose shapes contain k parts. Let $A(X) = \sum_{n,k \geq 0} A_{n,k}(X) u^k$. By considering the first occurrence of the letter r, prove that

$$A(X_r)(1 - x_r F(X_{r-1})) = (1 - x_r(1-u)) A(X_{r-1}).$$

It then follows that

$$A(X) = \frac{1-u}{1 - uE((1-u)X)}.$$

See Désarménien 1983.

Section 11.3

11.3.1 From Problem 11.2.1 and from Corollary 11.3.5, prove that the enumeration by imaj of permutations of $[n]$ with only one descent in position k is

$$\begin{bmatrix} n \\ k \end{bmatrix} - 1.$$

This is a particular case of a more general result proved in Gessel and Reutenauer 1993.

11.3.2 (The q-Euler numbers) Consider the q-exponential $e(t;q)$. The q-tangent and q-secant series may be defined by

$$\tan(t;q) = \frac{1}{i} \frac{e(it;q) - e(-it;q)}{e(it;q) + e(-it;q)}$$

and

$$\sec(x;q) = \frac{2}{e(it;q) + e(-it;q)}.$$

The q-Euler numbers $\mathrm{Eul}_n(q)$ (also called tangent and secant numbers) are the coefficients of the series expansions of those functions:

$$\sum_{n \geq 0} \mathrm{Eul}_n(q) \frac{t^n}{(q)_n} = \tan(t;q) + \sec(t;q).$$

Using Problem 11.2.2, prove that $\text{Eul}_n(q)$ enumerates alternating permutations of $[n]$ by imaj. (Alternating permutations are those whose shape is $(2,2,2,\ldots 2)$ or $(2,2,2,\ldots 1)$, depending on the parity of n.) By letting q tend to 1, one obtains André's classical interpretation of the Euler numbers as counting alternating permutations. See Désarménien 1983.

11.3.3 (The q-Eulerian polynomials) The q-Eulerian polynomials are defined by

$$\sum_{n\geq 0} A_n(u;q) \frac{t^n}{(q)_n} = \frac{1-u}{1 - ue((1-u)t;q)}.$$

From Problem 11.2.3, prove that the coefficient of u^k in $A_n(u;q)$ enumerates permutations of $[n]$ with $k-1$ descents (i.e. their shape is a composition with k parts) by imaj. This generalizes the classical interpretation of the Eulerian polynomials as generating functions for permutations according to the number of descents. See Désarménien 1983.

Section 11.4

11.4.1 (The q-Euler numbers again) According to Theorem 11.4.4, the q-Euler numbers also enumerate the alternating permutations of $[n]$ by number of inversions. This can be proved directly from the definitions of the q-tangent and q-secant series. The q-derivative of a series $f(t;q)$ is $D_q f(t;q) = (f(t;q) - f(tq;q))/t$. Prove that the q-tangent and q-secant series satisfy the following q-differential equations:

$$D_q(\tan(t;q) + \sec(t;q)) = 1 + \tan(t;q)(\tan(tq;q) + \sec(tq;q)).$$

It follows that the q-Euler numbers satisfy the following quadratic recurrence:

$$\text{Eul}_{n+1}(q) = \sum_{0 \leq j \leq (n-1)/2} \begin{bmatrix} n \\ 2j+1 \end{bmatrix} q^{n-2j-1} \, \text{Eul}_{2j}(+1\,q) \, \text{Eul}_{n-2j-1}(q).$$

By considering the position of $n+1$ in an alternating permutation of length $n+1$ and counting the inversions, it can be shown that the enumeration of permutations by number of inversions satisfies the same quadratic recurrence. See Désarménien 1982.

Section 11.5

11.5.1 Actually, Lemma 11.5.4 provides a characterization of Lyndon words. Prove that the following statements are equivalent:
(i) u is a Lyndon word;
(ii) for any word v, it is equivalent that $u > v$ and $u^\omega > v^\omega$.

Section 11.6

11.6.1 (The q-counting of cycles) The generating series of Lyndon words of length n on the alphabet X is

$$L_n(X) = \frac{1}{n} \sum_{d\mid n} \mu(d) \left(\sum_i x_i^d\right)^{n/d}.$$

This can be obtained by observing that any word of length n is some power of a primitive word, taking the commutative image and using Möbius inversion. By substituting powers of q for indeterminates, obtain the enumeration of n-cycles by imaj:

$$C_n(q) = \frac{(q)_n}{n} \sum_{d\mid n} \frac{\mu(d)}{(1 - q^d)^{n/d}}.$$

See Gessel and Reutenauer 1993.

11.6.2 (*Dérangements* and *désarrangements*) A *dérangement* is a permutation without fixed points. A *désarrangement* is a permutation whose first ascent is even (i.e. its shape starts with an odd number of 1's). A *dérangement* is a permutation with a given cycle structure (no 1-cycle) and a *désarrangement* is a permutation with a given shape.

If A is a subset of $[n-1]$, then the number of *dérangements* whose backsteps are the elements of A is equal to the number of *désarrangements* whose backsteps are the elements of A. Prove this by showing that the generating series of both types of permutations are equal, then applying Corollaries 11.3.4 and 11.6.3.

Then the numbers of *dérangements* and of *désarrangements* counted by imaj are equal. Their common value is

$$d_n(q) = [n]! \sum_{0 \le k \le n} \frac{(-1)^k q^{k(k-1)/2}}{[k]!},$$

where $[n]! = (q)_n/(1-q^n)$. This is the natural q-analogue of the number of *dérangements*. This problem has been considered in Désarménien and Wachs 1988, Désarménien and Wachs 1993 and Gessel and Reutenauer 1993.

11.6.3 (Involutions) An involution has cycles of length 1 or 2. Let $I_{n,k}(X)$ be the generating series for words on the alphabet N corresponding to involutions of $[n]$ with k fixed points. Prove that
$$\sum_{n,k} I_{n,k}(X)u^k = \prod_{i \in N}(1-ux_i)^{-1} \prod_{i<j \in N}(1-x_ix_j)^{-1}.$$

From this generating series, one can derive the generating series for the number $I_{n,k}$ of involutions of $[n]$ with k fixed points:
$$\sum_{n,k} I_{n,k}(X)u^k \frac{t^n}{(q)_n} = \prod_{0 \le i}(1-utq^i)^{-1} \prod_{0 \le i < j}(1-t^2q^iq^j)^{-1}.$$

See e.g. Désarménien and Foata 1985 and Gessel and Reutenauer 1993.

11.6.4 (A symmetry property) The generating series considered for words on the alphabet N with a given shape or with a given type are symmetric functions of the indeterminates. This translates into a symmetry property for the distribution of the backsteps on permutations on $[n]$ with a given shape or with a given type. More precisely, any subset $A = \{a_1 < a_2 < \cdots < a_k\}$ of $[n]$ can be encoded by a composition $c(A)$ of n, namely, $(a_1, a_2-a_1, \ldots, n-a_k)$. Then, given a composition **a** of n, the number of permutations of $[n]$, with a given shape or with a given type, such that their backsteps are a subset of $c^{-1}(\mathbf{a})$ does not depend on the order of the parts of **a**.

For example, with $n = 4$, to the partition $(2, 1, 1)$ correspond three compositions, $(2, 1, 1)$, $(1, 2, 1)$ and $(1, 1, 2)$. The corresponding sets A are respectively $\{2, 3\}$, $\{1, 3\}$ and $\{1, 2\}$. There are five permutations of $[4]$ of shape $(2, 2)$: 1324 (backstep 2), 1423 (backstep 3), 2314 (backstep 1), 3412 (backstep 2) and 2413 (backsteps 1 and 3). Among those permutations three have backsteps in each of the sets A. See Désarménien 1990.

Notes

The major ingredients for this chapter are q-calculus and symmetric functions. What is needed of the former can be found in Andrews

1976. An excellent introduction to symmetric functions is contained in Chapter 1 of Macdonald 1995. In it can also be found the essence of Corollary 11.3.5.

Actually, the determinant in Section 11.2 is a particular case of a Schur function on a flag of alphabets. This concept is due to Lascoux, and can be found in Lascoux 1974. If all alphabets are equal, it becomes the determinantal expression of a Schur function of ribbon shape. Corollary 11.3.5, as well as Corollary 11.6.4, can be extended to the case of finite alphabets. What arises is a double generating function for imaj and the number of descents. This is Theorem 4.1 of Désarménien and Foata 1985, which contains various applications of this theorem, in particular to involutions (see Problem 11.6.3). See also Désarménien and Foata 1991.

The key to Section 11.4, which is the use of Lehmer encoding together with a Schur function on a flag of alphabets, is due to Thibon. This leads to a new proof of Theorem 11.4.4, which was originally proved bijectively in Foata and Schützenberger 1970.

A proof of Theorem 11.5.1 can be found in Chapter 5 of Lothaire 1983. In Chapter 10 of the same reference, Foata's "first fundamental transformation" is described.

The Gessel normalization is to be found in Gessel and Reutenauer 1993, along with many enumerative results related to descent sets and cycle structure. This article contains different formulations of Sections 11.3, 11.5 and 11.6 in the setting of symmetric functions. This encoding had been exploited earlier in Désarménien and Wachs 1988 and Désarménien and Wachs 1993.

CHAPTER 12

Makanin's Algorithm

12.0. Introduction

A seminal result of Makanin 1977 states that the existential theory of equations over free monoids is decidable. Makanin achieved this result by presenting an algorithm which solves the satisfiability problem for word equations with constants. The satisfiability problem is usually stated for a single equation, but this is no loss of generality.

This chapter provides a self-contained presentation of Makanin's result. The presentation has been inspired by Schulz 1992a. In particular, we show the result of Makanin in a more general setting, due to Schulz, by allowing that the problem instance is given by a word equation $L = R$ together with a list of rational languages $L_x \subseteq A^*$, where $x \in \Omega$ denotes an unknown and A is the alphabet of constants. We will see that it is decidable whether or not there exists a solution $\sigma : \Omega \to A^*$ which, in addition to $\sigma(L) = \sigma(R)$, satisfies the rational constraints $\sigma(x) \in L_x$ for all $x \in \Omega$. Using an algebraic viewpoint, rational constraints mean to work over some finite semigroup, but we do not need any deep result from the theory of finite semigroups. The presence of rational constraints does not make the proof of Makanin's result much harder; however, the more general form is attractive for various applications.

In the following we explain the outline of the chapter; for some background information and more comments on recent developments we refer to the Notes.

The major step toward Makanin's result is to bound the exponent of periodicity, which is, by definition, the maximal number of direct repetitions of a primitive word in a solution of minimal length. A priori it is not at all clear why an upper bound for the exponent of periodicity is a key, since there are arbitrarily long words where the exponent of periodicity is 3. This means that the exponent of periodicity alone does not give any recursive bound on the length of a minimal solution. However, together with a deep combinatorial analysis in the situation of word equations, it does. The bound for the exponent of periodicity is calculated in Section 12.2 using the notion of p-stable normal form (Subsection 12.1.5) and some standard linear algebra.

Instead of working with word equations directly, it turns out to be more convenient to work with boundary equations. Systems of boundary equations are introduced in Section 12.3. In some sense they store the relative lengths of the variables in possible solutions. The important point is that a notion of convex chain can be defined (Subsection 12.3.4). This leads to a geometrical reflection on the problem. An upper bound for the exponent of periodicity yields an upper bound on the maximal length of clean convex chains (Proposition 12.3.15). As long as the convex chain condition is satisfied, the maximal length of convex chains yields an upper bound on the number of boundary equations (Corollary 12.3.16). The strategy of Makanin's algorithm is therefore as follows. A word equation is transformed into a system of boundary equations, which will satisfy the convex chain condition for trivial reasons. Then transformation rules are applied which maintain the convex chain condition (which is not trivial) and which either lead to a solution of the word equation or introduce more and more boundary equations. But, for the number of boundary equations, there is an upper bound provided by the exponent of periodicity. Hence, we can stop the procedure at some stage. The transformation rules (Subsection 12.3.5) are at the heart of Makanin's algorithm. The central idea is a left-to-right transport of positions in combination with a splitting of variables. It is not so much Makanin's algorithm which is complicated; the hard part is the termination proof, when to stop the procedure. Major steps are Proposition 12.3.15 and the proof that the transformations preserve the convex chain condition. Makanin's algorithm itself becomes the construction of a finite search graph: the vertices are systems of boundary equations and edges are transformation rules.

During our presentation we do not focus on necessary decidable conditions which might be used to prune the search graph. A good pruning strategy is of course extremely important for an implementation since the search graph tends to be huge. However, pruning doesn't help to understand the algorithm, nor does it seem to have any effect on the worst-case analysis. For the worst-case analysis we use standard notions of complexity theory as they can be found in the textbooks of Hopcroft and Ullman 1979 and Papadimitriou 1994. The final result of this chapter shows that Makanin's algorithm can be implemented in exponential space, Theorem 12.4.2.

Exponential space is not optimal for the satisfiability problem of word equations, since Plandowski 1999b has shown that the satisfiability problem of word equations can be decided in polynomial space. Plandowski's new approach is rather different from the material presented here; for example, an important ingredient of Plandowski's method is data compression in terms of exponential expressions, whereas we do not need any data compression here. Makanin's algorithm has many other nice features, and, since the equations are written in plain form, it seems to be easier to follow some strategy during the search for a solution. Experimental results indicate Makanin's algorithm is quite suitable for practical application.

12.1. Words and word equations

12.1.1. Basic notions

By $A = \{a, b, \ldots\}$ we mean an alphabet of constants and Ω is a set of *variables* (or *unknowns*) such that $A \cap \Omega = \emptyset$. Throughout this chapter we shall use the same symbol σ to denote a mapping $\sigma : \Omega \to A^*$ and its canonical extension to a homomorphism $\sigma : (A \cup \Omega)^* \to A^*$ leaving the letters of A invariant. The empty word (and also the unit element in other monoids) is denoted by ε. The length of a word w is denoted by $|w|$. We have $|\varepsilon| = 0$. The prefix relation of words is denoted by $u \leq v$, the proper prefix relation is $u < v$. As usual, the set of integers is $\mathbb{Z}$. The set of natural numbers is $\mathbb{N}$; these are the nonnegative integers. Lower-case Greek letters like α, β etc. are mostly used to denote natural numbers. By $\log \alpha$ we mean $\max\{1, \lceil \log_2 \alpha \rceil\}$.

A *word equation* is a pair $(L, R) \in (A \cup \Omega)^* \times (A \cup \Omega)^*$; it is written as $L = R$. A *system* of word equations is a set of equations $\{L_1 = R_1, \ldots, L_k = R_k\}$. A system where each variable occurs at most twice is called a *quadratic system*. A *solution* is a homomorphism $\sigma : (A \cup \Omega)^* \to A^*$ leaving the letters of A invariant such that $\sigma(L_i) = \sigma(R_i)$ for all $1 \leq i \leq k$. It is called *nonsingular*, if $\sigma(x) \neq \varepsilon$ for all $x \in \Omega$; otherwise it is called *singular*.

EXAMPLE 12.1.1. Let $A = \{a, b\}$ and $\Omega = \{x, y, z, u\}$. Consider the equation
$$xauzau = yzbxaaby.$$
This is a solvable quadratic equation. There are singular and nonsingular solutions. A possible nonsingular solution is given by
$$\sigma(x) = abb, \quad \sigma(y) = ab, \quad \sigma(z) = ba, \quad \sigma(u) = bab.$$

We have
$$abbababbaabab = \sigma(xauzau) = \sigma(yzbxaaby).$$

12.1.2. Solving quadratic systems

Using Nielsen transformations there is a simple strategy for solving quadratic systems. The strategy is as follows. Let $E = \{L_1 = R_1, \ldots, L_k = R_k\}$ be a system of word equations and assume that every variable $x \in \Omega$ occurs at most twice in the system. Let $\|E\| = \sum_{i=1}^{k} |L_i R_i|$ denote the denotational length of E. Using induction on $\operatorname{Card} \Omega$ we describe a nondeterministic decision algorithm which solves the question whether there is a solution in space $O(\|E\|)$. The case $\Omega = \emptyset$ is trivial, hence let $\Omega \neq \emptyset$. The first step is the guess whether there is a solution $\sigma: \Omega \to A^*$ such that $\sigma(x) = \varepsilon$ for some $x \in \Omega$. This is done by choosing some $\Omega' \subseteq \Omega$ and by replacing all occurrences of all $x \in \Omega'$ by the empty word. We obtain a new system E' over $\Omega \setminus \Omega'$ and recursively, if $\Omega' \neq \emptyset$, we decide in nondeterministic linear space whether E' has a solution. Thus, after this step we are looking for nonsingular solutions of E, only. We may assume that the first equation is of the form

$$\begin{array}{lll} \text{either} & x \cdots = a \cdots & \text{with } x \in \Omega,\ a \in A \\ \text{or} & x \cdots = y \cdots & \text{with } x \in \Omega,\ y \in \Omega,\ x \neq y. \end{array}$$

By symmetry (or a nondeterministic guess to interchange the roles of L_1 and R_1) we may write either $x = az$ or $x = yz$, where z is a new variable. Replacing the occurrences of x by az or yz respectively, we obtain a new system where x does not occur any more and z occurs at most twice. On the left of the first equation we may cancel either a or y, and then y also occurs at most twice. Hence we end up with a new system E' where the number of variables is the same as in E, every variable occurs at most twice and we have $\|E'\| \leq \|E\|$. Clearly, if E has a nonsingular solution, then E' is solvable including the possibility of a singular solution with $\sigma(z) = \varepsilon$. However, if E' is solvable, then E is also solvable. Now, let $\sigma: \Omega \to A^*$ be a nonsingular solution of E where $\sum_{x \in \Omega} |\sigma(x)|$ is minimal. Then we find a solution σ' for E' with $|\sigma'(z)| < |\sigma(x)|$ since $\sigma(y) \neq \varepsilon$. Thus, the length of a shortest solution has decreased. This shows that the nondeterministic procedure will find a solution, if there is any. The space requirement for this algorithm is linear, but its time complexity might be exponential. The exponential time bound is perhaps inevitable, because the satisfiability problem for quadratic word equations remains NP-hard.

The algorithm above has a convenient graphical representation which we show by another example: Consider $A = \{a, b, c\}$ and $\Omega = \{x, y, z\}$.

12.1. Words and word equations

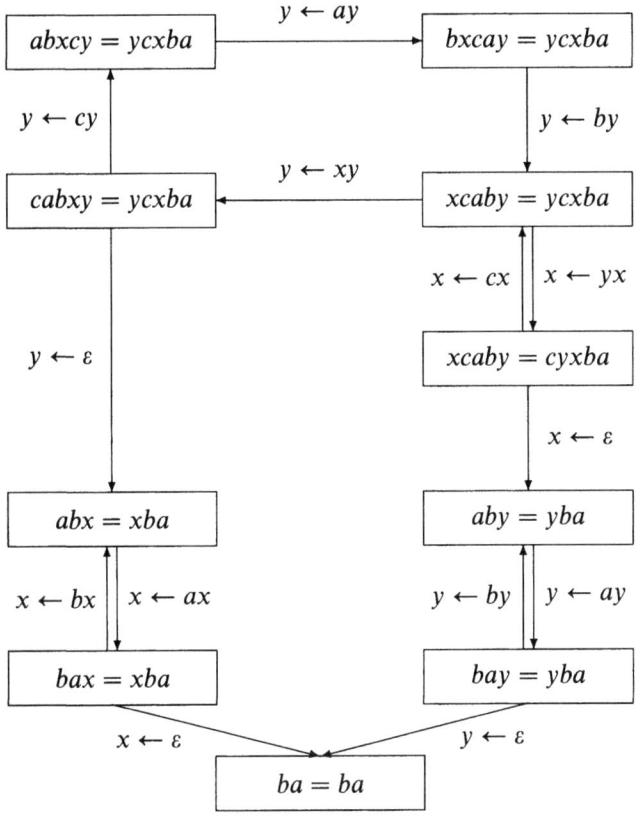

Figure 12.1. Solving the equation $abxcy = ycxba$.

Let the word equation be $abxcy = ycxba$. Running the algorithm leads to the graph as depicted in Figure 12.1. The arcs are labeled in such a way that we can reconstruct a solution by going backwards on a path from the initial equation to the trivial equation $ba = ba$. One of the paths has the following labels:

$$y \leftarrow ay, \; y \leftarrow by, \; x \leftarrow yx, \; x \leftarrow \varepsilon, \; y \leftarrow ay, \; y \leftarrow \varepsilon.$$

It corresponds to the minimal solution, where $\sigma(x) = a$ and $\sigma(y) = aba$. Nodes or arcs which cannot lead to any solution have been omitted in the picture; they are not drawn.

12.1.3. Combinatorial properties

Two words $y, z \in A^*$ are *conjugate*, if $xy = zx$ for some $x \in A^*$. The next proposition shows that in free monoids conjugates are obtained by transposition.

PROPOSITION 12.1.2. *Let $x, y, z \in A^*$ be words, $y, z \neq \varepsilon$. Then the following assertions are equivalent:*

(i) $xy = zx$,
(ii) $\exists r, s \in A^*, s \neq \varepsilon, \alpha \geq 0 : x = (rs)^\alpha r, \ y = sr, \ \text{and} \ z = rs$.

A word p is called *primitive*, if it cannot be written in the form $p = r^\alpha$ with $r \in A^+$ and $\alpha \neq 1$. In particular, a primitive word p is nonempty, $p \neq \varepsilon$.

PROPOSITION 12.1.3. *Let $p \in A^*$ be primitive and $p^2 = xpy$ for some $x, y \in A^*$. Then we have either $x = \varepsilon$ or $y = \varepsilon$ (but not both).*

Proofs of Propositions 12.1.2 and 12.1.3 can be found e.g. in Lothaire 1983, Section 1.3.

An overlapping of two words w_1 and w_2 is depicted by the following figure:

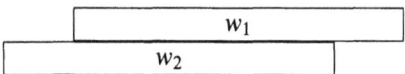

It says that the common border is an identical factor, i.e., $w_1 = xy$, $w_2 = zx$. Usually we mean $x \neq \varepsilon$ and sometimes the figure also indicates that both $y \neq \varepsilon$ and $z \neq \varepsilon$. But there will be no risk of confusion. For example, Proposition 12.1.3 can be rephrased by saying that the following picture is not possible for a primitive word.

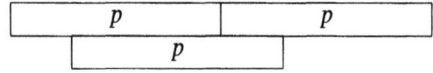

12.1.4. Domino towers

Every nonempty word $w \in A^+$ can be written in the form $w = (rs)^{h-1} r$ with $s \neq \varepsilon, h \geq 2$. This is trivial for $r = \varepsilon$ and $h = 2$. The more interesting case is when we have $r \neq \varepsilon$. Writing $w = (rs)^{h-1} r$ leads to an arrangement

12.1. Words and word equations

of the following shape:

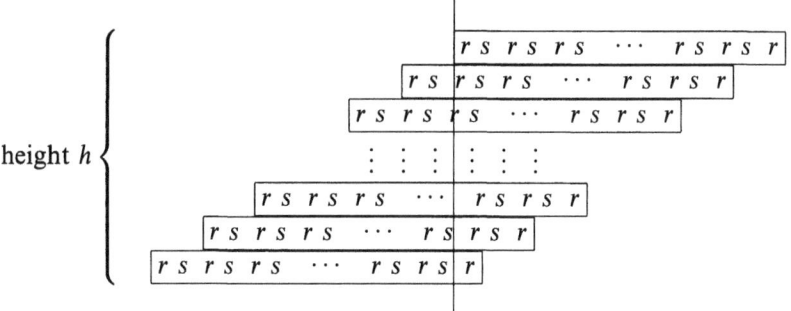

The position of the vertical line says that the upper left boundary is never to the right of the lower right boundary. The formal definition of such an arrangement also allows a less uniform shape. Let $h \geq 2$. We say that a nonempty word $w \in A^+$ can be arranged in a *domino tower of height h*, if there are words $x_1, \ldots, x_{h-1} \in A^*$ and nonempty words $y_1, \ldots, y_{h-1}, z_2, \ldots, z_h \in A^+$, such that

(i) $w = x_i y_i = z_{i+1} x_i$ for all $1 \leq i < h$,
(ii) $|z_2 \cdots z_h| \leq |w|$.

In the figure above we have $x_1 = \cdots = x_{h-1} = (rs)^{h-2}r$, $y_1 = \cdots = y_{h-1} = sr$, and $z_2 = \cdots = z_h = rs$. Note also that a domino tower of height 2 may degenerate as in the following figure.

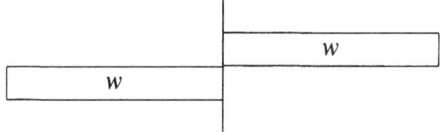

Let $w \in A^*$ be a word. The *exponent of periodicity* $\exp(w)$ is defined by
$$\exp(w) = \max\{\alpha \in \mathbb{N} \mid \exists r, s, p \in A^*, p \neq \varepsilon \colon w = rp^\alpha s\}.$$

LEMMA 12.1.4. *Let $h \geq 2$ and $w \in A^+$ be a nonempty word which can be arranged in a domino tower of height h. Then we have $\exp(w) \geq h-1$.*

Proof. Choose a domino tower and words x_i, y_i, z_i as in the definition above. Let $z = z_i \in \{z_2, \ldots, z_h\}$ be of minimal length, $x = x_{i-1}$, $y = y_{i-1}$. Then $(h-1)|z| \leq |w|$, and we have $xy = zx = w$. Hence y and z are conjugate and we may apply Proposition 12.1.2. We obtain $z = rs$ and $x = (rs)^\alpha r$ for some $\alpha \geq 0$ and $|r| < |z|$. Hence $w = z^{\alpha+1}r$ and therefore
$$(h-1)|z| \leq |w| < (\alpha+2)|z|.$$
Since $|z| > 0$ we see that $h-1 \leq \alpha+1 \leq \exp(w)$. ∎

12.1.5. Stable normal forms

Let $p \in A^+$ be a primitive word. The *p-stable normal form* of the word $w \in A^*$ is a shortest sequence (k is minimal)

$$(u_0, \alpha_1, u_1, \ldots, \alpha_k, u_k)$$

such that $k \geq 0$, $u_0, u_i \in A^*$, $\alpha_i \geq 0$ for $1 \leq i \leq k$, and the following three conditions are satisfied:
(i) $w = u_0 p^{\alpha_1} u_1 \cdots p^{\alpha_k} u_k$;
(ii) $k = 0$ if and only if p^2 is not a factor of w;
(iii) if $k \geq 1$, then

$$u_0 \in A^*p \setminus A^*p^2 A^*,$$
$$u_i \in (A^*p \cap pA^*) \setminus A^*p^2 A^* \text{ for } 1 \leq i < k,$$
$$u_k \in pA^* \setminus A^*p^2 A^*.$$

EXAMPLE 12.1.5. Let $p = aba$ and $w = ab(aba)^5 ba(aba)^4 ba$. Then the p-stable normal form of w is the sequence

$$(ababa, 3, ababa, 3, ababa).$$

PROPOSITION 12.1.6. *Let $p \in A^+$ be primitive. The p-stable normal form of $w \in A^*$ is uniquely defined. This means, if $(u_0, \alpha_1, u_1, \ldots, \alpha_k, u_k)$ and $(v_0, \beta_1, \ldots, \beta_\ell, v_\ell)$ are p-stable normal forms of the same word $w \in A^*$, then they are identical, i.e., we have $k = \ell$, $u_0 = v_0$, $u_i = v_i$, and $\alpha_i = \beta_i$ for $1 \leq i \leq k$.*

Proof. Assume that $(u_0, \alpha_1, u_1, \ldots, \alpha_k, u_k)$ and $(v_0, \beta_1, v_1, \ldots, \beta_\ell, v_\ell)$ are both p-stable normal forms of w. Since these are shortest sequences, the indices k and ℓ are both minimal, hence $k = \ell$.

For $k = 0$ we have $w = u_0 = v_0$, hence let $k = \ell \geq 1$.

We show first that $u_0 = v_0$. To see this, suppose by symmetry that $|u_0| \leq |v_0|$. Since $u_0 p \in A^* p^2$ and $v_0 \in (A^*p \setminus A^*p^2 A^*)$, we obtain that $u_0 \leq v_0 < u_0 p$. By Proposition 12.1.3 this yields $u_0 = v_0$.

Let w' denote the word $u_1 p^{\alpha_2} u_2 \cdots p^{\alpha_k} u_k$. A simple reflection using $u_1 \neq p$, Proposition 12.1.3, and $u_1 \in (A^*p \cap pA^*) \setminus A^*p^2 A^*$ shows that $p^{\alpha_1} w' \in p^{\alpha_1+1} A^* \setminus p^{\alpha_1+2} A^*$. This implies $\alpha_1 = \beta_1$ and $w' = v_1 p^{\beta_2} v_2 \cdots p^{\beta_k} v_k$. Since we have $w' \in pA^*$, we see that the first component of its p-stable normal form is in pA^*. Hence $(u_1, \alpha_2, u_2, \ldots, \alpha_k, u_k)$ is the p-stable normal form of w'. By induction we conclude that $(u_1, \alpha_2, u_2, \ldots, \alpha_k, u_k) = (v_1, \beta_2, v_2, \ldots, \beta_k, v_k)$. ∎

12.1.6. The existential theory of concatenation

The existential theory of equations over free monoids is decidable, i.e., the satisfiability of any propositional formula over word equations (with rational constraints) can be decided. This can be deduced from Makanin's result as follows. In a first step we may assume that all negations in a given formula are of type $L \neq R$. Due to the following proposition these negations can be eliminated.

PROPOSITION 12.1.7. *An inequality $L \neq R$ is equivalent with the following positive existential formula:*

$$\exists x\, \exists y\, \exists z : \bigvee_{a \in A}(L = Rax \vee R = Lax) \vee \bigvee_{a,b \in A,\, a \neq b}(L = xay \wedge R = xbz).$$

In a second step the formula (without negations) is written in disjunctive normal form. Then, for satisfiability, it is enough to see how a system of word equations can be transformed into a single word equation. The method is given in Proposition 12.1.8. It relies on the observation that if $ua \leq va, ub \leq vb, u, v \in A^*, a, b \in A$, and $a \neq b$, then we have $u = v$.

PROPOSITION 12.1.8. *Let $a, b \in A$ be distinct letters, $a \neq b$, and let $E = \{L_1 = R_1, \ldots, L_k = R_k\}$ be a system of word equations. Then the set of solutions of E is identical with the set of solutions of the following equation:*

$$L_1 a \cdots L_k a\, L_1 b \cdots L_k b = R_1 a \cdots R_k a\, R_1 b \cdots R_k b.$$

Sometimes it is useful to do the opposite of Proposition 12.1.8 and to split a single word equation into a system where all equations are of type $xy = z$ with $x, y, z \in A \cup \Omega$. This can be derived from the next proposition. Again its (simple) proof is left to the reader.

PROPOSITION 12.1.9. *Let $x_1 \cdots x_g = x_{g+1} \cdots x_d$ be a word equation with $1 \leq g < d$, $x_i \in A \cup \Omega$ for $1 \leq i \leq d$. Then the set of solutions is in canonical bijection with the set of solutions of the following system:*

$$\begin{aligned}
x_1 &= y_1, & x_{g+1} &= y_{g+1}, \\
y_1 x_2 &= y_2, & y_{g+1} x_{g+2} &= y_{g+2}, \\
&\vdots & &\vdots \\
y_{g-1} x_g &= y_g, & y_{d-1} x_d &= y_d, \\
& & y_g &= y_d.
\end{aligned}$$

In the system above, $y_1, \ldots, y_d$ denote new variables.

It is worth noting that a disjunction of word equations can be replaced by an existential formula in a single equation, too. The construction shown below has been taken from Karhumäki, Mignosi, and Plandowski 2000.

PROPOSITION 12.1.10. *Let $a, b \in A$ be distinct letters, $a \neq b$. A disjunction of two word equations is equivalent with a single word equation in two extra unknowns.*

Proof. Consider a disjunction

$$L_1 = R_1 \vee L_2 = R_2,$$

where $L_1, L_2, R_1, R_2 \in (A \cup \Omega)^*$. This is equivalent to the disjunction

$$L_1 R_2 = R_1 R_2 \vee R_1 L_2 = R_1 R_2.$$

Thus, for the construction we can start with a disjunction where the right-hand sides are equal: $L_1 = R \vee L_2 = R$.

It turns out that the word $P = L_1 L_2 R a L_1 L_2 R b \in (A \cup \Omega)^+$ is primitive. In fact, we need a sharper statement: Choose a primitive word $Q \in (A \cup \Omega)^+$ and some $\alpha \geq 1$ such that P is a prefix of Q^α. Then we have $|Q| > \frac{1}{2}|P|$. To see this, assume to the contrary that $|Q| \leq \frac{1}{2}|P|$. Since $a \neq b$ we have $|Q| < \frac{1}{2}|P|$ and Q is a prefix of $L_1 L_2 R$. But this is impossible due to Proposition 12.1.3. As a consequence, if P^2 is a factor of some word $P^2 W P^2$ where $|W| \leq \frac{1}{2}|P|$, then P^2 is either a prefix or a suffix of $P^2 W P^2$. (This can be seen from the statement above, again using Proposition 12.1.3, and by Proposition 12.1.2.)

Given this, let x, y be two extra unknowns. Then the disjunction $L_1 = R \vee L_2 = R$ is equivalent to the existential formula

$$\exists x \exists y : P^2 L_1 P^2 L_2 P^2 = x P^2 R P^2 y.$$

Indeed, if $L_1 = R \vee L_2 = R$ is solvable then we can satisfy the existential formula above. For the other direction let σ be a solution to the formula. Since $|\sigma(L_i)| \leq \frac{1}{2}|\sigma(P)|$, $i = 1, 2$, the first P^2 of the right-hand side matches either the first or the second P^2 on the left-hand side. If it matches the second one, we are done. Hence we may assume that the first P^2 of the right-hand side matches the first P^2 on the left-hand side. Now the second P^2 of the right-hand side cannot match the third P^2 on the left-hand side since $|\sigma(R)| < |\sigma(L_1 P^2 L_2)|$, so it matches the second one. The assertion follows. ∎

Let us look at the number of different constants which are used in a word equation. It is well known that the problem of solving word equations can be reduced to the case where only two constants appear:

12.1. Words and word equations

PROPOSITION 12.1.11. *Let $L = R$ be a word equation over a set of constants A and $B = \{a, b\}$ be a two-letter alphabet. Then we can construct (in polynomial time) a word equation over B which is solvable if and only if $L = R$ has a nonsingular solution.*

Proof. We may assume that $A = \{a_1, \ldots, a_k\}$ with $k > 2$. We define an injective homomorphism $\eta : (A \cup \Omega)^* \longrightarrow (B \cup \Omega)^*$ by $\eta(a_i) = ab^i a$ for $1 \leq i \leq k$ and $\eta(x) = axa$ for $x \in \Omega$. We obtain an equation $\eta(L) = \eta(R)$.

Clearly, if $L = R$ has a nonsingular solution $\sigma : \Omega \longrightarrow A^+$, then for all $x \in \Omega$ we can write $\eta(\sigma(x)) = a\tau(x)a$; and $\tau : \Omega \longrightarrow B^+$ is a nonsingular solution of $\eta(L) = \eta(R)$.

For the converse, let $\tau : \Omega \longrightarrow B^*$ be any solution of $\eta(L) = \eta(R)$. (Even if this solution is singular, we will produce a nonsingular solution of $L = R$.) Define $\sigma'(x) = a\tau(x)a$ for $x \in \Omega$, and modify η by defining $\eta'(a_i) = ab^i a$ and $\eta'(x) = x$ for $1 \leq i \leq k$ and $x \in \Omega$. Let $L' = \eta'(L)$, and $R' = \eta'(R)$. Then σ' is a nonsingular solution of $L' = R'$ such that $\sigma'(x) \in aB^*a$ for all $x \in \Omega$. Of course, we cannot guarantee that $\sigma'(x) \in \eta(A)^+$, it might happen that $\sigma'(x)$ contains factors of the form aaa or ab^+ab^+a or so. But such a *wrong* factor on one side of the equation must correspond to the same wrong factor on the other side, which must be again inside some piece corresponding to a variable. In order to formalize this idea we observe that the subset $\{\varepsilon\} \cup (aB^* \cap B^*a)$ is a free submonoid of B^*. The (infinite) basis is $\Sigma = \{a\} \cup aB^*a \setminus B^*aaB^*$. Hence $\sigma' \circ \eta' : \Omega \longrightarrow \Sigma^+$ is a nonsingular solution of the original equation $L = R$ if we identify $\eta'(A)$ with A. The only difference is that $\sigma'(x)$ may contain (finitely many) letters from $\Sigma \setminus \eta'(A)$. Hence for some finite set $C \subseteq \Sigma \setminus \eta'(A)$ we have $\sigma'(LR) \subseteq (\eta'(A) \cup C)^*$. Choosing any mapping $\rho : C \longrightarrow \eta'(A)^+$ we obtain a nonsingular solution $\sigma = \rho \circ \sigma'$, which can be identified with a nonsingular solution of $L = R$ using the following composition leaving the letters of A invariant:

$$(A \cup \Omega) \xrightarrow{\eta'} (\eta'(A) \cup \Omega)^* \xrightarrow{\sigma'} (\eta'(A) \cup C)^+ \xrightarrow{\rho} \eta'(A)^+ \xrightarrow{\eta'^{-1}} A^+. \quad \blacksquare$$

12.1.7. A single variable

A parametric description of the set of all solutions can be computed in polynomial time, if there is only one variable occurring in the equation. This serves as an example of why p-stable normal forms might be useful.

Let E be a set of word equations where exactly one variable x occurs, $\Omega = \{x\}$. By Proposition 12.1.8 we may assume that E is given by a single equation $L = R$ with $L, R \in (A \cup \{x\})^*$. The basic check is whether $\sigma(x) = \varepsilon$ yields the singular solution. It is therefore enough to consider only nonsingular solutions. Let us denote by $\mathscr{L}$ a list of pairs (p, r) where $p \in A^+$ is primitive and $r \in A^*$ is some prefix $r < p$. We say that $\mathscr{L}$ is *complete for the equation* $L = R$, if every nonsingular solution σ has the form $\sigma(x) = p^\alpha r$ for some $\alpha \geq 0$ and $(p, r) \in \mathscr{L}$.

Assume for a moment that a finite complete list $\mathscr{L}$ has already been computed in a first phase of the algorithm. Then we proceed as follows. For each pair $(p, r) \in \mathscr{L}$ we make a first test whether $\sigma(x) = r$ is a solution and a second test whether $\sigma(x) = pr$ is a solution. After that, we search (for this pair (p, r)) for solutions where $\sigma(x) = p^\alpha r$ with $\alpha \geq 2$. Replace all occurrences of x in the equation $L = R$ by the expression $pp^{\alpha-2}pr$, where α now denotes an integer variable. Thus, the problem is now to find solutions for α such that $\alpha \geq 2$. Using the symbolic expression we can factorize L and R in their p-stable normal forms:

$$L = u_0 p^{m_1 \alpha + n_1} u_1 \cdots p^{m_k \alpha + n_k} u_k,$$
$$R = v_0 p^{m'_1 \alpha + n'_1} v_1 \cdots p^{m'_\ell \alpha + n'_\ell} v_\ell.$$

Here $k, \ell \geq 0$ and $m_i, m'_j \in \mathbb{N}$, $n_i, n'_j \in \mathbb{Z}$ for $1 \leq i \leq k$ and $1 \leq j \leq \ell$. By Proposition 12.1.6 we have to verify $k = \ell$, $u_i = v_i$ for $0 \leq i \leq k$, and we have to solve a linear Diophantine system:

$$(m_i - m'_i)\alpha = n'_i - n_i \quad \text{for } 1 \leq i \leq k.$$

There are three cases. Either no or exactly one or all $\alpha \geq 2$ satisfy these equations.

It is clear that for each pair (p, r) the necessary computations can be done in polynomial time. In fact, using pattern-matching techniques it can be proved that linear time is enough for each pair (p, r). The performance of the algorithm therefore depends on an efficient computation of a short and complete list $\mathscr{L}$.

We may assume that $L = ux \cdots$ and $R = xv \cdots$, where $u \in A^+, v \in A^*$ and both words u and v are of maximal length. Let $p \in A^+$ be the primitive root of u, i.e., p is primitive and $u = p^e$ for some $e \geq 1$. If σ is a solution of $L = R$, then σ also solves an equation of type $ux = xw$ for some word $w \in A^+$. By Proposition 12.1.2 it is immediate that we have $\sigma(x) = p^\alpha r$ for some $\alpha \geq 0$ and $r < p$. Thus, the obvious method is to define the list $\mathscr{L}$ by all pairs (p, r) where $r < p$. We obtain a list $\mathscr{L}$ with $|p|$ elements.

There is an improvement of the algorithm due to Eyono Obono, Goralčik, and Maksimenko 1994 by observing that there is a complete list $\mathscr{L}$ of at most logarithmic length. This improvement uses a finer combinatorial analysis and it relies, in particular, on the following well-known fact:

12.1. Words and word equations

- Let $u, v, w \in A^+$ be primitive words such that $u^2 < v^2 < w^2$. Then we have $|u| + |v| \leq |w|$. In particular, a word $w \in A^*$ of length n has at most $O(\log n)$ distinct prefixes of the form pp where p is primitive.

For a proof of the fact see Lemma 8.1.14 or Crochemore and Rytter 1995, Lemma 10.

We outline the method of Eyono Obono et al. 1994: The set of nonsingular solutions is divided into two classes. The first class contains all solutions where $|\sigma(x)| \geq |u| - |v|$. (Of course, in the case $|u| \leq |v|$ all solutions satisfy this condition.) Let w be the prefix of the word vu such that $|w| = |u|$. If σ is a solution with $|\sigma(x)| \geq |u| - |v|$, then we have $u\sigma(x) = \sigma(x)w$. Let p be the primitive root of u and let q be the primitive root of w. Then $\sigma(x) = p^\alpha r$ for some $\alpha \geq 0$ and the unique prefix $r < p$ such that $p = rs$ and $q = sr$. If p and q are not conjugate, then there is no such solution. Otherwise, if p and q are conjugate, we include the unique pair (p, r) into $\mathscr{L}$. This pair covers all solutions where $|\sigma(x)| \geq |u| - |v|$.

Now, let σ be a nonsingular solution such that $0 \neq |\sigma(x)| < |u| - |v|$. This implies that R has the form $R = xvx \cdots$ and that $\sigma(x)v\sigma(x) < u\sigma(x)$. Hence $\sigma(x)v\sigma(x) < uu$ and $ww < vuu$, where w denotes the nonempty word $v\sigma(x)$. Let q be the primitive root of w, then we have $qq < vuu$.

There is a unique factorization $q = sr$ with $s < q$ such that $v \in q^*s$. The word rs also is primitive and we have $\sigma(x) = (rs)^\alpha r$ for some $\alpha \geq 0$. Therefore it is enough to compute the list of all primitive words q such that $qq < vuu$. If $v = \varepsilon$, then we add all pairs (q, ε) to $\mathscr{L}$. Otherwise, if $v \neq \varepsilon$, then we compute for each q the unique factorization $q = sr$ with $s \neq \varepsilon$ such that $v \in q^*s$. We add all pairs (rs, r) to $\mathscr{L}$. It follows from Crochemore 1981 that the list $\mathscr{L}$ can be computed in time $O(|LR| \log |LR|)$. The conclusion is that the solvability of an equation $L = R$ in one variable can be decided in time $O(|LR| \log |LR|)$. It is however not clear whether there is a linear-time algorithm.

12.1.8. Constraints over a semigroup

The input for Makanin's algorithm is an equation $L = R$ with $L, R \in (A \cup \Omega)^*$ together with rational languages $L_x \subseteq A^*$ for all variables $x \in \Omega$. We assume that the languages are specified by nondeterministic finite automata. If it happens that for some variable no rational constraint is defined, then we simply put $L_x = A^*$. We are looking for a solution $\sigma: \Omega \to A^*$ such that $\sigma(L) = \sigma(R)$ and $\sigma(x) \in L_x$ for all $x \in \Omega$. For notational convenience, henceforth we will not distinguish between variables and constants in the equation. Every constant $a \in A$ is replaced by a new variable x_a and the constraint $L_{x_a} = \{a\}$ for all $a \in A$. (For readability we shall use constants in examples however.) From now on the equation is given as

$$x_1 \cdots x_g = x_{g+1} \cdots x_d$$

with $x_i \in \Omega$. In order to exclude trivial cases we shall assume $1 \leq g < d$ whenever convenient. The number d is called the *denotational length* of the equation. It is enough to consider nonsingular solutions. Hence we shall assume that $\varepsilon \notin L_x$ for all $x \in \Omega$. Next we fix a finite semigroup S and a semigroup homomorphism $\varphi: A^+ \to S$ such that $L_x = \varphi^{-1}\varphi(L_x)$ for all $x \in \Omega$. For later purposes we require that φ is surjective. The semigroup S can be realized as the image $\varphi(A^+)$ of the canonical homomorphism to the direct product of the syntactical monoids with respect to L_x for $x \in \Omega$. Sometimes it is more convenient to work with monoids instead of semigroups. We denote by S^ε the monoid which is obtained by adjoining a unit element ε to S. We have $S^\varepsilon \setminus \{\varepsilon\} = S$ and the homomorphism φ is extended to a monoid homomorphism $\varphi: A^* \to S^\varepsilon$. We have $\varphi^{-1}(\varepsilon) = \{\varepsilon\}$ and $\varphi(A^+) = S$.

Given S we can compute constants $t(S) \geq 0$ and $q(S) > 0$ such that $s^{t(S)+q(S)} = s^{t(S)}$ for all $s \in S^\varepsilon$. In the following we actually use another constant $c(S)$, which is defined as the least multiple of $q(S)$ such that $c(S) \geq \max\{2, t(S)\}$. Note that this implies $s^{r+\alpha c(S)} = s^{r+\beta c(S)}$ for all $s \in S$ and $r \geq 0$ and $\alpha, \beta \geq 1$.

REMARK 12.1.12. Assume that each rational language L_x is specified by a (nondeterministic) finite automaton with r_x states, $x \in \Omega$. Let $r = \sum_{x \in \Omega} r_x$. Then we may choose the semigroup S such that

$$\text{Card } S \leq 2^{r^2} \text{ and } c(S) \leq r!$$

A proof for these bounds can be found in Markowsky 1977, where a more precise analysis is given. For the moment explicit upper bounds for Card S and $c(S)$ are not relevant. They are used only later (Subsection 12.4.2) when complexity issues are investigated.

12.2. The exponent of periodicity

This section provides an effective upper bound for the exponent of periodicity in a solution of minimal length of a given word equation (with rational constraints). For the decidability result any effective upper bound would be sufficient, but, due to its close relation to linear Diophantine equations and by techniques from linear optimization, one can be precise.

12.2. The exponent of periodicity

The upper bound for the exponent of periodicity is exponential in the input size, and this is essentially optimal. In the proof below, a rather detailed analysis is given hiding perhaps some basic ideas. In a first reading one is therefore invited to ignore the exact values. We shall use the notations as introduced in Subsection 12.1.8.

THEOREM 12.2.1. *Let $d \geq 1$ be a natural number, $\varphi: A^* \to S^\varepsilon$ a homomorphism, and $c(S) \geq 2$ as above. There is a computable number $e(c(S), d) \in c(S) \cdot 2^{O(d)}$ satisfying the following assertion.*

Given as instance a word equation $x_1 \cdots x_g = x_{g+1} \cdots x_d$ of denotational length d together with a solution $\sigma': \Omega \to A^*$, we can effectively find a solution $\sigma: \Omega \to A^*$ and a word $w \in A^*$ such that the following conditions hold:

(i) $\varphi\sigma'(x) = \varphi\sigma(x)$ *for all* $x \in \Omega$,
(ii) $w = \sigma(x_1 \cdots x_g) = \sigma(x_{g+1} \cdots x_d)$,
(iii) $\exp(w) \leq e(c(S), d)$.

Proof. For $g = 0$ or $g = d$, we have $\exp(w) = 0$, hence let $1 \leq g < d$.

Testing all words of length up to $|\sigma'(x_1 \cdots x_g)|$ we find a solution σ and a word w such that $w = \sigma(x_1 \cdots x_g) = \sigma(x_{g+1} \cdots x_d)$ is of minimal length among all solutions σ where $\varphi\sigma'(x) = \varphi\sigma(x)$ for all $x \in \Omega$. Recall that $x_1 \cdots x_g = x_{g+1} \cdots x_d$ is equivalent to the following system:

$$
\begin{array}{ll}
x_1 = y_1, & x_{g+1} = y_{g+1}, \\
y_1 x_2 = y_2, & y_{g+1} x_{g+2} = y_{g+2}, \\
\vdots & \vdots \\
y_{g-1} x_g = y_g, & y_{d-1} x_d = y_d, \\
& y_g = y_d.
\end{array}
$$

Note also that $\exp(w) = \exp(\sigma(y_g))$. After an obvious elimination of variables, the system above is equivalent to a system of $d - 2$ equations of type

$$xy = z, \quad x, y, z \in \Omega.$$

Choose a primitive word $p \in A^+$ such that $w = up^{\exp(w)}v$ for some $u, v \in A^*$. Consider an equation $xy = z$ from the system above and write the words $\sigma(x), \sigma(y), \sigma(z)$ in their p-stable normal forms:

$$\sigma(x): (u_0, r_1 + \alpha_1 c(S), u_1, \ldots, r_k + \alpha_k c(S), u_k);$$
$$\sigma(y): (v_0, s_1 + \beta_1 c(S), v_1, \ldots, s_\ell + \beta_\ell c(S), v_\ell);$$
$$\sigma(z): (w_0, t_1 + \gamma_1 c(S), w_1, \ldots, t_m + \gamma_m c(S), w_m).$$

The natural numbers $r_i, s_i, t_i, \alpha_i, \beta_i,$ and γ_i are uniquely determined by w, $c(S)$, and the requirement $0 \leq r_i, s_i, t_i < c(S)$.

Since w is a solution, there are many equations among the words and among the integers. For example, for $k, \ell \geq 2$ we have $u_0 = w_0$, $v_l = w_m$, $r_1 = t_1$, $\alpha_1 = \gamma_1$, etc. In order to be precise, we shall use

$$\alpha_1 = \gamma_1, \quad \ldots, \alpha_{k-1} = \gamma_{k-1},$$
$$\beta_2 = \gamma_{m-\ell+2}, \ldots, \quad \beta_\ell = \gamma_m.$$

We have no bound on k, ℓ, or m, but we have $|k + \ell - m| \leq 2$. What exactly happens depends on the p-stable normal form of the product $u_k v_0$. Since $u_k, v_0 \notin A^* p^2 A^*$, it is enough to distinguish nine cases. Here are the nine possible p-stable normal forms of $u_k v_0$, where $t \in \{0, 1\}$, $u_k, v_0 \in A^*$, and $u'_k, v'_0, w' \in A^+$:

$(u_k v_0)$, (p, t, p), (p, t, pv'_0),
$(u'_k p, t, p)$, $(u'_k p, t, pv'_0)$, $(p, 0, w', 0, p)$,
$(p, 0, w', 0, pv'_0)$, $(u'_k p, 0, w', 0, p)$, $(u'_k p, 0, w', 0, pv'_0)$.

The case $(p, 0, w', 0, p)$ can be produced, if p has an overlap as in $p = ababa$. Then we might have $u_k = pabab$, $v_0 = abap$, which yields $u_k v_0 = ppbap = pabpp$ and $abp = pba$. Hence the p-stable normal form $u_k v_0$ is $(p, 0, abp, 0, p)$. We may conclude that $w_{k+1} = abp$ and

$$t_k + \gamma_k c(S) = r_k + \alpha_k c(S) + 1, \quad t_{k+1} + \gamma_{k+1} c(S) = s_1 + \beta_1 c(S) + 1.$$

In particular $k + \ell = m$. If $r_k < c(S) - 1$, then $\alpha_k = \gamma_k$, otherwise $\alpha_k + 1 = \gamma_k$. Similarly, if $s_1 < c(S) - 1$, then $\beta_1 = \gamma_{k+1}$, otherwise $\beta_1 + 1 = \gamma_{k+1}$.

A p-stable normal form of type $(u'p, 0, w', 0, pv')$ with $u', v', w' \in A^+$ leads to $k + \ell = m + 2$ and $0 = \gamma_k = \gamma_{k+1}$. Let us consider another example. If $u_k v_0 = p^3$, then $k + \ell = m + 1$ and we have

$$r_k + s_1 + 3 + (\alpha_k + \beta_1)c(S) = t_k + \gamma_k c(S).$$

Since by assumption $c(S) \geq 2$, the case $u_k v_0 = p^3$ leads to the equation

$$\gamma_k - (\alpha_k + \beta_1) = c \text{ with } c \in \{0, 1, 2\}.$$

We have seen that there are various possibilities for $u_k v_0$. However, always the same phenomenon arises. First of all we obtain a bunch of trivial equations which can be eliminated by renaming. All equations of type $\gamma = 0$ are eliminated by substitution. Then, for each $xy = z$ either there are at most two equations of type $\gamma = \alpha + 1$ or there is one equation of type $\gamma - (\alpha + \beta) = c$ with $c \in \{0, 1, 2\}$. If there are two equations of type $\gamma = \alpha + 1$, then one of them is eliminated by substitution. So after renaming and substituting we end up with at most one nontrivial

12.2. The exponent of periodicity

equation having at most three variables. Proceeding this way through all $d-2$ word equations we have various interactions due to renaming and substitution. However, finally each equation $xy = z$ leads to at most one nontrivial equation with at most three variables. The type of this equation is

$$c_1\gamma + i_1 - c_2\alpha - i_2 - c_3\beta - i_3 = c$$

where we have $0 \leq i_1, i_2, i_3 \leq d-2$, $0 \leq c \leq 2$, $c_1, c_2, c_3 \in \{0, 1\}$. This can be written as

$$c_1\gamma - c_2\alpha - c_3\beta = c' \text{ with } |c'| \leq 2d - 2.$$

For the case $\alpha = \beta \neq \gamma$ and $c_1 = c_2 = c_3 = 1$ we obtain a coefficient -2, because then $\gamma - 2\alpha = c'$.

We have viewed the symbols $\alpha, \beta, \ldots$ as variables ranging over natural numbers. Going back to the solution σ, which is given by the word w, the symbols $\alpha_1, \ldots, \alpha_k, \beta_1, \ldots, \beta_\ell, \gamma_1, \ldots, \gamma_m$ represent concrete values. Some of them might still be zero. These are eliminated now. The reason is that they cannot be replaced by other values without risk of changing the image under φ. If $\delta \geq 1$ is a remaining value, i.e., a number greater than zero, then we replace it by $\delta = 1 + Z_\delta$ where now Z_δ denotes a variable over $\mathbb{N}$. For example an equation

$$\gamma - \alpha - \beta = c'$$

with $\alpha, \beta, \gamma \geq 1$ is transformed to a linear Diophantine equation with integer variables $Z_\alpha, Z_\beta, Z_\gamma \geq 0$ as follows:

$$Z_\gamma - Z_\alpha - Z_\beta = c' + 1 \text{ with } |c' + 1| \leq 2d - 1.$$

Putting all equations of type $xy = z$ together we obtain a (possibly) huge system of linear equations. After substitution and elimination of variables, we end up with a system of at most $d-2$ equations and n integer variables with $n \leq 3(d-2)$. The absolute values of the coefficients are bounded by 2 and those of the constants by $2d-1$. For each equation the sum of the squares of the coefficients is bounded by 5. The linear Diophantine system is defined by w and the word w provides a nonnegative integer solution.

What becomes crucial now is the converse: Every solution in nonnegative integers yields by backward substitution a word w'' and a solution $\sigma'' : \Omega \to A^*$ satisfying (i) and (ii) of the theorem. Therefore, since w was chosen of minimal length, the solution of the integer system given by w is a minimal solution with respect to the natural partial ordering of $\mathbb{N}^n$.

In this ordering we have $(\alpha_1,\ldots,\alpha_n) \leq (\beta_1,\ldots,\beta_n)$ if and only if $\alpha_i \leq \beta_i$ for all $1 \leq i \leq n$.

For $\vec{\alpha} = (\alpha_1,\ldots,\alpha_n) \in \mathbb{N}^n$ let $\|\vec{\alpha}\| = \max\{\alpha_i \mid 1 \leq i \leq n\}$. All we need is a recursive bound for the following value:

$e(d) = \max\{\|\vec{\alpha}\| \mid \vec{\alpha}$ is a minimal solution of a system of linear Diophantine equations with at most $d-2$ equations, $3(d-2)$ variables, where the absolute value of the coefficients is bounded by 2, the sum of the squares of the coefficients in each equation is bounded by 5, and the absolute values of constants are bounded by $2d-1\}$.

Obviously, there are only finitely many systems of linear Diophantine equations where the numbers of equations, variables, and the absolute values of coefficients and constants are bounded. For each system the set of minimal solutions is finite; this is a special case of Lemma A of Dickson 1913. Moreover the set of minimal solutions is effectively computable. Hence, the set of values of $\|\vec{\alpha}\|$ above is finite and effectively computable. Therefore $e(d)$ is computable. Since $e(d)+d-1 \geq \alpha_1,\ldots,\beta_1,\ldots$ for original values under the consideration above, we obtain a recursive upper bound for the exponent of periodicity. A much more precise statement is possible. It is known that $e(d) \in 2^{O(d)}$; see Remark 12.2.2. Hence we can state

$$\exp(w) \leq 2 + (c(S) - 1) + (e(d) + d - 1) \cdot c(S) \in c(S) \cdot 2^{O(d)}.$$

This proves the theorem. ∎

REMARK 12.2.2. The result on the exponent of periodicity $e(d)$ saying that it can be bounded by a singly exponential function is due to Kościelski and Pacholski 1996. The analysis given there is more accurate than the one presented here, and it leads to linear Diophantine systems having a slightly different structure. The article uses results of von zur Gathen and Sieveking 1978. They show that the exponent of periodicity of a minimal solution of a word equation of denotational length d (without rational constraints) is in $O(2^{1.07d})$. The introduction of rational constraints doesn't change the situation very much: it yields the factor $c(S)$, as is shown above. Therefore the actual result including rational constraints is

$$e(c(S), d) \in c(S) \cdot O(2^{1.07d}).$$

It is rather difficult to obtain this very good bound. However, a bound which is good enough to establish Theorem 12.2.1 is $e(d) \in O(2^{cd})$

12.2. The exponent of periodicity

for some constant c, say $c = 4$. Such a more moderate bound can be obtained using the present approach and some standard knowledge in linear algebra; see Problem 12.3.1.

EXAMPLE 12.2.3. Consider $c, n \geq 2$ and let $S = \mathbb{Z}/c\mathbb{Z}$ be the cyclic group of c elements. We give a rational constraint for the variable x_1 by defining

$$L_{x_1} = \{w \in A^+ \mid |w| \equiv 0 \pmod{c}\}.$$

The system is given by

$$x_1 = a^c, \quad x_2 = x_1^2, \quad \ldots, \quad x_n = x_{n-1}^2.$$

Its unique solution σ is $\sigma(x_i) = a^{c \cdot 2^{i-1}}$, $1 \leq i \leq n$. A transformation into a single equation according to Proposition 12.1.8 shows that $e(c(S), d) \in c(S) \cdot 2^{\Omega(d)}$. Thus, the assertion given in Theorem 12.2.1 is essentially optimal.

The following example shows that the length of a minimal solution can be very long although the exponent of periodicity is bounded by a constant.

EXAMPLE 12.2.4. Consider the following system of word equations:

$$x_0 = a, \quad\quad y_0 = b,$$
$$x_i = x_{i-1} y_{i-1}, \quad y_i = y_{i-1} x_{i-1} \text{ for } 1 \leq i \leq n.$$

The unique solution is the Thue–Morse word:

$$\sigma(x_n) = abbabaabbaababbabaababbaabbabaab \cdots \text{ for } n \geq 5.$$

We have $|\sigma(x_n)| = 2^n$, but $\exp(\sigma(x_n)) = 2$.

EXAMPLE 12.2.5. Consider the equation with rational constraints

$$axyz = zxay, \quad L_x = a^2 a^*, \quad L_y = \{a, b\}^* \setminus (a^* \cup b^*), \quad L_z = \{a, b\}^+.$$

A suitable homomorphism $\varphi: \{a, b\}^+ \to S$ is given by the canonical homomorphism onto the quotient semigroup of $\{a, b\}^+$, which is presented by the defining relations

$$a^2 = a^3, \ b = b^2, \ ab = ba = aab.$$

Thus, S is a semigroup with a zero, $0 = ab$; and S has four elements:

$$S = \{a, a^2, b, 0\}.$$

The constant $c(S) = 2$ fits the requirement $s^{r+c(S)} = s^{r+\alpha c(S)}$ for all $s \in S^\varepsilon$ and $r \geq 0$, $\alpha \geq 1$. It is not difficult to find a solution σ for the equation above, e.g. $\sigma(x) = a^2$, $\sigma(y) = ba^2$, and $\sigma(z) = a^3ba^2$. Now let α, β, γ, and δ be some integer variables and let u, v, and w be parametric words, which are described by the following a-stable normal forms:

$$u: (a, 2\alpha, a); \quad v: (ba, 2\beta, a); \quad w: (a, 1 + 2\gamma, aba, 2\delta, a).$$

In order to derive the system of linear Diophantine equations, we make a direct approach: We want to solve $auvw = wuav$. First we write $auvw$ as a sequence of a-stable normal forms:

$$((a), (a, 2\alpha, a), (ba, 2\beta, a), (a, 1 + 2\gamma, aba, 2\delta, a)).$$

The resulting a-stable normal form is

$$(a, 2\alpha + 1, aba, 2\beta + 2\gamma + 3, aba, 2\delta, a).$$

Now consider the right-hand side $wuav$. This yields

$$(a, 2\gamma + 1, aba, 2\alpha + 2\delta + 3, aba, 2\beta, a).$$

We obtain the linear Diophantine system

$$2\alpha + 1 = 2\gamma + 1,$$
$$2\beta + 2\gamma + 3 = 2\alpha + 2\delta + 3,$$
$$2\delta = 2\beta.$$

Going back to the equation we see that for all $\alpha \geq 0$ and $\beta \geq \alpha$ the mapping

$$\sigma(x) = a^{2+2\alpha}, \quad \sigma(y) = ba^{2+2\beta}, \quad \sigma(z) = a^{3+2\alpha}ba^{2+2\beta}$$

yields a solution of the equation $axyz = zxay$ satisfying the rational constraints.

12.3. Boundary equations

12.3.1. Linear orders over a semigroup

We introduce some concepts using the semigroup S which describes the rational constraints. Let us start with an informal explanation of the notions discussed in this subsection. Assume that $x_1 \cdots x_g = x_{g+1} \cdots x_d$, $1 \leq g < d$, $x_i \in \Omega$ for $1 \leq i \leq d$, is a solvable word equation with rational

12.3. Boundary equations

constraints and that there is a nonsingular solution σ such that $\sigma(x_i) = u_i$ for $1 \leq i \leq d$. The equation and the solution define a word $w \in A^+$ and two factorizations $w = u_1 \cdots u_g = u_{g+1} \cdots u_d$. The positions between the factors u_i and u_{i+1} for $1 \leq i < g$ or $g < i < d$ are called *cuts*. By convention, the first and the last position of w are also cuts, and then we have at most d cuts. Reading the word from cut to cut, we obtain a sequence $(w_1, \ldots, w_m)$ such that each u_i is a product of some w_k and such that $w = w_1 \cdots w_m$, $w_k \neq \varepsilon$, $1 \leq k \leq m$, $m < d$.

On an abstract level we can say that the sequence $(w_1, \ldots, w_m)$ refines the two sequences $(u_1, \ldots, u_g)$ and $(u_{g+1}, \ldots, u_d)$. Let us see what happens if we pass via the homomorphism φ to the finite semigroup S. Thus we replace the u_i and w_k by $p_i = \varphi(u_i)$ and $s_k = \varphi(w_k)$ respectively.

Two sequences $(p_1, \ldots, p_g) \in S^g$ and $(p_{g+1}, \ldots, p_d) \in S^{d-g}$ are refined to a single sequence $(s_1, \ldots, s_m) \in S^m$, $m < d$, such that each $p_i \in S$ is a product of some s_k. We shall say that $(s_1, \ldots, s_m)$ is a *common refinement* of $(p_1, \ldots, p_g)$ and $(p_{g+1}, \ldots, p_d)$.

However, for each d, there are only finitely many candidates for $(s_1, \ldots, s_m)$ with $m < d$. Hence, in a nondeterministic step, we can guess and fix such a sequence $(s_1, \ldots, s_m)$ being the φ-image of $(w_1, \ldots, w_m)$.

A basic technique of solving word equations is to split a variable. Working over the sequence $(s_1, \ldots, s_m) \in S^m$, a splitting of a variable $x = x'x''$ corresponds to a splitting of some s_i and a guess of $s', s'' \in S$ such that $s_i = s's''$. In this way the lengths of the sequences are increasing.

EXAMPLE 12.3.1. Consider the equation $xauzau = yzbxaaby$. The solution, which was given in Example 12.1.1, leads to the sequences $(abb, a, bab, ba, a, bab)$ and $(ab, ba, b, abb, a, a, b, ab)$, where $(ab, b, a, b, ab, b, a, a, b, ab)$ is a common refinement. This can be visualized by the following figure.

a	b	b	a	b	a	b	b	a	a	b	a	b
a	b	b	a	b	a	b	b	a	a	b	a	b
a	b	b	a	b	a	b	b	a	a	b	a	b

Passing to the semigroup $S = \{a, a^2, b, 0\}$ of Example 12.2.5, we could start to search for a solution with the sequence $(0, b, a, b, 0, b, a, a, b, 0) \in S^{10}$.

We now start the formal discussion of this section. The semigroup S and the homomorphism $\varphi: A^+ \longrightarrow S$ are given as in Subsection 12.1.8. An *S-sequence* is a sequence $(s_1, \ldots, s_m) \in S^m$, $m \geq 0$. A *representation* of $(s_1, \ldots, s_m)$ is a triple $(I, \leq, \varphi_I)$ such that $(I, \leq)$ is a totally ordered set of

$m+1$ elements and

$$\varphi_I : \{(i,j) \in I \times I \mid i \leq j\} \to S^\varepsilon$$

is a mapping satisfying for some order respecting bijection $\rho : I \tilde\to \{0, \ldots, m\}$ the condition

$$\varphi_I(i,j) = s_{\rho(i)+1} \cdots s_{\rho(j)} \in S^\varepsilon \text{ for all } i,j \in I, \ i \leq j.$$

We have $\varphi_I(i,j) = \varepsilon$ if and only if $i = j$, and we have $\varphi_I(i,k) = \varphi_I(i,j)\varphi_I(j,k)$ for all $i,j,k \in I$, $i \leq j \leq k$.

The *standard representation* of $(s_1, \ldots, s_m)$ is simply $(I, \leq, \varphi_I)$ where $I = \{0, \ldots, m\}$ and $\varphi_I(i,j) = s_{i+1} \cdots s_j$ for $i,j \in I, i \leq j$. Hence for the standard representation the bijection ρ is the identity.

In the following any representation $(I, \leq, \varphi_I)$ of some S-sequence is called a *linear order over S*.

REMARK 12.3.2. An S-sequence can be viewed as an abstraction of a linear order over S. In most cases we are interested in the abstract objects only, but if we work with them we have to pass to concrete representations. When counting linear orders over S (cf. Lemma 12.3.6), by convention, we count only standard representations.

Let $w = a_1 \cdots a_m \in A^*$, $a_i \in A$ for $1 \leq i \leq m$. The set $\{0, \ldots, m\}$ is the *set of positions* of w, and for $0 \leq i \leq j \leq m$ let $w(i,j)$ denote the factor $a_{i+1} \cdots a_j$. In particular, $w = w(0, m) = w(0, i)w(i, m)$ for all $0 \leq i \leq m$. The associated S-sequence of a word w is defined by $w_S = (\varphi(a_1), \ldots, \varphi(a_m))$. The notation w_S also refers to its standard representation $w_S = (\{0, \ldots, m\}, \leq, \varphi_w)$. The mapping φ_w is defined by $\varphi_w(i,j) = \varphi(w(i,j))$ for all $0 \leq i \leq j \leq m$.

Let s, s' be S-sequences, which are given by some representations $(I, \leq, \varphi_I)$ and $(I', \leq, \varphi_{I'})$. We say that s' is a *refinement* of s (or that s matches s'), if there exists an order respecting injective mapping $\rho : I \to I'$ such that $\varphi_I(i,j) = \varphi_{I'}(\rho(i), \rho(j))$ for all $i,j \in I$, $i \leq j$. We write either $s \leq s'$ or, more precisely, $s \leq_\rho s'$ and $(I, \leq, \varphi_I) \leq_\rho (I', \leq, \varphi_{I'})$ in this case.

REMARK 12.3.3. Let s, s' be S-sequences such that $s \leq s'$. Then we may choose concrete representations and a refinement $(I, \leq, \varphi_I) \leq_\rho (I', \leq, \varphi'_I)$ such that $\rho : I \to I'$ is an inclusion, i.e., $I \subseteq I'$ and φ_I is the restriction of $\varphi_{I'}$ to I.

Let s be an S-sequence and $(I, \leq, \varphi_I)$ some representation. A word $w \in A^*$ is called a *model* of s (of $(I, \leq, \varphi_I)$ resp.), if the associated S-sequence w_S is a refinement of s, i.e., $(I, \leq, \varphi_I) \leq_\rho w_S$ for some ρ.

12.3. Boundary equations

If w is a model of s, then we write $w \models s$ or $w \models (I, \leq, \varphi_I)$. By abuse of language, we make the following convention. As soon as we have chosen a word w as a model, we are free to view the set I as a subset of positions of w, i.e., ρ becomes an inclusion and therefore $\varphi_I(i,j) = \varphi(w(i,j))$ for all $i, j \in I$, $i \leq j$.

LEMMA 12.3.4. *Every S-sequence* $(s_1, \ldots, s_m)$ *has a model* $w \in A^*$.

Proof. Since φ is surjective, there are nonempty words $w_i \in A^+$ such that $s_i = \varphi(w_i)$ for all $1 \leq i \leq m$. Let $w = w_1 \cdots w_m$; then we have $w \models (s_1, \ldots, s_m)$. ∎

The lemma above will yield the positive termination step in Makanin's algorithm if there are no more variables. In the positive case we can eventually reconstruct some S-sequence such that some model w describes a solution of the word equation.

Let $i, j \in I$, $i \leq j$ be positions in a linear order over S. Then $[i, j]$ denotes the interval from i to j; this is a linear suborder over S which is induced by the subset $\{k \in I \mid i \leq k \leq j\}$. More generally, let $T \subseteq I$ be a subset; then we view $(T, \leq, \varphi_T)$ as a linear suborder of $(I, \leq, \varphi_I)$. In the following, $\min(T)$ and $\max(T)$ refer to the minimal, respectively to the maximal, element of a subset T of a linear order I.

Let $(I, \leq, \varphi_I)$ be a representation of some S-sequence, $T \subseteq I$ a nonempty subset, and $\ell^*, r^* \in I$ positions such that $\ell^* < r^*$.

An *admissible extension* of $(I, \leq, \varphi_I)$ by T at $[\ell^*, r^*]$ is given by a linear order $(I^*, \leq, \varphi_{I^*})$ and two refinements $(I, \leq, \varphi_I) \leq_\rho (I^*, \leq, \varphi_{I^*})$ and $(T, \leq, \varphi_T) \leq_{\rho^*} (I^*, \leq, \varphi_{I^*})$ such that the following two conditions are satisfied:

(i) $I^* = \rho(I) \cup \rho^*(T)$,
(ii) $\min(\rho^*(T)) = \ell^*$ and $\max(\rho^*(T)) = r^*$.

The intuition behind the last definition should be fairly clear. An admissible extension refines $(I, \leq, \varphi_I)$ by defining new positions between ℓ^* and r^* until T matches the enlarged interval $[\ell^*, r^*]$ in such a way that all new points have a corresponding point in T and such that $\min(T)$ is mapped to ℓ^* and $\max(T)$ is mapped to r^*. The other way round: Let $(I^*, \leq, \varphi_{I^*})$ denote an admissible extension of $(I, \leq, \varphi_I)$ by T at $[\ell^*, r^*]$; then we may view $I \subseteq I^*$, whence $T \subseteq I^*$. There is a subset $T^* \subseteq I^*$ representing the same S-sequence as T; and we have $I^* = I \cup T^*$, $\min(T^*) = \ell^*$, and $\max(T^*) = r^*$.

EXAMPLE 12.3.5. Let $(s_1, \ldots, s_6)$ be some S-sequence, $(I, \leq, \varphi_I)$ its standard representation, $\ell^* = 4$ and $r^* = 6$. Let $(I^*, \leq, \varphi_{I^*})$ represent an admissible extension of $(I, \leq, \varphi_I)$ by $\{0, 3, 4, 5\}$ at $[4, 6]$. Then we may assume $I^* = \{0, \ldots, 6\} \cup \{3^*, 4^*\}$. The ordering of I^* satisfies $0 < 1 < 2 < 3 < 4 < 5 < 6$ and $4 = 0^* < 3^* < 4^* < 5^* = 6$.

We may or may not have $5 \in \{3^*, 4^*\}$. Say we have $5 = 3^*$. Then the corresponding S-sequence has the form

$$(s_1, s_2, s_3, s_4, s_5, s_4, s_5)$$

such that $s_5 = s_1 s_2 s_3$ and $s_6 = s_4 s_5$.

The following figure represents this admissible extension.

```
     s1      s2     s3     s4     s5           s6
0 ——— 1 ——— 2 ——— 3 ——— 4 ——— 5 —————————— 6
                        ‖     ‖              ‖
                     0* —s1s2s3— 3* —s4— 4* —s5— 5*
```

LEMMA 12.3.6. *Given* $(I, \leq, \varphi_I)$, $T \subseteq I$, $\ell^*, r^* \in I$. *Then the list of all admissible extensions of* $(I, \leq, \varphi_I)$ *by* T *at* $[\ell^*, r^*]$ *is finite and effectively computable.*

Proof. Trivial, since the cardinality of an admissible extension is bounded by $\operatorname{Card} I + \operatorname{Card} T$. ∎

EXAMPLE 12.3.7. Consider the same situation as in Example 12.3.5. The number of admissible extensions by the subset $\{0, 3, 4, 5\}$ at the interval $[4, 6]$ is given as a sum $e_1 + e_2 + e_3$. The numbers e_1, e_2, and e_3 respectively are the numbers of admissible extensions with $4^* \leq 5$, with $3^* < 5 < 4^*$, and with $5 \leq 3^*$ respectively. We have

$$e_1 = \operatorname{Card}\{s \in S^\varepsilon \mid s_5 = s_1 s_2 s_3 s_4 s, \, s_5 = s s_6\},$$
$$e_2 = \operatorname{Card}\{(r, s) \in S \times S \mid s_4 = rs, \, s_5 = s_1 s_2 s_3 r, \, s_6 = s s_5\},$$
$$e_3 = \operatorname{Card}\{s \in S^\varepsilon \mid s_1 s_2 s_3 = s_5 s, \, s_6 = s s_4 s_5\}.$$

Note that $s_1 s_2 s_3 s_4 s_5 \neq s_5 s_6$ implies $e_1 + e_2 + e_3 = 0$. Thus, there is no admissible extension of $\{0, 3, 4, 5\}$ at $[4, 6]$ in this case.

12.3.2. From word equations to boundary equations

Let $x_1 \cdots x_g = x_{g+1} \cdots x_d$, $1 \leq g < d$, $x_i \in \Omega$ for $1 \leq i \leq d$, be a word equation with rational constraints $L_x \subseteq A^*$ such that, without restriction, $\varepsilon \notin L_x \neq \emptyset$ for all $x \in \Omega$. Recall that we fixed a homomorphism $\varphi : A^+ \to S$ to some finite semigroup S such that $\varphi^{-1}\varphi(L_x) = L_x$ for all

12.3. Boundary equations

$x \in \Omega$. Since the images $\varphi(L_x) \subseteq S$ are finite sets we can split into finitely many cases where in each case $\varphi(L_x)$ is a singleton. Thus, it is enough to consider a situation where the input is $x_1 \cdots x_g = x_{g+1} \cdots x_d$, $1 \leq g < d$, and the question is the existence of a nonsingular solution $\sigma: \Omega \to A^+$ satisfying $\psi = \varphi \circ \sigma$ for some fixed mapping $\psi: \Omega \to S$. The question will be reformulated in terms of boundary equations.

Let $n \geq 0$ and $\varphi: A^+ \to S$ be a homomorphism to a finite semigroup S.

(i) A *system of boundary equations* is specified by a tuple

$$\mathscr{B} = ((\Gamma, ^-), (I, \leq, \varphi_I), \text{left}, B)$$

where Γ is a set of $2n$ variables, $^-: \Gamma \to \Gamma$ is an involution without fixed points, i.e., $\bar{\bar{x}} = x$, $x \neq \bar{x}$, for all $x \in \Gamma$, the triple $(I, \leq, \varphi_I)$ is a linear order over S, $\text{left}: \Gamma \to I$ is a mapping, and B is a set of *boundary equations*. Every boundary equation $b \in B$ has the form $b = (x, i, \bar{x}, j)$ with $x \in \Gamma$, $i, j \in I$, such that $\text{left}(x) \leq i$ and $\text{left}(\bar{x}) \leq j$.

(ii) A *solution* of $\mathscr{B}$ is a model $w \models (I, \leq, \varphi_I)$, $w \in A^*$, such that

$$w(\text{left}(x), i) = w(\text{left}(\bar{x}), j) \text{ for all } (x, i, \bar{x}, j) \in B.$$

(Recall that if a word $w \in A^*$ is a model for $(I, \leq, \varphi_I)$, then we view I as a subset of positions of w. Hence it makes sense to write $w(p, q)$ for $p, q \in I$, $p \leq q$.)

(iii) If $\mathscr{B}$ is solvable, then the *exponent of periodicity* $\exp(\mathscr{B})$ of $\mathscr{B}$ is defined by

$$\exp(\mathscr{B}) = \min\{\exp(w) \mid w \text{ is a solution of } \mathscr{B}\}.$$

We shall not distinguish between *isomorphic* systems. In particular, we may always think that $\Gamma = \{x_1, \ldots, x_n, \bar{x_1}, \ldots, \bar{x_n}\}$ and that $(I, \leq, \varphi_I)$ is the standard representation of some S-sequence, $I = \{0, \ldots, m\}$ for some $n, m \geq 0$.

REMARK 12.3.8. If we have $n = 0$, then there are no variables, hence no boundary equations, and any model $w \models (I, \leq, \varphi_I)$ is a solution of $\mathscr{B}$. Therefore, if $n = 0$, then the system is solvable by Lemma 12.3.4.

We are now ready to pass from word equations to boundary equations. The formal description is rather technical. We will see an example later. Consider a word equation $x_1 \cdots x_g = x_{g+1} \cdots x_d$ and a mapping $\psi: \Omega \to S$. We are going to construct a system

$$\mathscr{B} = ((\Gamma, ^-), (I, \leq, \varphi_I), \text{left}, B)$$

of boundary equations having the following two properties.

(i) Let $\sigma:\Omega \to A^+$ be a solution of the word equation such that $\psi = \varphi \circ \sigma$, and let $v \in A^*$ be a word with $v = \sigma(x_1 \cdots x_g) = \sigma(x_{g+1} \cdots x_d)$. Then $w = vv$ is a solution of $\mathcal{B}$.

(ii) Let $w \models (I, \leq, \varphi_I)$ be a solution of $\mathcal{B}$. Then we have $w \in A^*vvA^*$ for some $v \in A^*$ and there is a solution of the word equation $\sigma:\Omega \to A^+$ such that $\psi = \varphi \circ \sigma$ and $v = \sigma(x_1 \cdots x_g) = \sigma(x_{g+1} \cdots x_d)$.

In order to define $\mathcal{B}$ we start with the S-sequence

$$(\psi(x_1), \ldots, \psi(x_d)).$$

Let $(I, \leq, \varphi_I)$ be some representation, $I = \{i_0, \ldots, i_d\}$, $i_0 \leq \cdots \leq i_d$. The next step is to define the pair $(\Gamma, ^-)$ and the mapping left: $\Gamma \to I$. The intuitive meaning of $(\Gamma, ^-)$ is that Γ is a new set of variables where the notion of *dual* is defined and that left indicates the leftmost position of a variable in a given solution. We formalize this concept by using some undirected graph. Let (V, E) be the undirected graph with vertex set $V = \{1, \ldots, d\}$ and edge set $E = \{(p, q) \in V \times V \mid x_p = x_q\}$. Clearly, each edge defines a variable, but now we have a canonical choice to define the dual of (p, q) to be (q, p).

The idea is now that for $v = \sigma(x_1, \ldots, x_g) = \sigma(x_{g+1}, \ldots, x_d)$ and $w = vv$ we can realize I as a subset of positions of w such that both $w \models (\psi(x_1), \ldots, \psi(x_d))$ and the following equations hold:

$$w(i_0, i_g) = w(i_g, i_d), \quad w(i_{p-1}, i_p) = w(i_{q-1}, i_q) \text{ for all } (p, q) \in E.$$

For the first equation we shall introduce below an extra variable x_0 (and its dual $\overline{x_0}$); in the other list of equations there is some redundancy since the edge relation in our graph is transitive. For $(p, q), (q, r) \in E$, we have by definition $(p, r) \in E$, but the equations $w(i_{p-1}, i_p) = w(i_{q-1}, i_q)$ and $w(i_{q-1}, i_q) = w(i_{r-1}, i_r)$ already imply $w(i_{p-1}, i_p) = w(i_{r-1}, i_r)$. Hence we do not need the edge (p, r) for the equation. To avoid this redundancy we let $F \subseteq E$ be a spanning forest of (V, E). This means $F = F^{-1}$, $F^* = E^*$, and (V, F) is an acyclic undirected graph. We have Card $F = 2(d - c)$, where c is the number of connected components of (V, E). For each $x = (p, q) \in F$ we define its dual and two positions left(x), right(x):

$$\overline{x} = (q, p), \; \text{left}(x) = i_{p-1}, \; \text{right}(x) = i_p.$$

Note that $x \neq \overline{x}$ and $\overline{\overline{x}} = x$ for all $x \in F$. Taking duals corresponds to edge reversing in (V, F). Define two extra elements x_0 and $\overline{x_0}$ with $\overline{\overline{x_0}} = x_0$ and define $\Gamma = \{x_0, \overline{x_0}\} \cup F$ and

$$\text{left}(x_0) = i_0, \; \text{right}(x_0) = i_g = \text{left}(\overline{x_0}), \; \text{right}(\overline{x_0}) = i_d.$$

12.3. Boundary equations

This defines the set Γ, the involution without fixed points $^-: \Gamma \to \Gamma$, and the mapping left: $\Gamma \to I$. The elements of Γ are called variables again.

The last step of the construction is to define the set B of boundary equations. It should be clear what to do. We define

$$B = \{(x, \text{right}(x), \overline{x}, \text{right}(\overline{x})) \mid x \in \Gamma\}.$$

We still have to verify the two properties above.

(i) Let $\sigma: \Omega \to A^+$ be a solution such that $\psi = \varphi \circ \sigma$, and let $w = vv$, where $v = \sigma(x_1 \cdots x_g) = \sigma(x_{g+1} \cdots x_d)$. The word w has positions $0 = i_0 < i_1 < \cdots < i_d$, where i_d is the last position and the following equations hold:

$$w(i_0, i_g) = w(i_g, i_d), \quad w(i_{p-1}, i_p) = \sigma(x_p) \text{ for } 1 \leq p \leq d.$$

In particular, $w \models (I, \leq, \varphi_I)$ and w is a solution of $\mathcal{B}$.

(ii) Let $w \models (I, \leq, \varphi_I)$ be a solution of $\mathcal{B}$. Without restriction we may view I as a subset of positions of w. Consider the factors $w(i_0, i_g)$ and $w(i_g, i_d)$. The boundary equation $(x_0, \text{right}(x_0), \overline{x_0}, \text{right}(\overline{x_0})) \in B$ implies $w(i_0, i_g) = w(i_g, i_d)$ and it follows that $w \in A^* vv A^*$ for $v = w(i_0, i_g)$. We define $\sigma: \Omega \to A^+$ by $\sigma(x_p) = w(i_{p-1}, i_p)$. Since $i_{p-1} < i_p$, this is a nonempty word. The elements $(x, \text{right}(x), \overline{x}, \text{right}(\overline{x})) \in B$ for $x = (p, q)$, $\overline{x} = (q, p)$, $(p, q) \in T$ imply $w(i_{p-1}, i_p) = w(i_{q-1}, i_q)$ whenever $x_p = x_q$. Hence σ is well defined. We have $\varphi\sigma(x_p) = \varphi w(i_{p-1}, i_p) = \psi(x_p)$ since $w \models (I, \leq, \varphi_I)$. Finally, $v = w(i_0, i_g) = w(i_g, i_d)$ implies $v = \sigma(x_1 \cdots x_g) = \sigma(x_{g+1} \cdots x_d)$.

Thus, the word equation with rational constraints given by the mapping ψ has a solution if and only if the system of boundary equations is solvable. The construction of the system $\mathcal{B}$ above can be performed in polynomial time (and logarithmic space). Due to this reduction, Makanin's result follows from Theorem 12.3.10. The assertion of this theorem is in fact equivalent to Makanin's result; see Lemma 12.3.12.

EXAMPLE 12.3.9. We assume that the equation is simply $xyxyz = zyxyx$ and that we ignore any constraints for a moment. Hence, $\sigma(x) = a$, $\sigma(y) = b$, and $\sigma(z) = aba$, i.e., the word $v = abababa$ solves the equation. The transformation which yields the system of boundary equations is based on the following picture. The first line represents the word $w = vv$ of length 14.

a	b	a	b	a	b	a	a	b	a	b	a	$\overline{b}$	$\overline{a}$
\multicolumn{7}{c}{x_0}	\multicolumn{7}{c}{$\overline{x_0}$}
x_1	x_2	x_3	x_4	\multicolumn{2}{c}{x_5}		\multicolumn{2}{c}{$\overline{x_5}$}		x_6	x_7	$\overline{x_6}$	$\overline{x_7}$
		\multicolumn{2}{c}{$\overline{x_1}\ \overline{x_2}$}							\multicolumn{2}{c}{$x_4\ x_3$}		

According to the picture above we may represent the equation by a system of word equations using a set of eight variables with their duals $\{x_0, \overline{x_0}, \ldots, x_7, \overline{x_7}\}$:

$$x_0 = x_1 x_2 x_3 x_4 x_5,$$
$$\overline{x_0} = \overline{x_5} x_6 x_7 \overline{x_6} \overline{x_7},$$
$$\overline{x_1} = x_3,$$
$$\overline{x_3} = x_7,$$
$$\overline{x_2} = x_4,$$
$$\overline{x_4} = x_6,$$
$$x_i = \overline{\overline{x_i}} \text{ for } 0 \leq i \leq 7.$$

The system looks more complicated than the original equation, but the pattern is straightforward from the picture. The word vv has positions $0, \ldots, 14$. We define left$(x_0) = 0$, left$(\overline{x_0}) = 7$, left$(x_1) = 0$, left$(\overline{x_1}) = 2$, left$(x_2) = 1$, left$(\overline{x_2}) = 3$, left$(x_3) = 2$, left$(\overline{x_3}) = 11$, left$(x_4) = 3$, left$(\overline{x_4}) = 10$, left$(x_5) = 4$, left$(\overline{x_5}) = 7$, left$(x_6) = 10$, left$(\overline{x_6}) = 12$, left$(x_7) = 11$, and left$(\overline{x_7}) = 13$.

The set B of boundary equations is defined by the following list:

$$(x_0, 7, \overline{x_0}, 14), (x_1, 1, \overline{x_1}, 3), (x_2, 2, \overline{x_2}, 4), (x_3, 3, \overline{x_3}, 12),$$
$$(x_4, 4, \overline{x_4}, 11), (x_5, 7, \overline{x_5}, 10), (x_6, 11, \overline{x_6}, 12), (x_7, 12, \overline{x_7}, 14).$$

Since there were no constraints, the linear order is just the pair $(\{0, \ldots, 14\}, \leq)$.

12.3.3. The main theorem

THEOREM 12.3.10. *It is decidable whether a system of boundary equations has a solution.*

The rest of this chapter is devoted to the proof of Theorem 12.3.10. An important step is done in the next proposition: we can bound the exponent of periodicity while searching for a solution.

PROPOSITION 12.3.11. *Given as instance a system of boundary equations $\mathcal{B}$, we can compute a number $e(\mathcal{B})$ having the property that if $\mathcal{B}$ is solvable, then we have $\exp(\mathcal{B}) \leq e(\mathcal{B})$.*

The proof of Proposition 12.3.11 could be based on the same techniques as presented in Section 12.2. However, for our purposes we prefer to prove Proposition 12.3.11 via a reduction to word equations.

LEMMA 12.3.12. *There is an effective reduction of the solvability of a system of boundary equations $\mathcal{B}$ to the satisfiability problem of some*

12.3. Boundary equations

word equation with rational constraints such that for all solutions $w \in A^*$ of the word equation we have $\exp(\mathcal{B}) \leq \exp(w)$.

Proof. Let $\mathcal{B} = ((\Gamma,^-), (I, \leq, \varphi_I), \text{left}, B)$ be a system of boundary equations. We may assume that the linear order $(I, \leq, \varphi_I)$ is the standard representation of its underlying S-sequence $s = (s_1, \ldots, s_m)$. Introduce new variables $y_1, \ldots, y_m$ with rational constraints $\psi(y_p) = s_p$, $1 \leq p \leq m$.

For each boundary equation $b = (x, i, \bar{x}, j) \in B$ we introduce a word equation

$$y_{\text{left}(x)+1} \cdots y_i = y_{\text{left}(\bar{x})+1} \cdots y_j.$$

This system of word equations with rational constraints is solvable if and only if $\mathcal{B}$ is solvable. Indeed, if $w \in A^*$ is a solution of $\mathcal{B}$, then, by definition, we have $(I, \leq, \varphi_I) \leq_\rho w_S$, and $\rho(I)$ is a subset of positions of w. All word equations

$$w(\rho(\text{left}(x)), \rho(i)) = w(\rho(\text{left}(\bar{x})), \rho(j))$$

are satisfied for $(x, i, \bar{x}, j) \in B$. Hence defining $\sigma(y_p) = w(\rho(p-1), \rho(p))$, $1 \leq p \leq m$, yields a solution of the system of word equations.

For the other direction let $\sigma(y_p) = v_p$, $1 \leq p \leq m$, be some solution of the system of word equations. Due to the rational constraints we have $\psi(y_p) = s_p$ and $v_p \neq \varepsilon$ for all $1 \leq p \leq m$. Therefore the word $v = \sigma(y_1) \cdots \sigma(y_m)$ solves $\mathcal{B}$.

Next, we transform the system of word equations into a single word equation $L = R$ using Proposition 12.1.8 and finally we reduce to the word equation $Ly_1 \cdots y_m = Ry_1 \cdots y_m$. The point is that if w is a solution of this equation, then some suffix v of w solves $\mathcal{B}$. Hence $\exp(\mathcal{B}) \leq \exp(v) \leq \exp(w)$. This yields Lemma 12.3.12. Now, let d be the denotational length of $Ly_1 \cdots y_m = Ry_1 \cdots y_m$. Then define the number $e(\mathcal{B}) = e(c(S), d)$, which has been given in Theorem 12.2.1. We can choose w such that $\exp(w) \leq e(c(S), d)$. This proves Proposition 12.3.11. ∎

12.3.4. The convex chain condition

Let $\mathcal{B} = ((\Gamma,^-), (I, \leq, \varphi_I), \text{left}, B)$ be a system of boundary equations. Henceforth, a boundary equation $b = (x, i, \bar{x}, j) \in B$ will also be called a *brick*. The variable x is called the *label* of the brick $b = (x, i, \bar{x}, j)$. Pictorially a brick is given as follows:

x	i
$\bar{x}$	j

The dual brick $\bar{b}$ of $b = (x, i, \bar{x}, j)$ is given by reversing the brick; it has label $\bar{x}$:

$\bar{x}$	j
x	i

We make the assumption that B is closed under duals (i.e., $b \in B$ implies $\bar{b} \in B$) and that there is at least one brick $b \in B$ having label x for all $x \in \Gamma$. Clearly, this is no restriction. For $x \in \Gamma$ let $B(x) \subseteq B$ be the subset of bricks with label x. Then $B(x) = \{(x, i_1, \bar{x}, j_1), \ldots, (x, i_r, \bar{x}, j_r)\}$ for some nonempty subset $\{i_1, \ldots, i_r\} \subseteq I$ such that $\text{left}(x) \le i_1 \le \cdots \le i_r$. The *right boundary* of x is defined by $\text{right}(x) = i_r$.

Before we continue, we make some additional assumptions on B. All of them are necessary conditions for solvability and easily verified.

Let $(x, i, \bar{x}, j), (y, i, \bar{y}, j), (y, i', \bar{y}, j') \in B$. Then we assume from now on

- $\text{left}(x) \le \text{left}(\bar{x})$ if and only if $i \le j$,

- $\varphi_I(\text{left}(x), i) = \varphi_I(\text{left}(\bar{x}), j)$,

- $\text{left}(x) \le \text{left}(y)$ if and only if $\text{left}(\bar{x}) \le \text{left}(\bar{y})$,

- $i \le i'$ if and only if $j \le j'$.

These assumptions imply that if $B(x) = \{(x, i_1, \bar{x}, j_1), \ldots, (x, i_r, \bar{x}, j_r)\}$ is given such that $\text{left}(x) \le i_1 \le \cdots \le i_r$, then we also have $\text{left}(\bar{x}) \le j_1 \le \cdots \le j_r$. In particular, $B(x)$ contains a brick $(x, \text{right}(x), \bar{x}, \text{right}(\bar{x}))$. The set $B(x)$ can be depicted as follows:

$$B(x) = \left\{ \begin{array}{|c|c|} \hline x & i_1 \\ \hline \bar{x} & j_1 \\ \hline \end{array}, \begin{array}{|c|c|} \hline x & i_2 \\ \hline \bar{x} & j_2 \\ \hline \end{array}, \ldots, \begin{array}{|c|c|} \hline x & \text{right}(x) \\ \hline \bar{x} & \text{right}(\bar{x}) \\ \hline \end{array} \right\}$$

In our pictures a brick $(x, i, \bar{x}, j)$ can be placed upon $(y, j', \bar{y}, \ell)$, if and only if $j = j'$. We obtain one out of three different shapes:

Which one of these cases occurs is determined by the function $\text{left}: \Gamma \to I$. The leftmost picture corresponds to $\text{left}(\bar{x}) < \text{left}(y)$, the picture in the middle corresponds to $\text{left}(\bar{x}) = \text{left}(y)$, the picture on the right means $\text{left}(\bar{x}) > \text{left}(y)$.

12.3. Boundary equations

Let $k \geq 1$. A *chain C of length k* is a sequence of bricks

$$C = ((x_1, i_1, \overline{x_1}, i_2), (x_2, i_2, \overline{x_2}, i_3), \ldots, (x_k, i_k, \overline{x_k}, i_{k+1})),$$

where $(x_p, i_p, \overline{x_p}, i_{p+1}) \in B$ for all $1 \leq p \leq k$.

For a chain C and a variable $x \in \Gamma$ we define the *x-length* $|C|_x$ of C to be the number of bricks in C having label x. Thus, the length of a chain C is the sum $\sum_{x \in \Gamma} |C|_x$.

A chain C is called *convex*, if for some index q with $1 \leq q \leq k$ we have

$$\text{left}(\overline{x_p}) \geq \text{left}(x_{p+1}) \text{ for } 1 \leq p < q,$$
$$\text{left}(\overline{x_p}) \leq \text{left}(x_{p+1}) \text{ for } q \leq p < k.$$

A convex chain C is called *clean*, if the bricks of C are pairwise distinct.

A brick $(x, i, \overline{x}, j)$ is *linked via a convex chain* to a brick $(x', i', \overline{x}', j')$, if there is a convex chain C of length k as above for some $k \geq 1$ such that $(x, i, \overline{x}, j) = (x_1, i_1, \overline{x_1}, i_2)$, and $(x', i', \overline{x}', j') = (x_k, i_k, \overline{x_k}, i_{k+1})$.

REMARK 12.3.13. If $C = (b_1, \ldots, b_k)$ is a convex chain, then its dual $\overline{C} = (\overline{b_k}, \ldots, \overline{b_1})$ and $(b_p, \ldots, b_q)$, $1 \leq p \leq q \leq k$ are convex chains. If $b_p = (x_p, i_p, \overline{x_p}, i_p)$ for some $1 < p < k$, then $(b_1, \ldots, b_{p-1}, b_{p+1}, \ldots b_k)$ is a convex chain. If $b_p = b_q$ for some $1 < p < q \leq k$, then $(b_1, \ldots, b_{p-1}, b_q, \ldots b_k)$ is also a convex chain. In particular, if two bricks are linked via a convex chain, then they are linked via some clean convex chain. The shortest chain linking two bricks to each other is always clean.

Let $F \subseteq I$ be a subset. A brick $(x, i, \overline{x}, j) \in B$ is called a *basis* or *foundation* with respect to F, if $j \in F$. We say that $\mathscr{B}$ satisfies the *convex chain condition* (with respect to F), if every brick $b \in B$ can be linked via some convex chain to some basis. The set F is also called the set of *final indices*.

In the following we concentrate on solvable systems and we need a few more notations. Let $\mathscr{B} = ((\Gamma, \bar{\ }), (I, \leq, \varphi_I), \text{left}, B)$ be a solvable system of boundary equations and $w \in \Gamma^*$ such that $w \models (I, \leq, \varphi_I)$ is a solution of $\mathscr{B}$. Since w is a solution we may assume that I is a subset of positions of w. For all $x \in \Gamma$ define a word $w(x) \in A^*$ by

$$w(x) = w(\text{left}(x), \text{right}(x)).$$

This also permits a notion of *w-length* for $x \in \Gamma$. We define

$$|x|_w = |w(x)|.$$

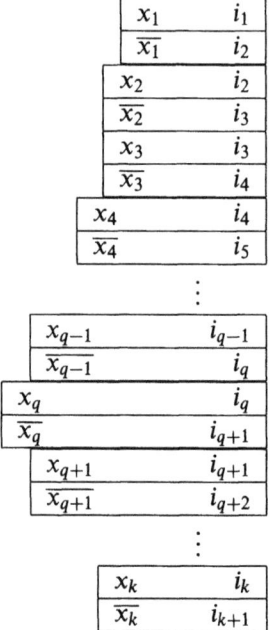

Figure 12.2. A convex chain.

Moreover, for each brick $b = (x, i, \bar{x}, j) \in B$ we also define its w-length by

$$|b|_w = |w(\text{left}(x), i)|.$$

For all $x \in \Gamma$ and $b \in B$ we have $w(x) = w(\bar{x})$, $|x|_w = |\bar{x}|_w$, $|b|_w = |\bar{b}|_w$, and $|b|_w \leq |x|_w$, if x is the label of b. A brick is uniquely determined by its label and its w-length $|b|_w$. A singly exponential bound on the number of bricks as given in the next lemma is due to Gutiérrez 1998a. The improvement on this number has been essential in order to obtain the singly exponential complexity bound in Theorem 12.4.2 below.

LEMMA 12.3.14. *Let $n, m, f \in \mathbb{N}$ and $\mathcal{B} = ((\Gamma, \bar{\ }), (I, \leq, \varphi_I), \text{left}, B)$ be a solvable system of boundary equations such that $w \models (I, \leq, \varphi_I)$ is a solution of $\mathcal{B}$. Let $\text{Card}\,\Gamma = 2n$, $F \subseteq I$, and $\text{Card}\,F = f$. Suppose that every brick $b \in B$ can be linked via a convex chain C to a basis with respect to F such that for each $x \in \Gamma$ the number of bricks in C having label x is at most m, i.e., $|C|_x \leq m$.*

12.3. Boundary equations

Then we can bound the size of B by

$$\text{Card } B \leq 2n \cdot f \cdot (2m+1)^n.$$

Proof. Consider a convex chain C of length k such that $|C|_x \leq m$ for all $x \in \Gamma$ and where the last brick is a basis:

$$C = ((x_1, i_1, \overline{x_1}, i_2), (x_2, i_2, \overline{x_2}, i_3), \ldots, (x_k, i_k, \overline{x_k}, i_{k+1})).$$

There are $2n$ possibilities for the label of the first brick. We shall calculate an upper bound for the number of possible w-lengths for the first brick $(x_1, i_1, \overline{x_1}, i_2)$. The length of the first brick is determined by the w-length of the last brick $(x_k, i_k, \overline{x_k}, i_{k+1})$ and by summing up the values $\text{left}(x_{i+1}) - \text{left}(\overline{x_i})$ for $i = k-1, \ldots, 1$; see Figure 12.2. Recall that $i \in I$ denotes a position in the solution w, hence $\text{left}(x_{i+1}) - \text{left}(\overline{x_i}) \in \mathbb{Z}$. So the w-length of the first brick is

$$i_{k+1} - \text{left}(\overline{x_k}) + \text{left}(x_k) - \text{left}(\overline{x_{k-1}}) + \cdots + \text{left}(x_2) - \text{left}(\overline{x_1}).$$

Then we can rearrange this sum in some formula of type

$$i_{k+1} - \text{left}(\overline{x_1}) + \sum_{x \in \Gamma} m_x \cdot \left(\text{left}(x) - \text{left}(\overline{x})\right)$$

where due to the hypothesis on C we have $-m \leq m_x \leq m$. The value $\text{left}(\overline{x_1})$ is uniquely determined by the label x_1 and i_{k+1} is a basis. Hence at most $f \cdot (2m+1)^n$ different values can be produced using these sums, when the label x_1 is fixed. Thus, at most

$$2n \cdot f \cdot (2m+1)^n.$$

different first bricks are possible. But this is also an upper bound for the number of bricks Card B by the convex chain condition. ∎

Every system of boundary equations $\mathcal{B}$ satisfies the convex chain condition with respect to the set I, trivially. Furthermore, if we construct $\mathcal{B}$ by starting from a word equation $x_1 \cdots x_g = x_{g+1} \cdots x_d$, $1 \leq g < d$, then we have Card $I \leq d$. The transformation rules below will increase neither the number $2n$ of variables nor the sum $2n + f$. They will increase the sizes of I and of B. However, Lemma 12.3.14 says that a large number of boundary equations (i.e., a large set of bricks) yields that there are long convex chains in order to satisfy the convex chain condition (pictorially: many bricks build *skyscrapers*). The next step is to show that long convex chains (or skyscrapers) lead to high domino towers and hence to a lower bound on the exponent of periodicity in any solution.

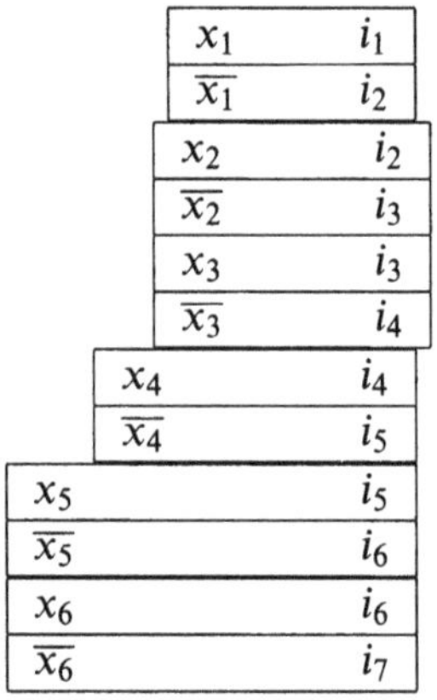

Figure 12.3. The upper part of a convex chain.

PROPOSITION 12.3.15. *Let $n, m \in \mathbb{N}$ and $\mathscr{B} = ((\Gamma, \bar{\ }), (I, \leq, \varphi_I), \text{left}, B)$ be a solvable system of boundary equations with Card $\Gamma = 2n$. Let $w \models (I, \leq, \varphi_I)$ be a solution of $\mathscr{B}$. Suppose that there is at least one clean convex chain such that $m \leq |C|_x$ for some $x \in \Gamma$. Then we have the following lower bound for the exponent of periodicity of the solution w:*

$$m \leq 2n \cdot (\exp(w) + 1) - 1.$$

Proof. The hypothesis implies $n \neq 0$, hence $w \neq \varepsilon$. The assertion is trivial for $m < 4n$. Hence let $n \geq 1$ and $m \geq 4n$. Define $h = \lceil \frac{m+1}{2n} \rceil$. We have $h \geq 3$. (Eventually h will be the height of some domino tower.)

Let $C = (b_1, \ldots, b_k)$ be a clean convex chain such that $m \leq |C|_x$ for some $x \in \Gamma$. Let $b_p = (x_{i_p}, i_p, \overline{x_{i_p}}, i_{p+1})$ for $1 \leq p \leq k$. Define $m' = \lceil \frac{m+1}{2} \rceil$; then by duality (replacing C by $\overline{C}$ and x by $\bar{x}$) we may assume that the label x occurs at least m' times in the upper part up to some k' where $k' \leq k$ such that

$$\text{left}(\overline{x_1}) \geq \text{left}(x_2), \quad \text{left}(\overline{x_2}) \geq \text{left}(x_3), \quad \ldots, \quad \text{left}(\overline{x_{k'-1}}) \geq \text{left}(x_{k'}).$$

This upper part of the chain C up to k' might look as in Figure 12.3.

In the following we need a suitable chain where the label of the last brick has minimal w-length. In order to find such a chain we scan $(b_1, \ldots, b_{k'})$ from right to left. We find a sequence of indices

$$0 = p_0 < p_1 < \cdots < p_{n'-1} < p_{n'} = k'$$

12.3. Boundary equations

such that $n' \leq n$ and for all q, j where $p_{j-1} < q \leq p_j$, $1 \leq j \leq n'$, we have

$$|x_q|_w \geq |x_{p_j}|_w.$$

This means that in each interval $[p_{j-1} + 1, p_j]$ the last label x_{p_j} has minimal w-length. By the pigeon-hole principle there is at least one index $j \in \{1, \ldots, n'\}$ such that the number of occurrences of the label x in the interval $[p_{j-1} + 1, p_j]$ is at least

$$\left\lceil \frac{m+1}{2n} \right\rceil.$$

We conclude that (after renaming) there are a clean convex chain $C = (b_1, \ldots, b_\ell)$ and a variable $x \in \Gamma$ having the following properties:

$$\begin{aligned}
|C|_x &= \left\lceil \tfrac{m+1}{2n} \right\rceil, \\
\text{left}(\overline{x_p}) &\geq \text{left}(x_{p+1}) \quad \text{for} \quad 1 \leq p < \ell, \\
|x_p|_w &\geq |x_\ell|_w \quad \text{for} \quad 1 \leq p \leq \ell.
\end{aligned}$$

Recall that $h = \left\lceil \frac{m+1}{2n} \right\rceil$. We have $h \geq 3$ and the label x occurs exactly h times in the clean convex chain C. By cutting off the sequence we may assume that x is the first label x_1.

This is the point where we switch from the chain to the sequence of words

$$(w(x_1), \ldots, w(x_\ell)).$$

We obtain a tower of words where $w(x_\ell)$ has minimal length and the word $w(x_1)$ occurs at least h times.

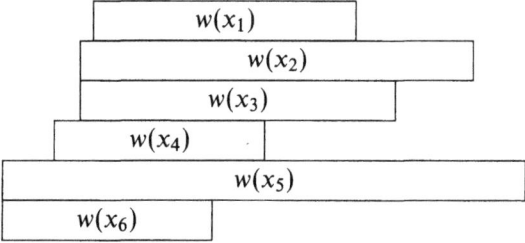

Define $v_p \in A^*$ to be the prefix of $w(x_p)$ of length $|w(x_\ell)|$ and let $u_p = w(\text{left}(x_p), i_p)$ for $1 \leq p \leq \ell$. Since $|u_p| \leq |w(\text{left}(x_\ell), i_\ell)| \leq |v_\ell| = |v_p|$, the word u_p is a prefix of v_p for all $1 \leq p \leq \ell$. The sequence $(v_1, \ldots, v_\ell)$ can be arranged in a tower of words which is already in better shape: all words v_p have equal length.

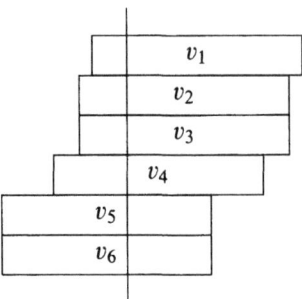

The vertical line corresponds to the factorization $v_p = u_p u'_p$ for $1 \leq p \leq \ell$.

Finally, let $\{q_1, q_2, \ldots, q_h\}$ be a set of the h indices where the bricks have label x_1. Since the convex chain leading to this tower is clean, we see that $u_{q_i} \neq u_{q_j}$ for all $1 \leq i, j \leq h$, $i \neq j$. (This is the only point where it is used that the chain is clean!) We obtain

$$0 \leq |u_{q_1}| < |u_{q_2}| < \cdots < |u_{q_h}|.$$

Moreover, we have $v_1 = v_{q_1} = v_{q_2} = \cdots = v_{q_h}$. We omit all other words in the tower above and we see that the word v_1 can be arranged in a domino tower of height h and $h \geq 2$. Applying Lemma 12.1.4 we obtain $h - 1 \leq \exp(w_1) \leq \exp(w)$. The assertion of the proposition follows. ∎

COROLLARY 12.3.16. *Let $\mathscr{B} = ((\Gamma, ^-), (I, \leq, \varphi_I), \text{left}, B)$ denote a solvable system of boundary equations which satisfies the convex chain condition with respect to some subset $F \subseteq I$. Let $\operatorname{Card} \Gamma = 2n$ and $\operatorname{Card} F = f$. Then we have*

$$\operatorname{Card} B \leq 2n \cdot f \cdot (4n \cdot (\exp(\mathscr{B}) + 1))^n.$$

If moreover $\operatorname{Card} \Gamma, \operatorname{Card} F \in O(d)$, and $\exp(\mathscr{B}) \in 2^{O(d + \log c(S))}$, then we have

$$\operatorname{Card} B \in 2^{O(d^2 + d \log c(S))}.$$

Proof. Let $2n = \operatorname{Card} \Gamma$, $f = \operatorname{Card} F$, and m be the maximal x-length of a clean convex chain, $x \in \Gamma$. By Remark 12.3.13 and Lemma 12.3.14 we have

$$\operatorname{Card} B \leq 2n \cdot f \cdot (2m + 1)^n.$$

Choose a solution w such that $\exp(w) \leq \exp(\mathscr{B})$. Proposition 12.3.15 yields

$$m \leq 2n \cdot (\exp(w) + 1) - 1.$$

12.3. Boundary equations

Putting things together we obtain

$$\text{Card } B \leq 2n \cdot f \cdot (4n \cdot (\exp(w) + 1))^n \leq 2n \cdot f \cdot (4n \cdot (\exp(\mathcal{B}) + 1))^n.$$

The result follows. ∎

12.3.5. Transformation rules

We are ready to define the (nondeterministic) transformation rules of Makanin's algorithm. If we apply a rule to a system $\mathcal{B} = ((\Gamma, \bar{\ }), (I, \leq, \varphi_I), \text{left}, B)$, then the new system is denoted by $\mathcal{B}' = ((\Gamma', \bar{\ }), (I', \leq, \varphi_{I'}), \text{left}', B')$. The transformation rules below will have the property that if $\mathcal{B} = ((\Gamma, \bar{\ }), (I, \leq, \varphi_I), \text{left}, B)$ satisfies the convex chain condition with respect to some subset $F \subseteq I$, then $\mathcal{B}'$ satisfies the convex chain condition with respect to some subset $F' \subseteq I'$ such that $\text{Card } \Gamma' + \text{Card } F' \leq \text{Card } \Gamma + \text{Card } F$. Thus, if we start with a system $\mathcal{B}_0$ where $\text{Card } \Gamma_0 = 2n_0$ and $\text{Card } I_0 \leq d$, then throughout the whole procedure the size of the set of final indices is smaller than or equal to $2n_0 + d$.

We say that a (nondeterministic) rule is *downward correct*, provided the following condition holds: if $w \in A^*$ is a solution of $\mathcal{B}$, then (for at least one nondeterministic choice) some suffix w' of w is a solution of $\mathcal{B}'$, and moreover either $\text{Card } \Gamma' < \text{Card } \Gamma$ or $|w'| < |w|$. Thus, applied to solvable systems at least one sequence of choices of downward correct rules leads to termination.

We say that a (nondeterministic) rule is *upward correct*, provided the following condition holds: if $w' \in A^*$ is a solution of $\mathcal{B}'$ (and $\mathcal{B}'$ is the result of any nondeterministic choice), then there is word $w \in A^*$, which is a solution of $\mathcal{B}$.

RULE 1. If there is some $x \in \Gamma$ with $\text{left}(x) = \text{right}(x)$, then cancel both bricks

$$(x, \text{right}(x), \bar{x}, \text{right}(\bar{x})) \text{ and } (\bar{x}, \text{right}(\bar{x}), x, \text{right}(x))$$

from B. Cancel x and $\bar{x}$ from Γ.

REMARK 12.3.17. Obviously Rule 1 is upward and downward correct since we have $w(i, i) = \varepsilon$ for all words w and all positions i of w. Hence the set of solutions is the same. In order to preserve the convex chain condition we introduce two new final indices. Let $x \in \Gamma$ be such that $\text{left}(x) = \text{right}(x)$ and assume that $x, \bar{x}$ are canceled by Rule 1. Define $F' = F \cup \{\text{left}(x), \text{left}(\bar{x})\}$. Consider a convex chain $C = (b_1, \ldots, b_m)$ where for some $1 < p \leq m$ the brick b_p has the form $b_p = (x, \text{right}(x), \bar{x}, \text{right}(\bar{x}))$.

Hence the brick b_p is canceled. However, the brick b_1 is linked to b_{p-1} via a convex chain and b_{p-1} is now a basis since right(x) = left$(x) \in F'$. Thus, if $\mathscr{B}$ satisfies the convex chain condition with respect to F, then the system $\mathscr{B}'$ (after an application of Rule 1) satisfies the convex chain condition with respect to F'. We have Card Γ' + Card $F' \leq$ Card Γ + Card F.

RULE 2. If there exists some $x \in \Gamma$ with left(x) = left$(\bar{x})$, then cancel all bricks $(x, j, \bar{x}, j)$ and $(\bar{x}, j, x, j)$ from B. Cancel x and $\bar{x}$ from Γ.

REMARK 12.3.18. Recall that for $(x, i, \bar{x}, j) \in B$ we have left(x) = left$(\bar{x})$ if and only if $i = j$. Thus, if left(x) = left$(\bar{x})$, then all bricks with label x have the form $(x, j, \bar{x}, j)$. Again, Rule 2 is obviously upward and downward correct. For the convex chain condition consider a convex chain $C = (b_1, \ldots, b_m)$ where $b_p = (x, j, \bar{x}, j)$ for some $1 < p \leq m$. If we have $p < m$, then $C' = (b_1, \ldots, b_{p-1}, b_{p+1}, \ldots, b_m)$ is a shorter convex chain linking b_1 with a basis. For $p = m$ we have $j \in F$. Hence b_{m-1} is also a basis.

RULE 3. Let $\ell = \min(I)$. If $\ell \notin$ left(Γ), then cancel the index ℓ from I. This means we replace the linear order over S by the induced suborder $(I', \leq, \varphi_{I'})$ where $I' = I \setminus \{\ell\}$.

REMARK 12.3.19. Clearly, the convex chain condition is not affected by this rule. Downward correctness is obvious, too. To see the upward correctness let $(I, \leq, \varphi_I)$ be given by the S-sequence $(s_1, \ldots, s_m)$ and let $w' \in A^*$ be a solution of the new system after an application of Rule 3 such that $\min(I')$ is the first position of w'. By definition of an S-sequence there is a nonempty word $u \in A^+$ with $\varphi(u) = s_1$. Then the first position of w' is not equal to the first position in the word uw', and uw' is a solution of $\mathscr{B}$. For later use notice that we can choose u such that $|u| \leq |S|$.

The next rule is very complex. It is the heart of the algorithm. Before we apply it to some system $\mathscr{B} = ((\Gamma,^-), (I, \leq, \varphi_I), \text{left}, B)$, we apply Rules 1, 2 and 3 as often as possible. In particular, we shall assume that left(x) < right(x), left$(x) \neq$ left$(\bar{x})$ for all $x \in \Gamma$, and that there exists some $x \in \Gamma$ with left(x) = $\min(I)$.

RULE 4. We divide Rule 4 into six steps.
    We need some notation. Define $\ell = \min(I)$ and $r = \max\{\text{right}(x) \mid x \in \Gamma, \text{left}(x) = \ell\}$. Note that $\ell \in$ left(Γ), hence $r \in I$ exists and we have $\ell < r$. Choose (and fix) some $x_o \in \Gamma$ with left$(x_o) = \ell$ and right$(x_o) = r$.

12.3. Boundary equations

Define $\ell^* = \text{left}(\overline{x_o})$ and $r^* = \text{right}(\overline{x_o})$. Define the *critical boundary* $c \in I$ by $c = \min\{c', r\}$ where

$$c' = \min\{\text{left}(x) \mid x \in \Gamma, \ r < \text{right}(x)\}.$$

Note that since $r < r^* = \text{right}(\overline{x_o})$, the minimum c' and hence the critical boundary c exist. We have $\ell < c \leq r < r^*$ and $c \leq \ell^* < r^*$. The ordering of r and ℓ^* depends on the system; it is of no importance.

Define the subset $T \subseteq I$ of *transport positions* by

$$T = \{i \in I \mid i \leq c\} \cup \{i \in I \mid \exists (x, i, \overline{x}, j) \in B : \text{left}(x) < c\}.$$

Note that $\min(T) = \ell$ and that $i \in T$ for all $(x_o, i, \overline{x_o}, j) \in B$. Moreover, since $\text{left}(x) < c$ implies $\text{right}(x) \leq r$, we have $\max(T) = r$.

STEP 1. Choose some admissible extension $(I^*, \leq, \varphi_{I^*})$ of $(I, \leq, \varphi_I)$ by T at $[\ell^*, r^*]$. By convention we identify I as a subset of I^*, whence $I \subseteq I^*$, and there is a subset $T^* \subseteq I^*$ with $\min(T^*) = \ell^*$ and $\max(T^*) = r^*$ and such that T^* is in order-respecting bijection with T. For each $i \in T$ the corresponding position in T^* is denoted by i^*. Having these notations we put a further restriction on the admissible extension: we consider only those admissible extensions where, first, $i < i^*$ for all $i \in T$ and, second, for all $(x, i, \overline{x}, j) \in B$ with $\text{left}(x) < c$ we require

$$\text{left}(x)^* = \text{left}(\overline{x}) \iff i^* = j,$$
$$\text{left}(x)^* < \text{left}(\overline{x}) \iff i^* < j.$$

In particular, for all bricks $(x_o, i, \overline{x_o}, j)$ we require $i^* = j$. If such an admissible extension is not possible, then Step 1 cannot be completed and Rule 4 is not applicable.

STEP 2. Introduce a new variable x_v and its dual $\overline{x_v}$. We define $\text{left}(x_v) = c$, $\text{left}(\overline{x_v}) = c^*$. For all $i \in T$ such that there is some $(x, i, \overline{x}, j) \in B$ with $\text{left}(x) < c \leq i$ introduce new bricks $(x_v, i, \overline{x_v}, i^*)$ and $(\overline{x_v}, i^*, x_v, i)$.

STEP 3. As long as there is a variable $x \in \Gamma$ with $\text{left}(x) < c$, replace $\text{left}(x)$ by $\text{left}'(x) = \text{left}(x)^*$ and replace all bricks $(x, i, \overline{x}, j), (\overline{x}, j, x, i) \in B$ by $(x, i^*, \overline{x}, j)$ and $(\overline{x}, j, x, i^*)$.

REMARK 12.3.20. To have some notation let x denote a variable before Step 3 and let x' be the corresponding variable after Step 3. Likewise let $b = (x, i, \overline{x}, j)$ denote a brick before Step 3 and let $b' = (x', i', \overline{x}', j)$ be the corresponding brick after Step 3. If $\text{left}(x) = \text{left}'(x')$, then sometimes we may still write $x = x'$. In particular, $x_v = x'_v$, $\overline{x}_v = \overline{x}'_v$, $\overline{x}_o = \overline{x}'_o$, but $x_o \neq x'_o$.

For $b = (x, i, \overline{x}, j)$ and $b' = (x', i', \overline{x}', j')$ there are four cases:
$$b' = (x', i^*, \overline{x}', j^*) \quad \text{if } \text{left}(x) < c, \quad \text{left}(\overline{x}) < c,$$
$$b' = (x', i^*, \overline{x}, j) \quad \text{if } \text{left}(x) < c, \quad c \leq \text{left}(\overline{x}),$$
$$b' = (x, i, \overline{x}', j^*) \quad \text{if } c \leq \text{left}(x), \quad \text{left}(\overline{x}) < c,$$
$$b' = (x, i, \overline{x}, j) \quad \text{if } c \leq \text{left}(x), \quad c \leq \text{left}(\overline{x}).$$

Note that after Step 3 all bricks $(x_o, i, \overline{x_o}, j) \in B$ have the form $(x'_o, i^*, \overline{x_o}', i^*)$.

STEP 4. Define as the new set of final indices
$$F' = \{i^* \in I^* \mid i < c \text{ and } i \in F\} \cup \{i \in F \mid c \leq i\}.$$

STEP 5. Cancel all bricks with label x'_o or $\overline{x_o}$, i.e., cancel all bricks of the form $(x'_o, i^*, \overline{x_o}, i^*)$ or $(\overline{x_o}', i^*, x'_o, i^*)$. Then cancel the variables $x_o, \overline{x_o}$.

STEP 6. Replace I^* by $I' = \{i \in I^* \mid c \leq i\}$ and consider the linear order $(I', \leq, \varphi_{I'})$ induced by $I' \subseteq I^*$.

After Step 6 the transformation rule is finished. The new system is denoted by $\mathscr{B}' = ((\Gamma', \bar{\ }), (I', \leq, \varphi_{I'}), \text{left}', B')$. We will show from Lemmas 12.3.25 to 12.3.28 that $\mathscr{B}'$ satisfies the convex chain condition with respect to F'. The first lemma is a trivial observation.

LEMMA 12.3.21. *We have* $\text{Card}\,\Gamma' = \text{Card}\,\Gamma$ *and* $\text{Card}\,F' \leq \text{Card}\,F$.

Proof. In Step 2 new variables x_v and $\overline{x}_v$ are introduced, but in Step 5 the variables x'_o and $\overline{x_o}$ are canceled. Hence $\text{Card}\,\Gamma' = \text{Card}\,\Gamma$. The set of final indices is changed in Step 4 in such a way that $\text{Card}\,F' \leq \text{Card}\,F$. ∎

The following lemma is used to bound the size of I during the transformation procedure. The lemma has a rather subtle proof.

LEMMA 12.3.22. *Let* $\beta' = \text{Card}\{(x', i', \overline{x}', j') \in B' \mid \text{left}'(x') < i'\}$ *and* $\beta = \text{Card}\{(x, i, \overline{x}, j) \in B \mid \text{left}(x) < i\}$. *Then we have*
$$2\,\text{Card}\,I' - \beta' \leq 2\,\text{Card}\,I - \beta.$$

Proof. The inequality can be destroyed either by a new position $i^* \in T^* \setminus I$ or by the cancellation of bricks $(x'_o, i^*, \overline{x_o}, i^*)$, $(\overline{x_o}, i^*, x'_o, i^*)$ in Step 5, where $\ell^* < i^*$. (Recall the definition of β and β' and that $\text{left}(x_o) = \ell$,

12.3. Boundary equations

left$'(x'_o) = \ell^*$.) The cancellation of these bricks involves again a position of type $i^* \in T^*$. Fortunately, if $(x'_o, i^*, \overline{x_o}, i^*)$ is canceled, where $\ell^* < i^*$, then $i^* = j$ for some $j \in I \setminus \{\ell\}$. In particular, i^* is not a new position and the two cases don't occur simultaneously. Therefore it is enough to find for each $i^* \in T^* \setminus \{\ell^*\}$ either two new bricks which are introduced in Step 2 or one position which is canceled in Step 6. Then the total balance will be negative or zero.

Let us consider the positions of type $i^* \in T^* \setminus \{\ell^*\}$ one by one. If $c^* < i^*$, then by the definition of T and Step 2 there are two new bricks $(x_v, i, \overline{x_v}, i^*), (\overline{x_v}, i^*, x_v, i) \in B'$ and we have left$(x_v) < i$, left$(\overline{x_v}) < i^*$. Next consider $i^* = c^*$. At least one position (namely ℓ) is canceled in Step 6. Next let $\ell^* < i^* < c^*$, i.e., $\ell < i < c$. The position i is canceled in Step 6. Hence we have the assertion of the lemma. ∎

LEMMA 12.3.23. *Rule 4 is downward correct.*

Proof. Let $w \in A^*$ be a solution of $\mathcal{B}$. Since $w \models (I, \leq, \varphi_I)$, we can view I as a subset of positions of w with $\ell = 0$. Let $w = vw'$ where $v = w(\ell, c)$. The word v is a nonempty prefix of $w(\ell, r)$. The word $w(\ell, r)$ is a prefix of w and at the same time another factor of w'; we have $w(\ell, r) = w(\ell^*, r^*)$ with $\ell < \ell^*$ due to the brick $(x_o, r, \overline{x_o}, r^*) \in B$. The set T is a subset of positions of $w(\ell, r)$, hence we find a corresponding subset T^* of positions of $w(\ell^*, r^*)$. The union $I \cup T^*$ leads to an admissible extension $(I^*, \leq, \varphi_I)$ such that, first, $i < i^*$ for all $i \in T$ and, second, $w(j, k) = w(j^*, k^*)$ for all $j, k \in T, j \leq k$. A careful but easy inspection of Rule 4 then shows that $w' \models (I', \leq, \varphi_{I'})$ and w' is a solution of $\mathcal{B}'$. ∎

LEMMA 12.3.24. *Rule 4 is upward correct.*

Proof. Let $w' \in A^*$ be a solution of $\mathcal{B}'$. Since $w' \models (I', \leq, \varphi'_I)$, we can view I' as a subset of positions of w' where c is the first position of w'. Define $v = w'(l^*, c^*)$ and let $w = vw'$. Then we have $w \models (I^*, \leq, \varphi_{I^*})$ such that $v = w(l, c) = w(l^*, c^*)$. With the help of the bricks $(x_v, i, \overline{x_v}, i^*)$ we conclude that $w(j, k) = w(j^*, k^*)$ for all $j, k \in T, j \leq k$. Therefore we have $w(\text{left}(x), i) = w(\text{left}(\overline{x}), j)$ for all $(x, i, \overline{x}, j) \in B$. Since $I \subseteq I^*$, we have $w \models (I, \leq, \varphi_I)$ and w is a solution of $\mathcal{B}$. ∎

Finally we show that Rule 4 preserves the convex condition. This is clear for Step 1; for the other steps we state lemmas.

LEMMA 12.3.25. *Step 2 preserves the convex chain condition with respect to the set F.*

Proof. The new bricks in Step 2 have the forms $(x_v, i, \overline{x_v}, i^*)$ and $(\overline{x_v}, i^*, x_v, i)$ for some $(x, i, \overline{x}, j) \in B$ with $\text{left}(x) < c = \text{left}(x_v) \leq i$. Since $(x, i, \overline{x}, j) \in B$ can be linked via a convex chain to some basis, it is enough to consider the following figure:

	x_v	i
	$\overline{x_v}$	i^*
	$\overline{x_v}$	i^*
	x_v	i
x		i
$\overline{x}$		j

∎

LEMMA 12.3.26. *Let $C = (b_1, \ldots, b_m)$ be a convex chain before Step 3 linking b_1 with b_m. Then after Step 3 there is a convex chain C' linking b'_1 with b'_m.*

Proof. Let us have a local look at the convex chain

$$C = (\ldots, (x, i, \overline{x}, j), (y, j, \overline{y}, k) \ldots).$$

By symmetry we may assume that $\text{left}(\overline{x}) \geq \text{left}(y)$. Pictorially this local part is then given by the following figure.

	x	i
	$\overline{x}$	j
y		j
$\overline{y}$		k

This is the situation before Step 3. After Step 3 let us denote the corresponding bricks by $(x', i', \overline{x}', j')$ and $(y', j'', \overline{y}', k')$. This yields the following figure.

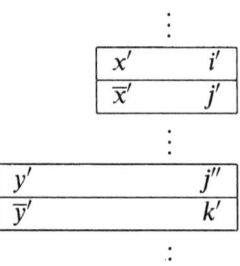

12.3. Boundary equations

The question is whether or not $j' = j''$. If $j' = j^*$ or $j'' = j$, then we have $j' = j''$, and the chain is not broken. Hence we have to consider the case $j' = j$ and $j'' = j^*$, only. This case is equivalent to

$$\text{left}(y) < c \le \text{left}(\bar{x}) \le j.$$

With the help of the brick $(x_v, j, \bar{x}_v, j^*)$, which was introduced in Step 2, we can repair the broken chain. We have

$$\text{left}(x_v) = c \le \text{left}(\bar{x}), \quad \text{left}'(y') < c^* = \text{left}(\bar{x}_v)$$

and we obtain the following figure:

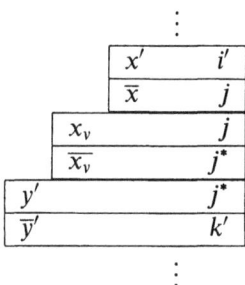

Doing this transformation wherever necessary we construct the convex chain C'. ∎

Note that C' constructed in the lemma above may contain many bricks of the forms $(x'_o, i^*, \bar{x}_o, i^*)$ and $(\bar{x}_o, i^*, x'_o, i^*)$. These bricks were canceled only later in Step 5. In fact their presence in the next lemma is very useful again.

LEMMA 12.3.27. *After Step 4 the convex chain condition is satisfied with respect to the set F'.*

Proof. Let b' be a brick after Step 3 and b the corresponding brick before Step 3. This brick b is linked before Step 3 via a convex chain to some basis $(x, i, \bar{x}, j)$ with $j \in F$. Lemma 12.3.26 states that after Step 3 the brick b' is linked via a convex chain to the corresponding brick $(x', i', \bar{x}', j')$. For $j < c$ we have $\text{left}(\bar{x}) < c$ and $j' = j^* \in F'$. Hence $(x', i', \bar{x}', j^*)$ is again a basis. For $j' = j$ we have $c \le j$ and therefore $j \in F'$. This also solves the case $j' = j$. The remaining case is $c \le j$ and $j' = j^*$. This means $\text{left}(\bar{x}) < c \le j$. By Step 2 there is a brick $(\bar{x}_v, j^*, x_v, j)$ and we have $\text{left}'(\bar{x}') < c^* = \text{left}(\bar{x}_v)$. We may put the brick $(x', i', \bar{x}', j^*)$ upon the basis $(\bar{x}_v, j^*, x_v, j)$. Since $j \in F \cap F'$, it is in fact a basis before and after Step 4. We obtain the following figure:

430   12. Makanin's Algorithm

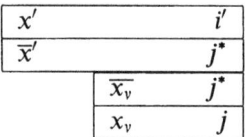

LEMMA 12.3.28. *Steps 5 and 6 preserve the convex chain condition with respect to the set F'.*

Proof. Step 5 is a special case of an application of Rule 2, likewise Step 6 is a special case of applications of Rule 3. In particular, the convex chain condition is preserved. ∎

The lemmas above yield the following proposition

PROPOSITION 12.3.29. *Rule 4 is upward and downward correct. It preserves the convex chain condition.*

EXAMPLE 12.3.30. Let $x_1 \cdots x_g = x_{g+1} \cdots x_d$ be a word equation, $1 \leq g < d$, such that the rational constraints are given by a mapping $\psi : \Omega \to S$. Let

$$\mathcal{B} = ((\Gamma, ^-), (I, \leq, \varphi_I), \text{left}, B)$$

be the result of the (log-space) reduction presented in Subsection 12.3.2. Recall that $(I, \leq, \varphi_I)$ represents the S-sequence

$$(\psi(x_1), \ldots, \psi(x_g), \psi(x_{g+1}), \ldots, \psi(x_d)).$$

We may assume that $(I, \leq, \varphi_I)$ is in its standard representation, $I = \{0, \ldots, d\}$. According to the reduction the set Γ contains two variables x_0 and $\overline{x_0}$ such that left$(x_0) = 0$, right$(x_0) = g =$ left$(\overline{x_0})$, and right$(\overline{x_0}) = d$. The set B contains at most d boundary equations (or bricks) among them there is the brick

x_0	g
$\overline{x_0}$	d

We have Card$I = d + 1$ and Card$\Gamma =$ Card$B \leq 2d$. If the word equation has a nonsingular solution satisfying the rational constraints, then $\exp(\mathcal{B}) \leq 2 \cdot e(c(S), d)$.

Rules 1 to 3 are not applicable to $\mathcal{B}$, but we can try Rule 4. Doing this we find

$$x_o = x_0, \quad l = 0, \quad c = g = r = l^*, \text{ and } c^* = g^* = r^* = d.$$

The set T of transport positions is $T = \{0, \ldots, g\}$.

12.3. Boundary equations

In Step 1 we have to choose some admissible extension of $(I, \leq, \varphi_I)$ by T at $[g, d]$. In general it is not clear that such an extension exists. Under the hypothesis that $x_1 \cdots x_g = x_{g+1} \cdots x_d$ has a nonsingular solution $\sigma: \Omega \to A^+$ with $\varphi \circ \sigma = \psi$ we can continue. Let $v = \sigma(x_1 \cdots x_g)$ and assume that v has minimal length among all solutions satisfying the rational constraints given by ψ. With the help of this word Step 1 can be completed: Define $w = vv$; then we have

$$w \models (\psi(x_1), \ldots, \psi(x_d)).$$

The set of positions of w is $\{0, \ldots, m, m+1, \ldots, 2m\}$ where $m = |v|$. The fact that w is a model of $(I, \leq, \varphi_I)$ is realized by an order-respecting injective mapping

$$\rho: \{0, \ldots, d\} \to \{0, \ldots, 2m\}.$$

Define $T^* = \{m + \rho(i) \mid 0 \leq i \leq g\}$ and $I^* = \rho(I) \cup T^*$. Since I^* is a subset of positions of w, this induces a linear suborder over S, which is denoted by $(I^*, \leq, \varphi_{I^*})$. We have $\operatorname{Card} I^* \leq d + g - 1$. After renaming we may assume $I^* = \{0, \ldots, d\} \cup T^*$ and $T^* = \{0^*, \ldots, g^*\}$ where $0^* = c = g$ and $c^* = g^* = d$. This completes Step 1 of Rule 4. Since in reality we usually do not know v, the choice of I^* is a nondeterministic guess!

The next steps in Rule 4 are deterministic. In Step 2 we introduce new variables x_v and $\overline{x_v}$ with $\operatorname{left}(x_v) = g = \operatorname{right}(x_v)$ and $\operatorname{left}(\overline{x_v}) = d = \operatorname{right}(\overline{x_v})$.

In Step 3 we transport the structure of the interval $[0, g]$ to $[0^*, g^*] = [g, d]$. If we still view I^* as a subset of positions of w, then this reflects a transport to the positions from the first to the second factor v in the word $w = vv$.

The definition of F' according to Step 4 is

$$F' = \{i \in I^* \mid g \leq i\}.$$

In Step 5 we cancel the bricks $(x_o, d, \overline{x_o}, d)$, $(\overline{x_o}, d, x_o, d)$ and the variables x_o, $\overline{x_o}$.

In Step 6 we replace I^* by $I' = F'$.

Rule 4 is finished. The cardinality of I' is bounded by d. Let $\mathcal{B}'$ denote the new system; then the word v is a solution, $v \models (I', \leq, \varphi_I(I'))$.

Since in the present situation $\operatorname{left}(x_v) = \operatorname{right}(x_v) = g$, Rule 1 is now applicable to $\mathcal{B}'$, it cancels the superfluous bricks $(x_v, g, \overline{x_v}, d)$, $(\overline{x_v}, d, x_v, g)$ and the variables x_v and $\overline{x_v}$. The new system after an application of Rule 1 is denoted by $\mathcal{B}'' = ((\Gamma_0'', \bar{}\,), (I_0'', \leq, \varphi_{I_0''}), \operatorname{left}_0'', B_0'')$. We have $\operatorname{Card} I'' \leq d$, $\operatorname{Card} \Gamma'' = \operatorname{Card} B'' \leq 2(d-1)$. It is now the word v which is a solution of $\mathcal{B}''$, hence $\exp(\mathcal{B}'') \leq \exp(v)$. Therefore we can choose $e(\mathcal{B}'') = e(c(S), d)$.

12.4. Proof of Theorem 12.3.10

12.4.1. Decidability

The proof of Theorem 12.3.10 is now a reduction to a reachability problem in some finite directed graph.

The instance is a system of boundary equations

$$\mathcal{B}_0 = ((\Gamma_0, \bar{\ }), (I_0, \leq, \varphi_{I_0}), \text{left}_0, B_0).$$

We may assume that $\mathcal{B}_0$ satisfies the assumptions made at the beginning of Subsection 12.3.4, because otherwise $\mathcal{B}_0$ is not solvable. For trivial reasons the system $\mathcal{B}_0$ satisfies the convex chain condition with respect to the set $F_0 = I_0$.

Let $2n_0 = \text{Card}\,\Gamma_0$ and $f_0 = \text{Card}\,F_0 = \text{Card}\,I_0$. In accordance with Proposition 12.3.11 choose a number $e(\mathcal{B}_0)$ such that either $\mathcal{B}_0$ is not solvable or $\exp(w) \leq e(\mathcal{B}_0)$ for some solution w of $\mathcal{B}_0$. Define an integer $\beta_{\max}$ by

$$\beta_{\max} = 2n_0 \cdot (2n_0 + f_0) \cdot (4n_0 \cdot (e(\mathcal{B}_0) + 1))^{n_0}.$$

Note that this value is defined just to fit Corollary 12.3.16 for a set of final indices having size at most $2n_0 + f_0$.

Now, consider a directed graph $\mathcal{G}$ (the search graph of Makanin's algorithm), which is defined as follows. The nodes of $\mathcal{G}$ are the systems of boundary equations $\mathcal{B} = ((\Gamma, \bar{\ }), (I, \leq, \varphi_I), \text{left}, B)$, where

$$\text{Card}\,\Gamma \leq 2n_0,$$
$$\text{Card}\,I \leq \frac{n_0 + 2}{2} \cdot \beta_{\max},$$
$$\text{Card}\,B \leq \beta_{\max}.$$

For systems $\mathcal{B}, \mathcal{B}' \in \mathcal{G}$ we define an arc from $\mathcal{B}$ to $\mathcal{B}'$ whenever, first, there is a transformation rule applicable to $\mathcal{B}$ and, second, $\mathcal{B}'$ is the result of the corresponding transformation. A system $\mathcal{B} \in \mathcal{G}$ with an empty set of variables is called a *terminal node*.

Clearly, $\mathcal{B}_0 \in \mathcal{G}$ and the search graph $\mathcal{G}$ has only finitely many nodes. Hence, it is enough to show the following claim: the system $\mathcal{B}_0$ has a solution if and only if there is a directed path in $\mathcal{G}$ from $\mathcal{B}_0$ to some terminal node.

The "if"-direction of the claim is trivial since all transformation rules are upward correct and since all terminal nodes are solvable by Lemma 12.3.4. For the "only if"-direction let $\mathcal{B}_0$ be solvable and let $w_0 \models (I_0, \leq, \varphi_{I_0})$ be a solution satisfying $\exp(w_0) \leq \exp(\mathcal{B}_0)$.

12.4. Proof of Theorem 12.3.10

Let $M \geq 0$ and assume that there is an inductively defined sequence of solvable systems $(\mathscr{B}_0, \mathscr{B}_1, \ldots, \mathscr{B}_M)$, $M \geq 0$, such that the following properties are satisfied for all $1 \leq k \leq M$:

- $\mathscr{B}_k = ((\Gamma_k, \bar{\ }), (I_k, \leq, \varphi_{I_k}), \text{left}_k, B_k)$ is the result of some transformation rule applied to $\mathscr{B}_{k-1}$,

- $\mathscr{B}_k$ has a solution $w_k \models (I_k, \leq, \varphi_{I_k})$ such that w_k is a suffix of w_{k-1},

- either $\operatorname{Card} \Gamma_k < \operatorname{Card} \Gamma_{k-1}$ or $|w_k| < |w_{k-1}|$,

- $\mathscr{B}_k$ satisfies the convex chain condition with respect to some subset $F_k \subseteq I_k$ with $\operatorname{Card} F_k + \operatorname{Card} \Gamma_k \leq 2n_0 + f_0$.

If $\mathscr{B}_M$ is a system of boundary equations without variables, then we stop. Otherwise, since $\mathscr{B}_M$ is solvable, a transformation rule is applicable. Consequently, the sequence can be continued by some solvable system $\mathscr{B}_{M+1}$ satisfying all the properties above. The third property however implies that $M \leq n_0 + |w_0|$. Hence, finally we must reach a system without variables. We may assume that this happens on reaching $\mathscr{B}_M$. Let us show that all $\mathscr{B}_k$ are nodes of $\mathscr{G}$ for all $0 \leq k \leq M$. This will imply the claim since then there is a directed path to $\mathscr{B}_M$, and $\mathscr{B}_M$ is a terminal node.

We have to verify $\operatorname{Card} \Gamma_k \leq 2n_0$, $\operatorname{Card} I_k \leq \frac{n_0+2}{2} \cdot \beta_{\max}$, and $\operatorname{Card} B_k \leq \beta_{\max}$.

The assertion $\operatorname{Card} \Gamma_k \leq 2n_0$ is trivial. The second property of the sequence implies $\exp(\mathscr{B}_k) \leq \exp(w_k) \leq \exp(w_0) \leq e(\mathscr{B}_0)$. By Corollary 12.3.16 and the fourth property we have $\operatorname{Card} B_k \leq \beta_{\max}$. The next lemma yields an invariant which will give the desired bound on the size of every I_k.

LEMMA 12.4.1. *For $0 \leq k \leq M$ define $\beta_k = \operatorname{Card}\{(x, i, \bar{x}, j) \in B_k \mid \text{left}_k(x) < i\}$. Then for all $1 \leq k \leq M$ we have*

$$2\operatorname{Card} I_k - \beta_k + \frac{\operatorname{Card} \Gamma_k}{2} \cdot \beta_{\max} \leq 2\operatorname{Card} I_{k-1} - \beta_{k-1} + \frac{\operatorname{Card} \Gamma_{k-1}}{2} \cdot \beta_{\max}.$$

Proof. Consider the rule which was applied to pass from $\mathscr{B}_{k-1}$ to $\mathscr{B}_k$. For Rule 1 or 2 we have

$$\operatorname{Card} \Gamma_k = \operatorname{Card} \Gamma_{k-1} - 2,$$
$$\operatorname{Card} I_k = \operatorname{Card} I_{k-1},$$
$$\beta_{k-1} - \beta_k \leq \beta_{\max}.$$

For Rule 3 we have

$$\text{Card } \Gamma_k = \text{Card } \Gamma_{k-1},$$
$$\text{Card } I_k = \text{Card } I_{k-1} - 1,$$
$$\beta_k = \beta_{k-1}.$$

Finally, for Rule 4 we have $\text{Card } \Gamma_k = \text{Card } \Gamma_{k-1}$ and Lemma 12.3.22 says

$$2\,\text{Card } I_k - \beta_k \leq 2\,\text{Card } I_{k-1} - \beta_{k-1}.$$

The assertion of the lemma follows. ∎

A consequence of Lemma 12.4.1 (and $\beta_k \leq \beta_{\max}$) is

$$2\,\text{Card } I_k \leq 2\,\text{Card } I_0 + (n_0 + 1)\beta_{\max} \text{ for all } 0 \leq k \leq M.$$

Since $\text{Card } I_0 \leq \frac{1}{2}\beta_{\max}$, we obtain $\text{Card } I_k \leq \frac{n_0+2}{2}\beta_{\max}$. Hence $\mathcal{B}_k \in \mathcal{G}$ for all $0 \leq k \leq M$. This proves Theorem 12.3.10, hence Makanin's result.

12.4.2. The complexity of Makanin's algorithm

Our estimations on the upper bounds of Makanin's algorithm are given by the size of the semigroup S and the maximal number of boundary equations $\beta_{\max}$ as defined in the preceding section.

A node $\mathcal{B} = ((\Gamma, \bar{}), (I, \leq, \varphi_I), \text{left}, B)$ of the search graph $\mathcal{G}$ is encoded as a binary string over $\{0, 1\}$ as follows. The code for $(\Gamma, \bar{})$ is simply the number n written in binary such that $|\Gamma| = 2n$. Thus, $O(\log n_0)$ bits are enough for this part. The linear order $(I, \leq, \varphi_I)$ is encoded by its underlying S-sequence. For this part $O(n_0 \beta_{\max} \log |S|)$ bits are used. The mapping $\text{left} : \Gamma \to I$ is encoded by using $O(n_0 \log(n_0 \beta_{\max}))$ bits. Finally, the set of bricks B can be encoded by using $O(\beta_{\max} \log(n_0 \beta_{\max}))$ bits. Note that $n_0 \leq \log \beta_{\max}$. It follows that there is effectively a constant $c \in \mathbb{N}$ such that every $\mathcal{B} \in \mathcal{G}$ can be described by a bit string of length equal to $c \cdot (\log |S| \cdot \beta_{\max} \cdot \log(\beta_{\max}))$. Up to some calculations performed over S this is the essential upper space bound for the nondeterministic procedure. It is at most exponential in the input size.

Consider the original question whether a given word equation $x_1 \cdots x_g = x_{g+1} \cdots x_d$, $1 \leq g < d$, with rational constraints has a solution. We may assume that each rational language $L_x \subseteq A^*$ is specified by a nondeterministic finite automaton with r_x states, $x \in \Omega$. Define $r = \sum_{x \in \Omega} r_x$; we are going to measure the complexity of Makanin's algorithm in terms of d and r. First, we choose a suitable semigroup S and a homomorphism $\varphi : A^+ \to S$. By Remark 12.1.12 we may assume that S satisfies Card $S \leq 2^{r^2}$ and $c(S) \leq r!$ By Theorem 12.2.1 choose a value $e(c(S), d) \in c(S) \cdot 2^{O(d)} \subseteq 2^{O(d+r \log r)}$ such that $e(c(S), d)$ is an upper bound for the exponent of periodicity. Transform the word equation (by a nondeterministic guess) into a system of boundary equations

$$\mathcal{B}_0 = ((\Gamma_0, \bar{\ }), (I_0, \leq, \varphi_{I_0}), \text{left}_0, B_0)$$

such that the word equation has a solution satisfying the rational constraints if and only if $\mathcal{B}_0$ is solvable. This is possible in such a way that, first, Card I_0, Card Γ_0, Card $B_0 \in O(d)$, and, second, if $\mathcal{B}_0$ is solvable, then

$$e(\mathcal{B}_0) \leq 2 \cdot e(c(S), d) \in 2^{O(d+r \log r)}.$$

More precisely, by Example 12.3.30 we can say Card $I_0 \leq d-1$, Card $\Gamma_0 = $ Card $B_0 \leq 2(d-1)$ and, if $\mathcal{B}_0$ is solvable, then $e(\mathcal{B}_0) \leq e(c(S), d)$.

Compute a value $\beta_{\max} \in 2^{\theta(d^2 + dr \log r)}$ such that $\beta_{\max}$ is an upper bound for the number of boundary equations of each node in the search graph $\mathcal{G}$. The value $\beta_{\max}$ can be taken large enough to perform all computations over the semigroup S and it can be taken small enough in order to solve the reachability problem in the search graph $\mathcal{G}$ in nondeterministic space NSPACE($2^{O(d^2 + dr \log r)}$). By Savitch's theorem (see e.g. Hopcroft and Ullman 1979) this is equal to DSPACE $\left(2^{O(d^2 + dr \log r)}\right)$. Hence we can state the final result of this chapter.

THEOREM 12.4.2. *The space requirement of Makanin's algorithm for word equations with rational constraints is at most exponential space. More precisely, we have the following complexity bound:*

$$\text{DSPACE}\left(2^{O(d^2 + dr \log r)}\right).$$

REMARK 12.4.3. The theorem above is an assertion on Makanin's algorithm and therefore it is no statement about the inherent complexity of the satisfiability problem for word equations. In fact, by Plandowski 1999b we know that the satisfiability problem for word equations can be solved in polynomial space; see the Notes below.

Problems

Section 12.1

12.1.1 Decide whether the solution *abbababbaabab* given in Example 12.1.1 is a nonsingular solution of minimal length.

12.1.2 Let $\Omega = \{x, y\}$ and $u, v \in A^*$ be words. Give necessary and sufficient conditions on u and v such that the equation $xu = vy$ is solvable.

12.1.3 Reduce the satisfiability problem of word equations to the satisfiability problem of systems of word equations where each variable occurs at most three times.

12.1.4 (Thierry Arnoux) Let $n > 0$. Consider the following word equation with rational constraints:

$$A = \{a, b\}, \quad \Omega = \{x_i \mid 0 \le i \le n\},$$
$$L_{x_0} = A^*, \quad L_{x_i} = aA^* \setminus (A^* b^i A^*) \text{ for } i > 0,$$
$$x_n ab^n x_n = a \, abx_0 ax_0 \, ab^2 x_1 abx_1 \cdots ab^n x_{n-1} ab^{n-1} x_{n-1}.$$

The denotational length of this equation is $d = n^2 + 5n + 4$. Show that there is only one solution satisfying the rational constraints, and that the length grows exponentially in n.

12.1.5 Show that the solvability of word equations becomes undecidable, if the constraints are allowed to be deterministic context-free languages.
(*Hint*: It is well known that the emptiness problem for intersections of deterministic context-free languages is undecidable.)

Section 12.2

12.2.1 Give a greedy algorithm to compute the p-stable normal form of a word $w \in A^*$. Modify the algorithm by pattern-matching techniques such that it runs in linear time.

12.2.2 Prove Propositions 12.1.7, 12.1.8, and 12.1.9. Show that the results remain true when there are rational constraints.

12.2.3 Show that the satisfiability problem of word equations without rational constraints is NP-hard.
(*Hint*: Show that the problem is NP-complete for systems of word equations, if there is exactly one constant, $A = \{a\}$. Use the fact that linear integer programming is NP-hard, even in unary notation.)

12.2.4 Let $L_x \subseteq A^*$ be a rational language. Describe the set of all solutions σ for an equation with only one unknown x under the constraint $\sigma(x) \in L_x$.

Section 12.3

12.3.1 An instance of a linear integer programming problem is given by an $m \times n$ matrix $D \in \mathbb{Z}^{m \times n}$ and a vector $c \in \mathbb{Z}^m$. Let $x \in \mathbb{N}^n$ be a minimal vector such that $Dx = c$. Assume that the sum over the squares of the coefficients in each row of D is in $O(1)$ and $\|c\| \in O(n^2)$. Show that there is a (small) constant c such that

$$\|x\| \in O(2^{cn}).$$

(*Hint*: The proof is a slight modification of the standard proof which shows that linear integer programming is NP-complete; see e.g. Hopcroft and Ullman 1979. Use Hadamard's inequality for an upper bound for the maximal absolute value over the determinants of square submatrices of D. Next, show that if $x \in \mathbb{N}^n$ is a minimal solution, then there is also a minimal solution $x' \in \mathbb{N}^n$ such that, first, the absolute value of at least one component can be bounded and, second, $\sum_{i=1}^n x_i \leq \sum_{i=1}^n x'_i$. Freeze by an additional equation one variable of x' to be a constant. Repeat the process until the homogeneous system $Dx = 0$ has only the trivial solution. Then apply Cramer's rule.

It should be noted that this method doesn't yield the best possible result. But it is good enough to establish that $e(d) \in 2^{O(d)}$, which was used in the proof of Theorem 12.2.1.)

Section 12.4

12.4.1 Consider the reduction in the proof of Lemma 12.3.12. Give an estimation for the length d of the word equation and thereby for an upper bound of $e(\mathscr{B})$. Define another reduction where the denotational length of the resulting word equation becomes smaller. This also improves the estimation for $e(\mathscr{B})$. Give a third estimation for $e(\mathscr{B})$ based on the techniques presented in Section 12.2.

(*Hint to the second part*: If a system contains two equations $x = x'$ and $xy = x'y'$, then the second one can be replaced by $y = y'$.)

12.4.2 According to Kościelski and Pacholski 1996, Theorem 4.8 the lower bound for $e(c(S), d)$ given in Example 12.2.3 can be refined. Consider the following equation with $k = 5$.

$$x_n a x_n b x_{n-1} b \cdots x_2 b x_1 = a x_n x_{n-1}^k b x_{n-2}^k b \cdots x_1^k b a^c.$$

Show that there is a unique solution. Derive from this solution a lower bound for the constant hidden in the notation $e(c(S), d) \in c(S) \cdot 2^{\Omega(d)}$. Why is $k = 5$ a good value? (*Hint*: Show first that $\sigma(x_i) \in a^*$ for all $1 \leq i \leq n$.)

Notes

A systematic study of equations in free monoids was initiated in the Russian school by A. A. Markov in the late 1950s in connection with Hilbert's Tenth Problem; see Hmelevskiĭ 1971, Makanin 1981. The connection is based on the fact that the set of matrices having nonnegative integer coefficients and determinant 1 form a free monoid inside the special linear group $SL_2(\mathbb{Z})$. Free generators are

$$a = \begin{pmatrix} 1 & 1 \\ 0 & 1 \end{pmatrix}, \qquad b = \begin{pmatrix} 1 & 0 \\ 1 & 1 \end{pmatrix}.$$

Let $L = R$ be a word equation over $\{a, b\}$ with $\Omega = \{x_1, \ldots, x_n\}$. Replace each variable $x_i \in \Omega$ by a matrix

$$\begin{pmatrix} \alpha_{i1} & \alpha_{i2} \\ \alpha_{i3} & \alpha_{i4} \end{pmatrix},$$

where α_{ij} denote variables over $\mathbb{N}$. Multiplying matrices corresponding to the words L and R yields an equation of the form

$$\begin{pmatrix} P_1 & P_2 \\ P_3 & P_4 \end{pmatrix} = \begin{pmatrix} Q_1 & Q_2 \\ Q_3 & Q_4 \end{pmatrix}.$$

The coefficients $P_1, \ldots, Q_4$ are polynomials in the α_{ij}. It is clear that the equation $L = R$ has a solution if and only if the following Diophantine system has a nonnegative solution:

$$\alpha_{i1}\alpha_{i4} - \alpha_{i2}\alpha_{i3} = 1, \quad i = 1, \ldots, n,$$
$$P_j = Q_j, \, j = 1, \ldots, 4.$$

The satisfiability problem of word equations becomes thereby a special instance of Hilbert's Tenth Problem: the satisfiability problem of Diophantine equations. The hope of Markov was to prove the unsolvability of Hilbert's Tenth Problem using this reduction. This hope failed; the unsolvability of Hilbert's Tenth Problem was shown in 1970 by Matiyasevich using an entirely different approach; see Matiyasevich 1993. The solvability of word equations is due to Makanin 1977. It is the subject of the present chapter. However, the reduction from word equations to

Diophantine equations is still very useful. For example it yields a simple proof of the Ehrenfeucht conjecture; see Chapter 13 for details.

A consequence of Makanin's result is the decidability of the existential theory of concatenation. The method is given in Subsection 12.1.6. The decidability of the existential theory is close to the borderline to undecidability. By Marchenkov 1982 and by Durnev 1995 it is known that the positive $\forall\exists^3$-theory of concatenation is unsolvable; see also the survey paper of Durnev 1997. Durnev 1974 and Büchi and Senger 1988 defined length predicates such that adding these predicates yields an undecidable existential theory of concatenation. The latter article also shows that equal length is not existentially definable by word equations. The decidability of word equations with an additional equal-length predicate is still an open problem. For more details about the expressibility of languages and relations by word equations see Karhumäki et al. 2000.

A few partial results about the decidability of word equations were known quite early. The fact that a disjunction of two equations can be replaced by a single equation was shown by Büchi in the mid-1960s, but his proof was published only much later; see Büchi and Senger 1986/7. In 1964 and 1967 Hmelevskiĭ found a positive solution for the cases with two and three variables respectively; see Hmelevskiĭ 1971. Other special cases were solved by Plotkin 1972 and Lentin 1972. In the case of two variables Charatonik and Pacholski 1993 analyzed Hmelevskiĭ's work by proving a polynomial-time bound on his algorithm. Their estimation about the degree of the polynomial was rather rough and extremely high. Ilie and Plandowski 2000 lowered the estimation on the degree down to 6 by giving a quadratic bound on the length of the minimal solution. In the case where each variable occurs at most twice, i.e., in the case of quadratic systems, there is a linear-time algorithm for the satisfiability problem, once the lengths for the solutions of variables are fixed and their binary representation is part of the input; see Robson and Diekert 1999. The linear space algorithm for this problem without fixing the lengths appeared in Matiyasevich 1968; the main result of that paper is however a quite different way to reduce word equations with additional conditions on equality of length of some words to Diophantine equations.

After Makanin presented his result in 1977 other questions became central. Makanin 1979 has shown that the rank of an equation is computable; see also Pécuchet 1981. The original article of Makanin is rather technical. Subsequently other presentations with various improvements were given; let us refer to Jaffar 1990, Schulz 1992a, 1993, Gutiérrez 1998b. The present chapter is along this line. A brief survey on equations in words can be found in the paper of Perrin 1989. Further material on equations in free monoids and, especially, on equations without constants is in the *Handbook of Formal Languages*; see Choffrut and Karhumäki 1997. There are two volumes in the Springer Lecture Notes series dedicated to word equations and related topics: Schulz 1992b and Abdulrab and Pécuchet 1993. Makanin's algorithm was implemented in 1987 at Rouen by Abdulrab; see Abdulrab and Pécuchet 1990.

The inherent complexity of the satisfiability problem of word equations with constants is not yet understood. The lower bound is NP-hardness, simply because linear integer programming (in unary notation) is a special instance and the latter problem is NP-hard. The satisfiability problem of word equations also remains NP-hard for a single quadratic equation. On the other hand, the exponent of periodicity is only linear for quadratic systems; see Diekert and Robson 1999, and it is believed that at least quadratic systems can be solved in NP. In fact, a conjecture of Plandowski and Rytter 1998 claims NP-completeness as the complexity bound for general word equations with constants. The development toward this conjecture over the past few years is somewhat unexpected since a first analysis of Makanin's algorithm done in the works of Jaffar and Schulz showed a 4-NEXPTIME result, only. By Kościelski and Pacholski 1996, Corollary 4.6 this went down to 3-NEXPTIME and then to 2-EXPSPACE during the work on the present chapter. The final version of this chapter uses another improvement due to Gutiérrez 1998a; see Lemma 12.3.14. It shows that the space requirement for Makanin's algorithm does not exceed EXPSPACE. This is the statement of Theorem 12.4.2. It is still the smallest space requirement for a full implementation of Makanin's algorithm.

However, in 1999 Plandowski found a new way for solving word equations which is independent of Makanin's work and which led to polynomial space. He obtained his result in two consecutive papers which both appeared in 1999: Plandowski 1999a showed that the satisfiability problem for word equations is in NEXPTIME. This is based on a result due to Plandowski and Rytter 1998, which shows that the minimal solution of a word equation is highly compressible in terms of Lempel–Ziv encodings and by a nontrivial combinatorial argument showing that the length of a minimal solution is at most doubly exponential in the denotational length of the equation. The NEXPTIME algorithm is to guess such an encoding of a minimal solution and to verify in deterministic polynomial time that the guess actually corresponds to some solution. Moreover, it is conjectured that the length of a minimal solution is at most exponential in the denotational length of the equation. If this is true, then the Lempel–Ziv encoding will have polynomial length and the satisfiability problem for word equation with constants will become NP-complete. So it might be that the trivial lower bound of NP-hardness already matches the upper bound, which is exactly the conjecture mentioned above. A counterexample to NP-completeness would imply the existence of a family of solvable word equations over a two-letter alphabet where the lengths of minimal solutions grow faster than an exponential function.

Plandowski 1999b showed that the satisfiability problem is in PSPACE. One important ingredient of his work is to use data compression in terms of exponential expressions. It is an interesting open problem whether the use of data compression could also lower the complexity bound in Makanin's method from exponential space down to polynomial space.

This chapter dealt with word equations having rational constraints. In this form the satisfiability problem becomes PSPACE-hard, simply because we may encode the intersection problem for rational languages, and the latter problem is known to be PSPACE-complete by Kozen 1977. Extending Plandowski's method Rytter has stated a PSPACE-completeness result for the satisfiability problem for word equations with rational constraints see Plandowski 1999b, Thm. 1.

Another surprising consequence of Plandowski's work is the dramatic improvement for solving equations over free groups. Let us first recall some background. Word equations in the framework of combinatorial group theory were introduced by Lyndon 1960; see Lyndon and Schupp 1977 for a standard reference. The corresponding notion of quadratic equation plays an important role in the classification of closed surfaces, and basic ideas how to solve quadratic equations go back to Nielsen 1918. The general satisfiability problem for equations with constants in free groups was shown to be decidable by Makanin 1982 and Makanin 1984. Razborov 1984 presented an algorithm which generates all solutions to a given equation. Let us also refer to the survey given by Razborov 1994. Makanin's method for group equations turned out to be even more complicated than in the word case, it is much more involved. Its complexity has been investigated by Kościelski and Pacholski 1998. The authors define the notion of abstract Makanin algorithm and they show that this abstract scheme is not primitive recursive. Therefore it was widely believed that the inherent complexity of the satisfiability problem in the group case is much higher than in the word case. However, there were hints that this was perhaps misleading: Using a result of Merzlyakov 1966 it has been shown by Makanin 1984 that the positive theory of equations in free groups is decidable whereas it was known to be undecidable in the word case. This contrast does not fit well to the assumption that the existential theory over free groups is much harder than over free monoids. And indeed, Gutiérrez 2000 achieved an extension of Plandowski's method such that it became applicable to the situation in free groups. As in the word case, the existential theory of equations in free groups is in PSPACE. Consequently, a nonprimitive recursive has been replaced by some polynomial space bounded algorithm. Finally, it became possible to cope with rational constraints in free groups. Diekert, Gutiérrez, and Hagenah 2001 have shown that the satisfiability problem for equations with rational constraints in free groups is PSPACE-complete, too.

An ongoing direction of research is to extend Makanin's result beyond free monoids and free groups. We briefly list some of the known results. For example, the main result of Diekert et al. 2001 is in fact a statement about free monoids with involution. This was used when the existential theory of equations in plain groups was shown to be decidable by Diekert and Lohrey 2001, thereby solving an open problem of Narendran and Otto 1997. According to Haring-Smith 1983 a group is called plain, if it is a free product of a finitely generated free group and finitely many finite groups. The class of plain groups is contained in the class of hyperbolic groups, which was introduced by Gromov 1987, and furthermore it is known that the existential theory of equations in torsion-free hyperbolic groups is decidable by Rips and Sela 1995. The intersection of plain groups and of torsion-free word hyperbolic groups is the class of free groups. It is strongly conjectured that the existential theory of equations is decidable in the whole class of hyperbolic groups.

On the other hand, if we move to free inverse semigroups, then the existential theory becomes undecidable; see Rozenblatt 1982, 1985. The situation improves if we wish to include partial commutation. Free partially commutative monoids are also called trace monoids. They are a tool to study some phenomena in concurrency theory; see Mazurkiewicz 1977 and Diekert and Rozenberg 1995 for a general reference. Matiyasevich 1997 has shown that the satisfiability problem of trace equations is decidable; see also Diekert, Matiyasevich, and Muscholl 1999. Diekert and Muscholl 2001 generalized this result to trace monoids with involution and the corresponding result in free partially commutative groups became a corollary. Free partially commutative groups are also calledgraph groups in mathematics; see e.g. Droms 1985, 1987a, 1987b.

The comments above show that the work on word equations led to remarkable results with progress all through the years and many connections to other fields. Makanin's deep insight in the combinatorics on words has been a basis and a source for an active area of research.

CHAPTER 13

Independent Systems of Equations

13.0. Introduction

The notion of a dimension, when available, is a powerful mathematical tool in proving finiteness conditions in combinatorics. An example of this is Eilenberg's equality theorem, which provides an optimal criterion for the equality of two rational series over a (skew) field. In this example a problem on words, i.e., on free semigroups, is first transformed into a problem on vector spaces, and then it is solved using the dimension property of those algebraic structures. One can raise the natural question: do sets of words possess dimension properties of some kind?

We approach this problem through systems of equations in semigroups. As a starting point we recall the well-known *defect theorem* (see Chapter 6), which states that if a set of n words satisfies a nontrivial relation, then these words can be expressed simultaneously as products of at most $n-1$ words. The defect effect can be seen as a weak dimension property of words. In order to analyze it further one can examine what happens when n words satisfy several independent relations, where independence is formalized as follows: a set E of relations on n words is *independent*, if E, viewed as a system of equations, does not contain a proper subset having the same solutions as E.

It is not difficult to see that a set of n words can satisfy two or more equations even in the case where the words cannot be expressed as products of fewer than $n-1$ words. This proposes an interesting problem: how many independent equations can a set of n words satisfy? In other words, how weak is the above dimension property? A partial answer is given in a fundamental result on words revealed in 1985, namely in the *compactness property* of free semigroups (known as Ehrenfeucht's conjecture): each independent set of equations of words is finite. This is the central theme of this chapter.

For finite systems of equations over a free semigroup, we show that there exist independent systems of $\Omega(n^3)$ equations in n variables. Moreover, in comparison with the defect theorem, we construct a set X of words with $\text{Card}(X) = n$ that satisfies $\Omega(n^2)$ independent relations, and still the words of X cannot be expressed as products of less than $n - 1$ words. That is, these relations cause the same defect effect as a single nontrivial relation.

Our central problem generalizes, in a natural way, to all semigroups S. In this setting we consider infinite systems of equations, i.e., pairs of words from a free semigroup of variables, and ask whether, for each such system, there exists a finite subsystem of equations that is equivalent to the given one, that is, whether the subsystem has exactly the same set of solutions in S as the original one.

The compactness property does not hold in all semigroups, an example being the bicyclic semigroup, while it does hold in some other than the free semigroups. In general, no characterization of semigroups satisfying the compactness property is known. This, however, changes if we consider varieties of semigroups or monoids. There is a nontrivial characterization in terms of ascending chains of congruences for a variety to satisfy the compactness property.

13.1. Sets and equations

Let A and Ξ be two finite sets, where the elements of Ξ are called *variables*. Let $(u, v) \in \Xi^+ \times \Xi^+$ be an equation, usually written as $u = v$. Its *solution* in the free semigroup A^+ (resp. in a semigroup S) is a morphism $\alpha: \Xi^+ \to A^+$ (resp. $\alpha: \Xi^+ \to S$) that satisfies $\alpha(u) = \alpha(v)$. Solutions of an (infinite) system of equations E are defined in the obvious way. Let $\text{Sol}(E)$ be the set of all solutions of a system E of equations. Two systems E and E' of equations are said to be *equivalent* if they have the same solutions, $\text{Sol}(E) = \text{Sol}(E')$. Further, a system E of equations is *independent*, if it is not equivalent to any of its proper subsystems.

For simplicity, we often write $x = w$ instead of $\alpha(x) = w$, when $x \in \Xi$ is a variable and α a morphism.

The *combinatorial rank* of a finite subset $X \subseteq A^+$ of words is defined by
$$r_c(X) = \min\{\text{Card}(Y) \mid X \subseteq Y^+\}.$$
Clearly, we have $r_c(X) \le \max\{\text{Card}(X), \text{Card}(A)\}$.

13.1. Sets and equations

EXAMPLE 13.1.1. Consider the following three systems of equations:

$$E_1: \quad xy = zx;$$
$$E_2: \quad xy^i = z^i x, \quad i = 1, 2, \ldots;$$
$$E_3: \quad xyz = zyx, \quad xy^2z = zy^2x.$$

Here E_1 and E_2 are equivalent, since $xy = zx$ implies, for $i \geq 1$,

$$xy^{i+1} = (xy)y^i = (zx)y^i = zxy^i,$$

which gives, by induction, that $xy^{i+1} = z^{i+1}x$. The system E_3 is independent, since $x = a$, $y = b$ and $z = aba$ is a solution of the first equation of E_3 that is not a solution of the second one, and $x = a$, $y = b$ and $z = abba$ is a solution of the second equation that is not a solution of the first one.

The morphisms $\alpha: \Xi^+ \to A^+$ are, in a natural way, in a one-to-one correspondence with the finite ordered subsets $X \subseteq A^+$ with $\operatorname{Card}(X) = \operatorname{Card}(\Xi)$. We exploit this by attaching to a finite ordered subset $X = \{w_1, w_2, \ldots, w_n\}$ a set $\Xi_X = \{x_1, x_2, \ldots, x_n\}$ of variables and a morphism $\alpha_X: \Xi_X^+ \to X^+$, for which $\alpha_X(x_i) = w_i$ for all i. Such a surjective morphism is a *presentation* of the semigroup X^+. Now we can view the set X of words as a solution of an equation $u = v$ over Ξ_X, if the morphism α_X is its solution. Further, the *set of relations* satisfied by X is defined as the kernel of the morphism α_X,

$$E(X) = \{(u, v) \in \Xi_X^+ \times \Xi_X^+ \mid \alpha_X(u) = \alpha_X(v)\}. \tag{13.1.1}$$

Clearly, $E(X)$ is a congruence of the free semigroup Ξ_X^+, that is, it is an equivalence relation and a subsemigroup of the direct product $\Xi_X^+ \times \Xi_X^+$.

A subsemigroup X^+ of a free semigroup A^+ is cancellative, and so is the semigroup $E(X) \subseteq \Xi_X^+ \times \Xi_X^+$. Therefore if X^+ satisfies the relations (u_1, v_1) and (u_1u_2, v_1v_2) (or (u_2u_1, v_2v_1)), it also satisfies (u_2, v_2). We say that a relation (u, v) is *reduced*, if it belongs to the minimal generating set $E_{\text{red}}(X)$ of the semigroup $E(X)$:

$$E_{\text{red}}(X) = (E(X) \setminus E(X)^2) \setminus \iota_{\Xi_X},$$

where $\iota_{\Xi_X} = \{(x, x) \mid x \in \Xi_X\}$ is the identity relation of Ξ_X. Clearly, if $(u, v) \in E_{\text{red}}(X)$, then the first variables in u and v are different, and so are the last ones. It is obvious that, as systems of equations over Ξ_X, $E(X)$ and $E_{\text{red}}(X)$ are equivalent. However, $E_{\text{red}}(X)$ need not be a minimal equivalent subsystem of $E(X)$.

EXAMPLE 13.1.2. Let $X = \{a, ab, ba\} \subseteq \{a,b\}^+$. Then it satisfies the relation (x_1x_3, x_2x_1), and, by the previous example, it also satisfies the relations $(x_1x_3^i, x_2^i x_1)$ for all $i \in \mathbb{N}$. It is not difficult to see that the latter are exactly the reduced relations satisfied by X^+, that is, $E_{\text{red}}(X) = \{(x_1x_3^i, x_2^i x_1) \mid i \in \mathbb{N}\}$. Indeed, the validity of this can be concluded from the finite automaton given in Figure 13.1 that seeks through all double X-factorizations of words in X^+.

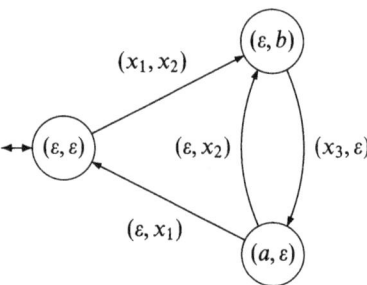

Figure 13.1. A finite automaton for the relations of X.

In general, such an automaton can be constructed as follows. The states of the automaton form a subset of the pairs $(u, \varepsilon), (\varepsilon, u)$, where u is a proper suffix of a word in X, and there is a transition

$$(u_1, v_1) \xrightarrow{(x,y)} (u_2, v_2)$$

if $u_1 \alpha_X(x) v_2 = v_1 \alpha_X(y) u_2$ for $x, y \in \Xi_X \cup \{\varepsilon\}$. The initial and the final state of the automaton is $(\varepsilon, \varepsilon)$. In our example, the automaton has been simplified after the construction.

The proof of the next theorem is left to the reader as an exercise. Example 13.1.2 serves as an illustration.

THEOREM 13.1.3. *The sets $E_{\text{red}}(X)$ and $E(X)$ for a finite set $X \subseteq A^+$ are rational relations.*

Thus the sets $E_{\text{red}}(X)$ and $E(X)$ are rather easy to compute. However, to compute $\text{Sol}(u, v)$ for a given equation $u = v$ may be very demanding; see Chapter 12.

13.2. The compactness property

In this section we shall prove a compactness result, which states that every system of equations in a free semigroup over a finite set of variables has an equivalent finite subsystem. In other words, each independent system of equations in a free semigroup over a finite set Ξ of variables is finite. This can be viewed, beside the defect theorem, as another positive dimension property of words.

13.2.1. The proof

In the proof of the compactness result we need Hilbert's basis theorem. For this, let $\mathbb{Z}[X]$ be the ring of polynomials with integer coefficients in a finitely many (commuting) variables X.

THEOREM 13.2.1 (Hilbert's basis theorem). *Let P_i, for $i \geq 1$, be polynomials in $\mathbb{Z}[X]$. There exists a finite subset $P_1, P_2, \ldots, P_t$ of these polynomials such that every P_i can be expressed as a linear combination*

$$P_i = P_1 Q_{i_1} + P_2 Q_{i_2} + \cdots + P_t Q_{i_t},$$

where $Q_{i_j} \in \mathbb{Z}[X]$.

We use Theorem 13.2.1 in the following form. Let $\{P_i = 0 \mid i \geq 1\}$ be a system of polynomial equations, where $P_i \in \mathbb{Z}[X]$. There exists a finite subsystem $\{P_i = 0 \mid i = 1, 2, \ldots, t\}$, every solution of which is a solution of the whole system of equations.

THEOREM 13.2.2. *Every independent system of equations in a free semigroup A^+ over a finite set Ξ of variables is finite.*

Proof. To take advantage of Hilbert's basis theorem, we transform each word equation $u = v$ into a polynomial equation in $\mathbb{Z}[X]$. For this, we use noncommuting matrices of integer polynomials.

Let $\Xi = \{x_1, x_2, \ldots, x_k\}$ be a fixed set of word variables. Since each A^+ can be embedded into B^+, where $B = \{a, b\}$, it is sufficient to solve equations over Ξ in B^+.

Consider the multiplicative semigroup $\mathbb{Z}^{2 \times 2}$ of 2×2 matrices over $\mathbb{Z}$. Let $\mathbb{F}$ be the subsemigroup of $\mathbb{Z}^{2 \times 2}$ of all matrices of the form

$$M = \begin{pmatrix} 2^m & n \\ 0 & 1 \end{pmatrix}, \qquad \text{where } 0 \leq n < 2^m. \tag{13.2.1}$$

CLAIM 1. *The semigroup $\mathbb{F}$ is free. In fact, the morphism $\mu:\{a,b\}^+ \to \mathbb{Z}^{2\times 2}$ defined by*

$$\mu(a) = \begin{pmatrix} 2 & 0 \\ 0 & 1 \end{pmatrix} \quad \text{and} \quad \mu(b) = \begin{pmatrix} 2 & 1 \\ 0 & 1 \end{pmatrix}$$

is injective, and onto $\mathbb{F}$.

We observe that $M_a = \mu(a)$ and $M_b = \mu(b)$ have the inverses

$$M_a^{-1} = \begin{pmatrix} 1/2 & 0 \\ 0 & 1 \end{pmatrix} \quad \text{and} \quad M_b^{-1} = \begin{pmatrix} 1/2 & -1/2 \\ 0 & 1 \end{pmatrix}$$

in rational entries. Further, for a matrix M in (13.2.1),

$$M_a^{-1} M = \begin{pmatrix} 2^{m-1} & n/2 \\ 0 & 1 \end{pmatrix} \quad \text{and} \quad M_b^{-1} M = \begin{pmatrix} 2^{m-1} & (n-1)/2 \\ 0 & 1 \end{pmatrix}.$$

This yields that $M_a^{-1} M \in \mathbb{F}$ if and only if n is even, and $M_b^{-1} M \in \mathbb{F}$ if and only if n is odd. Consequently, M_a and M_b generate $\mathbb{F}$, and each $M \in \mathbb{F}$ has a unique factorization $M = M_{a_1} M_{a_2} \ldots M_{a_k}$ in terms of the matrices M_a and M_b. This shows that $\mathbb{F}$ is freely generated by M_a and M_b, proving Claim 1.

Next we introduce, for each $x_i \in \Xi$, two commuting *integer variables* y_i and z_i, and denote $X = \{y_i, z_i \mid i = 1, 2, \ldots, k\}$. Further, for each i, let

$$M_i = \begin{pmatrix} y_i & z_i \\ 0 & 1 \end{pmatrix},$$

and let $\mathbb{M}(\Xi)$ be the subsemigroup of the multiplicative semigroup $\mathbb{Z}[X]^{2\times 2}$ generated by the matrices $M_1, M_2, \ldots, M_k$.

CLAIM 2. *The semigroup $\mathbb{M}(\Xi)$ is free. In fact, the morphism $\varphi: \Xi^+ \to \mathbb{M}(\Xi)$, defined by $\varphi(x_i) = M_i$, is an isomorphism.*

For this, consider any $M = M_{i_1} M_{i_2} \ldots M_{i_t} \in \mathbb{M}(\Xi)$. Then

$$M = \begin{pmatrix} y_{i_1} y_{i_2} \cdots y_{i_t} & m_{12} \\ 0 & 1 \end{pmatrix},$$

where the right upper corner entry $m_{12} = z_{i_1} + y_{i_1} z_{i_2} + \cdots + y_{i_1} y_{i_2} \cdots y_{i_{t-1}} z_{i_t}$. From the entry m_{12} we conclude, despite the fact that the unknowns y_j and z_j commute, that it determines the sequence $i_1, i_2, \ldots, i_t$ uniquely, and Claim 2 follows.

13.2. The compactness property

For each morphism $\alpha: \Xi^+ \to B^+$, let $\tilde{\alpha} = \mu\alpha\varphi^{-1}: \mathbb{M}(\Xi) \to \mathbb{Z}^{2\times 2}$ so that the following diagram commutes:

$$\begin{array}{ccc} \Xi^+ & \xrightarrow{\alpha} & B^+ \\ \varphi \downarrow & & \downarrow \mu \\ \mathbb{M}(\Xi) & \xrightarrow{\tilde{\alpha}} & \mathbb{Z}^{2\times 2} \end{array}$$

Denote for each word $w \in \Xi^+$,

$$\varphi(w) = \begin{pmatrix} P_1(w) & P_2(w) \\ 0 & 1 \end{pmatrix} \in \mathbb{Z}[X]^{2\times 2}.$$

Finally, let $\hat{\alpha}: \mathbb{Z}[X] \to \mathbb{Z}$ be the ring morphism which is defined by

$$\tilde{\alpha}(M_i) = \begin{pmatrix} \hat{\alpha}(y_i) & \hat{\alpha}(z_i) \\ 0 & 1 \end{pmatrix}.$$

It follows that

$$\tilde{\alpha}\varphi(w) = \begin{pmatrix} \hat{\alpha}(P_1(w)) & \hat{\alpha}(P_2(w)) \\ 0 & 1 \end{pmatrix}.$$

Now, a morphism $\alpha: \Xi^+ \to B^+$ is a solution of an equation $u = v$ over Ξ if and only if $\tilde{\alpha}$ is a solution of the matrix equation $\varphi(u) = \varphi(v)$. We conclude that α is a solution of $u = v$ if and only if the corresponding ring morphism $\hat{\alpha}: \mathbb{Z}[X] \to \mathbb{Z}$ is a solution of the system

$$\begin{cases} P_1(u) = P_1(v), \\ P_2(u) = P_2(v) \end{cases}$$

of integer equations, or equivalently, of the equation $e(u,v) = 0$, where

$$e(u,v) = (P_1(u) - P_1(v))^2 + (P_2(u) - P_2(v))^2.$$

Let then $E = \{(u_i, v_i) \mid i \geq 1\}$ be a system of word equations, and denote by $J = \{e(u_i, v_i) \mid i \geq 1\}$ the corresponding system of polynomial equations. By Hilbert's basis theorem, J has an equivalent finite subsystem, say $J_0 = \{e(u_i, v_i) \mid i = 1, 2, \ldots, t\}$. By the above, if α is a solution of the finite subsystem $E_0 = \{(u_i, v_i) \mid i = 1, 2, \ldots, t\}$, then the corresponding $\hat{\alpha}$ is a solution of J_0, and thus also of J, which gives that α is a solution of E. This shows that E_0 is equivalent to E, proving the theorem. ∎

The above deserves a special comment on the proof method. We have proved there a combinatorial result for semigroups (with one operation) concerning *noncommuting* variables by reducing it to a result for rings (with two operations) concerning *commuting* variables.

13.2.2. Applications

We give two applications of the above compactness theorem. The first of these solves the isomorphism problem for finitely generated subsemigroups of free semigroups. For the proof of this we need an effective restriction of Theorem 13.2.2; see Problem 13.2.3.

LEMMA 13.2.3. *Let $R \subseteq \Xi^+ \times \Xi^+$ be a rational relation (considered as a system of equations). Then an equivalent finite subsystem $R_0 \subseteq R$ can be effectively found.*

Now we state our first application.

THEOREM 13.2.4. *Let $X, Y \subseteq A^+$ be two finite sets of words. It is decidable whether the semigroups X^+ and Y^+ are isomorphic.*

Proof. Since computing the base of X^+ is effective, we may assume that X and Y are the bases of their semigroups X^+ and Y^+. Further, we can suppose that $\text{Card}(X) = n = \text{Card}(Y)$; otherwise X^+ and Y^+ are not isomorphic. Let Ξ be a set of variables with $\text{Card}(\Xi) = n$.

We consider the bijections $\varphi: X \to Y$. Note that there are only finitely many of these, since X and Y are finite. We need to decide whether the (unique) extension $\varphi: X^+ \to Y^+$ of φ is an isomorphism. For this, consider the representing morphisms, $\alpha_X: \Xi^+ \to X^+$ and $\alpha_Y = \varphi\alpha_X: \Xi^+ \to Y^+$ for X and Y. Now φ is an isomorphism if and only if it is injective, and this holds if and only if $E(X) = E(Y)$ as sets of relations defined in (13.1.1). By Theorem 13.1.3, $E(X)$ and $E(Y)$ are rational relations. Despite the fact that the equivalence problem for rational relation is *undecidable*, we can solve the present problem. Considering $E(X)$ and $E(Y)$ as systems of equations, Theorem 13.2.2 states that they have finite equivalent subsystems, and by Lemma 13.2.3, such subsystems, say $E_0(X) \subseteq E(X)$ and $E_0(Y) \subseteq E(Y)$, can be effectively constructed. It is now a simple task to check whether α_X is a solution of $E_0(Y)$. If the answer is positive, then α_X is a solution of $E(Y)$, and $E(X) = \ker(\alpha_X) \subseteq E(Y)$. If the answer is negative, then clearly $E(X) \neq E(Y)$. Similarly, if α_Y is a solution of $E_0(X)$, then $E(Y) \subseteq E(X)$. This proves the theorem. ∎

We emphasize that the above theorem reveals one of the rare cases where the isomorphism problem for infinite (finitely generated) semigroups is known to be decidable. Indeed, already for multiplicative matrix semigroups with nonnegative entries, it is undecidable.

As a second application of Theorem 13.2.2, we prove a decidability result for monoids of endomorphisms. By an *endomorphism* we mean a morphism $\alpha: A^+ \to A^+$ of a free semigroup A^+ into itself.

13.2. The compactness property

We first study test sets of languages, which constitute the original language-theoretic formulation of the compactness property of Ehrenfeucht's conjecture. Let $L \subseteq A^+$ be a language, i.e., a subset of A^+. We say that a subset $T \subseteq L$ is a *test set* for L, if for all morphisms $\alpha, \beta : A^+ \to B^+$

$$\alpha(u) = \beta(u) \text{ for all } u \in T \iff \alpha(u) = \beta(u) \text{ for all } u \in L.$$

From Theorem 13.2.2 we obtain

THEOREM 13.2.5. *Every set $L \subseteq A^+$ possesses a finite test set.*

Proof. For any alphabet B, let $\bar{B} = \{\bar{a} \mid a \in B\}$ denote its copy, and write for all words $u = a_1 a_2 \ldots a_k \in B^+$, $\bar{u} = \bar{a}_1 \bar{a}_2 \ldots \bar{a}_k \in \bar{B}^+$. Here the function $u \mapsto \tilde{u}$ is an isomorphism $B^+ \to \bar{B}^+$.

Let $L \subseteq B^+$, and define $E = \{(u, \tilde{u}) \mid u \in L\}$. By Theorem 13.2.2, when E is considered as a system of equations, it has an equivalent finite subsystem E_0. Let $T = \{u \mid (u, \tilde{u}) \in E_0\}$. Certainly, T is a finite subset of L. Consider morphisms $\alpha, \beta : B^+ \to A^+$ satisfying $\alpha(u) = \beta(u)$ for all $u \in T$. Let $\Xi = B \cup \bar{B}$ be our set of variables, and define a morphism $\gamma : \Xi^+ \to A^+$ such that $\gamma(x) = \alpha(x)$ and $\gamma(\tilde{x}) = \beta(x)$ for all $x \in B$. Now $\gamma(u) = \gamma(\tilde{u})$ for all $u \in T$, and hence γ is a solution of the finite system E_0, and therefore γ is a solution of E as well. Consequently, $\alpha(u) = \beta(u)$ for all $u \in L$, which shows that T is a finite test set of L. ∎

Using Makanin's result (see Chapter 12), one can prove (see Problem 13.2.4)

LEMMA 13.2.6. *Let $L_1, L_2 \subseteq A^+$ be finite sets such that $L_1 \subseteq L_2$. It is decidable whether L_1 is a test set of L_2.*

With the help of the above lemma, we now show

THEOREM 13.2.7. *Let H be a finitely generated monoid of endomorphisms of a free semigroup A^+, and let w be a word in A^+. It is decidable, for given endomorphisms α, β of A^*, whether $\alpha \gamma(w) = \beta \gamma(w)$ for all $\gamma \in H$.*

Proof. Let H be generated by $G = \{\gamma_1, \gamma_2, \ldots, \gamma_n\}$, and denote $L(H, w) = \{\gamma(w) \mid \gamma \in H\}$. We consider the *level sets*

$$L_p(H, w) = \{\gamma(w) \mid \gamma \in G^p\},$$

where G^p denotes the set of all compositions $\gamma_{i_j} \gamma_{i_{j-1}} \ldots \gamma_{i_1}$ of at most p morphisms from G. By Theorem 13.2.5, the set $L(H, w)$ has a finite test set, and, consequently, there exists an index p such that $L_p(H, w)$ is a test set for $L(H, w)$.

It follows that there exists an index q, and hence the least index q, such that $L_q(H, w)$ is a test set for $L_{q+1}(H, w)$ as well. We claim that $L_q(H, w)$ is a test set of $L(H, w)$. This is seen inductively as follows. Assume that for morphisms α and β, $\alpha\gamma(w) = \beta\gamma(w)$ for all $\gamma \in H^{q+i}$. Let $\gamma' \in H^i$, $\gamma_j \in G$ and $\gamma'' \in H^q$, and denote $\kappa = \gamma'\gamma_j\gamma'' \in H^{q+i+1}$. Now, by the assumption on α and β,

$$(\alpha\gamma')\gamma''(w) = \alpha\gamma'\gamma''(w) = \beta\gamma'\gamma''(w) = (\beta\gamma')\gamma''(w),$$

and, by the assumption on H^q, also $(\alpha\gamma')\gamma_j\gamma''(w) = (\beta\gamma')\gamma_j\gamma''(w)$, which shows that $\alpha\kappa(w) = \beta\kappa(w)$ as required.

Finally, by Lemma 13.2.6, the index q can be effectively found. Indeed, it is sufficient to check, for $i = 1, 2, \ldots$, whether $L_i(H, w)$ is a test set of $L_{i+1}(H, w)$. The existence of q guarantees that this checking will end in a positive answer. The test set $L_q(H, w)$ is finite, and therefore the decidability claim follows. ∎

In the *DT0L problem* we are given a word $w \in A^+$, two monoids H_1 and H_2 of endomorphisms of A^* with equally many generators $\{\alpha_1, \alpha_2, \ldots, \alpha_n\}$ and $\{\beta_1, \beta_2, \ldots, \beta_n\}$, respectively. We ask whether for all sequences $i_1, i_2, \ldots, i_k$ of indices,

$$\alpha_{i_k} \ldots \alpha_{i_1}(w) = \beta_{i_k} \ldots \beta_{i_1}(w). \tag{13.2.2}$$

This problem reduces to Theorem 13.2.7, since, as is easy to see, (13.2.2) holds for all sequences of indices if and only if $\alpha_i\alpha(w) = \beta_i\alpha(w)$ for all i and $\alpha \in H_1$.

COROLLARY 13.2.8. *The DT0L problem is decidable.*

Our proof for the DT0L problem is short. However, it is based on two deep results, namely Makanin's algorithm and the compactness property. Amazingly this is the only known proof for this problem, although its special case, the celebrated *D0L problem*, where H_1 and H_2 are both generated by a single morphism, has several different proofs.

Corollary 13.2.8 should be compared with the *DT0L language equivalence problem*, where we are given a word w, and two finitely generated monoids H_1 and H_2 of endomorphisms of a free monoid A^*, and we ask whether for all $\alpha \in H_1$ there exists $\beta \in H_2$ such that $\alpha(w) = \beta(w)$. This problem is known to be undecidable.

13.3. Independence of finite systems of equations

As we have mentioned, both the defect theorem and the compactness property formalize – although from a different perspective – a weak dimension property of words. A link between these two important results is found by considering independent systems of equations. At the same time this allows us to analyze how weak these dimension properties are.

From the point of view of the compactness property, it is natural to ask how large a (necessarily finite) independent system of equations in n variables can be. From the point of view of (extensions of) the defect theorem, a natural question is to ask whether two or more 'different' relations force a larger 'defect effect' than a single equation, or how many different relations still allow nonperiodic solutions, i.e., solutions of combinatorial rank at least 2. Here we formalize the notion of *different relations* as the independence of a system of equations and the *defect effect* of a system E of equations in n variables as the number $n - t$, where t is the maximal combinatorial rank of a solution of E.

The goal of this section is to search for answers to these questions. In other words, we want to construct as large as possible independent systems of equations in n variables in general, or requiring in addition that they still possess a solution of a certain rank. Recall that the set of equations was defined to be *independent*, if it was not equivalent to any of its proper subsets.

At this point we emphasize that for any solution $X = \alpha(\Xi)$ of a system E of equations, the (combinatorial) rank of X is a property of X, i.e., of *a* solution of E. The independence of E, in turn, is a property of *all* solutions of E. Therefore attempts to fulfill our goal face a problem of relating a particular solution of E to all solutions of it.

Using Makanin's algorithm (see Chapter 12, Theorem 13.1.3 and Problem 13.2.3), it is not difficult to conclude that the following holds.

THEOREM 13.3.1. *For each finite set $X \subseteq A^+$, one can effectively find an independent subset $E_I(X)$ of $E(X)$ which is equivalent to $E(X)$.*

From the point of view of our goal, Theorem 13.3.1 is not really helpful. Indeed, the set $E_I(X)$ need not be unique, and, moreover, to find $E_I(X)$ in practice is very difficult, as is illustrated by the challenging Problem 13.3.1.

We now define formally the central notion of this section. For this, let Ξ be a set of variables and denote $n = \text{Card}(\Xi)$. For $t \leq n$, define

$$D_t(n) = \sup\{\text{Card}(E) \mid E \text{ is an independent system over } \Xi \text{ having a solution of combinatorial rank at least } n - t\}.$$

Now, for all t with $1 \le t \le n-1$, $D_1(n) \le D_t(n) \le D_{n-2}(n)$, and therefore the two most natural choices for the parameter t are the values $t = 1$ and $t = n - 2$. The former corresponds to the case where the defect effect is minimal, that is, equal to 1, while the case $t = n - 2$ corresponds to the case where nonperiodic solutions are required to exist. The topic of this section is to search for lower bounds of the value $D_t(n)$.

13.3.1. In free semigroups

We start with an example.

EXAMPLE 13.3.2. Let $\Xi = \{x, y\} \cup \{p_i, q_i, z_i \mid i = 1, 2, \ldots, n\}$ be a set of variables, and let E be the following system of equations over Ξ:

$$E: \quad xp_j z_k q_j y = y p_j z_k q_j x \quad \text{for } j, k = 1, 2, \ldots, n.$$

Then $\text{Card}(E) = n^2$ and $\text{Card}(\Xi) = 3n + 2$. We claim that
(i) E has a solution of combinatorial rank $3n + 1$, and
(ii) E is independent.

Part (i) is easy to verify. Indeed, choose $x = y$, which makes the equations of E trivial, so that a required solution can be found over the free semigroup having $3n + 1$ generators.

The essential part is to prove (ii). For this, we have to show that, for each pair (j, k), there exists a solution of the system

$$E(j, k) = E \setminus \{xp_j z_k q_j y = y p_j z_k q_j x\}$$

which is not a solution of E. Here is such a solution:

$$\begin{aligned}
x &= b^2 ab, \\
y &= b, \\
p_t &= \begin{cases} ba & \text{if } t = j, \\ bab & \text{otherwise}, \end{cases} \\
z_\ell &= \begin{cases} bab^2 & \text{if } \ell = k, \\ b & \text{otherwise}, \end{cases} \\
q_t &= \begin{cases} ba & \text{if } t = j, \\ a & \text{otherwise}. \end{cases}
\end{aligned} \qquad (13.3.1)$$

How to find (13.3.1) is not obvious, but to verify that it is a required one is easy. Indeed, we compute for $t = j$ and $\ell = k$,

$$xp_j z_k q_j y = b^2 ab \cdot ba \cdots \ne b \cdot ba \cdot bab^2 \cdots = y p_j z_k q_j x.$$

13.3. Independence of finite systems of equations

Therefore (13.3.1) is not a solution of E. For the remaining cases, we compute as follows.

$t \neq j, \ell \neq k$: $b^2ab \cdot bab \cdot b \cdot a \cdot b = (bba)^3 b = b \cdot bab \cdot b \cdot a \cdot b^2 ab$.
$t \neq j, \ell = k$: $b^2ab \cdot bab \cdot bab^2 \cdot a \cdot b = (bba)^4 b = b \cdot bab \cdot bab^2 \cdot a \cdot b^2 ab$.
$t = j, \ell \neq k$: $b^2ab \cdot ba \cdot b \cdot ba \cdot b = (bba)^3 b = b \cdot ba \cdot b \cdot ba \cdot b^2 ab$.

Indeed, (13.3.1) is a solution of $E(j,k)$.

The above example yields

THEOREM 13.3.3. (i) $D_1(n) = \Omega(n^2)$ in A^+ with $\mathrm{Card}(A) = \infty$.
(ii) $D_{n-2}(n) = \Omega(n^3)$ in A^+ with $\mathrm{Card}(A) \geq 2$.

Proof. Part (i) of the claim follows directly from Example 13.3.2. Note that here we have to solve the equations over an infinitely generated free semigroup.

To prove part (ii), we modify Example 13.3.2 slightly by introducing n copies of x and y, say x_i and y_i for $i = 1, 2, \ldots, n$, and by setting

$$E': \quad x_i p_j z_k q_j y_i = y_i p_j z_k q_j x_i \quad \text{for } i, j, k = 1, 2, \ldots, n.$$

Accordingly we extend the solutions (13.3.1) by setting

$$x_t = \begin{cases} b^2 ab & \text{if } t = i, \\ a & \text{otherwise,} \end{cases}$$
$$y_t = \begin{cases} b & \text{if } t = i, \\ a & \text{otherwise.} \end{cases} \quad (13.3.2)$$

Then we have a system E' of cardinality n^3 over $5n$ variables, which, moreover, by the computations of Example 13.3.2, is independent. It contains nonperiodic solutions, namely, those specified by $x_i = y_i$. Hence, (ii) is also valid, and here indeed A can be binary. ∎

We remark that in part (i) of the preceding theorem we used an infinite generating set A. Indeed, this is unavoidable if the definition of $D_1(n)$ is based on the combinatorial rank. However, the definition can be based, for instance, on the *prefix rank*, which, for a subset $X \subseteq A^+$, is defined as the cardinality of the least prefix code Y such that $X \subseteq Y^+$. If $D_1(n)$ were defined using the prefix rank, then in part (i) we could choose a binary generating set A. This is due to the fact that countably generated free semigroups can be embedded into a two-generator one using a prefix code as an embedding.

13.3.2. In free monoids

Lower bounds of Subsection 13.3.1 can be improved if the equations are solved in free monoids instead of free semigroups. This is no surprise, since if the empty word is available, it is essentially easier to find nontrivial equations, and hence also independent systems of equations, having a given set as a solution.

EXAMPLE 13.3.4. Let

$$\Xi = \{y\} \cup \{x_i, p_i, q_i, x'_i, p'_i, q'_i, x''_i, p''_i, q''_i \mid i = 1, 2, \ldots, n\}$$

be a set of variables, and let E be the following system of equations:

$$E: \ yx_j p_k q_\ell x'_j p'_k q'_\ell x''_j p''_k q''_\ell = x_j p_k q_\ell x'_j p'_k q'_\ell x''_j p''_k q''_\ell y \quad \text{for } j, k, \ell = 1, 2, \ldots, n.$$

Then E is over $9n + 1$ variables, and it has n^3 equations. Let us denote by $e(j, k, \ell)$ the equation of E for the triple (j, k, ℓ). As in Example 13.3.2, we show that
  (i) E has a solution of combinatorial rank $9n$ in A^*, and
  (ii) E is independent.

Part (i) is again clear: fix $y = \varepsilon$, so that all equations of E become trivial over $\Xi \setminus \{y\}$. To prove part (ii), we fix an equation from E, say $e(j_0, k_0, \ell_0)$, and search for a solution of the system $E \setminus \{e(j_0, k_0, \ell_0)\}$ which is not a solution of the whole E. Such a solution is provided by

$$\begin{aligned} y &= ababa, \\ x_{j_0} &= p_{k_0} = q_{\ell_0} = ab, \\ x'_{j_0} &= p'_{k_0} = q'_{\ell_0} = a, \\ x''_{j_0} &= p''_{k_0} = q''_{\ell_0} = ba, \\ z &= \varepsilon \text{ for all others.} \end{aligned} \quad (13.3.3)$$

Indeed, this is not a solution of $e(j_0, k_0, \ell_0)$, since $ababa \cdot ab \ldots \neq ab \cdot ab \cdot ab \ldots$. On the other hand, for any triple $(j, k, \ell) \neq (j_0, k_0, \ell_0)$, we obtain one of the following identities when substituting (13.3.3) into $e(j, k, \ell)$:

$$ababa = ababa,$$
$$ababa \cdot ab \cdot a \cdot ba = ab \cdot a \cdot ba \cdot ababa,$$
$$ababa \cdot ab \cdot ab \cdot a \cdot a \cdot ba \cdot ba = ab \cdot ab \cdot a \cdot a \cdot ba \cdot ba \cdot ababa.$$

Note that in the above factorizations we have omitted factors that are equal to ε.

By the above, we can formulate

13.3. Independence of finite systems of equations

THEOREM 13.3.5. (i) $D_1(n) = \Omega(n^3)$ *in* A^* *with* $\mathrm{Card}(A) = \infty$.
(ii) $D_{n-2}(n) = \Omega(n^4)$ *in* A^* *with* $\mathrm{Card}(A) \geq 2$.

The proof of Theorem 13.3.5 is analogous to that of Theorem 13.3.3. Note also that the remark made after Theorem 13.3.3 applies to the present case.

13.3.3. In free groups

The problems of this section change drastically when free semigroups are replaced by free groups, although both of these satisfy the compactness property; see Example 13.5.5. First, instead of considering the values $D_t(n)$ in the free groups, it is more meaningful to consider only the maximal cardinality of independent systems of equations. Second, the independent systems can be unboundedly large in free groups.

THEOREM 13.3.6. *Let n be a positive integer. There exists an independent system E_n of n equations in a free group using six variables.*

Proof. Let $\Xi = \{x, y, z, \bar{x}, \bar{y}, \bar{z}\}$ be a set of variables. Denote $\bar{u} = \bar{x}_m \bar{x}_{m-1} \ldots \bar{x}_1$ for each $u = x_1 x_2 \ldots x_m$, and $\bar{\bar{x}} = x$, where $x_i \in \{x, y, z\}$. Let $[u, v] = \bar{u}\bar{v}uv$ correspond to the commutator word of u and v for $u, v \in \Xi^+$, and define inductively

$$[v_1, \ldots, v_{k+1}] = [[v_1, \ldots, v_k], v_{k+1}].$$

Let $v_k = \bar{z}^k x^k \bar{y} z^k$ for $k \geq 1$, and let $w_i = [v_1, v_2, \ldots, v_{i-1}, v_{i+1}, \ldots, v_n]$. For each $n \in \mathbb{N}$, consider the system E_n of equations,

$$E_n: \quad (w_i)^2 = w_i \quad \text{for } i = 1, 2, \ldots, n.$$

It is now easy to verify that for two generators a, b of a free group,

$$x = a, \ \bar{x} = a^{-1}, \ y = a^j, \ \bar{y} = a^{-j}, \ z = b, \ \bar{z} = b^{-1}$$

is a solution of $(w_i)^2 = w_i$, if $i \neq j$, but not of $(w_j)^2 = w_j$. This shows that E_n is independent. ∎

REMARK 13.3.7. For groups G one usually prefers *group equations* $w = \varepsilon$, where w is an element of the free group $\Xi^{(*)}$ generated by the variables in Ξ, that is, w is a word over the alphabet $\Xi \cup \Xi^{-1}$, where $\Xi^{-1} = \{x^{-1} \mid x \in \Xi\}$. In this case a solution $\alpha: \Xi^{(*)} \to G$ of an equation $w = \varepsilon$ is required to respect the inverses, $\alpha(x^{-1}) = \alpha(x)^{-1}$ for all $x \in \Xi$. It is straightforward to show (see Problem 13.3.5) that a group G satisfies the compactness property for group equations if and only if G satisfies it for the ordinary semigroup equations.

In the proof of the above theorem the equations $(w_i)^2 = w_i$ can now be replaced by the equations $w_i = \varepsilon$, and the variables $\bar{x}, \bar{y}, \bar{z}$ can be replaced by the expressions x^{-1}, y^{-1}, z^{-1} that respect the group inversion. With these modifications Theorem 13.3.6 can be rephrased using only three variables.

13.4. Semigroups without the compactness property

We now consider systems of equations in arbitrary semigroups. A semigroup S is said to satisfy the *compactness property*, if for every finite set Ξ of variables, each system $E \subseteq \Xi^+ \times \Xi^+$ of equations is equivalent in S to a finite subsystem $E' \subseteq E$, that is, if every independent system of equations is finite.

In this section we shall demonstrate, via several examples of semigroups, that the compactness property of the previous section does not hold in general.

EXAMPLE 13.4.1. Let $A = \{a, b\}$, and $\Xi = \{x, y, z\}$. The monoid $\text{Fin}(A^*)$ of all nonempty finite subsets of the free monoid A^* does not satisfy the compactness property. Indeed, the system E of equations

$$xy^ix = xz^ix \quad \text{for } i \geq 1$$

over three variables does not have an equivalent finite subsystem in $\text{Fin}(A^*)$. To see this, define, for each $n \geq 1$, a morphism $\sigma_n: \Xi^+ \to \text{Fin}(A^*)$ as follows:

$$\sigma_n(x) = \{a^j \mid 0 \leq j \leq 2n+2\},$$
$$\sigma_n(y) = \{a^iba^j \mid 0 \leq i+j < n, \text{ or}$$
$$\quad 0 \leq i \leq 2n+2 \text{ and } n+1 \leq j \leq 2n+2, \text{ or}$$
$$\quad n+1 \leq i \leq 2n+2 \text{ and } 0 \leq j \leq 2n+2\},$$
$$\sigma_n(z) = \{a^iba^j \mid 0 \leq i, j \leq 2n+2\}.$$

Now for all $i \geq 1$,

$$\sigma_n(xy^ix) \subseteq \sigma_n(xz^ix) = \{a^{r_1}ba^{r_2}b \ldots ba^{r_i} \mid 0 \leq r_k \leq 4n+4, \ k = 1, 2, \ldots, i\}.$$

We leave it as an exercise to show that $\sigma_n(xy^ix) = \sigma_n(xz^ix)$ for all $i < n-1$. However, $\sigma_n(xy^{n+1}x) \neq \sigma_n(xz^{n+1}z)$, since $(ba^n)^nb \in \sigma_n(xz^{n+1}x)$, but $(ba^n)^nb \notin \sigma_n(xy^{n+1}x)$. These show that E does not have a finite equivalent subsystem in $\text{Fin}(A^*)$.

13.4. Semigroups without the compactness property

EXAMPLE 13.4.2. Let **B** be the monoid of functions generated by $\alpha, \beta: \mathbb{N} \to \mathbb{N}$,

$$\alpha(n) = \max\{0, n-1\}, \qquad \beta(n) = n+1.$$

The product of **B** is the ordinary composition of functions. This monoid is called the *bicyclic monoid*, and it has a simple monoid presentation $\langle a, b \mid ab = 1 \rangle$.

Let $\mathrm{id}: \mathbb{N} \to \mathbb{N}$ be the identity function. Now $\alpha\beta = \mathrm{id}$, but $\beta\alpha \neq \mathrm{id}$. Let $\gamma_i = \beta^i \alpha^i$ for all $i \geq 0$. We have

$$\gamma_i(n) = \begin{cases} i & \text{if } n \leq i, \\ n & \text{if } n > i, \end{cases}$$

and we observe that $\gamma_i \gamma_j = \gamma_{\max\{i,j\}}$. In particular, each γ_i is an idempotent of **B**, i.e., $\gamma_i^2 = \gamma_i$. Consider the system $E \subseteq \Xi^+ \times \Xi^+$ consisting of the equations

$$x^i y^i z = z \qquad \text{for } i \geq 1$$

in the variables $\Xi = \{x, y, z\}$. For a fixed j, the morphism δ_j defined by $\delta_j(x) = \beta$, $\delta_j(y) = \alpha$ and $\delta_j(z) = \gamma_j$ is a solution of $x^i y^i z = z$ for all $i \leq j$, but δ_j is not a solution of $x^{j+1} y^{j+1} z = z$. We conclude that the system E does not have an equivalent finite subsystem, and therefore the bicyclic monoid does not satisfy the compactness property.

This example extends directly to the bicyclic semigroup (without the identity element).

EXAMPLE 13.4.3. In this example we give a finitely generated semigroup S such that it and its ideal I both satisfy the compactness property but the Rees quotient S/I does not.

Let $S = A^+$ for $A = \{a, b\}$, and define

$$I = A^* \{ab^k ab^j a \mid 1 \leq j \leq k\} A^*.$$

It is plain that I is an ideal of A^+, and that both A^+ and I, as a subsemigroup of A^+, satisfy the compactness property. Let $\Xi = \{x, y, z\}$ be a set of variables, and E the following system of equations:

$$E: \quad xy^k xz^j x = xy^k xz^j xx \qquad \text{for } k, j \geq 1.$$

Let $E' \subseteq E$ be any finite subsystem, and set

$$m = \max \{ j \mid xy^k xz^j x = xy^k xz^j xx \text{ in } E' \}.$$

Define a morphism $\alpha: \Xi^+ \to A^+/I$ by $\alpha(x) = a$, $\alpha(y) = b^m$, $\alpha(z) = b$. Now, by the definition of I, $\alpha(xy^k xz^j x) = \alpha(xy^k xz^j xx)$ in A^+/I for each $xy^k xz^j x = xy^k xz^j xx$ from E', but $\alpha(xyxz^{m+1}x) \neq \alpha(xyxz^{m+1}xx)$. We conclude that E' is not equivalent to E. This proves that A^+/I does not satisfy the compactness property.

We obtain other semigroups that do not satisfy the compactness property after we prove some necessary conditions for this property in the next section; see Examples 13.5.7 and 13.5.11.

13.5. Semigroups with the compactness property

In this section, we search for connections between the compactness property and some classical notions of semigroup theory. We shall give, apart from positive examples, some necessary conditions for a semigroup to guarantee that it satisfies the compactness property. These conditions provide examples of semigroups that defy the compactness property.

Also, it turns out that all the monoids in a *variety* satisfy the compactness property if and only if these monoids satisfy the maximal condition on congruences. Such a characterization does not hold for individual monoids (or semigroups). Indeed, the free semigroups do not satisfy the maximal condition on congruences, but, as we have seen, they do satisfy the compactness property. The bicyclic semigroup, on the other hand, is an example of a semigroup that does satisfy the maximal condition, but does not satisfy the compactness property.

13.5.1. An extension of the proof

As is shown in Theorem 13.5.1, the compactness property is preserved under some natural operations on semigroups. On the other hand, the second part of the theorem shows that the compactness property fails on some other basic operations.

THEOREM 13.5.1. (i) *The class of semigroups that satisfy the compactness property is closed under taking isomorphic images, subsemigroups and finite direct products.*

(ii) *The class of semigroups that satisfy the compactness property is not closed under taking morphic images or infinite direct products.*

Proof. The proof of (i) is fairly easy, and it is left to the reader as an exercise.

That the compactness property is not inherited by the morphic images (or by the quotients) holds simply because the free semigroup $\{a,b\}^+$ satisfies the compactness property, but, as we have seen, its morphic image **B** (the bicyclic semigroup) does not.

13.5. Semigroups with the compactness property

Consider the semigroup $F = \text{Fin}(A^*)$ of all nonempty finite subsets of A^* from Example 13.4.1. For each $k \geq 0$, define a relation θ_k on F by

$$X\theta_k Y \iff X \cup A^{[k]} = Y \cup A^{[k]},$$

where $A^{[k]} = \{w \in A^* \mid |w| \geq k\}$. It is immediate that θ_k is a congruence on F, that is, for all $X, Y, Z \in F$, $X\theta_k Y$ implies that also $ZX\theta_k ZY$ and $XZ\theta_k YZ$. Furthermore, the quotient F/θ_k is a finite semigroup, and therefore it satisfies the compactness property.

Let $S = \prod_{k=0}^{\infty}(F/\theta_k)$ be the direct product of these finite semigroups, and define a morphism $\alpha: F \to S$ by its projections,

$$\pi_k \alpha(X) = X\theta_k,$$

where π_k is the projection of S onto F/θ_k, and $X\theta_k \in F/\theta_k$ is the congruence class of X with respect to θ_k.

For any two distinct $X, Y \in F$, we have $X\theta_k \neq Y\theta_k$ for $k = \max\{\text{Card } X, \text{Card } Y\}$, and therefore the morphism α is an embedding of F into S. We conclude from Example 13.4.1 and part (i) of the present theorem that $S = \prod_k (F/\theta_k)$ does not satisfy the compactness property although each of the semigroups F/θ_k does so. ∎

EXAMPLE 13.5.2. By Theorem 13.2.2, finitely generated free semigroups satisfy the compactness property. For a countably generated free semigroup A^+, where $A = \{a_1, a_2, \ldots\}$, the mapping $\alpha: A^+ \to \{a, b\}^+$ defined by $\alpha(a_i) = a^i b$ for all i is an embedding of A^+ into the free semigroup $\{a, b\}^+$. Therefore the compactness property holds also for countably generated free semigroups.

EXAMPLE 13.5.3. A *trace monoid* M, often referred to as a *free partially commutative monoid*, is a monoid that has a presentation $\langle A \mid ab = ba \ ((a, b) \in R)\rangle$, where R is a symmetric relation on the finite set $A = \{a_1, a_2, \ldots, a_k\}$ of generators. A trace monoid M can be embedded into the k-fold direct product $P^{(k)} = \{a, b\}^* \times \cdots \times \{a, b\}^*$. To see this, let for each $i = 1, 2, \ldots, k$, v_i be the vector defined by the conditions

$$\pi_j(v_i) = \begin{cases} a & \text{if } i = j, \\ \varepsilon & \text{if } (a_i, a_j) \in R \text{ and } i \neq j, \\ b & \text{otherwise,} \end{cases}$$

where π_j denotes the jth projection of $P^{(k)}$ into $\{a, b\}^*$. Then it is plain that $v_j v_i = v_i v_j$ for $i \neq j$ if and only if $(a_i, a_j) \in R$, which shows that the monoid generated by the vectors v_i, for $i = 1, 2, \ldots, k$, is isomorphic to M. Now, the claim follows from Theorem 13.5.1.

The above examples are special cases of a general theorem which we now prove as an extension of Theorem 13.2.2.

THEOREM 13.5.4. *Let R be a commutative Noetherian ring R containing an identity element. If a semigroup S can be embedded in the multiplicative matrix semigroup $R^{n \times n}$, then it satisfies the compactness property.*

Proof. We give a detailed outline of the proof. Indeed, polynomials over such a commutative Noetherian ring R satisfy Hilbert's basis theorem, and therefore R can be used instead of $\mathbb{Z}$ in the proof of Theorem 13.2.2. In the case $n = 2$, for each variable x_i in $\Xi = \{x_1, x_2, \ldots, x_k\}$, we introduce a matrix

$$M_i = \begin{pmatrix} x_{i1} & x_{i2} \\ x_{i3} & x_{i4} \end{pmatrix},$$

where $x_{i1}, x_{i2}, x_{i3}, x_{i4}$ are commuting variables. Let the set of these new variables be $X = \{x_{ij} \mid i = 1, 2, \ldots, k, j = 1, 2, 3, 4\}$. The subsemigroup $\mathbb{M}$ of $R[X]^{2 \times 2}$ generated by the matrices M_i is a free semigroup (see Problem 13.2.2). If S is a semigroup such that there exists an embedding $\mu: S \to R^{2 \times 2}$, then the commuting diagram of the proof of Theorem 13.2.2 takes the form

$$\begin{array}{ccc} \Xi^+ & \xrightarrow{\alpha} & S \\ \varphi \downarrow & & \downarrow \mu \\ \mathbb{M} & \xrightarrow{\tilde{\alpha}} & R^{2 \times 2} \end{array}$$

The proof of Theorem 13.2.2 can now be easily modified for Theorem 13.5.4. The general case is treated in a similar way. ∎

As an illustration of the above, we provide

EXAMPLE 13.5.5. It is easy to see that the matrices

$$M_a = \begin{pmatrix} 1 & 2 \\ 0 & 1 \end{pmatrix} \quad \text{and} \quad M_b = \begin{pmatrix} 1 & 0 \\ 2 & 1 \end{pmatrix}$$

generate a free subgroup of $\mathbb{Z}^{2 \times 2}$. Further, as is well known, every finitely generated free group can be embedded into a free group generated by two elements, and therefore Theorem 13.5.4 yields that the finitely generated free groups satisfy the compactness property.

13.5.2. Necessary conditions

The next result is motivated by Example 13.4.2.

An element e of a semigroup S is an *idempotent*, if it satisfies $e^2 = e$. The idempotents of a semigroup S can be partially ordered by defining $e \le f$ if and only if $fe = e = ef$. We say that S satisfies the *chain condition on idempotents*, if each subset of idempotents of S contains a maximal and a minimal element, i.e., each chain $\cdots e_{i-1} < e_i < e_{i+1} \cdots$ of idempotents is finite.

THEOREM 13.5.6. *Let S be a finitely generated semigroup satisfying the compactness property. Then it satisfies the chain condition on idempotents.*

Proof. Let S be generated by n elements, and let $\mu \colon \Xi^+ \to S$ be a natural morphism onto S, where $\mathrm{Card}(\Xi) = n$.

Suppose first that $e_1 > e_2 > \cdots$ is an infinite descending chain of idempotents of S. Therefore $e_i e_j = e_{\max\{i,j\}}$ for all $i, j \ge 1$. Now, for each $i \ge 1$, let $w_i \in \Xi^+$ be a word such that $\mu(w_i) = e_i$, and denote $Y = \Xi \cup \{y\}$. Consider the system E of equations $w_i y = y$, with $i \in \mathbb{N}$, over Y. For each $j \ge 1$, let $\alpha_j \colon Y^+ \to S$ be a morphism such that $\alpha_j(x) = \mu(x)$ for $x \in \Xi$, and let $\alpha_j(y) = e_j$. Now $\alpha_j(w_i y) = e_i e_j$ for all i and j. Consequently, α_j is a solution of $w_i y = y$ for all i with $i \le j$, but α_j is not a solution to $w_{j+1} y = y$. We conclude that E does not have an equivalent finite subsystem in S.

The case of an infinite ascending chain $e_1 < e_2 < \cdots$ of idempotents is treated analogously. Hence our proof is complete. ∎

Theorem 13.5.6 yields immediately another example of semigroups that do not satisfy the compactness property.

EXAMPLE 13.5.7. The free inverse semigroups do not satisfy the chain condition on idempotents, and therefore the compactness property fails for these. Indeed, the *free monogenic inverse semigroup*, which is generated by one element as an inverse semigroup, has a semigroup presentation

$$FI_1 = \langle a, b \mid a = aba, b = bab, a^m b^{m+n} a^n = b^n a^{n+m} b^m, \ n, m \ge 1 \rangle.$$

Here $a^n b^n$ is an idempotent for $n \ge 1$, and $a^n b^n \cdot a^m b^m = a^n b^n = a^m b^m \cdot a^n b^n$, i.e., $a^n b^n \le a^m b^m$ for all $n \ge m$.

Next we look for a connection between the compactness property and certain types of congruences of semigroups. A congruence θ of a semigroup S is called *nuclear*, if it is induced by an endomorphism, i.e., if

$\theta = \ker(\alpha)$ for an endomorphism $\alpha: S \to S$. In other words, a congruence θ of S is nuclear, if the quotient S/θ is isomorphic to a subsemigroup of S.

LEMMA 13.5.8. *Let S be a semigroup with the compactness property, and let Ξ be a finite alphabet. Then each sequence $\alpha_i: \Xi^+ \to S$ of morphisms with $\ker(\alpha_i) \subset \ker(\alpha_{i+1})$, for $i = 1, 2, \ldots$, is finite.*

Proof. Assume that $(\alpha_i)_{i \geq 0}$ is an infinite sequence of morphisms such that $\ker(\alpha_i) \subset \ker(\alpha_{i+1})$. Consider a system $E = \{(u_i, v_i) \mid i = 1, 2, \ldots\}$ of equations, where $(u_i, v_i) \in \ker(\alpha_{i+1}) \setminus \ker(\alpha_i)$ for each i. Clearly, E has no equivalent finite subsystem in S. This proves the lemma. ∎

We say that a semigroup S satisfies the *maximal condition on nuclear congruences*, if each ascending chain $\theta_1 \subset \theta_2 \subset \cdots$ of its nuclear congruences is finite. We obtain our second necessary condition.

THEOREM 13.5.9. *Let S be a semigroup with the compactness property. Then the finitely generated subsemigroups of S satisfy the maximal condition on nuclear congruences.*

Proof. Suppose that S_0 is a finitely generated subsemigroup of S such that, for each $i \geq 1$, $\alpha_i: S_0 \to S_0$ is an endomorphism satisfying $\ker(\alpha_i) \subset \ker(\alpha_{i+1})$ for all $i \geq 1$. Let $\mu: \Xi^+ \to S_0$ be a natural morphism onto S_0. Consequently, $\ker(\alpha_i \mu) \subset \ker(\alpha_{i+1} \mu)$ for all $i \geq 1$, and the claim follows from Lemma 13.5.8. ∎

Theorem 13.5.9 can be used to formulate another necessary condition for the compactness property. A semigroup S is said to be *Hopfian*, if it is not isomorphic to a quotient S/θ for any of its nontrivial congruences θ. Equivalently, S is Hopfian, if every surjective endomorphism $\alpha: S \to S$ is an automorphism.

THEOREM 13.5.10. *Let S be a semigroup with the compactness property. Then every finitely generated subsemigroup of S is Hopfian.*

Proof. Assume that S_0 is a non-Hopfian finitely generated subsemigroup of S, and let $\alpha: S_0 \to S_0$ be a noninjective endomorphism onto S_0. Now the nuclear congruences $\theta_i = \ker(\alpha^i)$ for $i \geq 1$ form a properly ascending chain, and hence the claim follows by Theorem 13.5.9. ∎

Note that if a semigroup S satisfies the compactness property, then S itself need not be Hopfian, if it is infinitely generated. Indeed, a countably generated free semigroup A^+ satisfies the compactness property, but it is not Hopfian.

13.5. Semigroups with the compactness property

EXAMPLE 13.5.11. One of the simplest non-Hopfian semigroups is the so-called *Baumslag–Solitar group* which has a group presentation $G = \langle a, b \mid b^2 a = ab^3 \rangle$.

For simplicity, we use group equations introduced in Remark 13.3.7. Let $[g, h] = g^{-1}h^{-1}gh$ be the *commutator* of the elements $g, h \in G$, and denote by ε_G the identity element of G. The morphism $\alpha: G \to G$ defined by $\alpha(a) = a$ and $\alpha(b) = b^2$ is surjective, because $\alpha([a, b^{-1}]) = b$. However, it can be shown that $\alpha(g) = \varepsilon_G$ for $g = [a^{-1}ba, b] \neq \varepsilon_G$, and therefore α is not injective.

Let $\Xi = \{x, y\}$, and denote by $\Xi^{(*)}$ the free group generated by Ξ. Let $u = [x^{-1}yx, y]$. Define a (group) morphism $\beta: \Xi^{(*)} \to \Xi^{(*)}$ by $\beta(x) = x$ and $\beta(y) = [x, y^{-1}]$. We obtain a system E of equations $\{\beta^i(u) = \varepsilon \mid i \geq 1\}$, which has no equivalent finite subsystem. Now if $\gamma: \Xi^{(*)} \to G$ is defined by $\gamma(x) = a$ and $\gamma(y) = b$, then $\alpha^i\gamma(\beta^j(u)) = \varepsilon_G$ for $j < i$, but $\alpha^i\gamma(\beta^i(u)) \neq \varepsilon_G$.

13.5.3. The compactness property for varieties

For convenience, in the rest of this chapter we shall consider monoids and groups rather than semigroups.

Recall that a class $\mathscr{V}$ of monoids (resp. groups) is a *variety*, if it is closed under taking submonoids (resp. subgroups), morphic images, and arbitrary direct products. By Birkhoff's theorem, a variety of monoids becomes defined by a set of *identities* $u \equiv v$, which are equations $(u, v) \in \Xi^*$ such that every morphism $\alpha: \Xi^* \to M$ with $M \in \mathscr{V}$ is a solution of (u, v). Note that here the set Ξ of variables is allowed to be infinite, although the equations are required to be finite. For instance, the identity $x_1 x_2 \equiv x_2 x_1$ defines the variety of all commutative monoids.

A monoid M satisfies the *maximal condition on congruences*, if each set of congruences of M has a maximal element. The following general result is easy to prove using Zorn's lemma.

LEMMA 13.5.12. *The following conditions are equivalent for a monoid M.*
 (i) *M satisfies the maximal condition on congruences.*
 (ii) *Each ascending chain $\theta_1 \subset \theta_2 \subset \cdots$ of congruences of M is finite.*
 (iii) *For each congruence θ of M generated by a subset $E \subseteq \theta$ there exists a finite subset $E' \subseteq E$ such that E' generates θ.*

Using the above lemma, we obtain the following characterization.

THEOREM 13.5.13. *A variety $\mathscr{V}$ of monoids satisfies the compactness property if and only if each finitely generated monoid $M \in \mathscr{V}$ satisfies the maximal condition on congruences.*

Proof. We first recall that, for all $n \geq 1$, a variety $\mathscr{V}$ has a monoid V_n generated by an n-element subset B satisfying the following extension property: for any mapping $\gamma_B: B \to M$ with $M \in \mathscr{V}$, there exists a unique morphism $\gamma: V_n \to M$, which is an extension of γ_B, i.e., $\gamma(b) = \gamma_B(b)$ for all $b \in B$. Such a monoid V_n is a *free monoid of* $\mathscr{V}$. The extension property yields that each morphism $\alpha: \Xi^* \to M$, with $M \in \mathscr{V}$ and Card(Ξ) = n, can be factored as $\alpha = \beta\mu$, where $\mu: \Xi^* \to V_n$ is the natural morphism onto V_n and $\beta: V_n \to M$ is a morphism.

Suppose now that each $M \in \mathscr{V}$ satisfies the maximal condition on congruences. Let $E = \{(u_i, v_i) \mid i \geq 1\} \subseteq \Xi^* \times \Xi^*$ be a system of equations, where Card(Ξ) = n. Further, let θ be the congruence on V_n generated by the relation $\mu(E) = \{(\mu(u_i), \mu(v_i)) \mid i \geq 1\}$, i.e., θ is the smallest congruence on V_n containing $\mu(E)$. By assumption and Lemma 13.5.12, θ is generated by a finite subset E'' of $\mu(E)$. Clearly, $E'' = \mu(E')$ for a finite subset E' of E. Now, if $M \in \mathscr{V}$ and $\alpha = \beta\mu: \Xi^* \to M$ is a solution to E', then $E' \subseteq \ker(\alpha)$ and hence $E'' = \mu(E') \subseteq \ker(\beta)$, which implies that $\theta \subseteq \ker(\beta)$. In particular, $\mu(E) \subseteq \ker(\beta)$, and, consequently, $E \subseteq \ker(\alpha)$. Therefore M satisfies the compactness property.

To prove the converse, let $\mathscr{V}$ be a variety of monoids, and assume that $M \in \mathscr{V}$ is a finitely generated monoid that does not satisfy the maximal condition on congruences. Let $\theta_1 \subset \theta_2 \subset \cdots$ be an ascending chain of congruences of M, and let $\alpha_i: M \to M_i = M/\theta_i$, for each $i \geq 1$, be a surjective morphism with $\ker(\alpha_i) = \theta_i$. Each quotient M_i is in $\mathscr{V}$, since $\mathscr{V}$ is a variety. Let again $\mu: \Xi^* \to M$ be a natural morphism from the free monoid Ξ^* onto M.

We observe that the congruences $\theta_i' = \ker(\alpha_i \mu)$ of Ξ^*, for $i \geq 1$, form a properly ascending chain. For each $i \geq 2$, choose a pair $(u_i, v_i) \in \theta_i' \setminus \theta_{i-1}'$, and let $E = \{(u_i, v_i) \mid i \geq 2\}$. Consider the direct product $\prod_{i \geq 1} M_i$. Since $\mathscr{V}$ is a variety, also $\prod_{i \geq 1} M_i \in \mathscr{V}$. Let $\beta_i: M_i \to \prod_{i \geq 1} M_i$ be the natural embedding, and define $\gamma_i = \beta_i \alpha_i \mu: \Xi^* \to \prod_{i \geq 1} M_i$. Now $\gamma_i(u_j) = \gamma_i(v_j)$ for all $j \leq i$, but $\gamma_i(u_{i+1}) \neq \gamma_i(v_{i+1})$, and therefore E does not have an equivalent finite subsystem in $\prod_{i \geq 1} M_i$. ∎

Theorem 13.5.13 is interesting in the sense that it provides a nontrivial characterization when a variety, i.e., all monoids in that variety, satisfies the compactness property. No similar characterizations are known for individual monoids (or semigroups).

By Redei's theorem, the finitely generated commutative monoids

13.5. Semigroups with the compactness property

satisfy the maximal condition on congruences. Hence we have the following corollary of Theorem 13.5.13.

COROLLARY 13.5.14. *Every commutative monoid satisfies the compactness property.*

By Theorem 13.2.2 and Corollary 13.5.14, the compactness property holds in the extremes with respect to commutativity, namely, for the free semigroups as well as for the commutative semigroups. Both results are based on Hilbert's basis theorem. Note also that in neither of these cases need the semigroups be finitely generated.

As a difference between these cases we mention that it is not known whether arbitrary large independent systems of equations with n variables exist in free semigroups, while for commutative semigroups such are easy to find; see Problem 13.3.3.

The proof of Theorem 13.5.13 does not use the property that varieties are closed under taking submonoids. Since (monoid) morphic images and direct products of groups are groups, the proof applies also to groups. Further, in a group G, the congruence class containing the identity element is a normal subgroup of G that determines the congruence θ. Therefore we obtain

THEOREM 13.5.15. *A variety $\mathscr{V}$ of groups satisfies the compactness property if and only if each finitely generated group of $\mathscr{V}$ satisfies the maximal condition on normal subgroups.*

EXAMPLE 13.5.16. For groups Corollary 13.5.14 can be much improved. Let G be a group, and let $[a, b] = a^{-1}b^{-1}ab$ be the commutator of the elements $a, b \in G$. The *metabelian groups* form a variety defined by a single identity $[[x_1, x_2], [x_3, x_4]] \equiv \varepsilon$. Clearly, every abelian group is metabelian. Moreover, by Hall's theorem, every finitely generated metabelian group satisfies the maximal condition on normal subgroups. Therefore the metabelian groups satisfy the compactness property. This result is interesting, since the free semigroups can be embedded into a free metabelian group. This gives the original proof of Albert and Lawrence for Theorem 13.2.2 for free semigroups. However, the proof of Hall's theorem uses a variant of Hilbert's basis theorem.

EXAMPLE 13.5.17. The nilpotent groups satisfy the compactness property, although they do not form a variety. Indeed, the smallest variety that contains all nilpotent groups consists of all groups. However, each nilpotent group belongs to a variety (nilpotent groups of class n, for some n) that does satisfy the maximal condition on normal subgroups.

Problems

Section 13.1

13.1.1 If a system $E \subseteq \Xi^+ \times \Xi^+$ of equations does not have an equivalent finite subsystem (in a semigroup S, in general), show that E has an infinite subsystem $E' = \{(u_i, v_i) \mid i = 1, 2, \ldots\}$ ordered in such a way that for each j there exists a solution of the system $u_i = v_i$ for $i = 1, 2, \ldots, j$, which is not a solution of the equation $u_{j+1} = v_{j+1}$.

13.1.2 Show that the system $xy^i z = zy^i x$, $i = 1, 2, 3$, is dependent in Σ^*.

13.1.3 Prove Theorem 13.1.3.

*13.1.4 Classify the relations defined by three-generator subsemigroups of a free semigroup. (See Spehner 1976.)

Section 13.2

13.2.1 Let S be a semigroup, $\Xi = \{x, y\}$, and let $k \geq 0$ and $m > 0$ be two fixed integers. Denote $I = \{k + jm \mid j \geq 0\}$. Show that the system of equations $x^i = y^i$ ($i \in I$) in S is equivalent to its finite subsystem $x^i = y^i$ ($i \in \{k + jm \mid j < k\}$).

13.2.2 Let R be any commutative nontrivial ring with an identity $1 \ (\neq 0)$ element. Let $X = \{x_{ij} \mid 1 \leq j \leq k, \ 1 \leq i \leq 4\}$ be a set of commuting variables for R. Show that the matrix semigroup $\mathbb{M}(\Xi)$ of $R[X]^{2 \times 2}$, generated by the matrices

$$M_i = \begin{pmatrix} x_{i1} & x_{i2} \\ x_{i3} & x_{i4} \end{pmatrix},$$

is a free semigroup.

13.2.3 Prove Lemma 13.2.3: each rational relation $R \subseteq \Xi^+ \times \Xi^+$ has (as a system of equations) an equivalent finite subsystem $R_0 \subseteq R$ that can be effectively found.

*13.2.4 Prove Lemma 13.2.6: it is decidable whether L_1 is a test set of L_2 for finite sets $L_1 \subseteq L_2$. For this, one applies Makanin's result in proving that the equivalence of two finite systems of equations is decidable. (See Culik and Karhumäki 1983)

13.2.5 The famous $2n$-conjecture for D0L systems states that if $\alpha^i(w) = \beta^i(w)$ for all $i < 2n$ (where $\alpha, \beta : A^* \to A^*$ are morphisms with $\text{Card}(A) = n$ and $w \in A^*$ is a word), then $\alpha^i(w) = \beta^i(w)$ for all $i \geq 0$. Prove, by modifying the system of equations in Subsection 13.3.2, that such a conjecture does not hold for HD0L systems, that is, for morphic images of D0L systems. Indeed, show that if a function $f:\mathbb{N} \to \mathbb{N}$ satisfies

$$\gamma_1\alpha^i(w) = \gamma_2\beta^i(w) \text{ for all } i < f(n) \Rightarrow \gamma_1\alpha^i(w)$$
$$= \gamma_2\beta^i(w) \text{ for all } i \geq 0$$

for all words $w \in A^*$ and morphisms $\alpha, \beta, \gamma_1, \gamma_2 : A^* \to A^*$, where $\text{Card}(A) = n$, then $f(n) = \Omega(n^4)$. (See Plandowski 1995.)

*13.2.6 Show that the $2n$-conjecture holds for D0L systems over binary alphabets. (See Karhumäki 1981.)

Section 13.3

*13.3.1 Show that any language over a binary alphabet has a test set of cardinality 3. This problem on three variables is completely open! In particular, does there exist an independent system of three equations in three variables that has a solution of rank 2 over a free semigroup A^+? (For the binary case, see Ehrenfeucht, Karhumäki, and Rozenberg 1983b.)

13.3.2 Show that for all $n, m, p, q \geq 1$,

$$D_p(n) + D_q(m) \leq D_{p+q}(n+m).$$

Use this inequality to prove that for $0 < c < 1$, $D_{\lceil cn \rceil}(n) = \Omega(n^4)$ in free monoids, and $D_{\lceil cn \rceil}(n) = \Omega(n^3)$ in free semigroups.

13.3.3 Show that there are arbitrarily large independent systems of equations with n variables in the commutative semigroup $S = \{z \mid \exists n : z^n = 1\}$ of the complex roots of unity.

**13.3.4 Improve the lower bounds of Theorems 13.3.3 and 13.3.5. This is an open problem.

13.3.5 Show that a group G satisfies the compactness property for group equations $w = \varepsilon$ with $w \in \Xi^{(*)}$ if and only if G satisfies it for the semigroup equations $u = v$ with $u, v \in \Xi^+$.

13.3.6 Prove Theorem 13.3.1.

Section 13.4

*13.4.1 Two elements a and b of a semigroup S form an *inverse pair*, if $a = aba$ and $b = bab$. In this case, the elements ab and ba are idempotents of S. Show that if S contains an inverse pair a, b such that $ba < ab$ in the partial ordering of the idempotents, then the subsemigroup of S generated by a and b is isomorphic to the bicyclic monoid, and therefore S does not satisfy the compactness property. (See Petrich 1984, p. 432.)

13.4.2
1. Prove that a variety $\mathscr{V}$ of monoids satisfies the compactness property if and only if the (relative) free monoids of this variety satisfy the maximal condition on congruences.
2. Show that the bicyclic monoid **B** satisfies the maximal condition on congruences. Show also that **B** is a monoid that does not satisfy the compactness property, but all its proper quotients do.

Section 13.5

13.5.1 Prove the claim of Example 13.5.5.

13.5.2 For a semigroup S that is not a monoid, let S^ε be the monoid obtained from S by adding an identity element ε_S to it. Show that S satisfies the compactness property if and only if the monoid S^ε does so.

13.5.3 Show that a monoid S satisfies the compactness property if and only if each sequence $\alpha_i : \Xi^+ \to \prod^\infty S$ of morphisms with $\ker(\alpha_i) \subset \ker(\alpha_{i+1})$ is finite for all finite Ξ. (See Harju, Karhumäki, and Plandowski 1997b.)

13.5.4 Let us say that a class $\mathscr{S}$ of semigroups satisfies the compactness property *uniformly*, if each system of equations $E \subseteq \Xi^+ \times \Xi^+$ has a finite subsystem $E' \subseteq E$ such that E' is equivalent to E in all semigroups $S \in \mathscr{S}$. Show that if a variety $\mathscr{V}$ of monoids (or groups) satisfies the compactness property, then it satisfies it uniformly. In particular, if a variety $\mathscr{V}$ satisfies the compactness property, and S is a semigroup that is *locally* $\mathscr{V}$ (i.e., each finitely generated subsemigroup of S is in $\mathscr{V}$), then S satisfies the compactness property. (See Harju et al. 1997b.)

*13.5.5 Show that the finite semigroups generated by two elements do not satisfy the compactness property uniformly. (They do satisfy the compactness property, but not uniformly. For this one can use a result, due to Munn, stating that the finitely generated free inverse semigroups are residually finite. See Harju et al. 1997b.)

Notes

For a general source in the theory of semigroups and groups, we refer to Howie 1976 and Magnus et al. 1966, respectively. General references on combinatorics of words are Lothaire 1983 and Choffrut and Karhumäki 1997.

For the statement and the proof of the equality theorem, see Eilenberg 1974, or for a generalization using skew fields. For the defect theorem, see Chapter 6, and the references given there.

The compactness theorem, Theorem 13.2.2, was conjectured by A. Ehrenfeucht in the beginning of the 1970s in a language-theoretic setting; see Theorem 13.2.5. Its reformulation in systems of equations is due to Culik and Karhumäki 1983. Theorem 13.2.2 was proved independently by Albert and Lawrence 1985b and Guba 1986. The present proof follows the ideas of the proof by Guba 1986. As mentioned in the notes to Chapter 12, this techniques was originated by Markov in the 1950s. Hilbert's basis theorem is proved in many "standard" textbooks on algebra. For the proof of the basis theorem formulated for Noetherian rings, see e.g. Kostrikin and Shafarevich 1990 (p. 45) and, for a different proof, Cohn 1989 (p. 318). Albert and Lawrence 1985b proved the compactness property by embedding the free semigroups into free metabelian groups, for which a variant of Hilbert's basis theorem was proved by Hall 1954.

The applications in Subsection 13.2.2 are treated in Harju and Karhumäki 1986, Choffrut et al. 1997, Culik and Karhumäki 1983 and Culik and Karhumäki 1986. The undecidability of the isomorphism problem for semigroups of nonnegative integer matrices follows, for instance, from the undecidability of freeness for these semigroups; see Klarner, Birget, and Satterfield 1991. The isomorphism problem, as well as many other algorithmic problems on semigroups, is treated in a more general setting in Kharlampovich and Sapir 1995. For the decidability problems on iterated morphisms, especially the D0L and DT0L systems, see Rozenberg and Salomaa 1980. The isomorphism problem is open for subsemigroups of free semigroups A^+ generated by a regular set.

The results on the independent systems of equations for free semigroups and monoids in Section 13.3 are due to Karhumäki and Plandowski 1994. Theorem 13.3.6 for free groups was proved by Albert and Lawrence 1985a. Sizes of independent systems of equations in various semigroups have been studied by Karhumäki and Plandowski 1996.

Example 13.4.1 for the semigroup of all finite subsets of a free semigroup is due to Lawrence 1986. For the other examples in Section 13.4, see Harju et al. 1997b and Harju, Karhumäki, and Petrich 1997a.

The compactness property for free groups (see Example 13.5.5) was proven by Guba 1986 and De Luca and Restivo 1986. Since every free semigroup can be embedded into a free group, Theorem 13.2.2 follows from this result. We refer also to Stallings 1986 for a more general approach of the compactness property.

For the theory of varieties needed in Section 13.5, see e.g. Cohn 1981. The results concerning the compactness property in varieties are due to Albert and Lawrence 1985a and Harju et al. 1997b. A short proof of Redei's theorem, needed in Corollary 13.5.14 for commutative semigroups, is given by Freyd 1968. This proof is based on Hilbert's basis theorem.

We have treated the compactness property for groups only cursorily. For more information on this topic, see Baumslag, Myasnikov, and Roman'kov 1997, where the groups satisfying the compactness property are called *equationally Noetherian*. In particular, Baumslag et al. 1997 show that if a group G has a subgroup of finite index that satisfies the compactness property, then so does G. Also, the authors construct a large class of groups that do not satisfy the compactness property. Indeed, they show that if G is any nonabelian group and H is any infinite group, then their wreath product $G \wr H$ does not satisfy the compactness property.

References

Abdulrab, H. and Pécuchet, J.-P. (1990). Solving word equations, *J. Symb. Comput.*, **8**, 499–521.

Abdulrab, H. and Pécuchet, J.-P. (Eds.) (1993). *Word Equations and Related Topics (IWWERT 91)*, Vol. 677 of *Lect. Notes Comp. Sci.* Springer-Verlag.

Adian, S. I. (1979). *The Burnside Problem and Identities in Groups*, Vol. 95 of *Ergebnisse der Mathematik und ihrer Grenzgebiete*. Springer-Verlag.

Albert, M. H. and Lawrence, J. (1985a). The descending chain condition on solution sets for systems of equations in groups, *Proc. Edinburgh Math. Soc. (2)*, **29**, 69–73.

Albert, M. H. and Lawrence, J. (1985b). A proof of Ehrenfeucht's conjecture, *Th. Comput. Sci.*, **41**, 121–123.

Alessandri, P. (1996). *Codages des rotations et basses complexités*. Thèse de doctorat, Université Aix-Marseille 2.

Alessandri, P. and Berthé, V. (1998). Three distance theorems and combinatorics on words, *Enseign. Math.*, **44**, 103–132.

Allauzen, C. (1998). Une caractérisation simple des nombres de Sturm, *J. Th. Nombres Bordeaux*, **10**, 237–241.

Allouche, J.-P. (1994). Sur la complexité des suites infinies, *Bull. Belg. Math. Soc. Simon Stevin*, **1**, 133–143.

Allouche, J.-P. and Bousquet-Mélou, M. (1995). On the conjectures of Rauzy and Shallit for infinite words, *Comment. Math. Univ. Carolin.*, **36**, 705–711.

Allouche, J.-P., Cateland, E., Gilbert, W. J., Peitgen, H.-O., Shallit, J. O., and Skordev, G. (1997). Automatic maps in exotic numeration systems, *Th. Comput. Syst.*, **30**, 285–331.

Amir, A. and Benson, G. E. (1992). Two-dimensional periodicity and its applications, in *Third ACM-SIAM Symp. on Discr. Algorithms*, pp. 440–452.

Amir, A. and Benson, G. E. (1998). Two-dimensional periodicity in rectangular arrays, *SIAM J. Comput.*, **27**, 90–106.

Andrews, G. E. (1976). *The Theory of Partitions*, Vol. 2 of *Encyclopedia of Mathematics and Its Applications*. Addison-Wesley.

Apostolico, A. and Ehrenfeucht (1993). Efficient detection of quasiperiodicities in strings, *Th. Comput. Sci.*, **119**, 247–265.

Arnoux, P., Ferenczi, S., and Hubert, P. (1999). Trajectories of rotations, *Acta Arith.*, **87**, 209–217.

Arnoux, P., Mauduit, C., Shiokawa, I., and Tamura, J. (1994). Complexity of sequences defined by billiard in the cube, *Bull. Soc. Math. France*, **122**, 1–12.

Arnoux, P. and Rauzy, G. (1991). Représentation géométrique de suites de complexité $2n + 1$, *Bull. Soc. Math. France*, **119**, 199–215.

Aršon, S. E. (1937). Démonstration de l'existence des suites asymétriques infinies, *Mat. Sb. (N.S.)*, *2(44)*, 769–777. In Russian. French summary.

Assous, R. and Pouzet, M. (1979). Une caractérisation des mots périodiques, *Discrete Math.*, **25**, 1–5.

Avizienis, A. (1961). Signed-digit number representations for fast parallel arithmetic, *IRE Trans.*, EC-10, 389–400.

Baker, K. A., McNulty, G. F., and Taylor, W. (1989). Growth problems for avoidable words, *Th. Comput. Sci.*, **69**, 319–345.

Bandt, C. (1991). Self-similar sets. V. Integer matrices and fractal tilings of $\mathbb{R}^n$, *Proc. Amer. Math. Soc.*, **112**, 549–562.

Baumslag, G., Myasnikov, A., and Roman'kov, V. (1997). Two theorems about equationally noetherian groups, *J. Algebra*, **194**, 654–664.

Béal, M.-P. and Perrin, D. (1997). Symbolic dynamics and finite automata, in Rozenberg, G. and Salomaa, A. (Eds.), *Handbook of Formal Languages*, Vol. 2, pp. 463–503. Springer-Verlag.

Bean, D. R., Ehrenfeucht, A., and McNulty, G. F. (1979). Avoidable patterns in strings of symbols, *Pacific J. Math.*, **85**, 261–294.

Beatty, S. (1926). Problem 3173, *Amer. Math. Monthly*, **33**, 159. Solution **34**:1927, 159.

Bender, E. A., Patashnik, O., and Rumsey, J. H. (1994). Pizza slicing, Phi's and the Riemann Hypothesis, *Amer. Math. Monthly*, **101**(4), 307–317.

Berend, D. and Frougny, C. (1994). Computability by finite automata and Pisot bases, *Math. Syst. Th.*, **27**, 274–282.

Bergman, G. (1969). Centralizers in free associative algebras, *Trans. Amer. Math. Soc.*, **137**, 327–344.

Bergman, G. (1979). The class of free subalgebras of a free associative algebra is not closed under taking unions of chains, or pairwise intersection, unpublished manuscript.

Bernoulli, J. (1772). Sur une nouvelle espèce de calcul, in *Recueil pour les astronomes*, Vol. 1, pp. 255–284. Berlin.

Berstel, J. (1979a). Sur les mots sans carré définis par un morphisme, in Maurer, H. A. (Ed.), *Automata, Languages and Programming (ICALP 97)*, Vol. 71 of *Lect. Notes Comp. Sci.*, pp. 16–25. Springer-Verlag.

Berstel, J. (1979b). *Transductions and Context-Free Languages*. Teubner, Stuttgart.

Berstel, J. (1984). Some recent results on square-free words, in Fontet, M. and Mehlhorn, K. (Eds.), *Theoretical Aspects of Computer Science (STACS 84)*, Vol. 166 of *Lect. Notes Comp. Sci.*, pp. 14–25. Springer-Verlag.

Berstel, J. (1996). Recent results on Sturmian words, in Dassow, J. and Salomaa, A. (Eds.), *Developments in Language Theory II*, pp. 13–24. World Scientific.

Berstel, J. and Boasson, L. (1999). Partial words and a theorem of Fine and Wilf, *Th. Comput. Sci.*, **218**, 135–141.

Berstel, J. and De Luca, A. (1997). Sturmian words, Lyndon words and trees, *Th. Comput. Sci.*, **178**, 171–203.

Berstel, J. and Perrin, D. (1985). *Theory of Codes*. Academic Press.

Berstel, J., Perrin, D., Perrot, J.-F., and Restivo, A. (1979). Sur le théorème du défaut, *J. Algebra*, **60**, 169–180.

Berstel, J. and Pocchiola, M. (1993). A geometric proof of the enumeration formula for Sturmian words, *Inter. J. Algebra Comput.*, **3**, 349–355.

Berstel, J. and Pocchiola, M. (1996). Random generation of finite Sturmian words, *Discrete Math.*, **153**, 29–39.

Berstel, J. and Reutenauer, C. (1988). *Rational Series and Their Languages*. No. 12 in *EATCS Monographs on Theoretical Computer Science*. Springer-Verlag.

Berstel, J. and Séébold, P. (1994a). Morphismes de Sturm, *Bull. Belg. Math. Soc. Simon Stevin*, **1**, 175–189.

Berstel, J. and Séébold, P. (1994b). A remark on morphic Sturmian words, *Th. Inform. Appl.*, **28**, 255–263.

Berthé, V. (1996). Fréquences des facteurs des suites sturmiennes, *Th. Comput. Sci.*, **165**, 295–309.

Bertrand, A. (1977). Développements en base de Pisot et répartition modulo 1, *C. R. Acad. Sci. Paris, Sér. A*, **285**, 419–421.

Bertrand-Mathis, A. (1986). Développement en base θ, répartition modulo un de la suite $(x\theta^n)_{n\geq 0}$, langages codés et θ-shift, *Bull. Soc. Math. France*, **114**, 271–323.

Bertrand-Mathis, A. (1989). Comment écrire les nombres entiers dans une base qui n'est pas entière, *Acta Math. Hungar.*, **54**, 237–241.

Bès, A. (2000). An extension of the Cobham–Semenov theorem, *J. Symb. Logic*, **65**, 201–211.

Blanchard, F. (1989). β-expansions and symbolic dynamics, *Th. Comput. Sci.*, **65**, 131–141.

Blanchard, F. and Hansel, G. (1986). Systèmes codés, *Th. Comput. Sci.*, **44**, 17–49.

Borel, J.-P. and Laubie, F. (1991). Construction de mots de Christoffel, *C. R. Acad. Sci. Paris, Sér. A*, **313**, 483–485.

Borel, J.-P. and Laubie, F. (1993). Quelques mots sur la droite projective réelle, *J. Th. Nombres Bordeaux*, **5**, 23–52.

Boshernitzan, M. and Fraenkel, A. S. (1981). Nonhomogeneous spectra of numbers, *Discrete Math.*, **34**, 325–327.

Boshernitzan, M. and Fraenkel, A. S. (1984). A linear algorithm for nonhomogeneous spectra of numbers, *J. Algorithms*, **5**, 187–198.

Boyd, D. (1989). Salem numbers of degree four have periodic expansions, in de Coninck, J.-H. and Levesque, C. (Eds.), *Number Theory*, pp. 57–64. Walter de Gruyter.

Breslauer, D. (1995). An Observation on the Periodicity Structure of Strings, tech. rep., private communication.

Breslauer, D., Jiang, T., and Jiang, Z. (1997). Rotations of periodic strings and short superstring, *J. Algebra*, **24**, 340–353.

Brodal, G. S. and Pedersen, C. N. S. (2000). Finding maximal quasiperiodicities in strings, in Giancarlo, R. and Sankoff, D. (Eds.), *Combinatorial Pattern Matching (CPM 2000)*, Vol. 1848 of *Lect. Notes Comp. Sci.*, pp. 391–422. Springer-Verlag.

Brown, T. C. (1993). Descriptions of the characteristic sequence of an irrational, *Canad. Math. Bull.*, **36**, 15–21.

Bruckstein, A. M. (1991). Self-similarity properties of digitized straight lines, in Melter, R. A., Rosenfeld, A., and Bhattacharya, P. (Eds.), *Vision Geometry*, Vol. 119 of *Contemporary Mathematics*, pp. 1–20. Amer. Math. Soc.

Bruyère, V. (1992). Automata and codes with bounded deciphering delay, in Simon, I. (Ed.), *Latin American Theoretical INformatics (LATIN 92)*, Vol. 583 of *Lect. Notes Comp. Sci.*, pp. 99–107. Springer-Verlag.

Bruyère, V. and Hansel, G. (1997). Bertrand numeration systems and recognizability, *Th. Comput. Sci.*, **181**, 17–43.

Büchi, J. R. and Senger, S. (1986/7). Coding in the existential theory of concatenation, *Arch. Math. Logik Grundlag.*, **26**, 101–106.

Büchi, J. R. and Senger, S. (1988). Definability in the existential theory of concatenation and undecidable extensions of this theory, *Z. Math. Logik Grundlag. Math.*, **34**, 337–342.

Carlitz, L. (1954). q-Bernoulli and Eulerian numbers, *Trans. Amer. Math. Soc.*, **76**, 332–350.

Carlitz, L. (1959). Eulerian numbers and polynomials, *Math. Mag.*, **33**, 247–260.

Carlitz, L. (1975). A combinatorial property of q-Eulerian numbers, *Amer. Math. Monthly*, **82**, 51–54.

Carpi, A. and Luca, A. D. (2000). Periodic-like words, in Nielsen, M. and Rovan, B. (Eds.), *Mathematical Foundations of Computer Science (MFCS 2000)*, Vol. 1893 of *Lect. Notes Comp. Sci.*, pp. 264–274. Springer-Verlag.

Cartier, P. and Foata, D. (1969). *Problèmes combinatoires de commutation et réarrangements*. Vol. 85 of *Lect. Notes Math.* Springer-Verlag.

Cassaigne, J. (1993a). Counting overlap-free binary words, in Enjalbert, P., Finkel, A., and Wagner, K. W. (Eds.), *Theoretical Aspects of Computer Science (STACS 93)*, Vol. 665 of *Lect. Notes Comp. Sci.*, pp. 216–225. Springer-Verlag.

Cassaigne, J. (1993b). Unavoidable binary patterns, *Acta Inform.*, **30**, 385–395.

Cassaigne, J. (1994a). An algorithm to test if a given HD0L-language avoids a pattern, in *IFIP World Computer Congress (IFIP 94)*. Vol. 1: *Technology and Foundations*, pp. 459–464. Elsevier.

Cassaigne, J. (1994b). *Motifs évitables et régularités dans les mots*. Thèse de doctorat, Université Paris 6. LITP research report TH 94-04.

Cassaigne, J. (1996). Special factors of sequences with linear subword complexity, in Dassow, J. and Salomaa, A. (Eds.), *Developments in Language Theory II*, pp. 25–34. World Scientific.

Cassaigne, J. (1997). On a conjecture of J. Shallit, in Degano, P., Gorrieri, R., and Marchetti-Spaccamela, A. (Eds.), *Automata, Languages and Programming (ICALP 97)*, Vol. 1256 of *Lect. Notes Comp. Sci.*, pp. 693–704. Springer-Verlag.

Cassaigne, J. (1999). Limit values of the recurrence quotient of Sturmian sequences, *Th. Comput. Sci.*, **218**, 3–12.

Castelli, M. G., Mignosi, F., and Restivo, A. (1999). Fine and Wilf's theorem for three periods and a generalization of Sturmian words, *Th. Comput. Sci.*, **218**, 83–94.

Cauchy, A. (1840). Sur les moyens d'éviter les erreurs dans les calculs numériques, *C. R. Acad. Sci. Paris*, **11**, 789–798. Reprinted in A. Cauchy, *Oeuvres complètes*, 1ère série, Tome V, Gauthier-Villars, 1885, pp. 431–442.

Césari, Y. and Vincent, M. (1978). Une caractérisation des mots périodiques, *C. R. Acad. Sci. Paris, Sér. A*, **286**, 1175–1177.

Charatonik, W. and Pacholski, L. (1993). Word equations with two variables, in Abdulrab, H. and Pécuchet, J.-P. (Eds.), *Word Equations and Related Topics (IWWERT 91)*, Vol. 677 of *Lect. Notes Comp. Sci.*, pp. 43–56. Springer-Verlag.

Choffrut, C. and Culik, K. (1984). On extendibility of unavoidable sets, *Discr. Appl. Math.*, **9**, 125–137.

Choffrut, C., Harju, T., and Karhumäki, J. (1997). A note on decidability questions on presentations of word semigroups, *Th. Comput. Sci.*, **183**, 83–92.

Choffrut, C. and Karhumäki, J. (1997). Combinatorics of words, in Rozenberg, G. and Salomaa, A. (Eds.), *Handbook of Formal Languages*, Vol. 1, pp. 329–438. Springer-Verlag.

Christoffel, E. B. (1875). Observatio arithmetica, *Math. Ann.*, **6**, 145–152.

Chuan, W. (1997). α-words and characteristic sequences, *Discrete Math.*, **177**, 33–50.

Clarke, R. J. (1995). A short proof of a result of Foata and Zeilberger, *Adv. in Appl. Math.*, **16**, 129–131.

Clarke, R. J. (1997). Han's conjecture on permutations, *European J. Combin.*, **18**, 511–524.

Clarke, R. J. and Foata, D. (1994). Eulerian calculus, I: univariable statistics, *European J. Combin.*, **15**, 345–362.

Clarke, R. J. and Foata, D. (1995a). Eulerian calculus, II: an extension of Han's fundamental transformation, *European J. Combin.*, **16**, 221–252.

Clarke, R. J. and Foata, D. (1995b). Eulerian calculus, III: the ubiquitous Cauchy formula, *European J. Combin.*, **16**, 329–355.

Clarke, R. J., Steingrimsson, E., and Zeng, J. (1997a). The k-extensions of some new Mahonian statistics, *European J. Combin.*, **18**, 143–154.

Clarke, R. J., Steingrimsson, E., and Zeng, J. (1997b). New Euler–Mahonian statistics on permutations and words, *Adv. in Appl. Math.*, **18**, 237–270.

Clifford, A. H. and Preston, G. B. (1961). *The Algebraic Theory of Semigroups*, Vol. 1. Amer. Math. Soc.

Cobham, A. (1969). On the base-dependence of sets of numbers recognizable by finite automata, *Math. Syst. Th.*, **3**, 186–192.
Cohn, P. M. (1963). Rings with a weak algorithm, *Trans. Amer. Math. Soc.*, **109**, 332–356.
Cohn, P. M. (1978). Centralisateurs dans les corps libres, in Berstel, J. (Ed.), *Séries formelles*, École de printemps d'informatique théorique, pp. 45–54. LITP and ENSTA.
Cohn, P. M. (1981). *Universal Algebra* (revised edition). D. Reidel.
Cohn, P. M. (1985). *Free Rings and Their Relations* (second edition). No. 19 in *London Mathematical Monographs*. Academic Press.
Cohn, P. M. (1989). *Algebra*, Vol. 2 (second edition). J. Wiley and Sons.
Cohn, P. M. (1991). *Algebra*, Vol. 3. J. Wiley and Sons.
Conway, J. (1987). The weird and wonderful chemistry of audioactive decay, in Cover, T. and Gopinath, B. (Eds.), *Open Problems in Communications and Computation*. Springer-Verlag.
Coudrain, M. and Schützenberger, M.-P. (1966). Une condition de finitude des monoïdes finiment engendrés, *C. R. Acad. Sci. Paris, Sér. A*, **262**, 1149–1151.
Coven, E. M. (1974). Sequences with minimal block growth II, *Math. Syst. Th.*, **8**, 376–382.
Coven, E. M. and Hedlund, G. A. (1973). Sequences with minimal block growth, *Math. Syst. Th.*, **7**, 138–153.
Crisp, D., Moran, W., Pollington, A., and Shiue, P. (1993). Substitution invariant cutting sequences, *J. Th. Nombres Bordeaux*, **5**, 123–137.
Crochemore, M. (1981). An optimal algorithm for computing the repetitions in a word, *Inform. Proc. Lett.*, **12**, 244–250.
Crochemore, M., Hancart, C., and Lecroq, T. (2001). *Algorithmique du texte*. Vuibert.
Crochemore, M., Lerest, M., and Wender, P. (1983). An optimal test of finite unavoidable sets of words, *Inform. Proc. Lett.*, **16**, 125–137.
Crochemore, M. and Perrin, D. (1991). Two-way string-matching, *J. Assoc. Comput. Mach.*, **38**, 651–675.
Crochemore, M. and Rytter, W. (1994). *Text Algorithms*. Oxford University Press.
Crochemore, M. and Rytter, W. (1995). Squares, cubes, and time-space efficient string searching, *Algorithmica*, **13**, 405–425.
Culik, K. II and Karhumäki, J. (1983). Systems of equations over a free monoid and Ehrenfeucht's conjecture, *Discrete Math.*, **43**, 139–153.
Culik, K. II and Karhumäki, J. (1986). The equivalence of finite valued transducers (on HDT0L languages) is decidable, *Th. Comput. Sci.*, **47**, 71–84.
Currie, J. D. (1993). Open problems in pattern avoidance, *Amer. Math. Monthly*, **100**, 790–793.
Curzio, M., Longobardi, P., and Maj, M. (1983). Su di un problema combinatorio di teoria dei gruppi, *Atti Accad. Naz. Lincei Rend. Cl. Sci. Fis. Mat. Natur. (8)*, **74**, 136–142.
Curzio, M., Longobardi, P., Maj, M., and Robinson, D. J. S. (1985). A permutational property of groups, *Arch. Math. (Basel)*, **44**, 385–389.

Czumaj, A. and Gasieniec, L. (2000). On the complexity of determining the period of a string, in Giancarlo, R. and Sankoff, D. (Eds.), *Combinatorial Pattern Matching (CPM 2000)*, Vol. 1848 of *Lect. Notes Comp. Sci.*, pp. 412–411. Springer-Verlag.

Date, E., Jimbo, M., and Miwa, T. (1990). Representations of $U_q(gl_n)$ at $q = 0$ and the Robinson–Schensted correspondence, in Brink, L., Friedan, D., and Polyakov, A. M. (Eds.), *Physics and Mathematics of Strings, Memorial Volume of V. Knizhnik*. World Scientific.

de Bruijn, N. G. (1946). A combinatorial problem, *Nederl. Akad. Wetensch. Proc*, **49**, 758–764.

de Bruijn, N. G. (1981). Sequences of zeros and ones generated by special production rules, *Indag. Math.*, **43**, 27–37.

de Bruijn, N. G. (1989). Updown generation of Beatty sequences, *Indag. Math.*, **51**, 385–407.

De Luca, A. (1997a). Combinatorics of standard Sturmian words, in Mycielski, J., Rozenberg, G., and Salomaa, A. (Eds.), *Structures in Logic and Computer Science*, Vol. 1261 of *Lect. Notes Comp. Sci.*, pp. 249–267. Springer-Verlag.

De Luca, A. (1997b). Standard Sturmian morphisms, *Th. Comput. Sci.*, **178**, 205–224.

De Luca, A. (1997c). Sturmian words: structure, combinatorics, and their arithmetics, *Th. Comput. Sci.*, **183**, 45–82.

De Luca, A. and Mignosi, F. (1994). On some combinatorial properties of Sturmian words, *Th. Comput. Sci.*, **136**, 361–385.

De Luca, A. and Restivo, A. (1984). A finiteness condition for finitely generated semigroups, *Semigroup Forum*, **28**, 123–134.

De Luca, A. and Restivo, A. (1986). On a generalization of a conjecture of Ehrenfeucht, *Bull. EATCS*, **30**, 84–90.

De Luca, A. and Varricchio, S. (1990). A note on ω-permutable semigroups, *Semigroup Forum*, **40**, 153–157.

De Luca, A. and Varricchio, S. (1991a). Combinatorial properties of uniformly recurrent words and an application to semigroups, *Inter. J. Algebra Comput.*, **1**, 227–245.

De Luca, A. and Varricchio, S. (1991b). Finiteness and iteration conditions for semigroups, *Th. Comput. Sci.*, **87**, 315–327.

De Luca, A. and Varricchio, S. (1994). A finiteness condition for semigroups generalizing a theorem of Coudrain and Schützenberger, *Adv. in Math.*, **108**, 91–103.

De Luca, A. and Varricchio, S. (1999). *Finiteness and Regularity in Semigroups and Formal Languages*. Springer-Verlag.

Dean, R. (1965). A sequence without repeats on x, x^{-1}, y, y^{-1}, *Amer. Math. Monthly*, **72**, 383–385.

Dekking, F. M. (1976). On repetitions of blocks in binary sequences, *J. Combin. Th. Ser. A*, **20**, 292–299.

Denert, M. (1990). The genus zeta function of hereditary orders in central simple algebras over global fields, *Math. Comput.*, **54**, 449–465.

Derencourt, D. (1996). A three-word code which is not prefix–suffix composed, *Th. Comput. Sci.*, **163**, 145–160.

Désarménien, J. (1982). Un analogue des congruences de Kummer pour les q-polynômes d'Euler, *European J. Combin.*, **3**, 19–28.

Désarménien, J. (1983). Fonctions symétriques associées à des suites classiques de nombres, *Ann. Sci. Ecole Norm. Sup. (4)*, **16**, 271–304.

Désarménien, J. (1990). Conséquences énumératives de la symétrie des fonctions de Schur, *Séminaire Lotharingien de Combinatoire*, B25a.

Désarménien, J. and Foata, D. (1985). Fonctions symétriques et séries hypergéométriques basiques multivariées, *Bull. Soc. Math. France*, **113**, 3–22.

Désarménien, J. and Foata, D. (1991). Statistiques d'ordre sur les permutations colorées, *Discrete Math.*, **87**, 133–149.

Désarménien, J. and Wachs, M. (1988). Descentes sur les dérangements et mots circulaires, *Séminaire Lotharingien de Combinatoire*, B19a.

Désarménien, J. and Wachs, M. (1993). Descent classes of permutations with a given number of fixed points, *J. Combin. Th. Ser. A*, **64**, 311–328.

Devolder, J. (1993). *Codes, mots infinis et bi-infinis*. Thèse de doctorat, Université Lille 1.

Devolder, J. (1999). Generators with bounded deciphering delay for rational ω-languages, *J. Autom. Lang. Combin.*, **4**, 183–204.

Devolder, J., Latteux, M., Litovsky, I., and Staiger, L. (1994). Codes and infinite words, *Acta Cybernet.*, **11**, 241–256.

Devolder, J. and Timmerman, E. (1992). Finitary codes for biinfinite words, *Th. Inform. Appl.*, **26**, 363–386.

Dickson, L. E. (1913). Finiteness of the odd perfect and primitive abundant numbers with n distinct prime factors, *Amer. J. Math.*, **35**, 413–422.

Didier, G. (1997). Echange de trois intervalles et suites sturmiennes, *J. Th. Nombres Bordeaux*, **9**, 463–478.

Diekert, V., Gutiérrez, C., and Hagenah, C. (2001). The existential theory of equations with rational constraints in free groups is PSPACE-complete, in Ferreira, A. and Reichel, H. (Eds.), *Theoretical Aspects of Computer Science (STACS 01)*, Vol. 2010 of *Lect. Notes Comp. Sci.*, pp. 170–182. Springer-Verlag.

Diekert, V. and Lohrey, M. (2001). A note on the existential theory in plain groups, *Inter. J. Algebra Comput.* Special issue of the International Conference on Geometric and Combinatorial Methods in Group Theory and Semigroup Theory, Lincoln, Nebraska, May 2000.

Diekert, V., Matiyasevich, Yu., and Muscholl, A. (1999). Solving word equations modulo partial commutations, *Th. Comput. Sci.*, **224**, 215–235.

Diekert, V. and Muscholl, A. (2001). Solvability of equations in free partially commutative groups is decidable, in Orejas, F., Spirakis, P. G., and van Leeuwen, J. (Eds.), *Automata, Languages and Programming (ICALP 01)*, Vol. 2076 of *Lect. Notes Comp. Sci.*, pp. 543–554. Springer-Verlag.

Diekert, V. and Robson, J. M. (1999). Quadratic word equations, in Karhumäki, J. et al. (Eds.), *Jewels Are Forever – Contributions on Theoretical Computer Science in Honor of Arto Salomaa*, pp. 314–326. Springer-Verlag.

Diekert, V. and Rozenberg, G. (Eds.) (1995). *The Book of Traces*. World Scientific.

Do Long Van (1982). Codes avec des mots infinis, *RAIRO Inform. Th.*, **16**, 371–386.

Dorst, L. and Smeulders, A. W. M. (1991). Decomposition of discrete curves into piecewise straight segments in linear time, in Melter, R. A., Rosenfeld, A., and Bhattacharya, P. (Eds.), *Vision Geometry*, Vol. 119 of *Contemporary Mathematics*. Amer. Math. Soc, Providence, RI.

Droms, C. (1985). Graph groups, coherence and three-manifolds, *J. Algebra*, **106**(2), 484–489.

Droms, C. (1987a). Isomorphisms of graph groups, *Proc. Amer. Math. Soc.*, **100**, 407–408.

Droms, C. (1987b). Subgroup of graph groups, *J. Algebra*, **110**, 519–522.

Droubay, X., Justin, J., and Pirillo, G. (2001). Episturmian words and some constructions of de Luca and Rauzy, *Th. Comput. Sci.*, **255**, 539–553.

Droubay, X. and Pirillo, G. (1999). Palindromes and Sturmian words, *Th. Comput. Sci.*, **223**, 73–85.

Dulucq, S. and Gouyou-Beauchamps, D. (1990). Sur les facteurs des suites de Sturm, *Th. Comput. Sci.*, **71**, 381–400.

Duprat, J., Herreros, Y., and Kla, S. (1993). New complex representations of complex numbers and vectors, *IEEE Trans. Comput.*, **42**, 817–824.

Durand, F. (1998). A generalization of Cobham's theorem, *Th. Comput. Syst.*, **32**, 169–185.

Durnev, V. G. (1974). Equations on free semigroups and groups, *Mat. Zametki*, **16**, 717–724. In Russian. English translation: *Math. Notes, 16*, 1024–1028, 1975.

Durnev, V. G. (1995). Undecidability of the positive $\forall\exists^3$-theory of a free semigroup, *Sibirsk. Mat. Zh.*, **36**(5), 1067–1080. In Russian. English translation: *Siberian Math. J., 36*(5), 917–929, 1995.

Durnev, V. G. (1997). Studying algorithmic problems for free semi-groups and groups, in Adian, S. and Nerode, A. (Eds.), *Logical Foundations of Computer Science (LFCS 97)*, Vol. 1234 of *Lect. Notes Comp. Sci.*, pp. 88–101. Springer-Verlag.

Duval, J.-P. (1979). Périodes et répétitions des mots du monoïde libre, *Th. Comput. Sci.*, **9**, 17–26.

Duval, J.-P. (1982). Relationship between the period of a finite word and the length of its unbordered segments, *Discr. Math.*, **40**, 31–44.

Duval, J.-P. (1998). Périodes locales et propagation de périodes dans un mot, *Th. Comput. Sci.*, **204**, 87–98.

Duval, J.-P., Mignosi, F., and Restivo, A. (2001). Recurrence and periodicity in infinite words from local periods, *Th. Comput. Sci.* To appear.

Ehrenfeucht, A., Hausler, D., and Rozenberg, G. (1983a). On regularity of context-free languages, *Th. Comput. Sci.*, **27**, 311–332.

Ehrenfeucht, A., Karhumäki, J., and Rozenberg, G. (1983b). On binary equality languages and a solution to the test set conjecture in the binary case, *Th. Comput. Sci.*, **85**, 76–85.

Ehrenfeucht, A. and Rozenberg, G. (1978). Elementary homomorphisms and a solution to the D0L sequence equivalence problem, *Th. Comput. Sci.*, **7**, 169–184.

Ehrenfeucht, A. and Silberger, D. M. (1979). Periodicity and unbordered words, *Discrete Math.*, **26**, 101–109.

Eilenberg, S. (1974). *Automata, Languages and Machines*, Vol. A. Academic Press.

Ekhad, S. B. and Zeilberger, D. (1997). Proof of Conway's lost cosmological theorem, *Electronic Res. Announc. Amer. Math. Soc.*, **3**, 78–82.

Entringer, R. C., Jackson, D. E., and Schatz, J. A. (1974). On nonrepetitive sequences, *J. Combin. Th. Ser. A*, **16**, 159–164.

Epifanio, C., Koskas, M., and Mignosi, F. (1999). On a conjecture on bidimensional words, tech. rep. 78, Dipartimento di Matematica e Appl., Università di Palermo.

Evdokimov, A. A. (1968). Strongly asymmetric sequences generated by a finite number of symbols, *Dokl. Akad. Nauk SSSR*, *179*, 1268–1271. In Russian. English translation: *Soviet Math. Dokl.*, **9**, 536–539, 1968.

Eyono Obono, S., Goralčik, P., and Maksimenko, M. (1994). Efficient solving of the word equations in one variable, in Privara, I. et al. (Eds.), *Proc. of the 19th MFCS*, Vol. 841 of *Lect. Notes Comp. Sci.*, pp. 336–341. Springer-Verlag.

Fabre, S. (1994). Une généralisation du théorème de Cobham, *Acta Arith.*, **67**, 197–208.

Fagnot, I. (1997). Sur les facteurs des mots automatiques, *Th. Comput. Sci.*, **172**, 67–89.

Fedou, J.-M. and Rawlings, D. (1995). Adjacencies in words, *Adv. in Appl. Math.*, **16**, 306–218.

Ferenczi, S. (1995). Les transformations de Chacon: combinatoire, structure géométrique, liens avec les systèmes de complexité $2n + 1$, *Bull. Soc. Math. France*, **123**, 271–292.

Fine, N. J. and Wilf, H. S. (1965). Uniqueness theorem for periodic functions, *Proc. Amer. Math. Soc.*, **16**, 109–114.

Flatto, L., Lagarias, J. C., and Poonen, B. (1994). The zeta function of the beta-transformation, *Ergodic Th. Dynamical Syst.*, **14**, 237–266.

Foata, D. (1965). Étude algébrique de certains problèmes d'analyse combinatoire et du calcul des probabilités, *Publ. Inst. Statist. Univ. Paris*, **14**, 81–241.

Foata, D. (1995). Les distributions Euler-Mahoniennes sur les mots, *Discrete Math.*, **139**, 167–188.

Foata, D. and Han, G. N. (1997). Calcul basique des permutations signées, I. Longueur et nombre d'inversions, *Adv. in Appl. Math.*, **18**, 489–509.

Foata, D. and Han, G. N. (1998). Transformations on words, *J. Algorithms*, **28**, 172–191.

Foata, D. and Krattenthaler, C. (1995). Graphical major indices, II, *Séminaire Lotharingien de Combinatoire*, B34k, 16 pp.

Foata, D. and Schützenberger, M.-P. (1970). Major index and inversion number of permutations, *Math. Nachr.*, **83**, 143–159.

Foata, D. and Zeilberger, D. (1990). Denert's permutation statistic is indeed Euler–Mahonian, *Stud. Appl. Math.*, **83**, 31–59.

Fraenkel, A. S. (1982). How to beat your Wythoff games opponent on three fronts, *Amer. Math. Monthly*, **89**, 353–361.

Fraenkel, A. S. (1985). Systems of numeration, *Amer. Math. Monthly*, **92**, 105–114.

Fraenkel, A. S., Mushkin, M., and Tassa, U. (1978). Determination of $[n\theta]$ by its sequence of differences, *Canad. Math. Bull.*, **21**, 441–446.

Fraenkel, A. S. and Simpson, J. (1998). How many squares can a string contain?, *J. Combin. Th. Ser. A*, **82**, 112–120.

Freyd, P. (1968). Redei's finiteness theorem for commutative semigroups, *Proc. Amer. Math. Soc.*, **19**, 1003.

Frougny, C. (1992). Representation of numbers and finite automata, *Math. Syst. Th.*, **25**, 37–60.

Frougny, C. and Solomyak, B. (1996). On representation of integers in linear numeration systems, in Pollicott, M. and Schmidt, K. (Eds.), *Ergodic Theory of $\mathbf{Z}^d$-Actions*, Vol. 228 of *London Math. Soc. Lect. Note Series*, pp. 345–368. Cambridge University Press.

Gambaudo, J.-M., Lanford, O., and Tresser, C. (1984). Dynamique symbolique des rotations, *C. R. Acad. Sci. Paris, Sér. A*, **299**, 823–826.

Gantmacher, F. R. (1960). *Matrix Theory*. Chelsea Pub.

Garsia, A. M. (1979). On the 'maj' and 'inv' q-analogues of Eulerian polynomials, *Linear and Multilinear Algebra*, **8**, 21–34.

Garsia, A. M. and Gessel, I. M. (1979). Permutation statistics and partitions, *Adv. in Math.*, **31**, 288–305.

Gasper, G. and Rahman, M. (1990). *Basic Hypergeometric Series*, Vol. 35 of *Encyclopedia of Mathematics and Its Applications*. Cambridge University Press.

Gathen, J. von zur and Sieveking, M. (1978). A bound on solutions of linear integer equalities and inequalities, *Proc. Amer. Math. Soc.*, **72**, 155–158.

Gauss, C. F. (1900). *Werke*. Teubner.

Gazeau, J.-P. (1995). Pisot-cyclotomic integers for quasilattice, in Moody, R. (Ed.), *The Mathematics of Long-Range Aperiodic Order*, Vol. 489 of *NATO-ASI Series*, pp. 175–198. Kluwer.

Gelfand, I. M., Krob, D., Lascoux, A., Leclerc, B., Retakh, V. S., and Thibon, J.-Y. (1995). Noncommutative symmetric functions, *Adv. in Math.*, **112**, 218–348.

Gessel, I. M. (1977). *Generating functions and enumeration of sequences*. PhD thesis, Dept. Math., MIT.

Gessel, I. M. (1982). A q-analogue of the exponential formula, *Discr. Math.*, **40**, 69–80.

Gessel, I. M. and Reutenauer, C. (1993). Number of permutations with given cycle structure and descent set, *J. Combin. Th. Ser. A*, **64**, 189–215.

Giancarlo, R. and Mignosi, F. (1994). Generalizations of the periodicity theorem of Fine and Wilf, in *Trees in Algebra and Programming (CAAP 94)*, Vol. 787 of *Lect. Notes Comp. Sci.*, pp. 130–141. Springer-Verlag.

Gilbert, E. N. and Moore, E. F. (1959). Variable length binary encodings, *Bell Syst. Tech. J.*, **38**, 933–967.

Gilbert, W. J. (1981). Radix representations of quadratic fields, *J. Math. Anal. Appl.*, **83**, 264–274.

Gilbert, W. J. (1986). The fractal dimension of sets derived from complex bases, *Canad. Math. Bull.*, **29**, 495–500.

Gilbert, W. J. (1994). Gaussian integers as bases for exotic number systems, unpublished manuscript.

Golod, E. S. (1964). On nil-algebras and finitely approximable p-groups, *Izv. Akad. Nauk SSSR Ser. Mat.*, **28**, 273–276.

Goralčik, P. and Vaniček, T. (1991). Binary patterns in binary words, *Inter. J. Algebra Comput.*, **1**, 387–391.

Grabner, P., Kirschenhofer, P., and Prodinger, H. (1998). The sum-of-digits function for complex bases, *J. London Math. Soc. (2)*, **57**, 20–40.

Graham, R. L., Knuth, D., and Patashnik, O. (1989). *Concrete Mathematics*. Addison-Wesley.

Greene, C. (1974). An extension of Schensted's theorem, *Adv. in Math.*, **14**, 254–265.

Gromov, M. (1987). Hyperbolic groups, in Gersten, S. M. (Ed.), *Essays in Group Theory*, No. 8 in Math. Sci. Res. Inst. Publ., pp. 75–263. Springer-Verlag.

Grünbaum, B. (1972). *Arrangements and Spreads*, Vol. 10 of *Conference Board of the Math. Sciences Regional Conf. Ser. in Math.* Amer. Math. Soc.

Guba, V. S. (1986). The equivalence of infinite systems of equations in free groups and semigroups with finite subsystems, *Mat. Zametki*, **40**, 321–324. In Russian.

Guibas, L. and Odlyzko, A. (1981). Periods in strings, *J. Combin. Th. Ser. A*, **30**, 19–42.

Gutiérrez, C. (1998a). Satisfiability of word equations with constants is in exponential space, in *Foundations of Computer Science (FOCS 98)*, pp. 112–119. IEEE Computer Society Press.

Gutiérrez, C. (1998b). Solving equations in strings: on Makanin's algorithm, in Lucchesi, C. L. and Moura, A. V. (Eds.), *LATIN'98: Theoretical Informatics*, No. 1380 in Lect. Notes Comp. Sci., pp. 358–373. Springer-Verlag.

Gutiérrez, C. (2000). Satisfiability of equations in free groups is in PSPACE, in *Theory of Computing (STOC 2000)*, pp. 21–27. ACM Press.

Halava, V., Harju, T., and Ilie, L. (2000?). Periods and binary words, *J. Combin. Th. Ser. A*. To appear.

Hall, P. (1954). Finiteness conditions for soluble groups, *Proc. Amer. Math. Soc.*, **4**, 419–436.

Han, G. N. (1990a). Distribution Euler–mahonienne: une correspondance, *C. R. Acad. Sci. Paris, Sér. A*, **310**, 311–314.

Han, G. N. (1990b). Une nouvelle bijection pour la statistique de Denert, *C. R. Acad. Sci. Paris, Sér. A*, **310**, 493–496.

Han, G. N. (1994). Une transformation fondamentale sur les réarrangements de mots, *Adv. in Math.*, **105**, 26–41.

Han, G. N. (1995). The k-extension of a Mahonian statistic, *Adv. in Appl. Math.*, **16**, 297–305.

Hansel, G. (1998). Systèmes de numération indépendants et syndéticité, *Th. Comput. Sci.*, **204**, 119–130.

Haring-Smith, R. H. (1983). Groups and simple languages, *Trans. Amer. Math. Soc.*, **279**, 337–356.

Harju, T. and Karhumäki, J. (1986). On the defect theorem and simplifiability, *Semigroup Forum*, **33**, 199–217.

Harju, T., Karhumäki, J., and Petrich, M. (1997a). Compactness of equations on completely regular semigroups, in Mycielski, J., Rozenberg, G., and Salomaa, A. (Eds.), *Structures in Logic and Computer Science*, Vol. 1261 of *Lect. Notes Comp. Sci.*, pp. 268–280. Springer-Verlag.

Harju, T., Karhumäki, J., and Plandowski, W. (1997b). Compactness of systems of equations in semigroups, *Inter. J. Algebra Comput.*, **7**, 457–470.

Hashiguchi, K. (1986). Notes on finitely generated semigroups and pumping conditions for regular languages, *Th. Comput. Sci.*, **46**, 53–66.

Hawkins, D. and Mientka, W. (1956). On sequences which contain no repetition, *Math. Student*, **24**, 185–187.

Hmelevskiĭ, Yu. I. (1971). Equations in free semigroups, in Petrovskiĭ, I. G. (Ed.), *Trudy Mat. Inst. Steklov.*, Vol. 107. In Russian. English translation: *Proc. Steklov Inst. Math., 107*, Amer. Math. Soc., 1976.

Hollander, M. (1998). Greedy numeration systems and regularity, *Th. Comput. Syst.*, **31**, 111–133.

Holton, C. and Zamboni, L. Q. (2000). Initial powers in Sturmian words, preprint.

Honkala, J. (1988). A defect property of codes with unbounded delays, *Discr. Appl. Math.*, **21**, 261–264.

Hopcroft, J. E. and Ullman, J. D. (1979). *Introduction to Automata Theory, Languages, and Computation*. Addison-Wesley.

Hotzel, E. (1979). On finiteness conditions in semigroups, *J. Algebra*, **60**, 352–370.

Howie, J. M. (1976). *An Introduction to Semigroup Theory*. Academic Press.

Hubert, P. (1995). Complexité de suites définies par des billards rationnels, *Bull. Soc. Math. France*, **123**, 257–270.

Hubert, P. (1996). Propriétés combinatoires des suites définies par le billard dans les triangles pavants, *Th. Comput. Sci.*, **164**, 165–183.

Ilie, L. and Plandowski, W. (2000). Two-variable word equations, in Reichel, H. et al. (Eds.), *Theoretical Aspects of Computer Science (STACS 2000)*, Vol. 1770 of *Lect. Notes Comp. Sci.*, pp. 122–132. Springer-Verlag.

Istrail, S. (1977). On irreducible languages and non rational numbers, *Bull. Math. Soc. Sci. Math. R. S. Roumanie (N.S.)*, **21**, 301–308.

Ito, S. and Takahashi, Y. (1974). Markov subshifts and realization of β-expansions, *J. Math. Soc. Japan*, **26**, 33–55.

Iwanik, A. (1994). Cyclic approximations of irrational rotations, *Proc. Amer. Math. Soc.*, **121**, 691–695.

Jacobson, N. (1964). *Structure of Rings* (revised edition)., Vol. 37 of *American Math. Soc. Colloquium Publ.* Amer. Math. Soc.

Jaffar, J. (1990). Minimal and complete word unification, *J. Assoc. Comput. Mach.*, **37**(1), 47–85.

Julia, S. (1996). On ω-generators and codes, in *Automata, Languages and Programming (ICALP 96)*, Vol. 1099 of *Lect. Notes Comp. Sci.*, pp. 393–402. Springer-Verlag.

Jürgensen, H. and Konstantinidis, S. (1997). Codes, in Rozenberg, G. and Salomaa, A. (Eds.), *Handbook of Formal Languages*, Vol. 1, pp. 511–607. Springer-Verlag.

Justin, J. (2000). On a paper by Castelli, Mignosi, Restivo, *Th. Inform. Appl.*, **34**, 373–377.

Justin, J. and Pirillo, G. (1991). Shirshov's theorem and ω-permutability of semigroups, *Adv. in Math.*, **87**, 151–159.

Justin, J. and Pirillo, G. (1997). Decimation and Sturmian words, *Th. Inform. Appl.*, **31**, 271–290.

Karhumäki, J. (1981). On the equivalence problem for binary DOL systems, *Inform. and Control*, **50**, 276–284.

Karhumäki, J. (1985a). On three-element codes, *Th. Comput. Sci.*, **40**, 3–11.

Karhumäki, J. (1985b). A property of three-element codes, *Th. Comput. Sci.*, **41**, 215–222.

Karhumäki, J., Lepistö, A., and Plandowski, W. (1998a). Locally periodic infinite words and a chaotic behaviour, in *Automata, Languages and Programming (ICALP 98)*, pp. 421–430.

Karhumäki, J., Maňuch, J., and Plandowski, W. (1998b). On defect effect of bi-infinite words, in *Mathematical Foundations of Computer Science (MFCS 98)*, Vol. 1450 of *Lect. Notes Comp. Sci.*, pp. 674–682. Springer-Verlag.

Karhumäki, J., Mignosi, F., and Plandowski, W. (2000). The expressibility of languages and relations by word equations, *J. Assoc. Comput. Mach.*, **47**, 483–505.

Karhumäki, J. and Plandowski, W. (1994). On defect effect of many identities in free semigroups, in Păun, G. (Ed.), *Mathematical Aspects of Natural and Formal Languages*, pp. 225–232. World Scientific.

Karhumäki, J. and Plandowski, W. (1996). On the size of independent systems of equations in semigroups, *Th. Comput. Sci.*, **168**, 105–119.

Kashiwara, M. (1991). Crystallizing the q-analogue of universal enveloping algebras, *Comm. Math. Phys.*, **133**, 249–260.

Kashiwara, M. (1994). Crystallization of quantized enveloping algebras, *Sugaku Expositiones*, **7**, 99–115.

Kátai, I. (1994). Number systems in imaginary quadratic fields, *Ann. Univ. Sci. Budapest. Sect. Comput.*, **14**, 91–103.

Kátai, I. and Kovács, B. (1981). Canonical number systems in imaginary quadratic fields, *Acta Math. Acad. Sci. Hungar.*, **37**, 159–164.

Kátai, I. and Szabó, J. (1975). Canonical number systems, *Acta Sci. Math. (Szeged)*, **37**, 255–280.

Kenyon, R. W. (1992). Self-replicating tilings, in Walters, P. (Ed.), *Symbolic Dynamics and Its Applications*, Vol. 135 of *Contemporary Mathematics*, pp. 239–263. Amer. Math. Soc.

Kharlampovich, O. G. and Sapir, M. V. (1995). Algorithmic problems in varieties, *Inter. J. Algebra Comput.*, **5**, 379–602.

Klarner, D. A., Birget, J. C., and Satterfield, W. (1991). On the undecidability of the freeness of integer matrix semigroups, *Inter. J. Algebra Comput.*, **1**, 223–226.

Knuth, D. E. (1970). Permutations, matrices, and generalized Young tableaux, *Pacific J. Math.*, **34**, 709–727.

Knuth, D. E. (1973). *The Art of Computer Programming*. Vol. 3: *Sorting and Searching*. Addison-Wesley.

Knuth, D. E. (1988). *The Art of Computer Programming*. Vol. 2: *Seminumerical Algorithms* (second edition). Addison-Wesley.

Kolotov, A. T. (1978). Free subalgebras of free associative algebras, *Sibirsk. Mat. Zh.*, **19**, 328–335. In Russian. English translation: *Siberian Math. J.*, **19**, 229–234, 1978.

Kolpakov, R., and Kucherov, G. (1999a). Finding maximal repetitions in a word in linear time, in *Foundations of Computer Science (FOCS 99)*, pp. 596–604. IEEE Computer Society Press.

Kolpakov, R. and Kucherov, G. (1999b). On maximal repetitions in words, in Ciobanu, G. and Păun, G. (Eds.), *Foundations of Computation Theory (FCT 99)*, Lect. Notes Comp. Sci., pp. 374–385. Springer-Verlag.

Komatsu, T. (1996). A certain power series associated with a Beatty sequence, *Acta Arith.*, **76**, 109–129.

Komatsu, T. and Poorten, A. J. van der (1996). Substitution invariant Beatty sequences, *J. Appl. Math. Mech.*, **22**, 349–354.

Koplowitz, J., Lindenbaum, M., and Bruckstein, A. (1990). The number of digital straight lines on an $n \times n$ grid, *IEEE Trans. Inform. Th.*, **36**, 192–197.

Kościelski, A. and Pacholski, L. (1996). Complexity of Makanin's Algorithm, *J. Assoc. Comput. Mach.*, **43**(4), 670–684. Preliminary version in *Foundations of Computer Science (FOCS 90)*, pp. 824–829.

Kościelski, A. and Pacholski, L. (1998). Makanin's algorithm is not primitive recursive, *Th. Comput. Sci.*, **191**, 145–156.

Kostrikin, A. I. and Shafarevich, I. R. (1990). *Algebra*, Vol. I. Springer-Verlag.

Kozen, D. (1977). Lower bounds for natural proof systems, In *Foundations of Computer Science (FOCS 77)*, pp. 254–266. IEEE Computer Society Press.

Krob, D. and Thibon, J.-Y. (1997). Noncommutative symmetric functions IV: quantum linear groups and Hecke algebras at $q = 0$, *J. Algebraic Combin.*, **6**, 339–376.

Lascoux, A. (1974). Puissances extérieures, déterminants et cycles de Schubert, *Bull. Soc. Math. France*, **102**, 161–179.

Lascoux, A. (1991). Cyclic permutations on words, tableaux and harmonic polynomials, In *Proceedings of the Hyderabad Conference on Algebraic Groups*, pp. 323–347. Manoj Prakashan.

Lascoux, A., Leclerc, B., and Thibon, J.-Y. (1995). Crystal graphs and q-analogues of weight multiplicities for the root system A_n, *Lett. Math. Phys.*, **35**, 359–374.

Lascoux, A. and Schützenberger, M.-P. (1980). A new statistics on words, *Discr. Math.*, **6**, 251–255.

Lascoux, A. and Schützenberger, M.-P. (1981). Le monoïde plaxique, in De Luca, A. (Ed.), *Non-Commutative Structures in Algebra and Geometric Combinatorics*, Vol. 109 of *Quaderni de "La Ricerca Scientifica"*, pp. 129–156. Consiglio Nazionale delle Ricerche.

Lascoux, A. and Schützenberger, M.-P. (1988). Keys and standard bases, in Stanton, D. (Ed.), *Invariant Theory and Tableaux*, Vol. 19 of *The IMA Volumes in Mathematics and Its Applications*, pp. 125–144. Springer-Verlag.

Lawrence, J. (1986). The nonexistence of finite test set for set-equivalence of finite substitutions, *Bull. EATCS*, **28**, 34–37.

Leclerc, B. and Thibon, J.-Y. (1996). The Robinson–Schensted correspondence, crystal bases, and the quantum straightening at $q = 0$, *Electronic J. Combin.*, **3**, 249–272.

Leech, J. (1957). A problem on strings of beads, *Math. Gaz.*, **41**, 277–278.

Lentin, A. (1972). *Equations dans les monoïdes libres*. Gauthier-Villars.

Lentin, A. and Schützenberger, M.-P. (1967). A combinatorial problem in the theory of free monoids, in Bose, R. C. and Dowlings, T. E. (Eds.), *Combinatorial Mathematics*, pp. 112–144. North Carolina Press.

Lepistö, A. (1999). Relations between local and global periodicity of words, In *Automata, Languages and Programming (ICALP 99)*, Vol. 1644 of *Lect. Notes Comp. Sci.*, pp. 534–543. Springer-Verlag.

Lind, D. (1984). The entropies of topological Markov shifts and a related class of algebraic integers, *Ergodic Th. Dynamical Syst.*, **4**, 283–300.

Lind, D. and Marcus, B. (1995). *An Introduction to Symbolic Dynamics and Coding*. Cambridge University Press.

Linna, M. (1977). The decidability of the D0L prefix problem, *Int. J. Comput. Math.*, **6**, 127–142.

Lint, J. H. van and Wilson, R. M. (1992). *A Course in Combinatorics*. Cambridge University Press.

Litovsky, I. (1991). Prefix-free languages as ω-generators, *Inform. Proc. Lett.*, **37**, 61–65.

Littelmann, P. (1994). A Littlewood–Richardson rule for symmetrizable Kac–Moody algebras, *Invent. Math.*, **116**, 329–346.

Littelmann, P. (1996). A plactic algebra for semisimple Lie algebras, *Adv. in Math.*, **124**, 312–331.

Littlewood, D. E. (1950). *The Theory of Group Characters* (second edition). Oxford University Press.
Loraud, N. (1995). β-shift, systèmes de numération et automates, *J. Th. Nombres Bordeaux*, **7**, 473–498.
Lothaire, M. (1983). *Combinatorics on Words*, Vol. 17 of *Encyclopedia of Mathematics and Its Applications*. Addison-Wesley. Reprinted in the *Cambridge Mathematical Library*, Cambridge University Press, 1997.
Lovasz, L. and Marx, M. L. (1976). A forbidden substructure characterization of Gauss codes, *Bull. Amer. Math. Soc.*, **82**, 121–122.
Lunnon, W. F. and Pleasants, P. A. B. (1992). Characterization of two-distance sequences, *J. Austral. Math. Soc. Ser. A*, **53**, 198–218.
Lyndon, R. (1960). Equations in free groups, *Trans. Amer. Math. Soc.*, **96**, 445–457.
Lyndon, R. C. and Schupp, P. E. (1977). *Combinatorial Group Theory*. Springer-Verlag.
Macdonald, I. G. (1995). *Symmetric Functions and Hall Polynomials* (second edition). Clarendon Press.
MacMahon, P. A. (1913). The indices of permutations and the derivation therefrom of functions of a single variable associated with the permutations of any assemblage of objects, *Amer. J. Math.*, **35**, 314–321.
MacMahon, P. A. (1915). *Combinatory Analysis*. Cambridge University Press. Reprinted by Chelsea Pub. 1955.
MacMahon, P. A. (1978). *Collected Papers*, Vol. 1. The MIT Press. Edited by G. E. Andrews.
Magnus, W., Karrass, A., and Solitar, D. (1966). *Combinatorial Group Theory*. Interscience.
Makanin, G. S. (1976). On the rank of equations in four unknowns in a free semigroup, *Mat. Sb. (NS)*, **100**, 285–311.
Makanin, G. S. (1977). The problem of solvability of equations in a free semigroup, *Mat. Sb. (NS)*, *103*(2), 147–236. In Russian. English translation in: *Math. USSR-Sb.*, **32**, 129–198, 1977.
Makanin, G. S. (1979). Recognition of the rank of equations in a free semigroup, *Izv. Akad. Nauk SSSR Ser. Mat.*, *43*. In Russian. English translation: *Math. USSR-Izv.*, **14**, 499–545, 1980.
Makanin, G. S. (1981). Equations in a free semigroup, *Amer. Math. Soc. Transl. (II Ser.)*, **117**, 1–6.
Makanin, G. S. (1982). Equations in a free group, *Izv. Akad. Nauk SSSR Ser. Mat.*, *46*, 1199–1273. In Russian. English translation: *Math. USSR-Izv.*, **21**, 483–546, 1983.
Makanin, G. S. (1984). Decidability of the universal and positive theories of a free group, *Izv. Akad. Nauk SSSR Ser. Mat.*, **48**, 735–749. In Russian. English translation: *Math. USSR-Izv.*, **25**, 75–88, 1985.
Mantaci, S. and Karhumäki, J. (1999). Defect theorems for trees, *Fund. Inform.*, **38**, 119–133.
Mantaci, S. and Restivo, A. (1999). On the defect theorem for trees, in *Automata and Formal Languages, VIII*, Vol. 54 of *Publ. Math. Debrecen*, pp. 923–932.

Marchenkov, S. S. (1982). Unsolvability of positive ∀∃-theory of a free semi-group, *Sibirsk. Mat. Zh.*, **23**(1), 196–198. In Russian.

Markoff, A. (1882). Sur une question de Jean Bernoulli, *Math. Ann.*, **19**, 27–36.

Markowsky, G. (1977). Bounds on the index and period of a binary relation on a finite set, *Semigroup Forum*, **13**, 253–259.

Martin, J. C. (1971). Substitution minimal sets, *Amer. J. Math.*, **93**, 503–526.

Marx, M. L. (1969). The Gauss realizability problem, *Proc. Amer. Math. Soc.*, **22**, 610–613.

Matiyasevich, Yu. (1968). A connection between systems of word and length equations and Hilbert's Tenth Problem, *Zap. Nauchn. Sem. Leningrad. Otdel. Mat. Inst. Steklov. (LOMI)*, **8**, 132–144. In Russian. English translation: *Sem. Math. V. A. Steklov*, **8**, 61–67, 1970.

Matiyasevich, Yu. (1993). *Hilbert's Tenth Problem*. MIT Press.

Matiyasevich, Yu. (1997). Some decision problems for traces, in Adian, S. and Nerode, A. (Eds.), *Logical Foundations of Computer Science (LFCS 97)*, Vol. 1234 of *Lect. Notes Comp. Sci.*, pp. 248–257. Springer-Verlag.

Mazurkiewicz, A. (1977). Concurrent program schemes and their interpretations, DAIMI Rep. PB 78, Aarhus University.

Melançon, G. (1993). Constructions des bases standard des $K\langle A\rangle$-modules à droite, *Th. Comput. Sci.*, **117**, 255–272.

Melançon, G. (1996). Lyndon factorization of infinite words, in *Thirteenth Annual Symposium on Theoretical Aspects of Computer Science*, Vol. 1046 of *Lect. Notes Comp. Sci.*, pp. 147–154. Springer-Verlag.

Merzlyakov, Yu. I. (1966). Positive formulae over free groups, *Algebra i Logika*, **5**, 25–42. In Russian.

Michaux, C. and Villemaire, R. (1996). Presburger arithmetic and recognizability of sets of natural numbers by automata: new proofs of Cobham's and Semenov's theorems, *Ann. Pure Appl. Logic*, **17**, 251–277.

Mignosi, F. (1989). Infinite words with linear subword complexity, *Th. Comput. Sci.*, **65**, 221–242.

Mignosi, F. (1990). Sturmian words and ambiguous context-free languages, *Internat. J. Found. Comput. Sci.*, **1**, 309–323.

Mignosi, F. (1991). On the number of factors of Sturmian words, *Th. Comput. Sci.*, **82**, 71–84.

Mignosi, F. and Pirillo, G. (1992). Repetitions in the Fibonacci infinite word, *Th. Inform. Appl.*, **26**(3), 199–204.

Mignosi, F., Restivo, A., and Salemi, S. (1995). A periodicity theorem on words and applications, in Wiedermann, J. and Hajek, P. (Eds.), *Mathematical Foundations of Computer Science (MFCS 95)*, Vol. 969 of *Lect. Notes Comp. Sci.*, pp. 337–348. Springer-Verlag.

Mignosi, F., Restivo, A., and Salemi, S. (1998). Periodicity and the golden ratio, *Th. Comput. Sci.*, **204**, 153–167.

Mignosi, F. and Séébold, P. (1993). Morphismes sturmiens et règles de Rauzy, *J. Th. Nombres Bordeaux*, **5**, 221–233.

Morse, M. and Hedlund, G. A. (1938). Symbolic dynamics, *Amer. J. Math.*, **60**, 815–866.
Morse, M. and Hedlund, G. A. (1940). Symbolic dynamics II: Sturmian sequences, *Amer. J. Math.*, **61**, 1–42.
Morse, M. and Hedlund, G. A. (1944). Unending chess, symbolic dynamics and a problem in semigroups, *Duke Math. J.*, **11**, 1–7.
Narendran, P. and Otto, F. (1997). The word matching problem is undecidable for finite special string-rewriting systems that are confluent, in Degano, P., Gorrieri, R., and Marchetti-Spaccamela, A. (Eds.), *Automata, Languages and Programming (ICALP 97)*, Vol. 1256 of *Lect. Notes Comp. Sci.*, pp. 638–648. Springer-Verlag.
Néraud, J. (1990a). Elementariness of a finite set of words is co-NP-complete, *Th. Inform. Appl.*, **24**, 459–470.
Néraud, J. (1990b). On the deficit of a finite set of words, *Semigroup Forum*, **41**, 1–21.
Néraud, J. (1993). Deciding whether a finite set of words has rank at most two, *Th. Comput. Sci.*, **112**, 311–337.
Nielsen, J. (1918). Die Isomorphismen der allgemeinen, unendlichen Gruppe mit zwei Erzeugenden, *Math. Ann.*, **78**, 385–397.
Nilgens, S. (1991). *3-vermeidbare Muster über dem ternären Alphabet*. Diplomarbeit, Fachbereich Informatik, Universität Frankfurt.
Papadimitriou, C. H. (1994). *Computatational Complexity*. Addison-Wesley.
Parry, W. (1960). On the β-expansions of real numbers, *Acta Math. Acad. Sci. Hungar.*, **11**, 401–416.
Parvaix, B. (1997). Propriétés d'invariance des mots sturmiens, *J. Th. Nombres Bordeaux*, **9**, 351–369.
Parvaix, B. (1998). *Contribution à l'étude des mots sturmiens*, Vol. 25. Publications du Lacim.
Pécuchet, J.-P. (1981). Sur la détermination du rang d'une équation dans le monoïde libre, *Th. Comput. Sci.*, **16**, 337–340.
Pedersen, A. (1988). Solution of Problem E 3156, *Amer. Math. Monthly*, **95**, 954–955. Other solutions at the same reference.
Perrin, D. (1989). Equations in words, in Ait-Kaci, H. and Nivat, M. (Eds.), *Resolution of Equations in Algebraic Structures*, Vol. 2, pp. 275–298. Academic Press, New York.
Petrich, M. (1984). *Inverse Semigroups*. J. Wiley and Sons.
Petronio, C. (1994). Thurston's solitaire tilings of the plane, *Rend. Istit. Mat. Univ. Trieste*, **26**, 247–248.
Pirillo, G. (1997). Fibonacci numbers and words, *Discr. Math.*, **173**, 197–207.
Plandowski, W. (1995). *The Complexity of the Morphism Equivalence Problem for Context-Free Languages*. Ph.D. thesis, Department of Mathematics, Informatics and Mechanics, Warsaw University.
Plandowski, W. (1999a). Satisfiability of word equations with constants is in NEXPTIME, in *Theory of Computing (STOC 99)*, pp. 721–725. ACM Press.

Plandowski, W. (1999b). Satisfiability of word equations with constants is in PSPACE, in *Foundations of Computer Science (FOCS 99)*, pp. 495–500. IEEE Computer Society Press.

Plandowski, W. and Rytter, W. (1998). Application of Lempel–Ziv encodings to the solution of words equations, in Larsen, K. G. et al. (Eds.), *Automata, Languages and Programming (ICALP 1998)*, Vol. 1443 of *Lect. Notes Comp. Sci.*, pp. 731–742. Springer-Verlag.

Pleasants, P. A. (1970). Non-repetitive sequences, *Math. Proc. Cambridge Philos. Soc.*, **68**, 267–274.

Plotkin, G. (1972). Building in equational theories, *Mach. Intelligence*, **7**, 115–162.

Point, F. and Bruyère, V. (1997). On the Cobham–Semenov theorem, *Th. Comput. Syst.*, **30**, 197–220.

Rauzy, G. (1979). Echanges d'intervalles et transformations induites, *Acta Arith.*, **34**, 315–328.

Rauzy, G. (1983). Suites à termes dans un alphabet fini, in *Séminaire Théorie des Nombres, année 1982–1983*, pp. 25–01 to 25–16. Université Bordeaux 1. Exposé No 25.

Rauzy, G. (1985). Mots infinis en arithmétique, in Nivat, M. and Perrin, D. (Eds.), *Automata on Infinite Words*, Vol. 192 of *Lect. Notes Comp. Sci.*, pp. 165–171. Springer-Verlag.

Rauzy, G. (1988). Rotations sur les groupes, nombres algébriques et substitutions, in *Séminaire de Théorie des Nombres, année 1987–1988*. Université Bordeaux 1. Exposé No 21.

Rawlings, D. (1981). Generalized Worpitzky identities with applications to permutation enumeration, *European J. Combin.*, **2**, 67–78.

Razborov, A. A. (1984). On systems of equations in a free group, *Izv. Akad. Nauk SSSR Ser. Mat.*, **48**, 779–832. In Russian. English translation: *Math. USSR-Izv.*, 25, 115–162, 1985.

Razborov, A. A. (1994). On systems of equations in free groups, in *Combinatorial and Geometric Group Theory*, pp. 269–283. Cambridge University Press.

Régnier, M. and Mouchard, L. (2000). Periods and quasiperiods characterization, in Giancarlo, R. and Sankoff, D. (Eds.), *Combinatorial Pattern Matching (CPM 2000)*, Vol. 1848 of *Lect. Notes Comp. Sci.*, pp. 388–396. Springer-Verlag.

Reiner, V. (1993). Signed posets, *J. Combin. Th. Ser. A*, **62**, 324–360.

Rényi, A. (1957). Representations for real numbers and their ergodic properties, *Acta Math. Acad. Sci. Hungar.*, **8**, 477–493.

Restivo, A. and Reutenauer, C. (1984). On the Burnside problem for semigroups, *J. Algebra*, **89**, 102–104.

Restivo, A., Salemi, S., and Sportelli, T. (1989). Completing codes, *Th. Inform. Appl.*, **23**, 135–147.

Reutenauer, C. (1986). Mots de Lyndon et un théorème de Shirshov, *Ann. Sci. Math. Québec*, **10**, 237–245.

Reutenauer, C. (1993). *Free Lie Algebras*. No. 7 in *London Mathematical Monographs New Series*. Clarendon Press.

Ribenboim, P. (1965). *Théorie des valuations.* Presses de l'Université de Montréal.
Richomme, G. (1999a). Another characterization of Sturmian words (one more), *Bull. EATCS,* **67,** 173–175.
Richomme, G. (1999b). Test-words for Sturmian morphisms, *Bull. Belg. Math. Soc. Simon Stevin,* **6,** 481–489.
Rips, E. and Sela, Z. (1995). Canonical representatives and equations in hyperbolic groups, *Invent. Math.,* **120,** 489–512.
Rivals, E. and Rahmann, S. (2001). Combinatorics of periods in strings, tech. rep. 01-017, LIRMM, Montpellier. In Orejas, F., Spirakis, P. G., and van Leeuwen, J. (Eds.), *Automata, Languages and Programming (ICALP 01),* Vol. 2076 of *Lect. Notes Comp. Sci.,* pp. 615–626, Springer-Verlag.
Robinson, G. de B. (1938). On the representations of the symmetric group, *Amer. J. Math.,* **60,** 745–760.
Robson, J. M. and Diekert, V. (1999). On quadratic word equations, in Meinel, C. et al. (Eds.), *Theoretical Aspects of Computer Science (STACS 99),* Vol. 1563 of *Lect. Notes Comp. Sci.,* pp. 217–226. Springer-Verlag.
Rosaz, L. (1995). Unavoidable languages, cuts and innocent sets of words, *Th. Inform. Appl.,* **29,** 339–382.
Rosaz, L. (1998). Inventories of unavoidable languages and the word-extension conjecture, *Th. Comput. Sci.,* **201,** 151–170.
Rosenstiehl, P. (1976). Solution algébrique du problème de Gauss sur la permutation des points d'intersection d'une ou plusieurs courbes fermées du plan, *C. R. Acad. Sci. Paris, Sér. A,* **283,** 417–419.
Rosenstiehl, P. (1999). A new proof of the Gauss interlace conjecture, *Adv. in Appl. Math.,* **23,** 3–13.
Rote, G. (1994). Sequences with subword complexity $2n$, *J. Number Th.,* **46,** 196–213.
Roth, P. (1991). l-occurrences of avoidable patterns, in Choffrut, C. and Jantzen, M. (Eds.), *Theoretical Aspects of Computer Science (STACS 91),* Vol. 480 of *Lect. Notes Comp. Sci.,* pp. 42–49. Springer-Verlag.
Roth, P. (1992). Every binary pattern of length six is avoidable on the two-letter alphabet, *Acta Inform.,* **29,** 95–106.
Rozenberg, G. and Salomaa, A. (1980). *The Mathematical Theory of L Systems.* Academic Press.
Rozenblatt, B. V. (1982). Equations on a free inverse semigroup, *Ural. Gos. Univ. Mat. Zap.,* **13,** 117–120.
Rozenblatt, B. V. (1985). Diophantine theories of free inverse semigroups, *Siberian Math. J.,* **26,** 860–865.
Safer, T. (1998). Radix representations of algebraic number fields and finite automata, in *Theoretical Aspects of Computer Science (STACS 98),* Vol. 1373 of *Lect. Notes Comp. Sci.,* pp. 356–365. Springer-Verlag.
Salomaa, A. (1981). *Jewels of Formal Language Theory.* Computer Science Press.
Schensted, C. (1961). Longest increasing and decreasing sub-sequences, *Canad. J. Math.,* **13,** 179–191.

Schmidt, K. (1980). On periodic expansions of Pisot numbers and Salem numbers, *Bull. London Math. Soc.*, **12**, 269–278.
Schmidt, U. (1986). *Motifs inévitables dans les mots*. Thèse de doctorat, Université Paris 6. LITP research report 86-63.
Schmidt, U. (1989). Avoidable patterns on two letters, *Th. Comput. Sci.*, **63**, 1–17.
Schulz, K. U. (1992a). Makanin's algorithm for word equations: two improvements and a generalization, in Schulz, K. U. (Ed.), *Word Equations and Related Topics (IWWERT 90)*, Vol. 572 of *Lect. Notes Comp. Sci.*, pp. 85–150. Springer-Verlag.
Schulz, K. U. (Ed.) (1992b). *Word Equations and Related Topics (IWWERT 90)*, Vol. 572 of *Lect. Notes Comp. Sci.* Springer-Verlag.
Schulz, K. U. (1993). Word unification and transformation of generalized equations, *J. Autom. Reasoning*, **11**, 149–184.
Schützenberger, M.-P. (1956). Une théorie algébrique du codage, in *Séminaire Dubreil–Pisot, année 1955–1956*. Institut H. Poincaré. Exposé No 15.
Schützenberger, M.-P. (1963). Quelques remarques sur une construction de Schensted, *Math. Scand.*, **12**, 117–128.
Schützenberger, M.-P. (1964). On the synchronizing properties of certain prefix codes, *Inform. and Control*, **7**, 23–36.
Schützenberger, M.-P. (1966). On a question concerning certain free submonoids, *J. Combin. Th. Ser. A*, **1**, 437–442.
Schützenberger, M.-P. (1976). A property of finitely generated submonoids of free monoids, in Pollack, G. (Ed.), *Algebraic Theory of Semigroups*, pp. 545–576. North-Holland.
Schützenberger, M.-P. (1977). La correspondance de Robinson, in Foata, D. (Ed.), *Combinatoire et représentation du groupe symétrique*, No. 579 in Lect. Notes Math., pp. 59–113. Springer-Verlag.
Schützenberger, M.-P. (1978). Propriétés nouvelles des tableaux de Young, in *Séminaire Delange–Pisot–Poitou, année 1977–1978*. Secrétariat Mathématique. Exposé No 26.
Séébold, P. (1985). *Propriétés combinatoires des mots infinis engendrés par certains morphismes*. Thèse de 3ème cycle, Université Paris 6. LITP research report 85-14.
Séébold, P. (1991). Fibonacci morphisms and Sturmian words, *Th. Comput. Sci.*, **88**, 365–384.
Séébold, P. (1998). On the conjugation of standard morphisms, *Th. Comput. Sci.*, **195**, 91–109.
Semenov, A. L. (1977). The Presburger nature of predicates that are regular in two number systems, *Sibirsk. Mat. Zh.*, **18**, 403–418. In Russian. English translation: *Siberian Math. J.* 18, 289–299, 1977.
Series, C. (1985). The geometry of Markoff numbers, *Math. Intelligencer*, **7**, 20–29.
Shallit, J. O. (1994). Numeration systems, linear recurrences, and regular sets, *Inform. Comput.*, **113**, 331–347.
Shallit, J. O. and Breibart, Y. (1996). Automaticity I: properties of a measure of descriptional complexity, *J. Comput. Syst. Sci.*, **53**, 10–25.

Shirshov, A. I. (1957). On certain nonassociative nil rings and algebraic algebras, *Mat. Sb.*, **41**, 381–394.
Shyr, H. J. (1977). A strongly primitive word of arbitrary length and its applications, *Internat. J. Comput. Math.*, **6**, 165–170.
Shyr, H. J. (1991). *Free Monoids and Languages* (second edition). Hon Min.
Sidorov, N. A. and Vershik, A. M. (1993). Arithmetic expansions associated with rotations of the circle and continued fractions, *Algebra i Anal.*, **5**, 97–115. In Russian. English translation: *St Petersburg Math. J.*, **5**, 1121–1136, 1994.
Simon, I. (1980). Conditions de finitude des semi-groupes, *C. R. Acad. Sci. Paris, Sér. A*, **290**, 1081–1082.
Simon, I. (1988). Infinite words and a theorem of Hindman, *Rev. Mat. Apl.*, **9**, 97–104.
Skordev, D. and Sendov, B. (1961). On equations in words, *Z. Math. Logik Grundlag. Math.*, **7**, 289–297.
Solomyak, B. (1994). Conjugates of beta-numbers and the zero-free domain for a class of analytic functions, *Proc. London Math. Soc.*, **68**, 477–498.
Spehner, J.-C. (1975). Quelques constructions et algorithmes relatifs aux sous-monoïdes d'un monoïde libre, *Semigroup Forum*, **9**, 334–353.
Spehner, J.-C. (1976). *Quelques problèmes d'extension, de conjugaison et de présentation des sous-monoïdes d'un monoïde libre*. Thèse, Université Paris 7.
Staiger, L. (1986). On infinitary finite length codes, *Th. Inform. Appl.*, **20**, 483–494.
Stallings, J. R. (1986). Finiteness properties of matrix representations, *Ann. of Math. (2)*, **124**, 337–346.
Stanley, R. P. (1972). *Ordered Structures and Partitions*, Vol. 119 of *Mem. Amer. Math. Soc.* Amer. Math. Soc, Providence, RI.
Stanley, R. P. (1976). Binomial posets, Möbius inversion and permutation enumeration, *J. Combin. Th. Ser. A*, **20**, 307–320.
Stolarsky, K. B. (1976). Beatty sequences, continued fractions, and certain shift operators, *Canad. Math. Bull.*, **19**, 473–482.
Thue, A. (1906). Über unendliche Zeichenreihen, *Kra. Vidensk. Selsk. Skrifter, I. Mat. Nat. Kl.*, **1906**(7), 1–22.
Thue, A. (1912). Über die gegenseitige Lage gleicher Teile gewisser Zeichenreihen, *Kra. Vidensk. Selsk. Skrifter, I. Mat. Nat. Kl.*, **1912**(1), 1–67.
Thurston, W. P. (1989). Groups, tilings, and finite state automata, geometry supercomputer project research report GCG1, Geometry Supercomputer Project, University of Minnesota.
Tijdeman, R. (1996). On disjoint pairs of Sturmian bisequences, tech. rep. W96-02, Mathematical Institute, University of Leiden.
Treybig, L. (1968). A characterization of the double point structure of the projection of a polygonal knot in regular position, *Trans. Amer. Math. Soc.*, pp. 223–247.
Vandeth, D. (2000). Sturmian words and words with a critical exponent, *Th. Comput. Sci.*, **242**, 283–300.
Vaniček, T. (1989). *Unavoidable words*. Diploma thesis, Faculty of Math. and Physics, Charles University, Prague. In Czech.

Vardi, I. (1991). *Computational Recreations in Mathematica*. Addison-Wesley.
Venkov, B. A. (1970). *Elementary Number Theory*. Wolters-Noordhoff.
Wen, Z.-X. and Wen, Z.-Y. (1994a). Local isomorphisms of invertible substitutions, *C. R. Acad. Sci. Paris, Sér. A*, **318**, 299–304.
Wen, Z.-X. and Wen, Z.-Y. (1994b). Some properties of the singular words of the Fibonacci word, *European J. Combin.*, **15**, 587–598.
Zech, T. (1958). Wiederholungsfreie Folgen, *Z. Angew. Math. Mech.*, **38**, 206–209.
Zeilberger, D. (1980). A lattice walk approach to the counting of multiset permutation, *J. Math. Anal. Appl.*, **74**, 192–199.
Ziccardi, G. (1995). *Parole Sturmiane*, Laurea, Università degli Studi di Roma.
Zimin, A. I. (1979). Blocking sets of terms, In *Fifteenth All-Union Algebra Conference, Krasnoyarsk, Abstract of Reports, Part 1*, p. 63. In Russian.
Zimin, A. I. (1982). Blocking sets of terms, *Mat. Sb. (NS)*, **119**, 363–375. In Russian. English translation: *Math. USSR Sbornik*, 47, 353–364, 1984.

Index of Notation

A^*, 3
A^+, 3
$\mathscr{A}$, 12
$\mathscr{A}(X)$, 14
$A_c(t,q)$, 334
$\mathrm{Adh}(X)$, 224
A^∞, 6
$\mathrm{Alph}_Y(X)$, 204
Alph, 4
Amb_X, 212
$A^{\mathbb{N}}$, 6
A^ω, 7
$A(t,q;\mathbf{u})$, 337
$A^{\mathbb{Z}}$, 11
A^ζ, 12

BS, 367

$\mathbb{C}$, 1
Card, 1

$\mathscr{D}$, 154
den, 354
des, 331, 367

$E_{\mathrm{red}}(X)$, 445
$E(X)$, 445
exc, 331

$F(X)$, 7
$F(x)$, 4, 7, 11
$\mathrm{First}_Y(X)$, 204
$F_n(x)$, 7
f_X, 22

gcd, 5

$\mathscr{H}$, 154
$h(S)$, 31

imaj, 367

$K_{\lambda\mu}(q)$, 191
$k\langle\!\langle X \rangle\!\rangle$, 313
$k\langle X \rangle$, 313

$\mathrm{Last}_Y(X)$, 204
$\mathrm{Lg}(X)$, 208

maj, 331

$\mathbb{N}$, 1

$P(w)$, 166
$\mathrm{Pl}(A)$, 168
$\mathrm{Pref}(X)$, 212

$\mathbb{Q}$, 1

$\mathbb{R}$, 1
$r_c(X)$, 444
ρ_X, 23
$\mathrm{Row}(A)$, 187

σ_X, 214
$\mathrm{STab}(\lambda)$, 170
$\mathrm{std}(w)$, 173
$\mathrm{Suff}(X)$, 212
S_X, 26

$\mathrm{Tab}(A)$, 170
$\mathrm{Tab}(\lambda, A)$, 170

$u \triangleright v$, 165
$u \wedge v$, 212

$\tilde{w}$, 4
$w^{-1}X$, 14

X^*, 2, 5
X^+, 2

$X + Y$, 5
$X^{-1}Y$, 14
$x < y$, 4
X^∞, 7
X^ω, 7
$\widetilde{X}$, 4
X^ζ, 12

$\mathbb{Z}$, 1

General Index

adjacency graph, 116
algebra of formulas, 131
alphabet, 3
alphabetic order, 4
anti-ideal, 323
automaton, 12
    deterministic, 13
    finite, 13
    path, 12, 17
    state, 12
    trim, 13
    unambiguous, 13
automorphism
    free group, 97
    positive, 97
avoidability, 112
    absolute, 111
    deciding, 116
    of cubes, 113
    of overlap, 113
    of squares, 113
    on a fixed alphabet, 112, 123
    proof techniques, 114
avoidability index, 112, 123
    bound, 126
avoidable, weakly, 133
avoidable pattern, 111

Büchi automaton, 17
backstep, 367
balanced set of words, 48
balanced word, 48
base, 199, 203
basis, 417
Baumslag–Solitar group, 465
Beatty sequence, 105

Bergman's centralizer theorem, 320
beta-expansion, 236
beta-number, 238
bi-ideal, 157
    principal, 157
bi-ideal sequence, 135, 136, 150
    order of a, 136, 150
bicyclic monoid, 40, 459
biletters, 172
binoid, 9, 203
    finitary, 9, 203
    free, 203
    stable, 203
biword, 172
blocking term, 134
bordered word, 270
boundary
    critical, 425
    right, 416
brick, 415
    label, 415

cancellation property, 153
canonical epimorphism, 149
canonical factorization, 138
    inverse, 138
canonical representative, 149
Cantor space, 7
central word, 68
centralizer, 313
chain
    clean convex, 417
    convex, 417
    convex condition, 417
chain condition on idempotents,
    463

characteristic morphism, 90
characteristic word, 54, 62
charge, 188
chop, 132
Christoffel pair, 103
Christoffel word, 59, 103, 109
closed set, 8
cocharge, 187
code, 6, 198
    complete, 222
    maximal, 221
    ω-code, 201
    prefix, 6, 198
    prefix–suffix composed, 226
    suffix, 198
    three-element, 219
    with bounded decoding delay, 201
    ξ-code, 209
    $\mathbb{Z}$-code, 224
    ς-code, 224
coding morphism, 198, 203
Cohn's centralizer theorem, 313
column, 166
combinatorial rank, 444
commutator, 465
compact set, 8
compactness property, 458
complexity function, 7, 46
composable sets, 208
composition, 208
conjugate partition, 166
conjugate words, 4, 392
content of a polynomial, 318
coplactic class, 182
correctness
    downward, 423
    upward, 423
critical factorization theorem, 286
critical point, 285
cross-section theorem, 169
crystal graphs, 186
cut, 128
cutting sequence, 55
cyclage, 187

de Bruijn graph, 20
decisive word, 121
decoding delay, 201
defect theorem, 204, 206, 443
degree, 136
deletion morphism, 113
Denert statistic, 354
dependency graph, 205
derangement, 384
derived sequence, 159
descent, 331, 367
directive sequence, 75
divisibility of patterns, 112
D0L problem, 452
D0L-avoidable, 132
dominates, 165
domino tower, 393
DT0L problem, 452

Ehrenfeucht's conjecture, 444
empty pattern, 114, 117
encounter, 112
endomorphism, 450
entropy, 31
equation
    equivalent systems, 444
    independent system, 444
    solution, 444
    word, 389
equationally Noetherian, 472
equivalent patterns, 112
equivalent systems of equations, 444
Euclid's algorithm, 272
Eulerian circuit, 42, 44
evaluation, 174
even subshift, 27
excedence, 331
exponent of periodicity, 393, 411
extension, admissible, 409

factor, 3, 7, 149
    conservative, 21
    left, 149
    proper, 3
    right, 149

right special, 21
factor graph, 20, 81
factor-closed set, 6
factorial set, 6, 149
Ferrers diagram, 166, 356
Fibonacci numbers, 23, 280
Fibonacci numeration system, 252
Fibonacci word, 11, 47, 49, 51, 63, 64, 75, 76, 78, 82, 84, 98, 106, 136, 280
Fine and Wilf's theorem, 5, 272
Foata–Schützenberger theorem, 373
foundation, 417
free algebra, 322
free group, 96
    automorphism, 97
free inverse semigroups, 463
free partially commutative monoid, 442, 461
free Schur function, 177
free set, 116
frequency, 83
full shift, 26

Gauss code, 40
generating function, 22
Gessel normalization, 376
golden mean subshift, 26
golden ratio, 279
graph group, 442
Greene's plactic invariants, 169
group equation, 457

height of a word, 48, 65
Hilbert's basis theorem, 447
Hopfian semigroup, 464
hull
    free, 204
    ω-free, 206
    ζ-free, 209

idealizer, 317
idempotent, 156, 463
independent system of equations, 444
index, 77, 151
    bounded, 77
    final, 417
infinite word avoiding a pattern, 113, 120
initial cut, 128
inverse major index, 367
inverse pair, 469
irreducible pattern, 116
iteration property, 153
    right, 156

k-chop, 132
Kleene's theorem, 16
Knuth correspondence, 174
Knuth relations, 168
König's lemma, 8, 113
Kostka–Foulkes polynomials, 191

l-occurrence, 133
leader of a set, 315
left quotient, 14
left repetition, 289
left special word, 63
Lehmer encoding, 372
length
    denotational, 400
    of a morphism, 85
Levi's lemma, 313
lexicographic order, 4, 7, 61
linear numeration system, 251
linear order over a semigroup, 408
Littlewood–Richardson rule, 178
local ring, 325
locally characteristic morphism, 90
locally Sturmian morphism, 85
locked pattern, 132
Lyndon word, 5, 103, 144, 375

major index, 331, 367
maximal condition on congruences, 465
mechanical word, 53
metabelian group, 467
minimal condition on principal bi-ideals, 158
minimal generating set, 6, 9, 199, 203

minimal subshift, 30
model, 408
monoid, 2
    morphism, 2
    Sturm, 85
morphism, 2, 6
    characteristic, 90
    elementary, 226
    length, 85
    literal, 6
    locally characteristic, 90
    locally Sturmian, 85
    nonerasing, 6
    simplifiable, 226
    standard, 90
    Sturmian, 83
    trivial, 83
Morse minimal set, 31

n-division, 139
    inverse, 139
Nielsen transformation, 97
node, terminal, 432
nonerasing morphism, 112
nontrivial relation, 203
normalization, 232, 235, 252, 264
n-sequence, 137, 151
    inverse, 137
nuclear congruence, 463

occurrence
    l-occurrence, 133
    of a pattern, 112
order of an element, 151
overlap-free word, 114

p-stable normal form, 394
partial quotients, 75
partition, 166
path, 12, 17
    successful, 13, 18
pattern, 111, 112
    avoidable, 113, 118
        on A, 113
    binary, 124
    irreducible, 116, 118
    k-avoidable, 113

    k-unavoidable, 113
    locked, 132
    reducible, 116–118
    reduction, 116
    unary, 113, 124
    unavoidable, 113, 114, 117, 118
        on A, 113
    with constants, 111
pattern language, 111, 112
period, 4, 151, 270
permutation property, 152
Perron number, 245
Pieri rule, 195
Pisot number, 245
plactic monoid, 168
plactic Schur function, 177
plus operation, 2
polynomial, 313
    primitive, 318
position, 408
    transport, 425
power, 113
    avoidability, 113
prefix, 3
    proper, 3
prefix code, 6
prefix order, 4
prefix rank, 455
prefix-closed, 6
primitive matrix, 247
primitive polynomial, 318
primitive word, 4, 392

q-binomial coefficient, 332
q-calculus, 332
q-Euler numbers, 382
q-Eulerian polynomials, 358
quadratic system of equations, 389
quasi-ideal sequence, 159
quasi-power, 159

radix order, 4, 149
rank
    combinatorial, 226
    free, 226

General Index

ω-free, 226
  right-unitary, 226
rational power of a word, 270
rational relation, 2
rational set, 2, 16
Rauzy's rules, 109
reciprocity law, 139
recognized set, 13, 18
recognizing letter, 121
recurrence index, 31, 142
recurrent word, 30
reduced relation, 445
reduced word, 96
reduction of patterns, 116
refinement, 407
relation satisfied by a set, 445
repetition
  central, 283
  external, 283
  internal, 283
  left, 288
representation of a semigroup, 445
right special factor, 21
right special word, 46
ring
  local, 325
  valuation, 325
Robinson–Schensted correspondence, 171
rotation, 56
row, 165

Salem number, 245
Schützenberger's theorem, 221
Schensted's algorithm, 166
Schur function, 176, 359
semigroup, 2
  finitely generated, 149
  idempotent, 156, 463
  locally finite, 149
  morphism, 2
  permutable, 152
  recognizing, 15
  strongly periodic, 156
  syntactic, 15
  weakly permutable, 161

sequential transducer, 18
sesquipower, 114, 135, 136, 141
  order of a, 135, 136
set
  elementary, 207
  maximal elementary, 227
  recognizable, 13
  recognized by a semigroup, 15
  simplifiable, 207
set of patterns, 131
shape, 166, 367, 369
shift, 7, 11
Shirshov's theorem, 144
σ-deletion, 134
simple formula, 131
singular word, 106
slope of a word, 50, 53
solution
  nonsingular, 389
  of an equation, 444
  of boundary equations, 411
  of word equations, 389
  singular, 389
square, 111, 123
  avoidability, 113
square-free word, 111, 114
stable monoid, 329
standard morphism, 90
standard normalization, 369
standard pair, 64
  unordered, 89
standard representation, 408
standard sequence, 75
standard tableau, 170
standard word, 64
standardization, 173
star operation, 2, 5
state, 12
  initial, 12, 17
  terminal, 12, 17
strictly bispecial word, 80
Sturmian
  morphism, 83
    locally, 85
  word, 46

submonoid, 2
    free, 200
    right-unitary, 225
    stable, 200
subsemigroup, 2
subshift
    coded, 32
    finite type, 27
    sofic, 27
substitution, 6
suffix, 3, 7
    proper, 3, 7
synchronization lemma, 125
synchronizing letter, 120
synchronizing word, 114
system of boundary equations, 411

tableau, 165
test set, 451
three-distance theorem, 110
Thue–Morse word, 10, 113, 114, 405
totient function, 72, 81
trace monoid, 442, 461
transducer, 17
    sequential, 18

U-representation, 248
unambiguous product, 5
unambiguous rational set, 16
unambiguous star, 6
unambiguous union, 5
unavoidable pattern, 111
    bound on length, 126
unbordered word, 222, 270
uniform compactness property, 470
uniform morphism, 114
uniformly recurrent, 30

valuation ring, 325
variable, 111
    deletion from a pattern, 113
variety of monoids, 460, 465

weak algorithm, 315
well-partial-order, 160
word
    ambiguously covered, 212
    aperiodic, 9
    balance, 48
    border, 270
    central, 68
    characteristic, 54, 62
    Christoffel, 59
    conjugate, 4, 392
    eventually periodic, 9
    height, 48, 65
    incomparable, 212
    index, 77
    inversely n-divided, 139
    irreducible, 149, 150
    left special, 63
    mechanical, 53
        intercept, 53
        slope, 53
    morphic, 10
    n-divided, 135, 139
    ω-power-free, 143
    palindrome, 4
    period, 4
    periodic, 9
    p-power-free, 143
    primitive, 4, 392
    reduced, 96
    reducible, 149, 150
    reversal, 4
    right special, 46
    slope, 50
    standard, 64
    strictly bispecial, 80
    Sturmian, 46
    Thue–Morse, 405
    ultimately periodic, 9
word equation, 389

X-adic valuation, 313
X-decomposition, 224
X-factorization, 198, 201, 224

Yamanouchi word, 180
Young tableau, 165

Zimin algorithm, 116, 117
Zimin word, 115, 137

For EU product safety concerns, contact us at Calle de José Abascal, 56–1°,
28003 Madrid, Spain or eugpsr@cambridge.org.

www.ingramcontent.com/pod-product-compliance
Ingram Content Group UK Ltd.
Pitfield, Milton Keynes, MK11 3LW, UK
UKHW022033130925
462840UK00023B/840